Representations of Finite Groups

Representations of Finite Groups

Hirosi Nagao

Department of Mathematics
Osaka University
Osaka, Japan

Yukio Tsushima

Department of Mathematics
Osaka City University
Osaka, Japan

ACADEMIC PRESS, INC.
Harcourt Brace Jovanovich, Publishers
Boston San Diego New York
Berkeley London Sydney
Tokyo Toronto

Originally published as *Yugen-gun no Hyogen.*

ACADEMIC PRESS, INC.
1250 Sixth Avenue, San Diego, CA 92101

United Kingdom Edition published by
ACADEMIC PRESS INC. (LONDON) LTD.
24–28 Oval Road, London NW1 7DX

Library of Congress Cataloging-in-Publication Data

Nagao, Hiroshi, Date–
[Yūgen-gun no hyōgen. English]
Representations of finite groups/Hirosi Nagao, Yukio Tsushima.
p. cm.
Translation of: Yūgen-gun no hyōgen.
Bibliography: p.
Includes index.
ISBN 0-12-513660-9
1. Representations of groups. 2. Finite groups. I. Tsushima.
Yukio, Date– . II. Title.
QA171.N27 1988
512′.2—dc19 88-10416

Printed in the United States of America
89 90 91 92 9 8 7 6 5 4 3 2 1

To the memory
of
T. Nakayama

Contents

Preface to the English Edition

This is a translation of our book "Yugen-gun no Hyogen," published by Shokabo.

We have added a few advanced results to this edition as problems that are of interest in their own right.

We appreciate the quick decision of Academic Press to publish this book. Indeed, we were able to start the translation before the Japanese edition was published.

We would like to express our gratitude to S. Murakami, who suggested that we publish the translation with Academic Press, and to K. Uno for his careful reading of the English manuscript.

Hirosi Nagao
Yukio Tsushima
January 1989

Preface

The fundamental theory of complex (or ordinary) representations of finite groups was almost completed by Frobenius and Burnside. Schur later simplified the rather complicated theory of Frobenius to a considerable extent by using a lemma now called Schur's lemma. The methods used were based on matrix representations and characters. On the other hand, the study of algebras developed by E. Noether and others made it possible to consider the representations of groups from the ring- and module-theoretic points of view.

The representation of a group G over a field of prime characteristic p was first called a modular representation by Dickson, which is the origin of the present terminology. He pointed out that if p does not divide the order of G, then there is no essential difference betweeen ordinary and modular representations. But since then, there had been no substantial development made on the subject for almost thirty years.

In 1935 Brauer obtained a result on the number of the irreducible modular representations. Then followed the fundamental paper of Brauer and Nesbitt, which was the beginning of the full-scale study of modular representations. The development thereafter was mostly the result of the considerable efforts of Brauer, but significant contributions were also made by some Japanese mathematicians, including Nakayama and Osima. In the meantime, Green's theory of vertices shed new light on the module-theoretic study of modular representations. Since then, the subject has been actively studied, embracing some deep problems.

This book gives an account of the fundamentals of ordinary and modular representations, assuming some standard facts of algebra. We have tried to keep the explanations comprehensible and assumed as few prerequisites as possible. We hope this book will be used as a guide to this fascinating subject.

At present, Feit[2] is one of the textbooks that treats the subject most extensively. The contents of this book corresponds to Chapters I through V and part of Chapter X of Feit, with the addition of some new results. Let us briefly outline the chapter contents of our book.

Chapter 1 contains, as preparatory to the subsequent chapters, basic facts about rings and modules. Particularly necessary facts on discrete valuation rings are summarized in §13 for later use.

Chapter 2 provides the theory of algebras, which includes theories of simple algebras, Schur indices with representation-theoretic versions of them, crossed products, Frobenius algebras, etc.

Chapter 3 gives the classical theory of ordinary representations and some fundamental theorems of Brauer. It also includes a survey of the fundamental theory of modular representations, with the focus on Brauer characters.

Chapter 4 provides the module-theoretic representation theory due to Green and also includes some recent topics such as Burry–Carlson's theorem, Scott modules, etc.

Chapter 5 presents the fundamental results of Brauer on blocks and Fong's theory of covering, and includes some new approaches to them.

At the end of each chapter, we provide a selection of problems. In addition to some routine exercises each chapter includes some advanced results in order to supplement the text. Thus, the reader may obtain complete solutions after consulting the (brief) solutions presented at the end of this book. Of course, it causes no trouble in understanding the text to skip them.

In the Postscript, we shall give some remarks on the text and problems, with references to some facts we did not mention in this book.

As pointed out at the beginning, there are various approaches to the study of the representations of groups, such as character-theoretic, ring-theoretic, module-theoretic, etc. Each has its own advantages, and the different developments stimulate each other. We do not adhere to any particular point of view, but are rather inclined to put an emphasis on the consideration of characters, in contrast to other books now available. This is partly because we believe that our readers will be led to a deeper understanding by carrying out the calculations themselves.

Hirosi Nagao
Yukio Tsushima
June 1987

Acknowledgments

K. Iizuka, A. Watanabe, and K. Uno read through the manuscript for this book very carefully from the first draft to the proofs. We also asked Y. Yamamoto and N. Nobusawa for examinations of part of the manuscript. They gave us innumerable valuable comments and criticisms. In fact, we are not able to estimate how many errors, essential or minor, have been eliminated by virtue of their suggestions. We would like to express our gratitude to all of them.

We extend our thanks to N. Iwahori for giving us the opportunity to write this book and to T. Abiko and K. Endo of Shokabo, our publisher in Japan, as well as to the Shokabo editorial staff, including S. Hosoki, for their encouragement and constant support.

Around 1957, one of the authors, Nagao, had planned to write a book on representation theory with T. Nakayama. However Nakayama told him in his letter of November 4, 1959, that he had decided to suspend the plan because R. Brauer, who visited Japan in the spring of that year, had told Nakayama that he was planning to write a book on representation theory himself. By the way, one of the authors, Tsushima, was one of the last pupils of Nakayama. Therefore, we dedicate this book to him.

1 Rings and Modules

In this chapter we summarize fundamental results on rings and modules that will be needed in the subsequent chapters. We assume that the reader is familiar with the basic concepts of algebra, such as groups, rings, and fields (see also, for example, Hungerford [1]).

1. Definitions and Notations

1.1. General Notations

If a set Y is a subset of a set X, we write $Y \subset X$. So, it is possible that $Y = X$. If Y is a proper subset, and if it is necessary to emphasize it, we write $Y \subsetneqq X$, and we denote by $X - Y$ the complement of Y in X. If X is a finite set, we denote by $|X|$ or by $\# X$ the number of the elements in X, while we write $|X| = \infty$ to mean that X is an infinite set. If a map $f: X \to X'$ sends $x \in X$ onto $x' \in X'$, we write $f: X \to X'(x \mapsto x')$, or simply $f: x \mapsto x'$. For $g: X' \to X''$, $g \circ f$ denotes the composite map of f and g, i.e., $g \circ f: X \to X''(x \mapsto g(f(x)))$. If Y is a subset of X, we obtain a map $f|_Y: Y \to X'$ by restricting the domain of f to Y. In particular $\iota_Y: Y \to X(y \mapsto y)$ and

$\mathrm{id}_X: X \to X(x \mapsto x)$ are called the *inclusion map* and the *identity map*, respectively. Throughout this book

(1.1) *a group means a finite group.*

Let G be a group. We write $H \leq G$ if H is a subgroup of G and $H < G$ if H is a proper subgroup. $|G:H|$ denotes the index of H in G. We write $H \lhd G$ if H is a normal subgroup of G. For $x \in G$, $o(x)$ denotes the order of x. For a set π of prime numbers, π' is the complementary set of primes to π. A natural number m is called a π-number if every prime divisor of m lies in π. A natural number n is written as $n = n_\pi n_{\pi'}$, with n_π being a π-number and $n_{\pi'}$ a π'-number. Then n_π and $n_{\pi'}$ are called the *π-component* and the *π'-component* of n, respectively. We say that an element x of G is a *π-element* if $o(x)$ is a π-number. A subgroup H of G is called a *π-subgroup* if H consists of π-elements. Every $x \in G$ can be expressed uniquely as the commuting product of the π-element x_π and the π'-element $x_{\pi'}$;

$$x = x_\pi x_{\pi'} = x_{\pi'} x_\pi.$$

Here we have $x_\pi, x_{\pi'} \in \langle x \rangle$, and they are called the *π-part* and *π'-part* of x, respectively. If π consists of a single prime p, then π and π' will be replaced with p and p', respectively, in the above expressions. We denote by $\mathrm{Syl}_p(G)$ the set of Sylow p-subgroups of G.

We assume, unless contrary stated explicitly, that

(1.2) *all rings have identity.*

If A is a ring, the identity of A is denoted by 1_A (or simply by 1 if there is no fear of confusion). A ring homomorphism $f: A \to B$ will be supposed to send the identity of A onto that of B; $f(1_A) = 1_B$. If $u \in A$ has a multiplicative inverse, namely, if there exists u^{-1} of A such that $uu^{-1} = u^{-1}u = 1$, then it is called a *unit* of A. The set of units of A forms a group called the *group of units* of A and is denoted by $A^\times$. A is called a *division ring* if $A^\times = A - \{0\}$.

For a ring A, the set $M_n(A)$ of all $n \times n$ matrices over A forms a ring called the *full matrix ring* of degree n over A. A unit of $M_n(A)$ is said to be a *nonsingular matrix* and the group of units of $M_n(A)$ is denoted by $GL_n(A)$, which is called the *general linear group* of degree n over A.

By an *ideal* of a ring A we mean a two-sided ideal of it. For an ideal I of A, the factor ring A/I will be usually denoted by $\bar{A}$, and, if $a \in A$, then $\bar{a}$ will denote the residue class $a + I$. Similar notation will be used in factor groups, too.

We assume, following the usual terminology, that

(1.3) *a field means a commutative field.*

The characteristic of a field K is denoted by Char K.

We let **N**, **Z**, **Q**, **R**, and **C** denote the set of natural numbers, the ring of rational integers and the fields of rational numbers, real numbers, and complex numbers, respectively.

1.2. A-modules

Let A be a ring. An A-module is assumed to be unital, namely, if V is a right (respectively, left) A-module, then $v = v1$ (respectively, $v = 1v$) for all $v \in V$. We understand by V_A (respectively, ${}_AV$) that V is a right (respectively, left) A-module. Let B be another ring. An *(A, B)-bimodule* ${}_AV_B$ is a left A-module and a right B-module simultaneously, satisfying $(av)b = a(vb)$ for $a \in A$, $b \in B$, $v \in V$. For example A acts on itself via the multiplication from the right and left, making A into A_A, ${}_AA$ and even ${}_AA_A$.

If X is a subset of a right A-module V, then the set $\{a \in A; Xa = 0\}$ is a right ideal of A, which is called the *annihilator ideal* of X and denoted by $(0{:}X)$. Note that $(0{:}V)$ is an ideal of A. If it is zero, then V_A is said to be *faithful*. For example A_A is faithful.

Consider the set of symbols $A^\circ = \{a^\circ; a \in A\}$, and define the sum and the multiplication in A° as follows:

$$a^\circ + b^\circ = (a + b)^\circ, \qquad a^\circ b^\circ = (ba)^\circ$$

The A° becomes a ring, which is called the *opposite ring* of A.

A right A-module V is considered as a left A°-module with the action $a^\circ v = va$ for $a \in A$ and $v \in V$. Likewise, ${}_AW$ becomes a right A°-module.

If R is a commutative ring, every right R-module can be naturally considered as a left R-module and vice-versa because of the isomorphism $R \simeq R^\circ (a \mapsto a^\circ)$. So we simply call it an R-module. However, the action of an element of R will be written from the left, following the usual manner. In this sense, an additive group is considered as a **Z**-module. An R-homomorphism $f\colon V \to W$ is sometimes said to be an *R-linear map*.

In what follows, we shall mainly discuss right A-modules and make no specific comments on left modules if our discussions will be equally valid for them. Also we occasionally omit "A-" from "A-submodules" or "A-homomorphisms" if from the context A is clearly understood to be the ring acting on them.

We henceforth assume, unless otherwise stated explicitly, that

(1.4) *an A-module means a right A-module.*

Let $W_\lambda(\lambda \in \Lambda)$ be A-submodules of V_A. Then $\sum_{\lambda \in \Lambda} W_\lambda$ denotes the sum of $W_\lambda(\lambda \in \Lambda)$, i.e.,

$$\sum_{\lambda \in \Lambda} W_\lambda = \left\{ \sum_\lambda w_\lambda \quad \text{(finite sum);} \quad w_\lambda \in W_\lambda \right\}.$$

A subset X of V is said to *generate* V if $V = \sum_{v \in X} vA$. In that case, the elements of X are called *generators* of V. If there is a finite subset X that generates V, then V is said to be *finitely generated* over A. In particular, if V is generated by a single element, V is called a *cyclic module.*

Exercise 1.1. $vA \simeq A/(0{:}v)$ (A-isomorphic).

Let Y be a set of elements of V. Then Y is said to be *A-free* or *linearly independent* over A, provided that for any finite subset $\{v_1, \ldots, v_n\}$ of Y, the linear relation $\sum_{i=1}^n v_i a_i = 0$ holds only if $a_i = 0$ for all i. If V has a set of generators that is linearly independent over A, V is called an *A-free module,* and such a set of generators is called an (*A*-)*basis.* For example A_A is a free module with basis $\{1\}$.

If R is a commutative ring, every finitely generated free R-module V has a basis consisting of finite elements, and any two bases have the same number of elements. This number is called the *rank* of V over R and is denoted by $\operatorname{rank}_R V$ (cf. Hungerford [1], p. 186). If R is a field, then $\operatorname{rank}_R V = \dim_R V$, the *dimension* of V over R.

Exercise 1.2. Let D be a division ring and let V_D be finitely generated over D. Then V is D-free, and any two bases have the same number of elements. (This number is also denoted by $\dim_D V$.)

1.3. A-homomorphisms

Let $\sigma: V \to W$ be an A-homomorphism. Then $\operatorname{Im} \sigma = \{\sigma(v); v \in V\}$, $\operatorname{Ker} \sigma = \{v \in V; \sigma(v) = 0\}$ and $\operatorname{Coker} \sigma = W/\operatorname{Im} \sigma$ are called the *image, kernel,* and *cokernel* of σ, respectively. If $\operatorname{Ker} \sigma = 0$ (respectively, $\operatorname{Im} \sigma = W$), σ is called a *monomorphism* (respectively, an *epimorphism*). If σ is both monomorphic and epimorphic, then σ is called an *isomorphism* and we write $\sigma: V \xrightarrow{\sim} W$.

The set of all A-homomorphisms from V into W is denoted by $\mathrm{Hom}_A(V, W)$ (or $\mathrm{Hom}_A(V_A, W_A)$). This is an additive group with the addition

$$(\sigma + \tau)(v) = \sigma(v) + \tau(v),$$

where $\sigma, \tau \in \mathrm{Hom}_A(V, W)$ and $v \in V$.

$\mathrm{Hom}_A(V, V)$ is a ring, which is called the *(A-)endomorphism ring* of V and is denoted by $\mathrm{End}_A(V)$; the multiplication of two elements in $E = \mathrm{End}_A(V)$ is given by the composition of maps. V is considered as a left E-module via the natural action: $\sigma v = \sigma(v)$ for $\sigma \in E$ and $v \in V$. Actually, V becomes an (E, A)-bimodule.

Remark. *When V is a left A-module, we shall write $v\sigma$ to denote the image of $v \in V$ under $\sigma \in E$.* The composite of σ and τ of E is then given by $v(\sigma\tau) = (v\sigma)\tau$ for $v \in V$. Thus V is an (A, E)-bimodule.

Exercise 1.3.

(i) $\mathrm{Hom}_A(A, V_A) \simeq V(\sigma \mapsto \sigma(1))$.

(ii) $\mathrm{End}_A({}_AA)$ and $\mathrm{End}_A(A_A)$ are both isomorphic to A as rings.

1.4. Direct Sum Decompositions

Let $V = \sum_{\lambda \in \Lambda} V_\lambda$ be a sum of A-submodules $V_\lambda (\lambda \in \Lambda)$. Then we say V is a *direct sum* of them and write $V = \bigoplus_{\lambda \in \Lambda} V_\lambda$, provided every element v of V is written *uniquely* as a (finite) sum $v = \sum_\lambda v_\lambda (v_\lambda \in V_\lambda)$. Here each v_λ is said to be the *λ-coordinate* of v, and each V_λ is a *direct summand* of V. If $\Lambda = \{1, \ldots, n\}$ is a finite set, we write $V = V_1 \oplus \cdots \oplus V_n$.

For example V is A-free if and only if V is a direct sum of submodules of V, each of which is isomorphic to A_A.

Exercise 1.4. A sum $V = \sum_{\lambda \in \Lambda} V_\lambda$ is a direct sum if and only if the following holds:

$$\left(\sum_{\lambda \neq \mu} V_\lambda\right) \cap V_\mu = 0 \qquad (\forall \mu \in \Lambda).$$

$V_A(\neq 0)$ is said to be *indecomposable* if it is not expressed in a direct sum of two nonzero A-submodules. If $V = \bigoplus_\lambda V_\lambda$ is a direct sum of indecomposable A-submodules $V_\lambda(\lambda \in \Lambda)$, we call it an *indecomposable decomposition* of V and write

$$V = \bigoplus_{\lambda \in \Lambda} V_\lambda \text{ (an indecomposable decomposition).}$$

Each V_λ is called an *indecomposable component* of V.

We write $U|V$ if U_A is isomorphic to a direct summand of V, i.e., there exists a decomposition $V = W \oplus W'$ such that $U \simeq W$.

Exercise 1.5.

(i) $W|U,\ U|V \Rightarrow W|V$.

(ii) If $V = W \oplus W'$ and $U \supset W$, then $U = W \oplus (U \cap W')$.

Let $V = \bigoplus_{\lambda \in \Lambda} V_\lambda$ be a direct sum. For each $\lambda \in \Lambda$, define $\pi_\lambda\colon V \to V$ by $\pi_\lambda(v) = v_\lambda$, where v_λ is the λ-coordinate of v. Then $\pi_\lambda \in \operatorname{End}_A(V)$, and it is called the *projection* on V_λ (with regard to (w.r.t.) the decomposition above). The following equalities are evident:

$$\pi_\lambda \pi_\mu = \delta_{\lambda\mu}\pi_\lambda, \qquad \mathrm{id}_V = \sum_\lambda \pi_\lambda,$$

where $\delta_{\lambda\mu}$ denotes the Kronecker delta. (The second equality makes sense because there are only a finite number of λ such that $\pi_\lambda(v) \neq 0$ for each $v \in V$).

Let $V_\lambda(\lambda \in \Lambda)$ be A-modules. Then the cartesian product $\prod_{\lambda \in \Lambda} V_\lambda$ becomes an A-module, called the *direct product* of $\{V_\lambda; \lambda \in \Lambda\}$, with the following addition and A-operation:

$$(v_\lambda) + (v'_\lambda) = (v_\lambda + v'_\lambda)_\lambda, \qquad (v_\lambda)_\lambda a = (v_\lambda a)_\lambda \quad \text{for} \quad a \in A.$$

In $\prod_{\lambda \in \Lambda} V_\lambda$ the subset consisting of the elements with only a finite number of nonzero coordinates is an A-submodule, which is referred to as an (external) direct sum and is denoted by $\bigoplus_{\lambda \in \Lambda} V_\lambda$. For $\lambda \in \Lambda$ the subset $\{(v_\lambda)_\lambda; v_\mu = 0 \text{ for } \mu \neq \lambda\}$ is an A-submodule of $\bigoplus_{\lambda \in \Lambda} V_\lambda$, being isomorphic to V_λ. We identify this with V_λ and regard V_λ as a submodule of $\bigoplus_{\lambda \in \Lambda} V_\lambda$. Then $\bigoplus_{\lambda \in \Lambda} V_\lambda$ is just the direct sum of the submodules $V_\lambda(\lambda \in \Lambda)$. If Λ is a finite set, the direct product is the same as the direct sum.

Theorem 1.6. *Every A-module is a homomorphic image of some A-free module.*

Proof. Let $X = \{v_\lambda; \lambda \in \Lambda\}$ be a set of generators of V (take $X = V$, for instance). Consider a set of symbols $\{x_\lambda; \lambda \in \Lambda\}$ and let F be the free A-module generated by them, i.e., $F = \bigoplus_{\lambda \in \Lambda} x_\lambda A$ with $x_\lambda A \simeq A$. Then $f: F \to V(\sum_\lambda x_\lambda a_\lambda \mapsto \sum_\lambda v_\lambda a_\lambda)$ is a (well-defined) A-epimorphism. ■

1.5. Algebras

Let A be a ring. The set of elements of A commuting with every element of A is a commutative subring of A, which is called the *center* of A and is denoted by $Z(A)$, i.e., $Z(A) = \{z \in A; az = za \text{ for all } a \in A\}$.

Let R be a commutative ring. If we are given a ring homomorphism $f: R \to Z(A)$, we say A is an *algebra* over R or an R-algebra, and R is called the *coefficient ring* of A. If V is a module over an R-algebra A, then V is an R-module via the action: $rv = vf(r)$ for $r \in R$ and $v \in V$. Moreover, it holds that $r(va) = (rv)a (r \in R, a \in A, v \in V)$. In particular, if we view A as an R-module in this way, then the following holds:

$$r(ab) = (ra)b = a(rb) \qquad \text{for all} \quad r \in R, \quad a, b \in A. \tag{1.5}$$

Conversely, if a ring A is an R-module satisfying (1.5), then $f: R \to Z(A)(r \mapsto r1_A)$ is a ring homomorphism, which makes A into an R-algebra.

If $f: R \to Z(A)$ is a monomorphism, we can regard R as a subring of $Z(A)$ by identifying r with $f(r)$.

A is said to be a finitely generated (respectively, free) R-algebra if A is finitely generated (respectively, free) as an R-module.

For example an algebra A over a field K is K-free, and we regard K as a subring of $Z(A)$.

Lemma 1.7. *Suppose that A is an integral domain that contains a field K. If $\dim_K A$ is finite, then A is a field, i.e., a finite extension of K.*

Proof. Let a be a nonzero element of A and consider the K-linear map $f_a: A \to A(x \mapsto xa)$. This is a monomorphism because A is an integral domain. Thus we have $A = f(A)$ by comparing the dimensions. In particular, there exists $x \in A$ such that $xa = 1$. This proves that A is a field. ■

Example 1.8. Let R be a commutative ring and G be a finite group. Consider the R-free module $RG = \bigoplus_{x \in G} Rx$ with basis G and define the multiplication in RG by

$$\left(\sum_x \alpha_x x\right)\left(\sum_x \beta_x x\right) = \sum_x \gamma_x x,$$

where $\alpha_x, \beta_x \in R$ and $\gamma_x = \sum_{yz=x} \alpha_y \beta_z$. Then RG becomes an R-algebra, which is called the *group ring* of G over R. This is sometimes denoted by $R[G]$.

For $X \subset G$, we let $\hat{X} = \sum_{x \in X} x \in RG$.

2. Noetherian and Artinian Modules

In what follows, A will always denote a ring.

An A-module V is called a *Noetherian* (respectively, an *Artinian*) *module* if every nonempty set of A-submodules of V contains a maximal (respectively, minimal) element w.r.t. inclusion. In this case, we also say that the V satisfies the *maximum* (respectively, *minimum*) *condition* on submodules.

Theorem 2.1. *The following three conditions on V_A are equivalent.*

(1) *V is a Noetherian module.*

(2) *(Ascending chain condition.) For every infinite chain of A-submodules of V,*

(2.1) $$V_1 \subset V_2 \subset \cdots \subset V_n \subset \cdots,$$

there is a number m such that $V_m = V_{m+1} = \cdots$.

(3) *Every A-submodule of V is finitely generated over A.*

Proof.

(1) $\Rightarrow$ (2). There is a maximal element, say V_m, in the set $\{V_n; 1 \le n\}$. Then $V_m = V_{m+1} = \cdots$.

(2) $\Rightarrow$ (1). Suppose that V is not Noetherian. Then there exists a set $\mathscr{V} = \{V_\lambda, \lambda \in \Lambda\}$ of submodules of V that contains no maximal element. So if $V_{\lambda_1} \in \mathscr{V}$, then there is $V_{\lambda_2} \in \mathscr{V}$ such that $V_{\lambda_1} \subsetneqq V_{\lambda_2}$. Repeating this, we get an infinite chain of submodules

$$V_{\lambda_1} \subsetneqq V_{\lambda_2} \subsetneqq \cdots \subsetneqq V_{\lambda_n} \subsetneqq \cdots,$$

contradicting the assumption.

(2) ⇒ (3). Let W be an arbitrary A-submodule of V. Take $w_1 \in W$ and let $W_1 = w_1 A$. If $W_1 \subsetneqq W$, take $w_2 \subsetneqq W - W_1$, and let $W_2 = w_1 A + w_2 A$. Repeating this, we get a chain $W_1 \subsetneqq W_2 \subsetneqq \cdots \subsetneqq W_n \cdots$, which will terminate by assumption. Thus $W = W_n = w_1 A + w_2 A + \cdots + w_n A$ for some n.

(3) ⇒ (2). Let $W = \bigcup_{n=1}^{\infty} V_n$. As is easily seen, W is an A-submodule of V, and thus it is finitely generated by assumption; $W = w_1 A + w_2 A + \cdots + w_n A$. Then there exists V_m which contains all w_i, and thus $V_m = V_{m+1} = \cdots = W$. ■

An argument similar to the one in the proof of the equivalence (1) ⇔ (2) of the above theorem gives the following.

Exercise 2.2. The following two conditions on V_A are equivalent.

(1) V is an Artinian module.

(2) (Descending chain condition.) For every infinite chain of A-submodules of V,

$$V_1 \supset V_2 \supset \cdots \supset V_n \supset \cdots,$$

there is a number m such that $V_m = V_{m+1} = \cdots$.

Theorem 2.3. *Let W be an A-submodule of V. Then V is Noetherian (respectively, Artinian) if and only if both W and V/W are Noetherian (respectively, Artinian).*

Proof. We prove the result in the Noetherian case. The Artinian case will be treated similarly.

It is clear that if V is Noetherian, then so are both W and V/W. Given an infinite chain of A-submodules $V_1 \subset V_2 \subset \cdots \subset V_n \subset \cdots$, we get the following chains of submodules of W and V/W:

$$V_1 \cap W \subset V_2 \cap W \subset \cdots \subset V_n \cap W \subset \cdots, \tag{2.2}$$

$$(V_1 + W)/W \subset (V_2 + W)/W \subset \cdots \subset (V_n + W)/W \subset \cdots. \tag{2.3}$$

Since both W and V/W satisfy the ascending chain condition, there is a number m such that

$$V_m \cap W = V_{m+1} \cap W = \cdots,$$

$$V_m + W = V_{m+1} + W = \cdots,$$

whence it follows easily that $V_m = V_{m+1} = \cdots$. ■

Exercise 2.4. Suppose that $V = \sum_{i=1}^{n} V_i$ is a sum of the A-submodules V_i. Then V is Noetherian (respectively, Artinian) if and only if each V_i is Noetherian (respectively, Artinian). [Hint: Use induction on n.]

A is said to be a right *Noetherian* (respectively, *Artinian*) *ring* if A is Noetherian (respectively, Artinian) as a right A-module. A left Noetherian or a left Artinian ring can be defined similarly. If A is commutative, there is no distinction between the right and left versions. *Henceforth we shall simply refer to a right Noetherian (respectively, Artinian) ring as a Noetherian (respectively, Artinian) ring.*

Theorem 2.5. *If A is a Noetherian (respectively, an Artinian) ring, then every finitely generated A-module is Noetherian (respectively, Artinian). In particular if A is a finitely generated algebra over a commutative Noetherian (respectively, Artinian) ring R, then A is right and left Noetherian (respectively, Artinian).*

Proof. Let $V_A = v_1A + v_2A + \cdots + v_nA$. Then each v_iA, being a homomorphic image of A, is Noetherian. Hence, V is Noetherian by Exercise 2.4. To show the second half, note that A is a Noetherian R-module by the above. But since every A-submodule of A is necessarily an R-submodule, both A_A and ${}_AA$ satisfy the maximum condition on A-submodules. ■

A nonzero A-module V is said to be *irreducible* (or *simple*) if it has no proper A-submodule other than 0. A chain of A-submodules of V

$$V = V_0 \supset V_1 \supset \cdots \supset V_n = 0 \tag{2.4}$$

is said to be a *composition series* of V if each V_i/V_{i+1} is irreducible. And the simple modules

$$V/V_1,\ V_1/V_2, \ldots,\ V_{n-1}/V_n = V_{n-1}$$

are called the *composition factors* of the series.

Exercise 2.6. A non-zero module has a composition series if and only if it is both Noetherian and Artinian.

We mention the following fundamental theorem without proof (cf. Hungerford [1], p. 111).

Theorem 2.7 (Jordan–Hölder). *If V has a composition series, any two composition factors obtained from the two series of V are the same up to order of occurrence and isomorphism.*

If (2.4) is a composition series of V, then the integer n is called the *composition length* and each V_i/V_{i+1} is an *irreducible constituent* of V.

3. The Radical of a Ring

Before mentioning the definition of the radical of a ring A, we show the following lemma.

Lemma 3.1. *Let W be a proper A-submodule of V_A. If V/W is finitely generated over A, then there is a maximal A-submodule of V containing W. In particular, if V is finitely generated, then there always exists a maximal A-submodule of V containing W.*

Proof. Let $\bar{V} = V/W = \sum_{i=1}^{n} \bar{v}_i A$. Let $\mathscr{W}$ be the set of proper A-submodules of V containing W. Then $\mathscr{W}$ is an ordered set w.r.t. inclusion. We claim that an arbitrary totally ordered subset in $\mathscr{W}$, say $\{W_\lambda; \lambda \in \Lambda\}$, has an upper bound. Let $U = \bigcup_{\lambda \in \Lambda} W_\lambda$. If $U = V$, there exists W_{λ_i} for each i such that $v_i \in W_{\lambda_i}$. Thus, if W_α is the largest module of $\{W_{\lambda_1}, \ldots, W_{\lambda_n}\}$, then it contains all v_i and hence $W_\alpha = V$, a contradiction. Hence $U \neq V$, and U gives an upper bound for $\{W_\lambda; \lambda \in \Lambda\}$. Thus by Zorn's lemma, we get the first assertion. The second half is clear. ■

If we apply the above lemma to $V = A(=1A)$, we obtain the following.

Corollary 3.2. *For every proper right ideal I of A, there exists a maximal right ideal of A containing I.*

The intersection of all maximal right ideals of A is called the (Jacobson) *radical* of A and is denoted by $J(A)$.

Theorem 3.3. *The following holds for the radical of a ring A.*

(i) *$J(A)$ coincides with the intersection of all the annihilator ideals of irreducible right A-modules. Consequently, $J(A)$ is an ideal of A.*

(ii) *$J(A) \ni a \Leftrightarrow 1 - ax$ has a right inverse for all $x \in A$, in other words, there exists $b \in A$ such that $(1 - ax)b = 1$.*

(iii) *$J(A)$ is the largest one among the ideals I of A satisfying the following condition:*

$$(*) \qquad I \ni a \Rightarrow 1 - a \text{ is a unit of } A.$$

(iv) *$J(A)$ coincides with the intersection of all maximal left ideals of A.*

Proof.

(i) Let V_A be irreducible and $v \neq 0 \in V$. Then $V = vA \simeq A/(0:v)$. Since $(0:v)$ is a maximal right ideal of A, it follows that $(0:v) \supset J(A)$, i.e., $vJ(A) = 0$. Thus $J(A) \subset (0:V)$. On the other hand, if $a \in \bigcap_V (0:V)$, V ranging over the irreducible A-modules, then $(A/M)a = 0$ for every maximal right ideal M of A. Thus $a \in Aa \subset M$, and hence $a \in J(A)$.

(ii) ($\Rightarrow$) Let $a \in J(A)$ and suppose that $1 - ax$ has no right inverse for some $x \in A$. So $(1 - ax)A \neq A$, and hence there exists a maximal right ideal M such that $M \supset (1 - ax)A \ni 1 - ax$. Thus $M \ni (1 - ax) + ax = 1$, as $M \supset J(A) \ni ax$. This is a contradiction.

($\Leftarrow$). If $a \notin J(A)$, then $a \notin M$ for some maximal right ideal M of A. Thus $M + aA = A$ and hence $b + ax = 1$ for some $x \in A$ and $b \in M$. Therefore $1 - ax = b \in M$ has no right inverse.

(iii) If $a \in J(A)$, then $(1 - a)b = 1$ for some $b \in A$, and $b = 1 + ab$ has a right inverse; $bc = 1$ for some $c \in A$. Thus $c = ((1 - a)b)c = (1 - a)bc = 1 - a$, and hence $b(1 - a) = 1$. Namely, $1 - a$ is a unit of A and $J(A)$ satisfies $(*)$.

Let I be any ideal of A satisfying $(*)$. If $I \not\subset J(A)$, then there is a maximal right ideal M such that $I \not\subset M$, so $I + M = A$. Hence there exists $a \in I$, $b \in M$ such that $a + b = 1$. But then $1 - a(= b \in M)$ is not a unit of A, contradicting the assumption. Therefore $I \subset J(A)$.

(iv) If we start by adopting the intersection of all maximal left ideals as the definition of $J(A)$, we shall arrive at the same characterization of $J(A)$ as given in (iii), which is free from the right and left versions. This completes the proof of the theorem. ■

An element a of A is said to be *nilpotent* if $a^n = 0$ for some $n \in \mathbf{N}$. If so, then $1 - a$ is a unit of A, since $(1 - a)(1 + a + \cdots + a^{n-1}) = (1 + a + \cdots + a^{n-1})(1 - a) = 1$.

A right ideal of A is said to be a *nil right ideal* if every element of I is nilpotent. Also, I is said to be *nilpotent* if $I^n = 0$ for some n. Clearly, a nilpotent right ideal is a nil right ideal.

Theorem 3.4. *If I is a nil right ideal, then $I \subset J(A)$.*

Proof. If $a \in I$ and $x \in A$, $ax(\in I)$ is nilpotent and hence $1 - ax$ is a unit of A. Therefore $a \in J(A)$ by Theorem 3.3(ii). ■

The following holds for the radical of an Artinian ring.

Theorem 3.5. *If A is Artinian, then $J(A)$ is nilpotent.*

Proof. Consider the chain of ideals $J(A) \supset J(A)^2 \supset \cdots$. Then there is n such that $J(A)^n = J(A)^{n+1} = \cdots$. We deduce a contradiction assuming that $N = J(A)^n \neq 0$. Since $N = N^2 \neq 0$, the set of right ideals I of A such that $IN = N$ is not empty. Let I_0 be a minimal member of it. Then there is $a \in I_0$ such that $aN \neq 0$. Since $aN \subset I_0$ and $(aN)N = aN \neq 0$, it follows that $aN = I_0$ by the minimality of I_0. In particular there exists $b \in N$ such that $ab = a$, so $a(1 - b) = 0$ and hence $a = 0$, since $1 - b$ is a unit by Theorem 3.3(iii). This is a contradiction since $aN \neq 0$. ■

The next result is of fundamental importance, which we shall refer to as *Azumaya–Nakayama's lemma.*

Theorem 3.6 (Azumaya–Nakayama).

(i) *Let V an A-module and W be an A-submodule of V such that V/W is finitely generated over A. Then*

$$V = W + VJ(A) \Rightarrow V = W.$$

(ii) *If V is finitely generated over A, then the following holds:*

$$VJ(A) = V \Rightarrow V = 0.$$

Proof.

(i) If $W \subsetneqq V$, there exists a maximal A-submodule, say M, such that $M \supset W$ by Lemma 3.1. Since V/M is irreducible, we have $(V/M)J(A) = 0$, i.e., $VJ(A) \subset M$. Therefore $V = W + VJ(A) \subset M$, which is a contradiction.

(ii) Let $W = 0$ in (i). ■

Let us consider the radicals of certain rings associated with A.

Theorem 3.7. *Let I be an ideal of A that is contained in $J(A)$. Then $J(A/I) = J(A)/I$.*

Proof. Since I is contained in every maximal right ideal of A, the set of maximal right ideals of A/I is given by $\{M/I\}_M$, where M runs through all maximal right ideals of A. This proves the assertion. ■

A is called a *simple ring* if it has no proper ideal other than 0. Also A is said to be *semisimple* if $J(A) = 0$. Clearly, a simple ring is semisimple and $A/J(A)$ is semisimple by Theorem 3.7.

A nonzero element e of A is called an *idempotent* if $e^2 = e$. If so, then eAe is a ring with identity e.

Exercise 3.8. No idempotent e of A is contained in $J(A)$. [Hint: $e(1 - e) = 0$.]

Theorem 3.9. *The following holds for an idempotent e of A:*

$$J(eAe) = eJ(A)e = eAe \cap J(A).$$

Proof. It is clear that $eAe \cap J(A) = eJ(A)e$. If $a \in J(A)$, then $eae \in J(A)$, and hence there exists $b \in A$ such that

$$(1 - eae)b = 1, \qquad b(1 - eae) = 1.$$

Multiplying both sides of these equations by e, we get $(e - eae)ebe = ebe(e - eae) = e$. Therefore $e - eae$ is a unit of eAe. Since $eJ(A)e$ is an ideal of eAe, it follows that $eJ(A)e \subset J(eAe)$ by Theorem 3.3(iii).

To show that $J(eAe) \subset J(A)$, it suffices to prove that $VJ(eAe) = 0$ for any irreducible A-module V. This is clear if $Ve = 0$. Suppose that $Ve \neq 0$, and let W be any nonzero eAe-submodule of Ve. Then $V = WA$ because V is irreducible, and hence $Ve = WAe = WeAe = W$. This implies that Ve is irreducible as an eAe-module, and we have $0 = VeJ(eAe) = VJ(eAe)$ as desired. ∎

Theorem 3.10. *Let A be a finite-dimensional algebra over a field K. Then the following holds for the radical of the center of A:*

$$J(Z(A)) = J(A) \cap Z(A).$$

Proof. $J(A) \cap Z(A)$ is contained in $J(Z(A))$ since it is a nilpotent ideal of $Z(A)$. On the other hand, $J(Z(A))A$ is a nilpotent ideal of A and hence contained in $J(A)$. Therefore $J(Z(A)) \subset J(A) \cap Z(A)$. ∎

For a subset S of A, we let $M_n(S)$ denote the subset of $M_n(A)$ consisting of those matrices all of whose entries lie in S. Then the following holds.

Theorem 3.11. $J(M_n(A)) = M_n(J(A))$.

Proof. Let e_{ij} be the matrix in $M_n(A)$ with (i, j)-entry 1 as its unique nonzero entry. Then $J_i = \bigoplus_{j=1}^{n} J(A)e_{ij}$ is a right ideal of $M_n(A)$. Let $a = \sum_j \alpha_{ij}e_{ij}(\alpha_{ij} \in J(A))$ be an element of J_i. Since $(1 - \alpha_{ii})A = A$, there exists

$\beta_{ij} \in A$ such that $(1 - \alpha_{ii})\beta_{ij} = \alpha_{ij}$, i.e., $\beta_{ij} - \alpha_{ij} = \alpha_{ii}\beta_{ij}$. Then if we let $b = \sum_j -\beta_{ij}e_{ij}$, we find that $(1 - a)(1 - b) = 1$. So, since J_i is a right ideal of $M_n(A)$, $1 - ax$ has a right inverse for all $x \in A$. Thus $J_i \subset J(M_n(A))$ and $M_n(J(A)) = \bigoplus_i J_i \subset J(M_n(A))$.

Conversely, let $a = \sum_{i,j} \alpha_{ij}e_{ij}$ be any element of $J(M_n(A))$ and $\beta \in A$. Then $c_{1p}a(\beta e_{q1}) = (\alpha_{pq}\beta)e_{11} \in J(M_n(A))$, and hence there exists $c = \sum_{i,j} \gamma_{ij}e_{ij}$ such that $(1 - (\alpha_{pq}\beta)e_{11})(1 - c) = 1$. Comparing the coefficients of e_{11} on both sides, we get $(1 - \alpha_{pq}\beta)(1 - \gamma_{11}) = 1$, i.e., $1 - \alpha_{pq}\beta$ has a right inverse. So, $\alpha_{pq} \in J(A)$ and we conclude that $a \in M_n(J(A))$. ■

4. Idempotents

4.1. Idempotents and Direct Sum Decompositions of Ideals

Two idempotents e_1, e_2 of a ring A are said to be *orthogonal*, provided $e_1e_2 = e_2e_1 = 0$. More generally, n idempotents $e_1, e_2, \ldots, e_n$ are said to be orthogonal if any pair of them is orthogonal. In this case, $e = e_1 + \cdots + e_n$ is also an idempotent.

An idempotent e is said to be *primitive* if it is not expressed as a sum of two orthogonal idempotents. By an *idempotent decomposition* of e we mean the decomposition

$$e = e_1 + \cdots + e_n$$

into a sum of orthogonal idempotents e_i. If moreover all e_i are primitive, then this is called a *primitive idempotent decomposition* of e.

For example if $e \neq 1$ is an idempotent of A, then e and $1 - e$ are orthogonal and $1 = e + (1 - e)$ is an idempotent decomposition of 1.

Idempotent decompositions and direct sum decompositions of right ideals of A are related as follows.

Theorem 4.1. *Let e be an idempotent of A. If $e = e_1 + e_2 + \cdots + e_n$ is an idempotent decomposition, then*

$$eA = e_1A \oplus e_2A \oplus \cdots \oplus e_nA. \tag{4.1}$$

Conversely, if eA is a direct sum of right ideals

$$eA = I_1 \oplus I_2 \oplus \cdots \oplus I_n. \tag{4.2}$$

then there is an idempotent decomposition $e = e_1 + e_2 + \cdots + e_n$ *such that* $I_i = e_i A (1 \leq i \leq n)$.

In this way the idempotent decompositions of e are in bijective correspondence to the direct sum decompositions of $(eA)_A$.

Proof. Let $e = \sum_{i=1}^n e_i$ be an idempotent decomposition. Thus $ee_i = e_i$, and hence $e_i A \subset eA$. But this yields that $eA = \sum_{i=1}^n e_i A$ since $ea = \sum_{i=1}^n e_i a \in \sum_{i=1}^n e_i A$ for all $a \in A$. To see whether the sum is direct, let $ea = \sum_{i=1}^n e_i a_i$. Then $e_i a = e_i ea = e_i a_i$, so the expression is unique and (4.1) holds.

Next, assume (4.2) and write $e = e_1 + \cdots + e_n$, where $e_i \in I_i$. Then for any $a_i \in I_i \subset eA$, we have $a_i = ea_i = \sum_j e_j a_i$ and hence $e_j a_i = \delta_{ij} a_i$. Thus $e_i A = I_i$. Furthermore, if we let $a_i = e_i$, then the above also shows that $e_1, \ldots, e_n$ are orthogonal idempotents. ■

As an easy consequence of the above theorem, we have the following.

Theorem 4.2. *Let e be an idempotent of A. Then the following conditions are equivalent:*

(1) *e is primitive.*

(2) $(eA)_A$ *is indecomposable.*

(3) *e is a unique idempotent of eAe.*

Proof.

(1)$\Leftrightarrow$(2). This is clear by Theorem 4.1.

(1)$\Rightarrow$(3). If e' is an idempotent of eAe different from e, then $e = e' + (e - e')$ is an idempotent decomposition of e, and hence e is not primitive.

(3)$\Rightarrow$(1). If $e = e_1 + e_2$ is an idempotent decomposition, then $e_1 = ee_1 e \in eAe$, and hence e_1 is an idempotent of eAe different from e. ■

Theorem 4.3. *Let e, f be idempotents of A, and V be an A-module. Then the following holds.*

(i) $\mathrm{Hom}_A(eA, V) \simeq Ve$. *In particular* $\mathrm{Hom}_A(eA, fA) \simeq fAe$.

(ii) $\mathrm{End}_A(eA) \simeq eAe$ *as rings.*

Proof. If $\varphi \in \mathrm{Hom}_A(eA, V)$, then $\varphi(e) = \varphi(ee) = \varphi(e)e \in Ve$. We show that the map $g: \mathrm{Hom}_A(eA, V) \to Ve(\varphi \mapsto \varphi(e))$ is an isomorphism. Since $\varphi(ex) = \varphi(e)x$ for $x \in A$, φ is uniquely determined by $\varphi(e)$, and hence g is a monomorphism. To see that g is an epimorphism, let v be any element of Ve. Then $\psi: eA \to V(ex \mapsto vex = vx)$ defines an A-homomorphism such that $v = \psi(e) = g(\psi)$. Thus g is an epimorphism.

If in particular $V = eA$, then it holds, for σ, $\tau \in \mathrm{End}_A(eA)$, that $\sigma(\tau e) = \sigma(e(\tau e)) = (\sigma e)(\tau e)$. Therefore we have the ring isomorphism $g: \mathrm{End}_A(eA) \xrightarrow{\sim} eAe(\sigma \mapsto \sigma(e))$. ■

For right ideals generated by idempotents of A, we have the following result.

Theorem 4.4. *Let e, f be idempotents of A. Then the following are equivalent.*

(1) *$eA \simeq fA$ (A-isomorphic).*

(2) *$Ae \simeq Af$ (A-isomorphic).*

(3) *There exist $a \in fAe$, $b \in eAf$ such that $ab = f$, $ba = e$.*

Proof.

(1) ⇒ (3). By assumption, there is an A-isomorphism $\varphi: eA \to fA$. Set $\varphi(e) = a$ and $\varphi^{-1}(f) = b$. Then we see that $a = \varphi(ee) = \varphi(e)e = ae \in fAe$, and similarly $b \in eAf$. Thus, $f = \varphi(\varphi^{-1}(f)) = \varphi(b) = \varphi(eb) = \varphi(e)b = ab$, and similarly we have $e = ab$.

(3) ⇒ (1). Since $a = fae$ and $b = eaf$, both $\varphi_a: eA \to fA(ex \mapsto ax)$ and $\varphi_b: fA \to eA(fx \mapsto bx)$ are A-homomorphisms such that $\varphi_b \circ \varphi_a = \mathrm{id}_{eA}$, $\varphi_a \circ \varphi_b = \mathrm{id}_{fA}$. Namely, φ_a is an A-isomorphism.

The equivalence of (2) and (3) can be proved similarly. ■

Theorem 4.5. *Suppose that I is an ideal of A contained in $J(A)$ and let $\bar{A} = A/I$. Then, for two idempotents e, f of A, an A-isomorphism $eA \simeq fA$ holds if and only if an $\bar{A}$-isomorphism $\bar{e}\bar{A} \simeq \bar{f}A$ holds.*

Proof. We need only to show that $eA \simeq fA$ assuming that $\bar{e}\bar{A} \simeq \bar{f}\bar{A}$. By the above theorem, there exist $\bar{a}, \bar{b} \in \bar{A}$ such that $\bar{f}\bar{a}\bar{e} = \bar{a}$, $\bar{e}\bar{b}\bar{f} = \bar{b}$, $\bar{a}\bar{b} = \bar{f}$, $\bar{b}\bar{a} = \bar{e}$.

We may assume that $fae = a$, $ebf = b$ by replacing a and b with fae and ebf, respectively. Let $ab = f - c(c \in I)$. Then, since $ab \in fAf$ and $c \in fIf \subset fJ(A)f = J(fAf)$, it follows that ab is a unit of fAf by Theorem 3.3(iii). This implies that the composite map $\varphi_{ab}: fA \to fA$ of $\varphi_b: fA \to eA(fx \mapsto bx)$ and $\varphi_a: eA \to fA(ex \mapsto ax)$ is an A-isomorphism. Likewise ba is a unit of eAe, and $\varphi_{ba}: eA \to eA$ is an A-isomorphism. Therefore φ_a gives a desired isomorphism $eA \xrightarrow{\sim} fA$. ∎

Two idempotents e and f of A are said to be *equivalent* and are written $e \simeq f$ if $eA \simeq fA$. We denote by pi(A) the set of primitive idempotents of A and by $\widetilde{\text{pi}}(A)$ a complete set of representatives of the equivalence classes of pi(A).

In general there does not always exist a primitive idempotent decomposition of an idempotent, nor is it uniquely determined even if it did exist. However, the following holds for commutative rings.

Theorem 4.6. *Let A be a commutative ring. If there exists a primitive idempotent decomposition*

$$1 = e_1 + e_2 + \cdots + e_n \tag{4.3}$$

of the identity, then we have

(i) pi(A) $= \{e_1, e_2, \ldots, e_n\}$. *In other words,* (4.3) *is a unique primitive idempotent decomposition of* 1.

(ii) *Every idempotent of A is a sum of certain e_i's.*

Proof. Let e be an idempotent of A. Then $e = ee_1 + ee_2 + \cdots + ee_n$ and $(ee_i)(ee_j) = \delta_{ij}(ee_i)$. If $ee_i \neq 0$, then $e_i = ee_i + (1 - e)e_i$. But since e_i is primitive, it must be $e_i = ee_i$ and hence $e = e_{i_1} + e_{i_2} + \cdots + e_{i_r}$, where $\{i_1, i_2, \ldots, i_r\} = \{i; ee_i \neq 0\}$. In particular, if e is primitive, we have $r = 1$ and $e = e_i$ for some i. ∎

An idempotent in the center $Z(A)$ of A is called a *central idempotent.* It is said to be *central primitive* if it is primitive in $Z(A)$. An idempotent decomposition in $Z(A)$ is said to be a *central idempotent decomposition.*

Theorem 4.7. *Let e be a central idempotent and let* $e = e_1 + e_2 + \cdots + e_n$ *be a central idempotent decomposition. Then*

$$eA = e_1 A \oplus e_2 A \oplus \cdots \oplus e_n A$$

holds, as (A, A)*-bimodules. Conversely, if*

$$eA = I_1 \oplus I_2 \oplus \cdots \oplus I_n \tag{4.4}$$

is a direct sum of ideals I_i, *there exists a central idempotent decomposition* $e = e_1 + e_2 + \cdots + e_n$ *such that* $I_i = e_i A$.

In this way the central idempotent decompositions of e are in bijective correspondence to the direct sum decompositions of ${}_A(eA)_A$.

In particular, e is central primitive if and only if eA is indecomposable as an (A, A)*-bimodule.*

Proof. The first statement is clear by Theorem 4.1 since eA and $e_i A$ are ideals of A. If (4.4) holds, then $e = e_1 + e_2 + \cdots + e_n$ and $I_i = e_i A$ by Theorem 4.1. Since $xe = ex$ for all $x \in A$, it follows that $xe_i = e_i x$ by comparing the ith coordinates of $\sum_i xe_i = \sum_i e_i x$. Thus e_i is central. The remaining assertions are now obvious. ■

Let

$$A = I_1 \oplus I_2 \oplus \cdots \oplus I_n \tag{4.5}$$

be a direct sum of ideals I_i of A. We see that $I_i I_j = 0$ whenever $i \neq j$, since $I_i I_j \subset I_i \cap I_j = 0$. Therefore, if we express $a, b \in A$ as

$$a = \sum_{i=1}^{n} a_i, \qquad b = \sum_{i=1}^{n} b_i \quad (a_i, b_i \in I_i),$$

then

$$a + b = \sum_{i=1}^{n} (a_i + b_i), \qquad ab = \sum_{i=1}^{n} (a_i b_i). \tag{4.6}$$

Moreover, if we write $1 = \sum_i e_i (e_i \in I_i)$, then each I_i is a ring with identity e_i.

Now, given rings $A_1, A_2, \ldots, A_n$, consider the direct sum

$$A = A_1 \oplus A_2 \oplus \cdots \oplus A_n \tag{4.7}$$

and define the multiplication in A as in (4.6). Then A becomes a ring called the *direct sum* of the rings $A_1, A_2, \ldots, A_n$.

Exercise 4.8. The following hold for the direct sum of rings (4.7).

(i) $Z(A) = Z(A_1) \oplus Z(A_2) \oplus \cdots \oplus Z(A_n)$.

(ii) $J(A) = J(A_1) \oplus J(A_2) \oplus \cdots \oplus J(A_n)$.

4.2. The Lifting Idempotent Theorem

In this subsection we assume that

(4.8) *I is a nilpotent ideal of A.*

We show certain relationships between the idempotents of A and those of $\bar{A} = A/I$.

Theorem 4.9. *Given an idempotent $\bar{c}$ of $\bar{A}$, there exists an idempotent e of A satisfying the following two conditions*:

(a) $\bar{e} = \bar{c}$.

(b) $e = f(c)$, *where $f(x)$ is a polynomial in x over $\mathbf{Z}$ with $f(0) = 0$.*

Proof. If $\bar{a} = \bar{c}$, then $a^2 - a \in I$ and $(a^2 - a)^n = 0$ for some n. Denote by $n(a)$ the minimum among those n. Let $X = \{a \in A;\ \bar{c} = \bar{a},\ a = f(c)$ for some $f(x) \in \mathbf{Z}[x]$ with $f(0) = 0\}$. This is not empty, because it contains c. Choose $e \in X$ so that $n(e)$ is minimal. We show that $n(e) = 1$. Suppose by way of contradiction that $n(e) > 1$ and let $t = e^2 - e (\in I)$, $e' = e - 2et + t$. We see readily that $e' \in X$ and $(e')^2 - e' = 4t^3 - 3t^2$. But it then follows that $n(e') \leq (n(e) + 1)/2 < n(e)$, a contradiction. ■

In general we say that an idempotent $\bar{c}$ of $\bar{A}$ can be lifted to A if there exists an idempotent e of A such that $\bar{e} = \bar{c}$. This e is called a *lift* of $\bar{c}$. The next result states more generally that any idempotent decomposition in $\bar{A}$ *lifts* to that in A.

Theorem 4.10. *Let $\bar{c}$ be an idempotent of $\bar{A}$ and let*

$$\bar{c} = \bar{c}_1 + \bar{c}_2 + \cdots + \bar{c}_n$$

be an idempotent decomposition. Let furthermore $e = f(c)$ *be a lift of* $\bar{c}$ *with* $f(x) \in \mathbf{Z}[x]$ *such that* $f(0) = 0$. *Then there exist orthogonal idempotents* e_1, $e_2, \ldots, e_n$ *of* A *satisfying the following conditions*:

(a) $\bar{e}_i = \bar{c}_i (i = 1, \ldots, n)$.

(b) $e = e_1 + e_2 + \cdots + e_n$.

(c) *There exist* $f_i(x_0, x_1, \ldots, x_n)(1 \leq i \leq n)$ *in the (noncommutative) polynomial ring* $\mathbf{Z}[x_0, x_1, \ldots, x_n]$ *with* $f_i(0, 0, \ldots, 0) = 0$ *such that* $e_i = f_i(c, c_1, \ldots, c_n)$.

Proof. We prove by induction on n. By the above theorem, we may assume that $n > 1$. Since $\overline{ec_1e} = \bar{c}_1$ is an idempotent of $\bar{A}$, there is an idempotent e_1 of A such that $\bar{e}_1 = \bar{c}_1$ and $e_1 = f_1(ec_1e)$ for some $f_1(x) \in \mathbf{Z}[x]$ with $f_1(0) = 0$. Then $e' = e - e_1$ is an idempotent orthogonal to $e_1 \in eAe$. If we apply the inductive hypothesis to $\bar{e}' = \bar{c}_2 + \bar{c}_3 + \cdots + \bar{c}_n$, we have orthogonal idempotents $e_2, e_3, \ldots, e_n$ of A such that $e' = e_2 + e_3 + \cdots + e_n$ and $e_i = g_i(e', c_2, \ldots, c_n)$ for some polynomial g_i over $\mathbf{Z}$ with $g_i(0, \ldots, 0) = 0 (2 \leq i \leq n)$. In particular, part (c) follows at once. Also, each $e_i (2 \leq i \leq n)$ is orthogonal to e_1 as $e_i = e_i e' = e' e_i$, and thus $e = e_1 + e_2 + \cdots + e_n$ is an idempotent decomposition. ■

As an easy consequence of the above result, we have

Theorem 4.11. *An idempotent* e *of* A *is primitive if and only if* $\bar{e}$ *is primitive in* $\bar{A}$.

Proof. If $e = e_1 + e_2$ is an idempotent decomposition, then $\bar{e} = \bar{e}_1 + \bar{e}_2$ is an idempotent decomposition since $\bar{e}_i \neq 0 (i = 1, 2)$.

Conversely, if $\bar{e} = \bar{c}_1 + \bar{c}_2$ is an idempotent decomposition in $\bar{A}$, there exists an idempotent decomposition $e = e_1 + e_2$ in A. ■

Combining the above theorem with Theorem 4.5, we get the following.

Theorem 4.12. *There is a bijection* $\widetilde{\mathrm{pi}}(A) \leftrightarrow \widetilde{\mathrm{pi}}(\bar{A}) (e \mapsto \bar{e})$.

5. Endomorphism Rings

We begin with the following fundamental result, which is known as *Schur's lemma.*

Theorem 5.1. *Let V, W be irreducible A-modules.*

(i) *If* $\mathrm{Hom}_A(V, W) \ni \varphi \neq 0$, *then φ is an isomorphism. In particular, we have* $\mathrm{Hom}_A(V, W) = 0$ *if V and W are not A-isomorphic.*

(ii) $\mathrm{End}_A(V)$ *is a division ring.*

Proof.

(i) Since W is irreducible and $\mathrm{Im}\,\varphi \neq 0$, it follows that $\mathrm{Im}\,\varphi = W$. Also, we have $\mathrm{Ker}\,\varphi = 0$ because $\mathrm{Ker}\,\varphi \subsetneq V$ and V is irreducible. Therefore φ is an isomorphism.

(ii) If $\mathrm{End}_A(V) \ni \varphi \neq 0$, then φ is an isomorphism, i.e., a unit of $\mathrm{End}_A(V)$. ■

We have the following result for the endomorphism ring of a module, which is a direct sum of isomorphic modules.

Theorem 5.2. *Let*

$$V_A = V_1 \oplus V_2 \oplus \cdots \oplus V_n \tag{5.1}$$

be a direct sum of A-modules and suppose that $V_1, V_2, \ldots, V_n$ are isomorphic to each other. Then $\mathrm{End}_A(V)$ is isomorphic to the full matrix ring of degree n over $\mathrm{End}_A(V_1)$.

Proof. Let $\pi_i\colon V \to V_i$ be the projection and fix an isomorphism $\theta_i\colon V_1 \simeq V_i$. For $\sigma \in \mathrm{End}_A(V)$, let $\varphi_{ij}(\sigma) = \theta_i^{-1}\pi_i\sigma\theta_j$. Then $\varphi_{ij} \in E_1 = \mathrm{End}_A(V_1)$, and we have a map $\varphi\colon \mathrm{End}_A(V) \to M_n(E_1)(\sigma \mapsto (\varphi_{ij}(\sigma)))$. It is easy to see that φ is a ring homomorphism. We show that φ is an isomorphism.

If $\varphi(\sigma) = 0$, then $\pi_i\sigma\theta_j(V_1) = \pi_i\sigma(V_j) = 0$ for all i, j and hence $\sigma(V_j) = 0$, i.e., $\sigma = 0$. Therefore φ is a monomorphism. To see that φ is an epimorphism, let (σ_{ij}) be an arbitrary element of $M_n(E_1)$ and $\iota_i\colon V_i \to V$ be the inclusion map. If we set $\sigma = \sum_{i,j} \iota_i\theta_i\sigma_{ij}\theta_j^{-1}\pi_j$, then $\sigma \in \mathrm{End}_A(V)$ and it follows that $\varphi(\sigma) = (\sigma_{ij})$. ■

Using Schur's lemma, we have the following special case of the above theorem.

Corollary 5.3. *If each V_i is irreducible in the above theorem, then* $\mathrm{End}_A(V)$ *is isomorphic to the full matrix ring of degree n over the division ring* $\mathrm{End}_A(V_1)$.

There is a close connection between direct sum decompositions of V and idempotent decompositions of id_V in $\mathrm{End}_A(V)$. Namely, we have

Theorem 5.4. *Let V be an A-module and put* $E = \mathrm{End}_A(V)$.

(i) *Let*

$$V = V_1 \oplus V_2 \oplus \cdots \oplus V_n \tag{5.2}$$

be a direct sum of A-submodules V_i and ε_i be the projection on V_i. Then

$$\mathrm{id}_V = \varepsilon_1 + \varepsilon_2 + \cdots + \varepsilon_n \tag{5.3}$$

is an idempotent decomposition of id_V

Conversely, if the idempotent decomposition (5.3) *of* id_V *is given, then V decomposes as in* (5.2) *with $V_i = \varepsilon_i V$, and ε_i gives the projection on V_i.*

Moreover we have a decomposition

$$E = E_1 \oplus E_2 \oplus \cdots \oplus E_n \qquad (E_i = \varepsilon_i E) \tag{5.4}$$

of E into the direct sum of right ideals E_i. In this way, the direct-sum decompositions of V, idempotent decompositions of id_V*, and decompositions of E into the direct sum of right ideals are in bijective correspondence to each other.*

(ii) *If ε_1, ε_2 are idempotents of E, then*

$$\mathrm{Hom}_A(\varepsilon_1 V, \varepsilon_2 V) \simeq \varepsilon_2 E \varepsilon_1.$$

In particular, $\mathrm{End}_A(\varepsilon_1 V) \simeq \varepsilon_1 E \varepsilon_1$ *as rings.*

(iii) *Let ε_1, ε_2 be as above. Then the following holds:*

$$\varepsilon_i V \simeq \varepsilon_2 V \quad (A\text{-isomorphic}) \quad \Leftrightarrow \varepsilon_1 E \simeq \varepsilon_2 E \quad (E\text{-isomorphic}).$$

Proof.

(i) This can be shown in a similar way as in Theorem 4.1.

(ii) If $\sigma \in E$, we have an A-homomorphism $\varphi(\varepsilon_2 \sigma \varepsilon_1)\colon \varepsilon_1 V \to \varepsilon_2 V (\varepsilon_1 v \to \varepsilon_2 \sigma \varepsilon_1 v)$. Thus we have a homomorphism $\varphi\colon \varepsilon_2 E \varepsilon_1 \to$

$\mathrm{Hom}_A(\varepsilon_1 V, \varepsilon_2 V)$, which is clearly a monomorphism. To show that it is an epimorphism, let $\varepsilon_1' = \mathrm{id}_V - \varepsilon_1$. Then $V = \varepsilon_1 V \oplus \varepsilon_1' V$, and any $\sigma_1 \in \mathrm{Hom}_A(\varepsilon_1 V, \varepsilon_2 V)$ is extended to $\sigma \in \mathrm{Hom}_A(V, V) = E$, i.e., $\sigma(v) = \sigma_1(\varepsilon_1 v)$. It is clear that $\varphi(\varepsilon_2 \sigma \varepsilon_1) = \sigma_1$.

(iii) By Theorem 4.4, $\varepsilon_1 E \simeq \varepsilon_2 E$ if and only if there exist $\sigma, \tau \in E$ such that $(\varepsilon_2 \sigma \varepsilon_1)(\varepsilon_1 \tau \varepsilon_2) = \varepsilon_2$, $(\varepsilon_1 \tau \varepsilon_2)(\varepsilon_2 \sigma \varepsilon_1) = \varepsilon_1$. But the latter condition asserts that $\varphi(\varepsilon_2 \sigma \varepsilon_1)\colon \varepsilon_1 V \to \varepsilon_2 V$ is an isomorphism having $\varphi(\varepsilon_1 \tau \varepsilon_2)\colon \varepsilon_2 V \to \varepsilon_1 V$ as its inverse. This proves our assertion. ∎

As a direct consequence of Theorem 5.4(i), we have the following.

Theorem 5.5. *V_A is indecomposable if and only if id_V is a unique idempotent of $\mathrm{End}_A(V)$.*

The following fact holds for modules with composition series. It is referred to as *Fitting's lemma.*

Lemma 5.6 (Fitting). *Suppose that V_A has a composition series. Then, for any $\sigma \in \mathrm{End}_A(V)$, there exists an integer $n \geq 0$ such that*

(a) $V = \sigma^n V \oplus \mathrm{Ker}\, \sigma^n$,

(b) $\sigma^n V = \sigma^{n+1} V = \cdots$, $\mathrm{Ker}\, \sigma^n = \mathrm{Ker}\, \sigma^{n+1} = \cdots$.

Proof. Consider the following two chains of A-submodules of V

$$\sigma V \supset \sigma^2 V \supset \cdots; \qquad \mathrm{Ker}\, \sigma \subset \mathrm{Ker}\, \sigma^2 \subset \cdots.$$

By assumption, there exists $n \geq 0$ such that $\sigma^n V = \sigma^{n+1} V = \cdots$, and $\mathrm{Ker}\, \sigma^n = \mathrm{Ker}\, \sigma^{n+1} = \cdots$. To show (a), we first claim that $\sigma^n V \cap \mathrm{Ker}\, \sigma^n = 0$. In fact, if $\sigma^n V \cap \mathrm{Ker}\, \sigma^n \ni v$, then $v = \sigma^n u$ for some $u \in V$, and hence $0 = \sigma^n v = \sigma^{2n} u$. Therefore $u \in \mathrm{Ker}\, \sigma^{2n} = \mathrm{Ker}\, \sigma^n$ and $v = \sigma^n u = 0$. Next, let v be any element of V. Since $\sigma^n v \in \sigma^n V = \sigma^{2n} V$, we have $\sigma^n v = \sigma^{2n} w$ for some $w \in V$. Hence $v - \sigma^n w \in \mathrm{Ker}\, \sigma^n$ and $v \in \sigma^n V + \mathrm{Ker}\, \sigma^n$. Therefore $V = \sigma^n V + \mathrm{Ker}\, \sigma^n$, which is a direct sum from the above. ∎

We say that A is a *local ring*, provided the set of nonunits of A forms an ideal of A.

Theorem 5.7. *The following conditions on a ring A are equivalent.*

(1) *A is a local ring.*

(2) *The set of nonunits of A coincides with the radical $J(A)$, i.e., $J(A)$ is a unique maximal right (or left) ideal of A.*

(3) *$A/J(A)$ is a division ring.*

Proof. Let M be the set of nonunits of A.

(1) $\Rightarrow$ (2). If M is an ideal of A, then M contains every proper right ideal of A. Hence M is a unique maximal right ideal of A and coincides with $J(A)$.

(2) $\Rightarrow$ (3). If $a \notin J(A)$, then a is a unit of A, and hence $\bar{a}$ is a unit of $\bar{A} = A/J(A)$. This implies that $\bar{A}$ is a division ring.

(3) $\Rightarrow$ (1). If $a \notin J(A)$, then by assumption there exist $b \in A$, $c \in J(A)$ such that $ab = 1 - c$, which has a right inverse by Theorem 3.3(ii). Hence a has a right inverse. Similarly, a has a left inverse, i.e., a is a unit of A. Therefore M coincides with $J(A)$, an ideal of A. ■

Exercise 5.8. Let A be a local ring. If $\sum_{i=1}^{n} a_i = 1$, then some a_i is a unit of A.

The following is one of the characteristic properties of a local ring.

Theorem 5.9. *If A is a local ring, then A has no idempotent other than the identity.*

Proof. If A has an idempotent e different from 1, then both e and $1 - e$ are nonunits as $e(1 - e) = 0$. So by assumption they belong to $J(A)$, but this is impossible since then $1 = e + (1 - e) \in J(A)$. ■

Now the following criterion for the indecomposability holds for modules with composition series.

Theorem 5.10. *Suppose that V_A has a composition series. Then*

$$V \text{ is indecomposable} \Leftrightarrow \operatorname{End}_A(V) \text{ is a local ring.}$$

Proof. The implication ($\Leftarrow$) follows from Theorems 5.9 and 5.5. To show the converse, let $\sigma \in E = \operatorname{End}_A(V)$. Since V is indecomposable, we see from Fitting's lemma that there exists some $n \geq 0$ such that either $V = \sigma^n V$, $\operatorname{Ker} \sigma^n = 0$ or $V = \operatorname{Ker} \sigma^n$ holds. In other words, σ^n is either a unit or nilpotent. Thus the set of nonunits of E coincides with the set, say M, of nilpotent elements of E. It remains only to show that M is an ideal of E. Let $\sigma \in M$ and let m be the smallest integer such that $\sigma^m = 0$. Then, since $\sigma^{m-1}(\sigma\rho) = 0 = (\rho\sigma)\sigma^{m-1}$ for any $\rho \in E$ and $\sigma^{m-1} \neq 0$, it follows that both $\sigma\rho$ and $\rho\sigma$ are nonunits and hence belong to M. Let $\sigma, \rho \in M$. If $\sigma + \rho = \tau$ is a unit, then $\tau^{-1}\sigma = 1 - \tau^{-1}\rho$ is a unit since $\tau^{-1}\rho \in M$ is nilpotent, but this is impossible as shown above. Therefore $\sigma + \rho \in M$ and M is an ideal of E. ■

6. The Krull-Schmidt-Azumaya Theorem

The following theorem is a generalization by Azumaya [1] of the Krull-Schmidt theorem, which was originally proved for modules with composition series. We shall henceforth refer to it as the *K-S-A theorem.*

Theorem 6.1 (Krull-Schmidt-Azumaya). *Suppose that V_A is a finite direct sum of indecomposable submodules*

$$V = V_1 \oplus V_2 \oplus \cdots \oplus V_n \tag{6.1}$$

and that $\operatorname{End}_A(V_i)$ is a local ring for all $V_i (1 \leq i \leq n)$. Then for any indecomposable decomposition

$$V = W_1 \oplus W_2 \oplus \cdots \oplus W_m, \tag{6.2}$$

we have the following:

(i) $n = m$.

(ii) *There exists a permutation* $p = \begin{pmatrix} i \\ p(i) \end{pmatrix}$ *on* $\{1, 2, \ldots, n\}$ *such that* $V_i \simeq W_{p(i)} (1 \leq i \leq n)$ *and*

$$V = W_{p(1)} \oplus \cdots \oplus W_{p(r)} \oplus \left(\bigoplus_{j=r+1}^{n} V_j \right) \tag{6.3}$$

for any r with $1 \leq r \leq n$ (namely, one can replace the first r modules on the right-hand side of (6.1) *with $W_{p(1)}, \ldots, W_{p(r)}$).*

Proof. Let $\pi_i\colon V \to V_i$ be the projection and let $\sigma \in \mathrm{End}_A(V)$. First we claim that if $\pi_1 \circ \sigma|_{V_1}\colon V_1 \to V_1$ is an isomorphism, then

$$V = \sigma V_1 \oplus V_2 \oplus \cdots \oplus V_n, \qquad \sigma V_1 \simeq V_1$$

holds. Indeed, since $\sigma|_{V_1}\colon V_1 \to \sigma V_1$ and $\pi_1|_{\sigma V_1}\colon \sigma V_1 \to V_1$ are both isomorphisms by assumption, we have $\sigma V_1 \cap \mathrm{Ker}\,\pi_1 = 0$. Also, for any $v \in V$, there exists v'_1 of σV_1 such that $\pi_1 v = \pi_1 v'_1$. Hence $v - v'_1 \in \mathrm{Ker}\,\pi_1$, and it follows that

$$V = \sigma V_1 + \mathrm{Ker}\,\pi_1 = \sigma V_1 \oplus \mathrm{Ker}\,\pi_1 = \sigma V_1 \oplus V_2 \oplus \cdots \oplus V_n,$$

as claimed. Now, if $\sigma_j \in \mathrm{End}_A(V)$ denotes the projection on W_j w.r.t. the decomposition (6.2), then $\mathrm{id}_V = \sum_{j=1}^m \sigma_j$, and so $\mathrm{id}_{V_1} = \sum_j \pi_1 \circ (\sigma_j|_{V_1})$ in $\mathrm{End}_A(V_1)$. This implies that $\pi_1 \circ (\sigma_j|_{V_1})\colon V_1 \simeq V_1$ is an isomorphism for some j since $\mathrm{End}_A(V_1)$ is a local ring. Thus, if we let $j = p(1)$, we have from the above,

$$V = \sigma_{p(1)} V_1 \oplus V_2 \oplus \cdots \oplus V_n, \qquad \sigma_{p(1)} V_1 \simeq V_1. \tag{6.4}$$

From this and $\sigma_{p(1)} V_1 \subset W_{p(1)}$, we have $\sigma_{p(1)} V_1 | W_{p(1)}$, and hence $\sigma_{p(1)} V_1 = W_{p(1)}$ by the indecomposability of $W_{p(1)}$.

Using the same argument starting with (6.4) and V_2 in place of (6.1) and V_1, respectively, we find $p(2)(\neq p(1))$ such that

$$V = W_{p(1)} \oplus W_{p(2)} \oplus V_3 \oplus \cdots \oplus V_n, \qquad W_{p(2)} \simeq V_2.$$

By repeating this, we finally obtain $V = W_{p(1)} \oplus \cdots \oplus W_{p(n)}$ and $n \leq m$. If $n < m$, then (6.2) implies that $W_{p(1)} \oplus \cdots \oplus W_{p(n)} \subsetneq V$, which is impossible. Therefore $n = m$ and (6.3) holds. ■

The following result holds especially for modules with maximum or minimum condition.

Theorem 6.2. *Let V be an A-module.*

(i) *If V is Neotherian or Artinian, then V is expressible in a finite direct sum of indecomposable submodules.*

(ii) *If V has a composition series, then the K–S–A theorem holds for any two indecomposable decompositions of V.*

Proof.

(i) Suppose the contrary; thus we have a decomposition $V = V_1 \oplus W_1$, and our statement is false for one of them, say W_1. Then we also have a decomposition $W_1 = V_2 \oplus W_2$, with our statement being false for W_2. By continuing this, we get the following descending or ascending chain of submodules:

$$W_1 \supsetneqq W_2 \supsetneqq W_3 \supsetneqq \cdots,$$

$$V_1 \subsetneqq V_1 \oplus V_2 \subsetneqq V_1 \oplus V_2 \oplus V_3 \subsetneqq \cdots,$$

which contradicts the assumption.

(ii) This is immediate from Theorem 5.10. ■

7. Completely Reducible Modules

Let V be an A-module. We say that V is *completely reducible* if V is expressed as a direct sum of irreducible A-submodules. And any such expression is said to be an *irreducible decomposition* of V. It is clear that $VJ(A) = 0$ for such V.

Theorem 7.1. *The following three conditions on V_A are equivalent.*

(1) *V is completely reducible.*

(2) *V is a sum of irreducible A-submodules.*

(3) *Any A-submodule of V is a direct summand of V.*

Proof.

(1) ⇒ (2). This is trivial.

(2) ⇒ (3). Suppose that $V = \sum_{\lambda \in \Lambda} V_\lambda$ is a sum of irreducible submodules $V_\lambda (\lambda \in \Lambda)$ and let W be any A-submodule of V. Let $\mathscr{S} = \{S \subset \Lambda;\ W \cap (\sum_{s \in S} V_s) = 0\}$. If $\mathscr{S}$ is empty, namely $W \cap V_\lambda \neq 0$ for all V_λ, then $W \supset V_\lambda$ for all V_λ since V_λ is irreducible, and thus $W = V$. If $\mathscr{S}$ is nonempty, we see easily that $\mathscr{S}$ is an inductively ordered set w.r.t. inclusion. Consequently $\mathscr{S}$ contains a maximal element, say S_0, by Zorn's lemma. Thus, $W \cap \sum_{s \in S_0} V_s = 0$. Let $U = W \oplus \sum_{s \in S_0} V_s$. We show that $V = U$. In fact, if there exists V_λ such that $U \cap V_\lambda = 0$, then $\lambda \notin S_0$ and $S_0 \cup \{\lambda\} \in \mathscr{S}$, contradicting the maximality of S_0. Therefore $U \supset V_\lambda$ for all λ, and we get $U = V$.

(3) ⇒ (1). First we claim that every nonzero A-submodule W of V contains an irreducible submodule. Let w be a nonzero element of W. Then the set of A-submodules of W not containing w is an inductively ordered set w.r.t. inclusion. Hence it contains a maximal element, say W_1. Since W_1 is a direct

summand of V and $W \supset W_1$, $W = W_1 \oplus W_2$ for some W_2. We assert that W_2 is irreducible. If not, then by the same reason as above, W_2 is expressible as a direct sum: $W_2 = W'_2 \oplus W''_2$ ($W'_2, W''_2 \neq 0$), and the maximality of W_1 implies that $(W_1 \oplus W'_2) \cap (W_1 \oplus W''_2) \ni w$. But this is a contradiction since $(W_1 \oplus W'_2) \cap (W_1 \oplus W''_2) = W_1$. Therefore W_2 is irreducible as asserted.

Now, let $\{V_\lambda; \lambda \in \Lambda\}$ be the set of all irreducible A-submodules of V, and let $\mathscr{T} = \{T \subset \Lambda; \sum_{t \in T} V_t$ is a direct sum$\}$. Then $\mathscr{T}$ is an inductively ordered set w.r.t. inclusion; therefore it contains a maximal element, say T_0. If $U = \bigoplus_{t \in T_0} V_t$ is different from V, then from the assumption, $V = U \oplus W$ for some $W \neq 0$. However, as was claimed above, W contains some V_λ, and hence $T_0 \cup \{\lambda\} \in \mathscr{T}$, which contradicts the maximality of T_0. Therefore $U = V$, and the proof is complete. ■

Exercise 7.2. If W is a submodule of a completely reducible module V, then W and V/W are also completely reducible.

Theorem 7.3. *Let V be a completely reducible A-module and let*

$$V = \bigoplus_{i \in I} \sum_{\lambda \in \Lambda_i} V_{i\lambda}$$

be an irreducible decomposition of V such that $V_{i\lambda}$ and $V_{k\mu}$ are isomorphic to each other if and only if $i = k$. Set $U_i = \bigoplus_{\lambda \in \Lambda_i} V_{i\lambda}$ and let V_i be any one of $\{V_{i\lambda}; \lambda \in \Lambda_i\}$, then the following holds.

(i) $V = \bigoplus_{i \in I} U_i$.

(ii) *Any irreducible A-submodule W of V is isomorphic to some V_i, in which case it holds that $W \subset U_i$. Consequently each U_i is the sum of all irreducible A-submodules of V isomorphic to V_i.*

(iii) $\mathrm{End}_A(V)U_i = U_i$.

(iv) *If $I = \{1, 2, \ldots, m\}$ is a finite set and $|\Lambda_i| < \infty$ for all i, then*

$$\mathrm{End}_A(V) \simeq \mathrm{End}_A(U_1) \oplus \cdots \oplus \mathrm{End}_A(U_m) \quad \textit{(ring isomorphism)},$$

and each $\mathrm{End}_A(U_i)$ is the full matrix ring of degree $|\Lambda_i|$ over the division ring $\mathrm{End}_A(V_i)$.

Proof.

(i) This is clear.

(ii) Let $\pi_{i\lambda}: V \to V_{i\lambda}$ be the projection. Then there exist i, λ such that $\pi_{i\lambda} W \neq 0$, whence it follows that $W \simeq V_{i\lambda} \simeq V_i$ by Schur's lemma. Moreover, if $k \neq i$, then $\pi_{k\mu} W = 0$ as $W \not\simeq V_{k\mu}$. This proves that $W \subset U_i$.

(iii) If $\sigma \in \mathrm{End}_A(V)$, then $\sigma V_{i\lambda}$ is either 0 or isomorphic to $V_{i\lambda}$. Hence we have $\sigma V_{i\lambda} \subset U_i$ from (ii). This yields that $\sigma U_i \subset U_i$.

(iv) For $\sigma \in \mathrm{End}_A(V)$, let σ_i be the restriction of σ to U_i. Note that $\sigma_i \in \mathrm{End}_A(U_i)$, since $\mathrm{Im}\, \sigma_i \subset U_i$ by (iii). Then the map $\sigma \mapsto (\sigma_1, \ldots, \sigma_m)$ gives a desired isomorphism. ∎

In the above theorem, the expression (i) is called the *homogenous decomposition* of V, and each U_i is called the *homogenous component* of V. The homogenous decomposition of V is independent of the choice of an irreducible decomposition of V.

8. Artinian Rings

8.1. The Structure of Artinian Rings

To begin with, we study the structure of semisimple Artinian rings.

Theorem 8.1. *The following conditions on a ring A are equivalent.*

(1) *A is a semisimple Artinian ring.*

(2) *A_A is completely reducible.*

(3) *Every A-module V is completely reducible.*

Proof.

(1) ⇒ (2). Let $\{M_\lambda; \lambda \in \Lambda\}$ be the set of maximal right ideals of A. The set of right ideals that are expressible as finite intersections of M_λ has a minimal element, say $L = M_{\lambda_1} \cap \cdots \cap M_{\lambda_r}$, by the minimum condition on A_A. If $L \neq 0$, then, since $J(A) = \bigcap_\lambda M_\lambda = 0$, there exists M_λ such that $L \supsetneqq L \cap M_\lambda$. This contradicts the minimality of L, and hence L must be 0. Consequently, the map

$$A \to (A/M_{\lambda_1}) \oplus \cdots \oplus (A/M_{\lambda_n})(a \mapsto (a + M_{\lambda_1}, \ldots, a + M_{\lambda_r}))$$

is a monomorphism, and it follows that A_A is completely reducible since each $A/M_{\lambda i}$ is irreducible.

(2) ⇒ (3). If $A = \sum_\lambda I_\lambda$ is a sum of irreducible A-submodules, then $V = \sum_{v \in V} \sum_\lambda vI_\lambda$. Hence V is completely reducible, because each vI_λ is isomorphic to I_λ unless it is zero.

(3) ⇒ (1). Since A_A is completely reducible, it follows that $0 = AJ(A) = J(A)$, i.e., A is semisimple. If $A = \bigoplus_{\lambda \in \Lambda} I_\lambda$ is an irreducible decomposition, it must be a finite sum. Indeed, if $1 \in I_1 \oplus \cdots \oplus I_n (i \in \Lambda)$, then we have $A = I_1 \oplus \cdots \oplus I_n$. Hence A_A has a composition series $A \supset I_1 \oplus \cdots \oplus I_{n-1} \supset \cdots \supset I_1 \supset 0$ and thus A is Artinian. ■

From the above theorem, we obtain the following.

Theorem 8.2. *Let A be an Artinian ring. Then*

$$V_A \text{ is completely reducible} \quad \Leftrightarrow \quad VJ(A) = 0.$$

Proof. The implication (⇒) is clear. To show the converse, let $\bar{A} = A/J(A)$. Then V can be naturally regarded as an $\bar{A}$-module. Since $\bar{A}$ is semisimple Artinian, $V_{\bar{A}}$ is completely reducible, or, which is the same thing, V is completely reducible as an A-module. ■

As observed above, every completely reducible module over A may be considered as a module over the semisimple ring $\bar{A} = A/J(A)$.

That a right ideal of a ring A is irreducible as an A-module means simply that it is a minimal right ideal of A.

Let A be a semisimple Artinian ring and let

$$A = \bigoplus_{i=1}^{k} \bigoplus_{\lambda=1}^{n_i} e_{i\lambda} A \tag{8.1}$$

be an irreducible decomposition, where $1 = \sum_{i,\lambda} e_{i\lambda}$ is a primitive idempotent decomposition of the identity, and we assume that $e_{i\lambda} A \simeq e_{j\mu} A$ if and only if $i = j$.

If we set $A_i = \bigoplus_\lambda e_{i\lambda} A$, then we have the homogenous decomposition

$$A = A_1 \oplus A_2 \oplus \cdots \oplus A_k. \tag{8.2}$$

Since every element of $\mathrm{End}_A(A_A)$ may be interpreted as the left multiplication by some element of A, each A_i is an ideal of A by Theorem 7.3 (iii), so that (8.2) is a direct sum as rings. Also, $D_i = \mathrm{End}_A(e_{i1}A) \simeq e_{i1}Ae_{i1}$ is a division ring and $A_i \simeq M_{n_i}(D_i)$. Then by the exercise below each A_i is a simple ring. This is called a *simple component* of A.

Exercise 8.3. Let A be an arbitrary ring. Then every ideal of the full matrix ring $M_n(A)$ can be expressed in the form $M_n(I)$, where I is an ideal of A. In particular, A is simple if and only if $M_n(A)$ is so. [Hint: For an ideal L of $M_n(A)$, let I be the set of (1,1)-entries of L. Show that I is an ideal of A and $L = M_n(I)$.]

From the above observations, we obtain the following two theorems.

Theorem 8.4 (Wedderburn). *The following conditions on a ring A are equivalent.*

(1) *A is a simple Artinian ring.*

(2) *A is a finite direct sum of mutually isomorphic minimal right ideals.*

(3) *$A \simeq M_n(D)$, where D is a division ring.*

Proof.

(1) ⇒ (2). Since A is simple, we have $k = 1$ in the homogenous decomposition (8.2) of A.

(2) ⇒ (3). By assumption, A_A has a unique homogenous component, i.e., A is a simple ring.

(3) ⇒ (1). This is clear from Exercise 8.3. ■

Theorem 8.5. *The following conditions on a ring A are equivalent.*

(1) *A is a semisimple (right) Artinian ring.*

(2) *A is a semisimple left Artinian ring.*

(3) *$A \simeq M_{n_1}(D_1) \oplus \cdots \oplus M_{n_k}(D_k)$, where each D_i is a division ring.*

Proof.

(1) ⇒ (3). This has already been observed in the above.

(3) ⇒ (1). Since each $M_{n_i}(D_i)$ is a simple right Artinian ring, A is also Artinian, and $J(A) = 0$ by Exercise 4.8(ii).

(2) ⇒ (3). This can be shown quite similarly to the above. ■

Let A be a semisimple Artinian ring and V_A be irreducible. By using the notations in (8.1) and (8.2), we have, for $v \neq 0 \in V$, that $V = vA = \sum_{i,\lambda} ve_{i\lambda}A = ve_{i\lambda}A \simeq e_{i1}A$ for some i, λ. It then follows that $VA_i = V$ and

$VA_j = 0$ if $j \neq i$. Thus V can be actually considered as an A_i-module. In that case, we say V belongs to the simple component A_i.

Summarizing the above, we get the following theorem.

Theorem 8.6. *Let A be a semisimple Artinian ring. Then the number of the representatives of nonisomorphic irreducible A-modules equals the number of the simple components of A, and under the notation of (8.1), $\{e_{11}A, \ldots, e_{k1}A\}$ gives a complete set of representatives for the isomorphism classes of irreducible A-modules.*

In the matrix ring $A = M_n(D)$ over a ring D, we denote by e_{ij} the element of A, all of whose entries are 0 except the (i, j)-entry, which is $1 (1 \leq i, j \leq n)$. These matrices are called *matrix units* of A. The following facts will be evident:

$$de_{ij} = e_{ij}d \qquad (\text{for all } d \in D), \qquad e_{ij}e_{kl} = \delta_{jk}e_{il},$$

$$1 = \sum_{i=1}^{n} e_{ii} \qquad (\text{an idempotent decomposition}),$$

$$A = \bigoplus_{i,j=1}^{n} e_{ij}D = \bigoplus_{i=1}^{n} e_{ii}A, \qquad \operatorname{End}_A(e_{ii}A) \simeq e_{ii}Ae_{ii} \simeq D.$$

If D is a division ring in the above (hence A is a simple ring), each $e_{ii}A$ is indecomposable by Theorem 5.5 and hence is irreducible in our situation. Therefore n is equal to the composition length of A_A, and D is isomorphic to the endomorphism ring of a (unique) irreducible A-module. Namely, we have shown the following result.

Theorem 8.7. *Let D_1, D_2 be division rings. Then*

$$M_{n_1}(D_1) \simeq M_{n_2}(D_2) \Leftrightarrow n_1 = n_2,\ D_1 \simeq D_2.$$

A division ring D is said to belong to a simple ring A, provided $A \simeq M_n(D)$ holds for some n.

Exercise 8.8

(i) A completely reducible module is Artinian if and only if it is Noetherian.

(ii) If A is an Artinian ring, then A is Noetherian. [Hint: Show that A_A has a composition series by applying (i) to the factor modules J^{i-1}/J^i of the series $A = J^\circ \supset J \supset J^2 \supset \cdots \supset J^n = 0$, where $J = J(A)$.]

The above exercise yields the following theorem.

Theorem 8.9. *Let A be an Artinian ring and V be a finitely generated A-module. Then*

$$V \text{ is indecomposable} \quad \Leftrightarrow \quad \mathrm{End}_A(V) \text{ is a local ring.}$$

In particular, if $e \in \mathrm{pi}(A)$, then eAe is a local ring.

Proof. Since V has a composition series by Theorem 2.5, the result follows from Theorem 5.10. ■

Throughout the remainder of this section,

A is assumed to be an Artinian ring and $\bar{A}$ denotes $A/J(A)$.

Since $\bar{A}$ is semisimple Artinian, it has an irreducible decomposition

$$\bar{A} = \bigoplus_{i=1}^{k} \bigoplus_{\lambda=1}^{n_i} \bar{e}_{i\lambda}\bar{A}. \tag{8.3}$$

Also, it is expressible as the direct sum of simple rings

$$\bar{A} = \bigoplus_{i=1}^{k} \bar{A}_i \qquad \left(\bar{A}_i = \bigoplus_{\lambda=1}^{n_i} \bar{e}_{i\lambda}\bar{A} \simeq M_{n_i}(D_i) \quad \text{with division ring } D_i\right). \tag{8.4}$$

Here we assume that $1 = \sum_{i,\lambda} e_{i\lambda}$ is a primitive idempotent decomposition and $e_{i\lambda} \simeq e_{j\mu}$ if and only if $i = j$ (cf. Section 4). Then it holds that

$$A = \bigoplus_{i=1}^{k} \bigoplus_{\lambda=1}^{n_i} e_{i\lambda}A \quad \text{(an indecomposable decomposition)} \tag{8.5}$$

and $e_{i\lambda}A \simeq e_{j\mu}A$ if and only if $i = j$.

Let $e_i = e_{i1}$ for brevity. As was noted in Theorem 8.6, every irreducible A-module can actually be considered as an $\bar{A}$-module. From this we obtain the following theorem.

Theorem 8.10. *With the notation above we have the following.*

(i) *The set* $\{\bar{e}_1\bar{A}, \ldots, \bar{e}_k\bar{A}\}$ *gives a complete set of representatives for the isomorphism classes of irreducible A-modules. Hence the number of representatives of nonisomorphic irreducible A-modules is equal to that of the simple components of* $\bar{A}$.

(ii) *If V is an irreducible A-module, then*

$$V \simeq \bar{e}_i\bar{A} \Leftrightarrow Ve_i \neq 0,$$

in which case we have $A/(0\colon V) \simeq \bar{A}_i$.

Proof.

(i) This follows immediately from Theorem 8.6.

(ii) The implication ($\Rightarrow$) is clear, since $\bar{e}_i\bar{A}\bar{e}_i \neq 0$. On the other hand, if $Ve_i \neq 0$, then V, as an irreducible $\bar{A}$-module, belongs to $\bar{A}_i$ and hence $V \simeq \bar{e}_i\bar{A}$. The second assertion follows from $(0\colon V)/J(A) = \bigoplus_{j \neq i} \bar{A}_j$. ■

Every indecomposable A-module that is isomorphic to an indecomposable component of A_A is called a *principal indecomposable A-module.* So, if $e \in \mathrm{pi}(A)$, then eA is a principal indecomposable A-module, and we see from the K–S–A theorem and (8.5) that the set $\{e_1A, \ldots, e_kA\}$ gives a complete set of representatives for the isomorphism classes of principal indecomposable A-modules.

Because A is also a Noetherian ring by Exercise 8.8(ii), if V_A is finitely generated, V has a composition series by Theorem 2.5. For $e \in \mathrm{pi}(A)$, the number of the composition factors that are isomorphic to $\bar{e}\bar{A}$ in a fixed composition series of V is said to be the *multiplicity* of $\bar{e}\bar{A}$ in V. For this, the following theorem holds.

Theorem 8.11. *If* $e \in \mathrm{pi}(A)$, *then the multiplicity of* $\bar{e}\bar{A}$ *in* V_A *is equal to the composition length of* Ve *as an* eAe*-module.*

Proof. Let $V = V_0 \supset V_1 \supset \cdots \supset V_n = 0$ be a composition series of V_A and consider the following series of eAe-modules

$$Ve = V_0e \supset V_1e \supset \cdots \supset V_ne = 0. \tag{8.6}$$

We know from Theorem 8.10(ii) that $V_i/V_{i+1} \simeq \bar{e}\bar{A} \Leftrightarrow (V_i/V_{i+1})e \neq 0$, and if this is the case, then

$$(V_i/V_{i+1})e = (V_ie + V_{i+1})/V_{i+1} \simeq V_ie/V_ie \cap V_{i+1} = V_ie/V_{i+1}e \simeq \bar{e}\bar{A}\bar{e},$$

which is an irreducible eAe-module. Consequently we obtain from (8.6) a composition series of $(Ve)_{eAe}$ by deleting repeated modules, so its length is equal to $\#\{i;\, V_i/V_{i+1} \simeq \bar{e}\bar{A}\}$. ■

We denote the multiplicity of $\bar{e}_j\bar{A}$ in e_iA by c_{ij}, which is called the *Cartan invariant* of A. And the $k \times k$ matrix (c_{ij}) is called the *Cartan matrix* of A.

From the above theorem, we get the following result.

Theorem 8.12. $c_{ij} \neq 0 \Leftrightarrow e_iAe_j \neq 0$.

8.2. Blocks

By the minimum condition on A_A, there exists a central primitive idempotent decomposition of the identity:

$$1 = \varepsilon_1 + \varepsilon_2 + \cdots + \varepsilon_n.$$

Corresponding to this, we have the following indecomposable decomposition of A as an (A, A)-bimodule

$$A = B_1 \oplus B_2 \oplus \cdots \oplus B_m (B_i = \varepsilon_i A = A\varepsilon_i). \tag{8.7}$$

Each B_i is called a *block* of A and ε_i the *block idempotent* of B_i. Also, the decomposition (8.7) is said to be the *block decomposition* of A, and an A-module V is said to *belong* to B_i if $V\varepsilon_i = V$.

Theorem 8.13. *Every indecomposable module V_A belongs to some block of A.*

Proof. Since $V\varepsilon_i$ is an A-submodule of V and $V = V\varepsilon_1 \oplus \cdots \oplus V\varepsilon_m$, it follows from the indecomposability of V that $V = V\varepsilon_i$ for some i. ■

In particular, the following holds for principal indecomposable modules.

Theorem 8.14. *Let $e \in \mathrm{pi}(A)$. Then eA belongs to a block B_i if and only if $eA \subset B_i$.*

Proof. If $eA\varepsilon_i = eA$, then it follows that $eA \subset B_i$ as $e = e\varepsilon_i \in B_i$. The converse is clear. ■

We say that two idempotents e, f of A are *linked* and write $e \sim f$ provided there exists a sequence of primitive idempotents

$$e = e^{(0)}, e^{(1)}, \ldots, e^{(r)} = f$$

such that $e^{(i)}A$ and $e^{(i+1)}A$ have an irreducible constituent in common for each $i (0 \leq i \leq r-1)$. Clearly the relation of linkage is an equivalence relation.

Theorem 8.15. *Let $I_1, \ldots, I_n$ be the equivalence classes of* $\mathrm{pi}(A)$ *by the relation "$\sim$". Then under the notation in* (8.7), *it holds that $n = m$, and after a suitable arrangement of indices, we have*

$$B_i = \sum_{e \in I_i} eA.$$

In particular for $e, f \in \mathrm{pi}(A)$, eA and fA belong to the same block if and only if $e \sim f$.

Proof. Let $D_i = \sum_{e \in I_i} eA$. If $e \in I_i$, $f \in I_j$ and $i \neq j$, then eA has no irreducible constituent isomorphic to $\bar{f}\bar{A}$. Thus, we have $eAf = 0$ by Theorem 8.12, and $D_iD_j = 0$ whenever $i \neq j$. Since $A = \sum_{j=1}^{n} D_j$, it follows that $AD_i = (\sum_j D_j)D_i \subset D_iD_i \subset D_i$, namely, D_i is an ideal of A. If $1 = \sum_j e_j (e_j \in D_j)$, then we have, for $a_i \in D_i$, $a_i = a_i(\sum_j e_j) = a_ie_i$ and similarly $a_i = e_ia_i$. Namely, e_i is the identity of D_i. Also, if $\sum_j a_j = 0 (a_j \in D_j)$, then $0 = (\sum_j a_j)e_i = a_ie_i = a_i$. Therefore we get the direct sum

$$A = D_1 \oplus \cdots \oplus D_n.$$

On the other hand, if eA and fA have an irreducible constituent isomorphic to $\bar{e}'\bar{A}$ in common, where $e' \in \mathrm{pi}(A)$, then $eAe' \neq 0$ and $fAe' \neq 0$. In particular, if $e' \in B_j$, we have that $eA, fA \subset B_j$. Thus, we readily find that eA and fA are contained in the same block whenever $e \sim f$. Therefore each D_i is contained in some B_j, whence we have $D_i | B_j$. But this forces $D_i = B_j$ since ${}_A(B_i)_A$ is indecomposable. ■

Let $\widetilde{\text{pi}}(A) = \{e_1, \ldots, e_k\}$ and C_i be the Cartan matrix of B_i. We know that $c_{ij} = 0$ if $e_i A$ and $e_j A$ belong to different blocks. Hence we can arrange the indices so that the Cartan matrix of A is of the form

$$C = \begin{pmatrix} C_1 & & & 0 \\ & C_2 & & \\ & & \ddots & \\ 0 & & & C_m \end{pmatrix}.$$

But no C_i splits into the form

$$C_i = \left(\begin{array}{c|c} * & 0 \\ \hline 0 & * \end{array}\right).$$

8.3. The Loewy and Socle Series

We maintain that A is an Artinian ring. Let $J = J(A)$ and $\bar{A} = A/J$.

Lemma 8.16. *Let V be an A-module. Then VJ coincides with the minimal submodule W of V such that V/W is completely reducible.*

Proof. Since $(V/VJ)J = 0$, V/VJ is completely reducible by Theorem 8.2. On the other hand, if $U \subset V$ and V/U is completely reducible, then it follows that $0 = (V/U)J = (VJ + U)/U$, namely, $VJ \subset U$. ■

For an A-module V, VJ is called the *radical* of V, whereas V/VJ is the *head* of V, which is denoted by $\text{hd}(V)$.

Let m be an integer such that $VJ^{m-1} \neq 0$, $VJ^m = 0$. Then the series of completely reducible A-modules

$$(8.8) \qquad V/VJ, \qquad VJ/VJ^2, \ldots, VJ^{m-1}/VJ^m = VJ^{m-1}$$

is said to be the *Loewy series* of V and m is the *length* of it. Each VJ^{i-1}/VJ^i is called the *ith head* of V. If we have an irreducible decomposition

$$VJ^{i-1}/VJ^i = X_{i1} \oplus \cdots \oplus X_{ir_i},$$

we usually denote the Loewy series by

$$(8.9)\qquad \begin{matrix} X_{11} \dots X_{1r_1} \\ X_{21} \dots X_{2r_2} \\ \vdots \\ X_{m1} \dots X_{mr_m} \end{matrix}$$

If $e \in \mathrm{pi}(A)$, then $(eA)J = eJ = eA \cap J$. Therefore we have

$$\mathrm{hd}(eA) = eA/eA \cap J \simeq (eA + J)/J = \bar{e}\bar{A}.$$

Thus, we have shown part(i) of the theorem below.

Theorem 8.17. *Let $e \in \mathrm{pi}(A)$. Then we have the following.*

(i) $\mathrm{hd}(eA) \simeq \bar{e}\bar{A}$, *which is irreducible. Consequently, eJ is a unique maximal submodule of eA.*

(ii) *If V_A is finitely generated and* $\mathrm{hd}(V) \simeq \bar{e}\bar{A}$, *then V is a homomorphic image of eA.*

Proof.

(ii) Let $\bar{v} \in \bar{V} = V/VJ$ be mapped onto $\bar{e}$ by the above isomorphism and consider the A-homomorphism $\varphi\colon eA \to V(ea \mapsto vea)$. Since $\overline{ve} = \bar{v}$, it follows that $veA + VJ = V$ and hence $veA = V$ by the Azumaya–Nakayama lemma. Therefore φ is an epimorphism. ■

The sum of all irreducible A-submodules of V_A is denoted by $\mathrm{soc}(V)$ and is called the *socle* of V. Clearly $\mathrm{soc}(V)$ is the largest completely reducible A-submodule of V.

Set $S_0(V) = 0$ and $S_1(V) = \mathrm{soc}(V)$. And define $S_i(V)$ inductively by

$$\mathrm{soc}(V/S_{i-1}(V)) = S_i(V)/S_{i-1}(V).$$

Each $S_i(V)/S_{i-1}(V)$ is said to be the ith *socle* of V.

Theorem 8.18. $\mathrm{soc}(V) = \{v \in V; vJ = 0\}$. *More generally, we have*

$$S_i(V) = \{v \in V;\ vJ^i = 0\}.$$

Proof. Let $T_i = \{v \in V; vJ^i = 0\}$, and let $S_i = S_i(V)$ for brevity. It is easy to see that $S_1 = T_1$ by Theorem 8.2. We show that $S_i = T_i$ by induction on i. So, assume that $S_{i-1} = T_{i-1}$. Then, since $S_i J \subset S_{i-1}$, we have $S_i \subset T_i$. On the other hand, from $T_i J \subset T_{i-1} = S_{i-1}$ it follows that $(T_i/S_{i-1})J = 0$, namely, T_i/S_{i-1} is completely reducible. Therefore $T_i \subset S_i$. ■

We see from the above theorem that $S_m(V) = V$ if $VJ^m = 0$. Let $n(\leq m)$ be the integer such that $S_{n-1}(V) \neq V$, $S_n(V) = V$. Then we have the sequence of completely reducible A-modules

(8.10) $$S_1(V), S_2(V)/S_1(V), \ldots, S_n(V)/S_{n-1}(V).$$

This is called the *socle series* of V, and n is the *length* of it. Also we express the socle series of V in the form such as (8.9), starting with an irreducible decomposition of $S_1(V)$ at the bottom.

Theorem 8.19. *The Loewy and socle length of V_A are equal. If this common integer is denoted by m, then*

$$VJ^i \subset S_{m-i}(V) \qquad (i = 0, 1, \ldots, m-1).$$

Proof. Let (8.8) and (8.10) be the Loewy and socle series of V, respectively, and $S_i = S_i(V)$. Then, as noted above, we have $n \leq m$. Also, we have $VJ^i \subset S_{n-i}$. In fact, if we assume that $VJ^{i-1} \subset S_{n-i+1}$, then $VJ^i \subset S_{n-i+1}J \subset S_{n-i}$. In particular, it follows that $VJ^n = S_0 = 0$ and thus $m \leq n$. ■

Lemma 8.20. *If W is an A-submodule of V, then $S_i(W) = W \cap S_i(V)$.*

Proof. This is clear from Theorem 8.18. ■

Theorem 8.21. *Let $e \in \mathrm{pi}(A)$. Then the following two conditions on V_A are equivalent.*

(1) *$\bar{e}\bar{A}$ is isomorphic to some irreducible constituent of the ith socle of V.*

(2) *There exists an A-submodule W of V with Loewy length i such that* $\mathrm{hd}(W) \simeq \bar{e}\bar{A}$.

Proof. Let $S_i = S_i(V)$ for brevity.

(1) ⇒ (2). By assumption, there exists $v \in S_i - S_{i-1}$ such that $v \equiv ve$ mod S_{i-1}. Let $W = veA$. Then $\mathrm{hd}(W) \simeq \bar{e}\bar{A}$. Moreover, since $WJ^i = 0$ and $veJ^{i-1} \neq 0$, W has the Loewy length i.

(2) ⇒ (1). It follows from $WJ^i = 0$ that $W \subset S_i$. Since $(W + S_{i-1})/S_{i-1} \simeq W/W \cap S_{i-1} = W/S_{i-1}(W) = \mathrm{hd}(W) \simeq \bar{e}\bar{A}$, $\bar{e}\bar{A}$ is isomorphic to some irreducible constituent of S_i/S_{i-1}. ■

We call V_A a *uniserial* module provided every ith head of V is irreducible.

Exercise 8.22. V is uniserial if and only if V has a unique composition series. If this is the case, then the Loewy and socle series of V are identical.

9. Hom and ⊗

Let A be a ring. A diagram consisting of A-modules and A-homomorphisms is said to be a diagram (in the category) of A-modules.

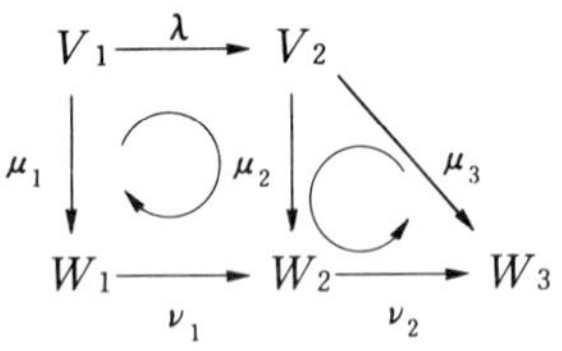

A diagram of A-modules is said to be *commutative* if every possible composition of maps is independent of path. For example, the above diagram is commutative if $\mu_2 \circ \lambda = \nu_1 \circ \mu_1$ and $\nu_2 \circ \mu_2 = \mu_3$. The commutativity will be indicated by drawing $\circlearrowright$ or $\circlearrowleft$ as above.

9.1. $Hom_A(V, W)$

We shall describe some properties of $\mathrm{Hom}_A(V, W)$, the additive group consisting of A-homomorphisms from V into W. The following theorem is obvious.

Theorem 9.1. *Let V, W, etc., denote A-modules.*

(i) *Given an A-homomorphism $\lambda: V \to V'$, we have a homomorphism* $\lambda^*: \mathrm{Hom}_A(V', W) \to \mathrm{Hom}_A(V, W)(\sigma \mapsto \sigma \circ \lambda)$. *If $\mu: V' \to V''$ is a homomorphism, then $(\mu \circ \lambda)^* = \lambda^* \circ \mu^*$.*

(ii) *Given an A-homomorphism* $\lambda: W \to W'$, *we have a homomorphism* $\lambda_*: \mathrm{Hom}_A(V, W) \to \mathrm{Hom}_A(V, W')(\sigma \mapsto \lambda \circ \sigma)$. *If* $\mu: W' \to W''$ *is a homomorphism, then* $(\mu \circ \lambda)_* = \mu_* \circ \lambda_*$.

A sequence of A-homomorphism

$$\cdots \xrightarrow{\lambda_{i-2}} V_{i-1} \xrightarrow{\lambda_{i-1}} V_i \xrightarrow{\lambda_i} V_{i+1} \xrightarrow{\lambda_{i+1}} \cdots$$

is said to be *exact* provided $\mathrm{Im}\, \lambda_{i-1} = \mathrm{Ker}\, \lambda_i$ for all i. By a *short exact sequence*, we mean the *exact sequence* of the form

$$0 \to V' \xrightarrow{\lambda} V \xrightarrow{\mu} V'' \to 0.$$

This means that λ is a monomorphism and μ is an epimorphism. Thus,

$$V' \simeq \lambda(V'), \qquad V/\lambda(V') \simeq V''$$

holds.

A monomorphism $\lambda: V' \to V$ is said to *split* if $\lambda(V')$ is a direct summand of V. Also an epimorphism $\mu: V \to V''$ is said to split if $\mathrm{Ker}\, \mu$ is a direct summand of V.

Lemma 9.2. *Let* $\lambda: V \to W$ *and* $\mu: W \to V$ *be homomorphisms such that* $\lambda \circ \mu = \mathrm{id}_W$. *Then the following holds.*

$$V = \mathrm{Ker}\, \lambda \oplus \mu(W), \qquad \mu(W) \simeq W.$$

Proof. For $v \in V$, we have $\lambda(v - \mu(\lambda(v))) = \lambda(v) - \lambda \circ \mu \circ \lambda(v) = 0$. So $v - \mu(\lambda(v)) \in \mathrm{Ker}\, \lambda$, and hence $V = \mathrm{Ker}\, \lambda + \mu(W)$. To show that the sum is direct, let $v \in \mathrm{Ker}\, \lambda \cap \mu(W)$. Then $v = \mu(w)$ for some $w \in W$, and hence $0 = \lambda(v) = \lambda \circ \mu(w) = w$. Therefore, $v = 0$, i.e., $\mathrm{Ker}\, \lambda \cap \mu(W) = 0$. The second assertion is trivial since μ is a monomorphism. ■

Theorem 9.3.

(i) *Let* $\lambda: V' \to V$ *be a monomorphism. Then the following two conditions are equivalent.*

(1) λ *splits.*

(2) *There exists a homomorphism* $\lambda': V \to V'$ *such that* $\lambda' \circ \lambda = \mathrm{id}_{V'}$.

(ii) *Let* $\mu\colon V \to V''$ *be an epimorphism. Then the following two conditions are equivalent.*

(3) μ *splits.*
(4) *There exists a homomorphism* $\mu'\colon V'' \to V$ *such that* $\mu \circ \mu' = \mathrm{id}_{V''}$.

Proof.
(i) (1) ⇒ (2). Let $V = \lambda(V') \oplus W$. For $v \in V$, there exist a uniquely determined $v' \in V'$ and $w \in W$ such that $v = \lambda(v') + w$. Define λ' by $\lambda'(v) = v'$, and (2) follows.

(2) ⇒ (1). This is clear from Lemma 9.2.

(ii) (3) ⇒ (4). If $V = \operatorname{Ker}\mu \oplus W$, then $\mu|_W\colon W \to V''$ is an isomorphism. Then (4) follows by setting $\mu' = (\mu|_W)^{-1}$.

(4) ⇒ (3). This is clear from Lemma 9.2. ∎

A short exact sequence

(9.1) $$0 \to V' \xrightarrow{\lambda} V \xrightarrow{\mu} V'' \to 0$$

is said to split if λ, or equivalently μ, splits.

The following property is usually referred to as the *left exactness* of Hom_A.

Theorem 9.4. *Suppose that the following two sequences of A-modules are exact:*

(9.2) $$V' \xrightarrow{\lambda_1} V \xrightarrow{\lambda_2} V'' \to 0,$$

(9.3) $$0 \to W' \xrightarrow{\mu_1} W \xrightarrow{\mu_2} W''.$$

Then the following sequences are both exact:

(9.4) $$0 \to \mathrm{Hom}_A(V'', W) \xrightarrow{\lambda_2^*} \mathrm{Hom}_A(V, W) \xrightarrow{\lambda_1^*} \mathrm{Hom}_A(V', W),$$

(9.5) $$0 \to \mathrm{Hom}_A(V, W') \xrightarrow{(\mu_1)_*} \mathrm{Hom}_A(V, W) \xrightarrow{(\mu_2)_*} \mathrm{Hom}_A(V, W'').$$

Proof. We show the exactness of (9.4). To show that λ_2^* is a monomorphism, let $\sigma \in \mathrm{Hom}_A(V'', W)$ and suppose that $\lambda_2^*(\sigma) = \sigma \circ \lambda_2 = 0$. Then $\sigma(\lambda_2(V)) = \sigma(V'') = 0$, namely $\sigma = 0$. So, it remains to show that $\operatorname{Im}\lambda_2^* = \operatorname{Ker}\lambda_1^*$. Since

$\lambda_1^* \circ \lambda_2^* = (\lambda_2 \circ \lambda_1)^* = 0$, it is clear that $\operatorname{Im} \lambda_2^* \subset \operatorname{Ker} \lambda_1^*$. If $\sigma \in \operatorname{Ker} \lambda_1^*$, it follows from $\sigma \circ \lambda_1 = 0$ that $\lambda_1(V') = \operatorname{Ker} \lambda_2 \subset \operatorname{Ker} \sigma$. Consequently, for a given $v'' \in V''$, if we choose $v \in V$ so that $\lambda_2(v) = v''$, then $\sigma(v)$ does not depend on the choice of v. Define $\tau \in \operatorname{Hom}_A(V'', W)$ by $\tau(v'') = \sigma(v)$. Then we have $\lambda_2^*(\tau) = \tau \circ \lambda_2 = \sigma$. ■

Exercise 9.5

(i) Prove the exactness of (9.5).

(ii) If λ_2 splits in (9.2), then λ_2^* splits in (9.4).

(iii) If μ_1 splits in (9.3), then $(\mu_1)_*$ splits in (9.5).

Theorem 9.6. *The following holds for the direct sum and the direct product of A-modules.*

(i) $\operatorname{Hom}_A(\bigoplus_{i \in I} V_i, W) \simeq \prod_{i \in I} \operatorname{Hom}_A(V_i, W)$.

(ii) $\operatorname{Hom}_A(V, \prod_{i \in I} W_i) \simeq \prod_{i \in I} \operatorname{Hom}_A(V, W_i)$.

Proof.

(i) Let $\iota_i: V_i \to \bigoplus_i V_i$ be the inclusion map. If we set $\lambda_i = \lambda \circ \iota_i$ for an A-homomorphism $\lambda: \bigoplus_i V_i \to W$, then the map $\lambda \mapsto (\lambda_i)_i$ provides the desired isomorphism.

(ii) Let $\pi_i: \prod_i W_i \to W_i$ be the projection. If we set $\mu_i = \pi_i \circ \mu$ for an A-homomorphism $\mu: V \to \prod_i W_i$, then the map $\mu \to (\mu_i)_i$ provides the desired isomorphism. ■

Remark. If I is a finite set, we usually write $\bigoplus_{i \in I}$ for $\prod_{i \in I}$ in the above theorem.

If A, B are rings, then $\operatorname{Hom}_A({}_BV_A, W_A)$ is a right B-module with the action defined by $(\sigma b)(v) = \sigma(bv)$ for $b \in B$, $\sigma \in \operatorname{Hom}_A({}_BV_A, W_A)$ and $v \in V$. Likewise, $\operatorname{Hom}_A(V_A, {}_BW_A)$ is a left B-module with the action defined by $(b\sigma)(v) = b(\sigma(v))$.

For example $\operatorname{Hom}_A({}_AA_A, V_A)$ is a right A-module and $\varphi: \operatorname{Hom}_A(A, V) \to V (f \mapsto f(1_A))$ is an A-isomorphism.

9.2. $V \otimes_A W$

For V_A and ${}_AW$, let F be the free $\mathbf{Z}$-module with basis $V \times W = \{(v, w); v \in V, w \in W\}$: $F = \bigoplus_{v \in V, w \in W} \mathbf{Z}(v, w)$. Consider the submodule I of F generated by the following elements:

$$(9.6) \qquad \begin{cases} (v + v', w) - (v, w) - (v', w), \\ (v, w + w') - (v, w) - (v, w'), \\ (va, w) - (v, aw) \qquad (a \in A). \end{cases}$$

Then the factor module F/I is denoted by $V \otimes_A W$ and is called the *tensor product* of V and W over A.

If $v \otimes w$ denotes the image of (v, w) under the natural map $\varphi_0: F \to F/I$, then the following holds by definition:

$$\begin{cases} (v + v') \otimes w = v \otimes w + v' \otimes w, \\ v \otimes (w + w') = v \otimes w + v \otimes w', \\ (va) \otimes w = v \otimes (aw). \end{cases}$$

Clearly, $V \otimes_A W$ is generated by $\{v \otimes w; v \in V, w \in W\}$ over $\mathbf{Z}$, namely, every $x \in V \otimes_A W$ is expressed as $x = \sum_{i=1}^{n} v_i \otimes w_i$. But this expression is not unique in general.

For ${}_{\mathbf{Z}}T$, a map $\varphi: V \times W \to T$ is said to be *(A-)bilinear* if it satisfies the following three conditions:

$$\begin{cases} \varphi(v + v', w) = \varphi(v, w) + \varphi(v', w), \\ \varphi(v, w + w') = \varphi(v, w) + \varphi(v, w'), \\ \varphi(va, w) = \varphi(v, aw) \qquad (a \in A). \end{cases}$$

The pair (T, φ) is said to have the *universality* property (w.r.t. V_A and ${}_AW$) if the following holds:

For any ${}_{\mathbf{Z}}S$ and a bilinear map $f: V \times W \to S$, there exists a *unique* homomorphism $g: T \to S$ such that $g \circ \varphi = f$.

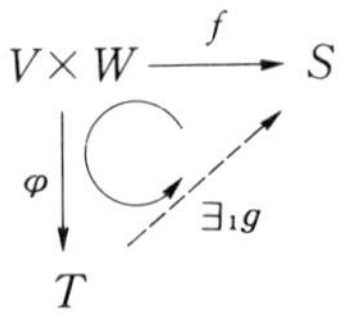

Theorem 9.7. *Let $\varphi_0: V \times W \to V \otimes_A W((v, w) \mapsto v \otimes w)$ be the natural map. Then the pair $(V \otimes_A W, \varphi_0)$ has the universality property.*

Proof. Let F, I be as above and let $f\colon V \times W \to S$ be a bilinear map. Then f extends to a homomorphism $f_1\colon F \to S$. Since f is bilinear, it follows that $f_1(I) = 0$. So f_1 induces $\bar{f}_1\colon F/I = V \otimes_A W \to S(v \otimes w \mapsto f(v, w))$ such that $\bar{f}_1 \circ \varphi_0 = f$. The uniqueness assertion is clear, since $V \otimes_A W$ is generated by $\varphi_0(V \times W) = \{v \otimes w;\, v \in V,\, w \in W\}$. ■

Theorem 9.8. *If both* (T, φ) *and* (T', φ') *have the universality property* w.r.t. V_A *and* ${}_AW$, *then there is a unique isomorphism* $g\colon T \xrightarrow{\sim} T'$ *such that* $g \circ \varphi = \varphi'$.

Proof. By the universality property of (T, φ), there exists a homomorphism $g\colon T \to T'$ such that $g \circ \varphi = \varphi'$.

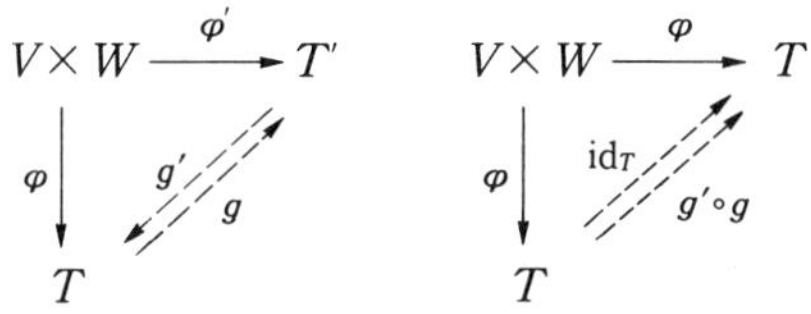

Likewise there exists $g'\colon T' \to T$ such that $g' \circ \varphi' = \varphi$. Then it follows that $(g' \circ g) \circ \varphi = g' \circ \varphi' = \varphi$. But since $\mathrm{id}_T \circ \varphi = \varphi$, we have $g' \circ g = \mathrm{id}_T$ by the uniqueness condition. Similarly, we have $g \circ g' = \mathrm{id}_{T'}$, and thus g is an isomorphism. ■

The above two theorems assert that the tensor product $V \otimes_A W$ is characterized by the universality property. This is useful in studying the tensor product.

For example, if B is a second ring and W is an (A, B)-bimodule, $V \otimes_A W$ becomes a right B-module with the action defined by $(v \otimes w)b = v \otimes wb$ for $b \in B$.

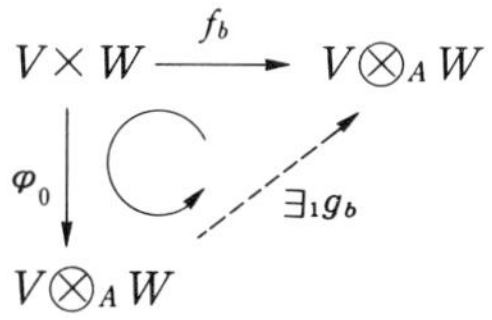

In fact, $f_b\colon V \times W \to V \otimes_A W((v, w) \mapsto v \otimes (wb))$ is a bilinear map and hence induces $g_b\colon V \otimes_A W \to V \otimes_A W$ such that $g_b(v \otimes w) = v \otimes wb$. Thus the above action is well defined. Similarly, if V is a (B, A)-bimodule, then $V \otimes_A W$ is a left B-module.

Theorem 9.9. *The following A-isomorphisms hold for V_A and ${}_AW$.*

(i) $V \otimes_A A \simeq V \quad (v \otimes a \mapsto va)$.

(ii) $A \otimes_A W \simeq W \quad (a \otimes w \mapsto aw)$.

Proof. (i) Note that $f: V \times A \to V((v, a) \mapsto va)$ is a bilinear map. Thus, by Theorem 9.7, there exists a homomorphism $g: V \otimes_A A \to V$ such that $g(v \otimes a) = va$, which is clearly an A-homomorphism. On the other hand, the map $h: V \to V \otimes_A A(v \mapsto v \otimes 1)$ is also an A-homomorphism, and $h \circ g = \mathrm{id}_{V \otimes A}$, $g \circ h = \mathrm{id}_V$ holds. Therefore g is an isomorphism.

The proof of part(ii) is similar. ■

Theorem 9.10. *Let $\lambda: V_A \to V'_A$, $\mu: {}_AW \to {}_AW'$ be A-homomorphisms. Then there exists a unique homomorphism $f: V \otimes_A W \to V' \otimes_A W'$ such that $f(v \otimes w) = \lambda(v) \otimes \mu(w)$.*

Proof. Let $\varphi_0(v, w) = v \otimes w$, and $g(v, w) = \lambda(v) \otimes \mu(w)$.

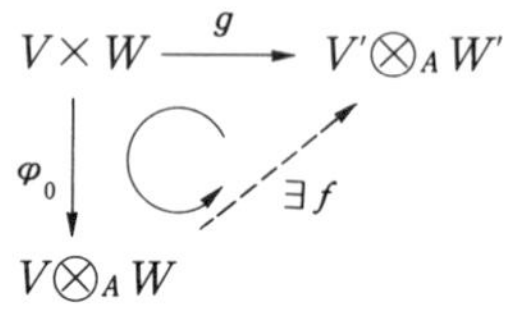

Since g is bilinear, the assertion follows from Theorem 9.7. ■

We denote the homomorphism f in the above theorem by $\lambda \otimes \mu$.

Exercise 9.11. Let $V_A \xrightarrow{\lambda} V'_A \xrightarrow{\lambda'} V''_A$ and ${}_AW \xrightarrow{\mu} {}_AW' \xrightarrow{\mu'} {}_AW''$ be A-homomorphisms. Then we have

$$(\lambda' \otimes \mu') \circ (\lambda \otimes \mu) = (\lambda' \circ \lambda) \otimes (\mu' \circ \mu).$$

The following property is usually referred to as the *right exactness* of $\otimes_A$.

Theorem 9.12. *Suppose that the following two sequences of A-modules are exact*:

(9.7) $$V'_A \xrightarrow{\lambda_1} V_A \xrightarrow{\lambda_2} V''_A \to 0,$$

(9.8) $${}_AW' \xrightarrow{\mu_1} {}_AW \xrightarrow{\mu_2} {}_AW'' \to 0.$$

Then both of the following two sequences are also exact:

(9.9) $$V' \otimes_A W \xrightarrow{\lambda_1 \otimes \mathrm{id}_W} V \otimes_A W \xrightarrow{\lambda_2 \otimes \mathrm{id}_W} V'' \otimes_A W \to 0,$$

(9.10) $$V \otimes_A W' \xrightarrow{\mathrm{id}_V \otimes \mu_1} V \otimes_A W \xrightarrow{\mathrm{id}_V \otimes \mu_2} V \otimes_A W'' \to 0.$$

Proof. We show the exactness of (9.9). Since λ_2 is an epimorphism, so is $\lambda_2 \otimes \mathrm{id}_W$. Also, it holds that $\mathrm{Im}(\lambda_1 \otimes \mathrm{id}_W) \subset \mathrm{Ker}(\lambda_2 \otimes \mathrm{id}_W)$ as $(\lambda_2 \otimes \mathrm{id}_W) \circ (\lambda_1 \otimes \mathrm{id}_W) = (\lambda_2 \circ \lambda_1) \otimes \mathrm{id}_W = 0$. Hence $\lambda_2 \otimes \mathrm{id}_W$ induces a homomorphism $\overline{\lambda_2 \otimes \mathrm{id}_W}: \overline{V \otimes_A W} = V \otimes_A W/\mathrm{Im}(\lambda_1 \otimes \mathrm{id}_W) \to V'' \otimes_A W(\overline{v \otimes w} \mapsto \lambda_2(v) \otimes w)$. Let us show that this is an monomorphism, which will complete the proof of our assertion.

Define $f: V'' \times W \to \overline{V \otimes_A W}$ by $f(v'', w) = \overline{v \otimes w}$, where $v \in V$ is chosen so that $\lambda_2(v) = v''$. We claim that f is well defined. If $\lambda_2(v_1) = v''$, then $\lambda_2(v_1 - v) = 0$ and hence $v_1 - v = \lambda_1(v')$ for some $v' \in V'$. Thus $v_1 \otimes w - v \otimes w = \lambda_1(v') \otimes w \in \mathrm{Im}(\lambda_1 \otimes \mathrm{id}_W)$. Clearly, f is bilinear, so there is a homomorphism $g: V'' \otimes W \to \overline{V \otimes_A W}$ such that $g(v'' \otimes w) = f(v'', w) = \overline{v \otimes w}$ with $\lambda_2(v) = v''$. Then it holds that $g \circ \overline{\lambda_2 \otimes \mathrm{id}_W}(\overline{v \otimes w}) = g(\lambda_2(v) \otimes w) = \overline{v \otimes w}$, i.e., $\overline{\lambda_2 \otimes \mathrm{id}_W}$ is a monomorphism. The exactness of (9.10) is proved similarly. ∎

Exercise 9.13. If λ_2 splits in (9.7), so does $\lambda_2 \otimes \mathrm{id}_W$ in (9.9). A similar statement is true for μ_2 and $\mathrm{id}_V \otimes \mu_2$.

Theorem 9.14. *The following isomorphism holds for V_A and ${}_AW_i (i \in I)$*:

$$V \otimes_A \left(\bigoplus_{i \in I} W_i \right) \simeq \bigoplus_{i \in I} (V \otimes_A W_i).$$

Proof. Since $\varphi\colon V \times (\bigoplus_i W_i) \to \bigoplus_i (V \otimes_A W_i)$ $((v, \sum_i w_i) \mapsto \sum_i v \otimes w_i)$ is bilinear, there exists a homomorphism $\bar{\varphi}\colon V \otimes_A (\bigoplus_i W_i) \to \bigoplus_i (V \otimes_A W_i)$ such that $\bar{\varphi}(v \otimes (\sum_i w_i)) = \sum_i v \otimes w_i$. On the other hand, if $\iota_j\colon W_j \to \bigoplus_i W_i$ is the inclusion map, then $\psi = \bigoplus_i (\mathrm{id}_V \otimes \iota_i)\colon \bigoplus_i (V \otimes W_i) \to V \otimes_A (\bigoplus_i W_i)$ gives the inverse map of $\bar{\varphi}$. Therefore $\bar{\varphi}$ is an isomorphism. ■

A statement similar to the above one also holds when V is a direct sum of A-modules.

As an application of the above theorem, we have the following.

Theorem 9.15.

(i) *Let $W = \bigoplus_{i \in I} Aw_i$ be an A-free module. Then every $u \in V \otimes_A W$ is expressed uniquely as $u = \sum_i v_i \otimes w_i (v_i \in V)$. Namely, if we put $\{v \otimes w_i;\ v \in V\} = V \otimes w_i$, then the following holds:*

$$V \otimes_A W = \bigoplus_{i \in I} V \otimes w_i.$$

(ii) *Let R be a commutative ring and suppose that $V = \bigoplus_{i \in I} Rv_i$, $W = \bigoplus_{j \in J} Rw_j$ be R-free modules. Then it holds that*

$$V \otimes_R W = \bigoplus_{i, j} R(v_i \otimes w_j),$$

which is again R-free.

Proof.

(i) We see that $V \otimes_A W = \bigoplus_i (V \otimes_A Aw_i) = \bigoplus_i (V \otimes w_i)$. Moreover, since ${}_A A \simeq {}_A(Aw_i)$, it follows that $\varphi_i\colon V = V \otimes_A A \to V \otimes w_i (v \mapsto v \otimes w_i)$ is an isomorphism and statement (i) holds.

(ii) We have

$$V \otimes_R W = \textstyle\bigoplus_{i,j} (Rv_i) \otimes_R (Rw_j) = \bigoplus_{i,j} (Rv_i) \otimes w_j = \bigoplus_{i,j} R(v_i \otimes w_j),$$

which is R-free. ■

The tensor products of more than two modules are defined similarly. For example, let A, B be rings and let U_A, ${}_A V_B$, ${}_B W$ be given. Let F be the free

Z-module with basis $U \times V \times W$, and I be the submodule of F generated by the following elements:

$$\begin{cases} (u + u', v, w) - (u, v, w) - (u', v, w), \\ (u, v + v', w) - (u, v, w) - (u, v', w), \\ (u, v, w + w') - (u, v, w) - (u, v, w'), \\ (ua, v, w) - (u, av, w) \qquad (a \in A), \\ (u, vb, w) - (u, v, bw) \qquad (b \in B). \end{cases}$$

The factor module F/I, denoted by $U \otimes_A V \otimes_B W$, is said to be the tensor product of U, V and W. Also, $u \otimes v \otimes w$ denotes the residue class $(u, v, w) + I$.

As in the case of the tensor product of two modules, the tensor product $U \otimes_A V \otimes_B W$ is also characterized by universality property.

Exercise 9.16.

(i) Describe the universality of $U \otimes_A V \otimes_B W$.

(ii) Prove that $(U \otimes_A V) \otimes_B W \simeq U \otimes_A (V \otimes_B W) \simeq U \otimes_A V \otimes_B W$.

Let I be an ideal of a ring A, $\bar{A} = A/I$ and for V_A, let $\bar{V} = V/VI$.

Theorem 9.17.

(i) *The following A-isomorphism holds for* V_A:

$$V \otimes_A A/I \xrightarrow{\sim} V/VI \qquad (v \otimes \bar{a} \mapsto \overline{va}).$$

(ii) *Let* V_A, ${}_AW$ *be such that* $VI = 0$, $IW = 0$. *Then* V *and* W *can be considered as a right and a left* $\bar{A}$-*module, respectively, and the following isomorphism holds*:

$$V \otimes_A W \xrightarrow{\sim} V \otimes_{\bar{A}} W \qquad (v \otimes w \mapsto v \otimes w).$$

Proof.

(i) For $v \in V$ and $\bar{a} \in \bar{A}$, $\overline{va}$ is determined independently of the choice of a representative a of $\bar{a}$. So, the map $\varphi: V \times \bar{A} \to \bar{V}((v, \bar{a}) \mapsto \overline{va})$ is well defined. Since φ is bilinear, it gives rise to a homomorphism $\tilde{\varphi}: V \otimes_A \bar{A} \to \bar{V}$ such that $\tilde{\varphi}(v \otimes \bar{a}) = \overline{va}$, which is an A-homomorphism. On the other hand, we can define $\psi: \bar{V} \to V \otimes_A \bar{A}$ by $\psi(\bar{v}) = v \otimes \bar{1}$. In fact, if $\bar{v} = \bar{v'}$, then $v' = v + \sum_i v_i x_i (v_i \in V, x_i \in I)$ and hence $v' \otimes \bar{1} = v \otimes \bar{1} + \sum_i v_i \otimes \bar{x}_i = v \otimes \bar{1}$. Thus $\psi(\bar{v})$ is independent of the choice of representatives of $\bar{v}$; and it is clear that ψ is the inverse of $\tilde{\varphi}$, and so $\tilde{\varphi}$ is an isomorphism.

(ii) Define the action of $\bar{a} \in \bar{A}$ on V by $v\bar{a} = va$ for each $v \in V$. Note that this is well defined and V becomes an $\bar{A}$-module by this action. Similarly, W is a left $\bar{A}$-module. Also it holds that

$$V \otimes_A W \simeq V \otimes_A (\bar{A} \otimes_{\bar{A}} W) \simeq (V \otimes_A \bar{A}) \otimes_{\bar{A}} W \simeq V/VI \otimes_{\bar{A}} W = V \otimes_{\bar{A}} W.$$

■

10. Projective and Injective Modules

10.1. Definitions and Schanuel's Lemma

An A-module P is said to be *projective* if, for any exact sequence $W \xrightarrow{\mu} W'' \to 0$ and any A-homomorphism $f: P \to W''$, there exists an A-homomorphism $g: P \to W$ such that $\mu \circ g = f$:

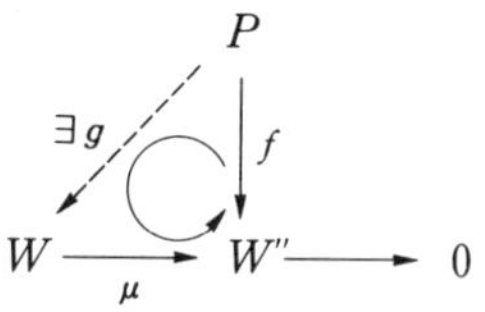

This is equivalent to saying that given an exact sequence

$$0 \to W' \xrightarrow{\lambda} W \xrightarrow{\mu} W'' \to 0,$$

the following is also exact:

$$0 \to \mathrm{Hom}_A(P, W') \xrightarrow{\lambda_*} \mathrm{Hom}_A(P, W) \xrightarrow{\mu_*} \mathrm{Hom}_A(P, W'') \to 0.$$

For example, every free A-module is projective. In fact, if $P = \bigoplus_{i \in I} x_i A$ is free in the above diagram, we define $g: P \to W$ by setting $g(x_i) = w_i$, where w_i is any element of $\mu^{-1}(f(x_i))$. Then we have $\mu \circ g = f$.

The above implies that any A-module V is a homomorphic image of some projective module P (cf. Theorem 1.6 of this chapter). Moreover, if V is finitely generated, then P may be assumed to be finitely generated.

Exercise 10.1. The following equivalence holds for $P = \bigoplus_{i \in I} P_i$.

$$P \text{ is projective} \Leftrightarrow P_i \text{ is projective for all } i \in I.$$

Theorem 10.2. *The following three conditions on P_A are equivalent.*

(1) *P is projective.*

(2) *An exact sequence $0 \to V' \xrightarrow{\lambda} V \xrightarrow{\mu} P \to 0$ always splits.*

(3) *P is isomorphic to a direct summand of some free A-module.*

Proof.

(1) $\Rightarrow$ (2). By the assumption, the diagram below can be completed to be a commutative one.

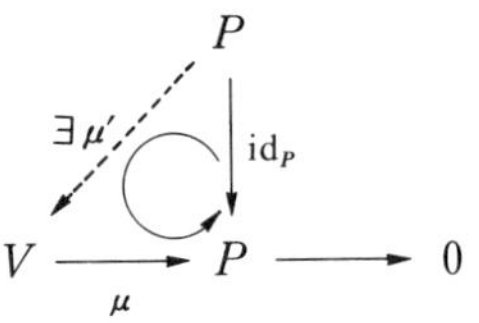

Namely, there exists an A-homomorphism $\mu'\colon P \to V$ such that $\mu \circ \mu' = \mathrm{id}_P$, and so the exact sequence in (2) splits.

(2) $\Rightarrow$ (3). There exists a free A-module F and an epimorphism $f\colon F \to P$. Then the exact sequence

$$0 \to \operatorname{Ker} f \to F \to P \to 0$$

splits, i.e. $P|F$.

(3) $\Rightarrow$ (1). This is clear by Exercise 10.1. ■

Theorem 10.3. *Let P_A be projective. Then, for an exact sequence,*

$$0 \to {}_AW' \xrightarrow{\lambda} {}_AW \xrightarrow{\mu} {}_AW'' \to 0,$$

the following is also exact:

$$0 \to P \otimes_A W' \xrightarrow{\mathrm{id}_P \otimes \lambda} P \otimes_A W \xrightarrow{\mathrm{id}_P \otimes \mu} P \otimes_A W'' \to 0.$$

Proof. We need only to show that $\mathrm{id}_P \otimes \lambda$ is a monomorphism. Assume first that P is A-free: $P = \bigoplus_{i \in I} x_i A$. Then by definition $\mathrm{id}_P \otimes \lambda$ is the map such that $(\mathrm{id}_P \otimes \lambda)(\sum_i x_i \otimes w_i') = \sum_i x_i \otimes \lambda(w_i')$. So, it is a monomorphism by Theorem 9.15.

For a general P, we may assume that $F = P \oplus P'$ for some free module F. Then $F \otimes_A W' = (P \otimes_A W') \oplus (P' \otimes_A W')$ and $F \otimes_A W' \to F \otimes_A W$ is a monomorphism by the above. Therefore $P \otimes_A W' \to P \otimes_A W$ is also a monomorphism. ■

The next result gives another characterization of semisimple Artinian rings.

Theorem 10.4. *The following two conditions are equivalent.*

(1) *A is a semisimple Artinian ring.*
(2) *Every A-module is projective.*

Proof.
(1) ⇒ (2). Any irreducible A-module is isomorphic to a direct summand of A_A by Theorem 8.6 and hence projective. This implies that every A-module is projective because it is a direct sum of irreducible A-modules.

(2) ⇒ (1). We show that every V_A is completely reducible (cf. Theorem 8.1). Let W be any A-submodule of V. Then, since V/W is projective, W is a direct summand of V. Thus V is completely reducible by Theorem 7.1. ■

Let I be an ideal of a ring A and $\bar{A} = A/I$. Then, for V_A, $\bar{V} = V/VI$ can be considered as an $\bar{A}$-module. Also, given an A-homomorphism $\varphi: V \to W$, we have an A-homomorphism $\bar{\varphi}: \bar{V} \to \bar{W}(\bar{v} \mapsto \overline{\varphi(v)})$.

Exercise 10.5. Suppose that P be a projective module.

With the notation as above, let $\lambda: \bar{P} \to \bar{V}$ be an $\bar{A}$-homomorphism. Then there exists an A-homomorphism $\varphi: P \to V$ such that $\bar{\varphi} = \lambda$.

Now, given A-modules V, X, X' and A-homomorphisms λ, λ' as in the diagram below,

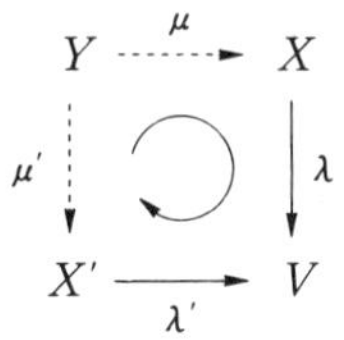

we see that

$$Y = \{(x, x'); \lambda(x) = \lambda'(x')\}$$

is an A-submodule of $X \oplus X'$, which is called the *pullback* of (λ, λ'). The above diagram can be completed to a commutative one if we set

$$\mu: Y \to X((x, x') \mapsto x),$$

$$\mu': Y \to X'((x, x') \mapsto x').$$

Note that if λ and λ' are both epimorphisms, then so are μ and μ'.

The next theorem is referred to as *Schanuel's lemma.*

Theorem 10.6. *Let*

$$0 \to K \to P \xrightarrow{\lambda} V \to 0,$$

$$0 \to K' \to P' \xrightarrow{\lambda'} V \to 0$$

be exact sequences of A-modules with P, P′ projective. Then the following isomorphism holds:

$$K' \oplus P \simeq K \oplus P'.$$

Proof. Let $Y = \{(p, p') \in P \oplus P';\ \lambda(p) = \lambda'(p')\}$ be the pullback of (λ, λ'). Then the diagram below is commutative and μ, μ' are epimorphisms.

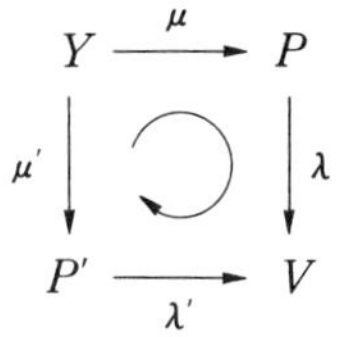

Since

$$\text{Ker } \mu = \{(0, p');\ \lambda'(p') = 0\} \simeq K'$$

$$\text{Ker } \mu' = \{(p, 0);\ \lambda(p) = 0\} \simeq K,$$

we have the following two exact sequences:

$$0 \to K' \to Y \xrightarrow{\mu} P \to 0,$$

$$0 \to K \to Y \xrightarrow{\mu'} P' \to 0.$$

But both of them split as P, P' are projective, and hence we have

$$K' \oplus P \simeq Y \simeq K \oplus P'. \quad \blacksquare$$

In the sense that all the arrows in the diagrams of A-modules are reversed, the dual notion of projective modules are the so-called *injective modules*. Here we mention the definition and give some elementary properties of them.

An A-module Q is said to be *injective* if, for any diagram of A-modules with concrete lines,

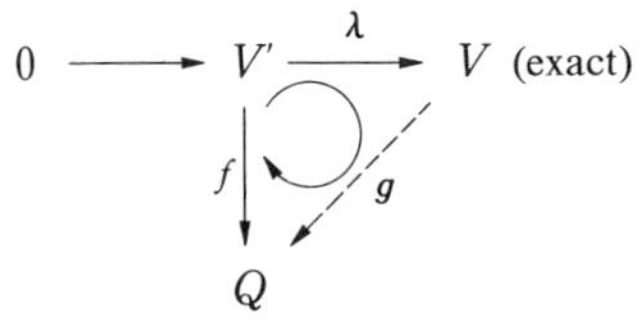

there exists an A-homomorphism $g: V \to Q$ such that $g \circ \lambda = f$. This is equivalent to saying that if

$$0 \to V' \to V \to V'' \to 0$$

is a short exact sequence of A-modules, then the following sequence is also exact:

$$0 \to \operatorname{Hom}_A(V'', Q) \to \operatorname{Hom}_A(V, Q) \to \operatorname{Hom}_A(V', Q) \to 0.$$

Exercise 10.7. If Q_A is injective, then an exact sequence

$$0 \to Q \to V \to V'' \to 0$$

always splits. (Remark: The converse is also true by Theorem 12.6 in Section 12.)

Exercise 10.8. A direct product of A-modules $Q = \prod_{i \in I} Q_i$ is injective if and only if each Q_i is injective.

Given A-modules V, X, X' and A-homomorphisms λ, λ' as in the diagram below, we let

$$Y = X \oplus X'/\{(\lambda(v), -\lambda'(v)); v \in V\}.$$

and call it the *pushout* of (λ, λ').

$$\begin{array}{ccc} V & \xrightarrow{\lambda} & X \\ \lambda' \downarrow & & \downarrow \mu \\ X' & \xrightarrow{\mu'} & Y \end{array}$$

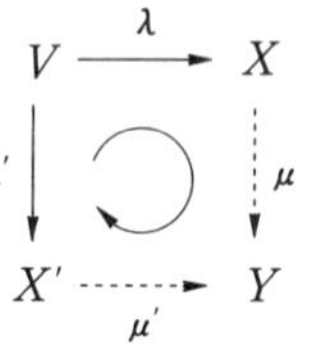

If we denote by $\overline{(x, x')}$ the residue class in Y containing $(x, x') \in X \oplus X'$, then $\overline{(\lambda(v), 0)} = \overline{(0, \lambda'(v))}$ holds. Thus, if we set

$$\mu: X \to Y(x \mapsto \overline{(x, 0)}), \qquad \mu': X' \to Y(x' \mapsto \overline{(0, x')}),$$

then the above diagram is commutative. Note that if λ and λ' are both monomorphisms, then so are μ and μ'.

The following result is a dual version of Schanuel's lemma.

Exercise 10.9. If

$$0 \to V \xrightarrow{\lambda} Q \to L \to 0,$$
$$0 \to V \xrightarrow{\lambda'} Q' \to L' \to 0$$

are exact with Q, Q' injective, then the following isomorphism holds.

$$Q \oplus L' \simeq Q' \oplus L.$$

10.2. Projective Covers and Injective Hulls

Exercise 10.10. The following holds for A-modules and A-homomorphisms.

(i) Let $f: U \to V$ be an epimorphism. Then the following two conditions are equivalent.

(1) A homomorphism $g: X \to U$ must be an epimorphism whenever the composite map $X \xrightarrow{g} U \xrightarrow{f} V$ is an epimorphism.
(2) If $W \subsetneqq U$, then $f(W) \subsetneqq V$.

(ii) Let $h: U \to V$ be a monomorphism. Then the following two conditions are equivalent.

(1) A homomorphism $k: V \to Y$ must be a monomorphism whenever the composite map $U \xrightarrow{h} V \xrightarrow{k} Y$ is a monomorphism.
(2) If $0 \neq W \subset V$, then $h(U) \cap W \neq 0$.

The epimorphism f and monomorphism h that satisfy the conditions in (i) and (ii), respectively, are said to be *essential*.

Let V be an A-module. If there exists a projective A-module P with an essential epimorphism $f: P \to V$, then P or the diagram $P \xrightarrow{f} V$ is called a *projective cover* of V.

Also, if there is an injective A-module Q with an essential monomorphism $h: V \to Q$, then Q or the diagram $V \xrightarrow{h} Q$ is called an *injective hull* of V.

A projective cover is unique up to natural isomorphism. Namely, we have the following.

Theorem 10.11. *If a projective cover of V_A exists, it is unique in the following sense: If $P \xrightarrow{f} V$, $P' \xrightarrow{f'} V$ are both projective covers of V, then there exists an isomorphism $\theta: P \xrightarrow{\sim} P'$ such that $f' \circ \theta = f$.*

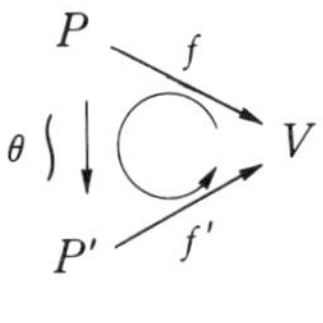

Proof. Since P is projective, there is $\theta: P \to P'$ such that $f' \circ \theta = f$. We see that θ is an epimorphism since f' is essential. Also, by the projectivity of P', there exists $\varphi: P' \to P$ such that $\theta \circ \varphi = \mathrm{id}_{P'}$. Then it holds that $f \circ \varphi = f' \circ \theta \circ \varphi = f'$, and hence φ is an epimorphism because f is essential. It then follows from $\theta \circ \varphi = \mathrm{id}_{P'}$ that θ is a monomorphism. ■

Exercise 10.12. If $V \xrightarrow{h} Q$, $V \xrightarrow{h'} Q'$ are injective hulls of V, then there exists $\theta: Q \xrightarrow{\sim} Q'$ such that $\theta \circ h = h'$.

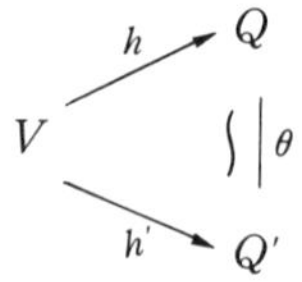

Given V_A, we let $P(V)$ and $I(V)$ denote a projective cover and an injective hull of V, respectively, if they exist.

The following is an easy consequence of the Azumaya–Nakayama lemma.

Lemma 10.13. *Let $f: U_A \to V_A$ be an epimorphism. If U is finitely generated and* $\mathrm{Ker}\, f \subset UJ(A)$, *then f is essential.*

Proof. Let $W \subset U$ be such that $f(W) = V$. Then $U = \mathrm{Ker}\, f + W = UJ(A) + W$, whence it follows that $U = W$. ■

From the above lemma we obtain the following.

Theorem 10.14. *Let I be an ideal of A contained in $J(A)$, and let P be a finitely generated projective module. Then P is a projective cover of P/PI. In particular, if e is an idempotent of A, eA is a projective cover of eA/eI.*

A ring A is said to be a *semiperfect ring* if it satisfies the following two conditions:

(10.1) *$\bar{A} = A/J(A)$ is an Artinian ring.*

(10.2) *Any idempotent of $\bar{A}$ lifts to that of A.*

For example an Artinian ring is semiperfect.

As will be shown later, every module has an injective hull. But there does not always exist a projective cover. For semiperfect rings, we have the following result.

Theorem 10.15. *Let A be a semiperfect ring and let $J = J(A)$, $\bar{A} = A/J$.*

(i) *Every finitely generated A-module V has a projective cover P which is finitely generated over A and satisfies $P/PJ \simeq V/VJ$.*

(ii) *Let $f: U \to V$ be an epimorphism and suppose that U is finitely generated and projective. Then the following equivalence holds.*

$$f \text{ is essential} \quad \Leftrightarrow \quad \operatorname{Ker} f \subset UJ.$$

(iii) *Under the assumption of* (i), *$P = P(V)$ is also a projective cover of $\bar{V} = V/VJ$.*

Proof.

(i) Let $\{\bar{e}_i\bar{A};\ 1 \le i \le k\}$ be a complete set of representatives of non-isomorphic irreducible $\bar{A}$-modules, where $e_i \in \mathrm{pi}(A)$. Write $\bar{V} = V/VJ \simeq \bigoplus_{i=1}^{k} m_i\bar{e}_i\bar{A}$ with $m_i \ge 0$. Let $P = \bigoplus_{i=1}^{k} m_i e_i A$. Then P is a finitely generated projective module such that $\bar{P} = P/PJ \simeq \bar{V}$.

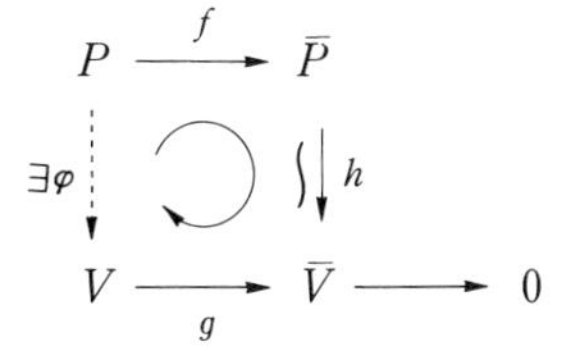

Let f, g be the natural homomorphisms in the above diagram. Then there exists $\varphi: P \to V$ such that $g \circ \varphi = h \circ f$ by the projectivity of P. So $g(\varphi(P)) = h(f(P)) = h(\bar{P}) = \bar{V}$, and hence $V = \varphi(P) + \operatorname{Ker} g = \varphi(P) + VJ$, whence it follows that $V = \varphi(P)$ by the Azumaya–Nakayama lemma. Thus φ is an epimorphism. Also, since $\operatorname{Ker} \varphi \subset \operatorname{Ker} f = PJ$, φ is essential by Lemma 10.13. Therefore P is a projective cover of V.

(ii) We need only to show the implication ($\Rightarrow$). Let $P \overset{\varphi}{\to} V$ be the same as in the proof of (i).

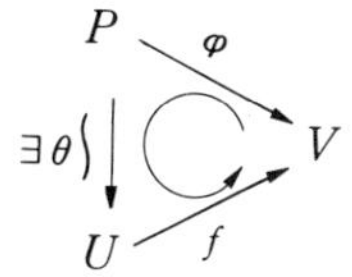

If f is essential, that is, if U is a projective cover of V, then there exists an isomorphism $\theta: P \xrightarrow{\sim} U$ such that $f \circ \theta = \varphi$. Then $\text{Ker } f = \theta(\text{Ker } \varphi) \subset \theta(PJ) = UJ$.

(iii) The projective cover P of V that is constructed in the proof of (i) is also a projective cover of $\bar{V}$, as is easily seen. ■

The next result states that both $P(V)$ and $I(V)$ are *additive*.

Lemma 10.16. *Let V_i be A-modules ($1 \leq i \leq n$) and let $V = \bigoplus_{i=1}^{n} V_i$.*

(i) *If $P_i \xrightarrow{f_i} V_i$ is a projective cover for each i, then $\bigoplus_{i=1}^{n} P_i$ is a projective cover of V.*

(ii) *If $V_i \xrightarrow{g_i} Q_i$ is an injective hull for each i, then $\bigoplus_{i=1}^{n} Q_i$ is an injective hull of V.*

Proof. It suffices to prove the assertion when $n = 2$.

(i) We show that $f = f_1 \oplus f_2: P = P_1 \oplus P_2 \to V$ is essential. Let $W \subset P$ be such that $f(W) = V$. Then we have $P = L_1 + L_2 + W$, where $L_i = \text{Ker } f_i$. From this we obtain $P_1 = L_1 + ((W + L_2) \cap P_1)$ and hence $P_1 \subset W + L_2$ because f_1 is essential. This implies that $P = L_2 + W$ and thus $P_2 = L_2 + (W \cap P_2)$, whence we have $P_2 \subset W$, since f_2 is essential. Therefore we conclude $P = W$.

(ii) We may assume that both g_1 and g_2 are the inclusion maps. So, it suffices to show that $vA \cap V \neq 0$ for any $v(\neq 0) \in Q_1 \oplus Q_2$. Write $v = v_1 + v_2$ with $v_i \in Q_i$. If $v_1 \neq 0$, then there exists $a \in A$ such that $v_1 a(\neq 0) \in V_1$. If $v_2 a \in V_2$, then $va \in V$. If $v_2 a \notin V_2$, then there exists $b \in A$ such that $v_2 ab(\neq 0) \in V_2$. Thus $vab = v_1 ab + v_2 ab$ is a nonzero element of V. ■

10.3. Heller Operators

Let V be an A-module. We denote by $\Omega(V)$ the kernel of a projective cover $P(V) \xrightarrow{f} V$ of V and by $\Omega^{-1}(V)$ the cokernel $I(V)/g(V)$ of an injective hull $V \xrightarrow{g} I(V)$. So we have the following two exact sequences

$$0 \to \Omega(V) \to P(V) \to V \to 0,$$

$$0 \to V \to I(V) \to \Omega^{-1}(V) \to 0,$$

and $\Omega(V)$, $\Omega^{-1}(V)$ are unique up to isomorphism. We call Ω, Ω^{-1} the *Heller operators*. Set $\Omega^0(V) = V$ and define $\Omega^n(V)$, $\Omega^{-n}(V)$ for $n > 0$ inductively by

$$\Omega^n(V) = \Omega(\Omega^{n-1}(V)), \qquad \Omega^{-n}(V) = \Omega^{-1}(\Omega^{-(n-1)}(V)).$$

Remark. If P is projective, then $P \xrightarrow{\mathrm{id}_P} P$ is a projective cover, and hence we have $\Omega(P) = 0$. Likewise $\Omega^{-1}(Q) = 0$ if Q is injective.

By Lemma 10.16 we have the following.

Theorem 10.17.

$$\Omega\left(\bigoplus_{i=1}^{n} V_i\right) \simeq \bigoplus_{i=1}^{n} \Omega(V_i), \qquad \Omega^{-1}\left(\bigoplus_{i=1}^{n} V_i\right) \simeq \bigoplus_{i=1}^{n} \Omega^{-1}(V_i).$$

Theorem 10.18. *Let $P(V) \xrightarrow{f} V$ be a projective cover of V_A. If P is a projective A-module with epimorphism $g: P \to V$, then there exists an A-module P' such that*

$$P \simeq P(V) \oplus P', \qquad \operatorname{Ker} g \simeq \Omega(V) \oplus P'.$$

Proof. Let $K = \operatorname{Ker} g$. We have the following two exact sequences

$$\begin{array}{ccccccccc} 0 & \longrightarrow & K & \longrightarrow & P & \xrightarrow{g} & V & \longrightarrow & 0 \\ & & & & \vdots \downarrow \lambda & & \wr\!\downarrow \mathrm{id}_V & & \\ 0 & \longrightarrow & \Omega(V) & \longrightarrow & P(V) & \xrightarrow[f]{} & V & \longrightarrow & 0. \end{array}$$

By the projectivity of P, there exists $\lambda: P \to P(V)$ such that $f \circ \lambda = g$, and λ is an epimorphism because f is essential. Hence there exists $\mu: P(V) \to P$ such that $\lambda \circ \mu = \mathrm{id}_{P(V)}$, whence it follows that $P = \operatorname{Ker} \lambda \oplus \mu(P(V))$, $\mu(P(V)) \simeq P(V)$. Let $P' = \operatorname{Ker} \lambda$, so $P \simeq P' \oplus P(V)$. Since $K = \operatorname{Ker} g \supset P' = \operatorname{Ker} \lambda$, we have $K = P' \oplus (K \cap \mu(P(V)))$. To complete the proof, it suffices to show that $\Omega(V) \simeq K \cap \mu(P(V))$. If $x \in P(V)$, then

$$\mu(x) \in K \Leftrightarrow g \circ \mu(x) = 0 \Leftrightarrow f \circ \lambda \circ \mu(x) = f(x) = 0 \Leftrightarrow x \in \Omega(V).$$

Since μ is a monomorphism, this gives $\Omega(V) \simeq K \cap \mu(P(V))$. ■

As a dual to the above theorem, we have

Theorem 10.19. *Let $V \xrightarrow{f} I(V)$ be an injective hull of V_A. If Q is an injective module with monomorphism $g: V \to Q$, then there exists an A-module Q' such that*

$$Q \simeq I(V) \oplus Q', \qquad \operatorname{Coker} g \simeq \Omega^{-1}(V) \oplus Q'.$$

Proof. Let $L = \operatorname{Coker} g$. We have the following two exact sequences

$$\begin{array}{ccccccccc} 0 & \longrightarrow & V & \xrightarrow{f} & I(V) & \longrightarrow & \Omega^{-1}(V) & \longrightarrow & 0 \\ & & \wr\!\downarrow \mathrm{id}_V & & \vdots\, \lambda & & & & \\ 0 & \longrightarrow & V & \xrightarrow[g]{} & Q & \longrightarrow & L & \longrightarrow & 0. \end{array}$$

By the injectivity of Q, there exists $\lambda: I(V) \to Q$ such that $\lambda \circ f = g$, and λ is a monomorphism because f is essential. Thus, there exists $\mu: Q \to I(V)$ such that $\mu \circ \lambda = \mathrm{id}_{I(V)}$, whence it follows that $Q = \operatorname{Ker} \mu \oplus \lambda(I(V))$, $\lambda(I(V)) \simeq I(V)$. Let $Q' = \operatorname{Ker} \mu$, so $Q \simeq I(V) \oplus Q'$. Since $g(V) = \lambda(f(V)) \subset \lambda(I(V))$, we have $L = Q/g(V) \simeq Q' \oplus (\lambda(I(V))/g(V))$. But under the isomorphism $\lambda: I(V) \xrightarrow{\sim} \lambda(I(V))$, $f(V)$ is mapped onto $g(V)$, and hence we have $\Omega^{-1}(V) = I(V)/f(V) \simeq \lambda(I(V))/g(V)$, and thus $L \simeq Q' \oplus \Omega^{-1}(V)$. ■

11. Change of Rings

11.1. Certain Connections Between Hom and Tensor

We begin with the following theorem.

Theorem 11.1. *Let A, B be rings. Then for X_B, ${}_BY_A$, Z_A the following isomorphism holds:*

(11.1) $$\operatorname{Hom}_A(X \otimes_B Y, Z) \simeq \operatorname{Hom}_B(X, \operatorname{Hom}_A(Y, Z)).$$

Proof. For an A-homomorphism $f: X \otimes_B Y \to Z$ and $x \in X$ consider the map $f_x: Y \to Z (y \mapsto f(x \otimes y))$. Obviously f_x is an A-homomorphism, and so we obtain a map $\varphi(f)\colon X \to \mathrm{Hom}_A(Y, Z)(x \mapsto f_x)$, which is a B-homomorphism, as can easily be seen. Since $\varphi(f_1 + f_2) = \varphi(f_1) + \varphi(f_2)$ holds, φ gives a homomorphism from the left-hand side of (11.1) into the right-hand side of it.

Next, let $g\colon X \to \mathrm{Hom}_A(Y, Z)$ be a B-homomorphism. Then the map $g'\colon X \times Y \to Z((x, y) \mapsto g(x)(y))$ is B-bilinear. Hence there exists a homomorphism $\psi(g)\colon X \otimes_B Y \to Z$ such that $\psi(g)(x \otimes y) = g'(x, y) = g(x)(y)$. Since $g(x)$ is an A-homomorphism, $\psi(g)$ is also. Thus we obtain a homomorphism ψ from the right-hand side of (11.1) into the left-hand side of it. And we see that $\psi \circ \varphi$ and $\varphi \circ \psi$ are the identity maps. Namely, φ is an isomorphism with $\psi = \varphi^{-1}$. ■

The following corollary will be needed to show the existence of injective hulls in the next section.

Corollary 11.2. *If ${}_BX_A$ is projective as a B-module and Y_A is injective, then the right B-module* $\mathrm{Hom}_A({}_BX_A, Y_A)$ *is injective.*

Proof. Let an exact sequence $0 \to V'_B \xrightarrow{\lambda} V_B$ be given. Then the sequence $0 \to V' \otimes_B X \xrightarrow{\mu} V \otimes_B X$ is exact since ${}_BX$ is projective, and from the above theorem we have the following commutative diagram:

$$\begin{array}{ccccc}
\mathrm{Hom}_A(V \otimes_B X, Y) & \xrightarrow{\mu^*} & \mathrm{Hom}_A(V' \otimes_B X, Y) & \longrightarrow & 0 \\
\wr\downarrow & & \wr\downarrow & & \\
\mathrm{Hom}_B(V, \mathrm{Hom}_A(X, Y)) & \xrightarrow{\lambda^*} & \mathrm{Hom}_B(V', \mathrm{Hom}_A(X, Y)) & \longrightarrow & 0.
\end{array}$$

Here the upper sequence is exact by the injectivity of Y, and hence the lower sequence is exact. Therefore, $\mathrm{Hom}_A(X, Y)$ is B-injective. ■

If a ring homomorphism $\chi\colon B \to A$ is given, then any V_A becomes a B-module, with the action of $b \in B$ being defined by $vb = v\chi(b)$ for $v \in V$. In particular, A itself can be considered to be A_B and ${}_BA$.

For example, if B is a subring of A with identity in common and if $\chi\colon B \to A$ is the inclusion map, then V_B in the above sense is just the B-module obtained from V_A by restricting the operation of A to B.

We get the following theorem from Theorem 11.1.

Theorem 11.3. *Given a ring homomorphism $\chi: B \to A$ and V_A, W_B, the following isomorphisms hold.*

(i) $\mathrm{Hom}_B(W, V_B) \simeq \mathrm{Hom}_A(W \otimes_B A, V)$.

(ii) $\mathrm{Hom}_B(V_B, W) \simeq \mathrm{Hom}_A(V, \mathrm{Hom}_B(A_B, W))$.

Proof.

(i) Let $X_B = W_B$, ${}_BY_A = {}_BA_A$, $Z_A = V_A$ in (11.1). Then the result follows from the B-isomorphism $\mathrm{Hom}_A(A, V) \simeq V$.

(ii) Exchange the rolls of A and B in (11.1) and let $X_A = V_A$, ${}_AY_B = {}_AA_B$, $Z_B = W_B$. Thus we have

$$\mathrm{Hom}_B(V \otimes_A A, W) \simeq \mathrm{Hom}_A(V, \mathrm{Hom}_B(A, W)),$$

whence the assertion follows from the B-isomorphism $V \otimes_A A \simeq V$. ■

11.2. Tensor Products of Algebras and Change of Coefficient Rings

Let R be a commutative ring and A be an R-algebra. Any A-module V becomes an (R, A)-bimodule if we define the action of $r \in R$ on V by $rv = v(r1_A)$.

For V_A and ${}_AW$, $V \otimes_A W$ is an R-module since V is an (R, A)-bimodule; $r(v \otimes w) = (rv) \otimes w$. Since the latter element equals $v \otimes (wr)$, we obtain the same R-module structure on $V \otimes_R W$ by regarding W as ${}_AW_R$. Similarly, $\mathrm{Hom}_A(V, W)$ is an R-module with the action defined by $(rf)(v) = f(rv) = r(f(v))$ for $f \in \mathrm{Hom}_A(V, W)$ and $v \in V$.

Now, let A and B be R-algebras. Recall that $A \otimes_R B$ is the factor module of the $\mathbf{Z}$-free module $F = \bigoplus_{a \in A, b \in B} \mathbf{Z}(a, b)$ modulo the submodule generated by the following elements: $(a + a', b) - (a, b) - (a', b)$, $(a, b + b') - (a, b) - (a, b')$, $(ra, b) - (a, rb)$. Define the multiplication by $(a, b)(a', b') = (aa', bb')$ and extend this linearly to all the elements of F. Then F is a ring and I is an ideal of it. Therefore, $F/I = A \otimes_R B$ is a ring with $1_A \otimes 1_B$ as identity, which is also an R-algebra. We call $A \otimes_R B$ the *tensor product of R-algebras A* and B.

For V_A and W_B, $V \otimes_R W$ becomes an $A \otimes_R B$-module. In fact, given $(a, b) \in A \times B$, the map $f(a, b): V \times W \to V \otimes_R W((v, w) \mapsto (va) \otimes (wb))$ is bilinear, and hence there exists a homomorphism $g(a, b): V \otimes_R W \to V \otimes_R W$ such that $v \otimes w \mapsto (va) \otimes (wb)$. This defines a bilinear map $g: A \times B \to \mathrm{End}_R(V \otimes_R W)$, which yields a homomorphism $h: A \otimes_R B \to \mathrm{End}_R(V \otimes_R W)$ such that $h(a \otimes b): v \otimes w \mapsto (va) \otimes (wb)$. Define the action of $a \otimes b$ via h, that is,

$$(v \otimes w)(a \otimes b) = (va) \otimes (wb).$$

With this action, $V \otimes_R W$ becomes an $A \otimes_R B$-module.

Lemma 11.4. *The following algebra isomorphisms hold for R-algebras A and B.*

(i) $M_n(A) \otimes_R B \simeq M_n(A \otimes_R B)$.
(ii) $M_n(A) \otimes_R M_m(B) \simeq M_{nm}(A \otimes_R B)$.

Proof.
(i) Let $e_{ij}(1 \leq i, j \leq n)$ be the matrix units of $M_n(A)$. Then $M_n(A) \otimes_R B \simeq \bigoplus_{i,j} (e_{ij} \otimes 1_B)(A \otimes_R B)$, which can be viewed as the full matrix ring over $A \otimes_R B$ with $\{e_{ij} \otimes 1_B\}$ as the matrix units.
(ii) By (i), the left-hand side is isomorphic to $M_n(A \otimes_R M_m(B)) \simeq M_n(M_m(A \otimes_R B)) \simeq M_{nm}(A \otimes_R B)$. ■

Let A, B be R-algebras and V, W be A-modules. For given $b \in B$ and $f \in \mathrm{Hom}_A(V, W)$, the map

$$\alpha'(b, f)\colon B \otimes_R V \to B \otimes_R W \; (c \otimes v \mapsto bc \otimes f(v))$$

is a $B \otimes_R A$-homomorphism as is easily seen. And the map

$$\alpha'\colon B \times \mathrm{Hom}_A(V, W) \to \mathrm{Hom}_{B \otimes_R A}(B \otimes_R V, B \otimes_R W)$$

is R-bilinear. Consequently, we have an R-homomorphism

$$\text{(11.2)} \qquad \alpha\colon B \otimes_R \mathrm{Hom}_A(V, W) \to \mathrm{Hom}_{B \otimes_R A}(B \otimes_R V, B \otimes_R W)$$

such that $\alpha(b \otimes f)(c \otimes v) = bc \otimes f(v)$. We shall show that this is an isomorphism under certain circumstances. Before doing so, we need some preparations.

An A-module V is said to be *finitely presented* if there exists an exact sequence

$$F_1 \to F_0 \to V \to 0$$

with F_0, F_1 free A-modules of finite rank.

Exercise 11.5. If A is Noetherian, then every finitely generated A-module is finitely presented.

An A-module V is said to be *A-flat* if, for any exact sequence,

$$0 \to W' \xrightarrow{\lambda} W,$$

the following sequence is also exact:

$$0 \to V \otimes_A W' \xrightarrow{\mathrm{id}_V \otimes \lambda} V \otimes_A W.$$

For example a projective module is always flat by Theorem 10.3.

Exercise 11.6. Suppose that the following commutative diagram with exact rows is given.

$$\begin{array}{ccccccc} 0 \longrightarrow & X & \longrightarrow & X' & \longrightarrow & X'' \\ & \alpha \downarrow & & \alpha' \downarrow & & \alpha'' \downarrow \\ 0 \longrightarrow & Y & \longrightarrow & Y' & \longrightarrow & Y'' \end{array}$$

Then the following holds.
(i) If α' is a monomorphism, so is α.
(ii) If α' and α'' are both isomorphisms, so is α.

Now we are ready to prove the following theorem.

Theorem 11.7. *Let A, B be R-algebras and V, W be A-modules. If B is R-flat and V_A is finitely presented, then the α defined in* (11.2) *is an isomorphism*:

$$\alpha\colon B \otimes \mathrm{Hom}_A(V, W) \xrightarrow{\sim} \mathrm{Hom}_{B \otimes A}(B \otimes V, B \otimes W),$$

where $\otimes$ is the abbreviation of $\otimes_R$.

Proof. If $V = A$, then both sides of the above are isomorphic to $B \otimes W$, and hence α is an isomorphism, as is easily seen. Consequently α is an isomorphism if V is a free A-module of finite rank.

By assumption, there are free A-modules, say F_0, F_1, of finite rank and an exact sequence

$$F_1 \to F_0 \to V \to 0.$$

From this, we obtain the following exact sequence:

$$0 \to \mathrm{Hom}_A(V, W) \to \mathrm{Hom}_A(F_0, W) \to \mathrm{Hom}_A(F_1, W).$$

For X_A, let $X' = B \otimes X$. Then we have the following isomorphisms from the above:

$$\alpha_i\colon B \otimes \mathrm{Hom}_A(F_i, W) \simeq \mathrm{Hom}_{A'}(F_i', W') \qquad (i = 0, 1).$$

Since B is R-flat, we get the following commutative diagram with exact rows:

$$\begin{array}{ccccccc}
0 \to & B \otimes \mathrm{Hom}_A(V, W) & \to & B \otimes \mathrm{Hom}_A(F_0, W) & \to & B \otimes \mathrm{Hom}_A(F_1, W) \\
& \alpha \downarrow & & \alpha_0 \downarrow & & \alpha_1 \downarrow \\
0 \to & \mathrm{Hom}_{A'}(V', W') & \to & \mathrm{Hom}_{A'}(F_0', W') & \to & \mathrm{Hom}_{A'}(F_1', W')
\end{array}$$

Therefore, α is an isomorphism by Exercise 11.6. ■

Now, we apply the above theorem to the case where R is a principal ideal domain and B is a commutative R-algebra. Before doing it, we collect here some necessary facts concerning modules and matrices over principal ideal domains.

Let V be a module over an integral domain R. An element v of V is said to be a *torsion element* if there exists a nonzero element r of R such that $rv = 0$. The set of all torsion elements of V is an R-submodule of V, which is denoted by $T(V)$. V is said to be *torsion-free* if $T(V) = 0$.

For the proof of the following two theorems, see Hungerford [1], Chapter IV.

Theorem 11.8. *Let R be a principal ideal domain.*

(i) *Let F be a free R-module of rank n and W be a nonzero R-submodule of F. Then W is a free R-module of rank not greater than n. In fact, there is a basis $\{v_1, \ldots, v_n\}$ of F such that $\{e_1v_1, \ldots, e_rv_r\}$ is a basis of W, with $e_i \in R$ satisfying $(e_i) \supset (e_{i+1})$ for each i $(1 \leq i \leq r-1)$.*

In particular, if $R = \mathbf{Z}$, then the R-rank of W equals n if and only if $|F/W| < \infty$.

(ii) *If V_R is finitely generated, then it is a direct sum of cyclic submodules: $V = Rv_1 \oplus \cdots \oplus Rv_n$.*

(iii) *The following two conditions are equivalent for a finitely generated R-module V.*

(1) *V is R-free.*

(2) *V is torsion-free.*

Theorem 11.9. *Let $C = (c_{ij})$ be an $m \times n$ matrix over a principal ideal domain R. Then there exist $P \in GL_m(R)$ and $Q \in GL_n(R)$ such that*

$$PCQ = \left(\begin{array}{ccc|c} e_1 & & 0 & \\ & \ddots & & 0 \\ 0 & & e_r & \\ \hline & 0 & & 0 \end{array}\right), \qquad (e_i) \supset (e_{i+1}) \quad (1 \leq i \leq r-1).$$

Moreover, the ideals $(e_1), \ldots, (e_r)$ are uniquely determined by C.

The above elements $e_1, \ldots, e_r$ are called the *elementary divisors* of the matrix C.

We also need the following theorem.

Theorem 11.10. *Let V be a module over a principal ideal domain R. If V is torsion-free, then V is R-flat.*

Note that in the above theorem any finitely generated R-submodule of V is R-free by Theorem 11.8(iii) and hence R-flat. Therefore, the theorem above is an immediate consequence of the following lemma.

Lemma 11.11. *Let A be a ring and V be an A-module. If any finitely generated A-submodule of V is A-flat, then V is also A-flat.*

Proof. Let $0 \to {}_AW' \xrightarrow{\lambda} {}_AW$ be an exact sequence. We need to show that the sequence $0 \to V \otimes_A W' \xrightarrow{\mathrm{id}_V \otimes \lambda} V \otimes_A W$ is exact. Going back to the definition of $V \otimes_A W$, let $F(V)$ be the free $\mathbf{Z}$-module with basis $V \times W$ and $I(V)$ be the submodule of $F(V)$ generated by the elements as in (9.6). Thus, $V \otimes_A W = F(V)/I(V)$.

Now let $u = \sum_{i=1}^r v_i \otimes w_i' \in V \otimes_A W'$ be such that $(\mathrm{id}_V \otimes \lambda)(u) = 0$. Then the element $u_o = \sum_{i=1}^r (v_i, \lambda(w_i'))$ of $F(V)$ lies in $I(V)$, and so it is expressed as a finite sum of elements of the form in (9.6). Consider the elements of V involved in such an expression and let V' be the A-submodule of V generated by those elements together with $\{v_i; 1 \leq i \leq n\}$. Then it is clear that $u' = \sum_{i=1}^r v_i \otimes w_i'$, viewed as an element of $V' \otimes_A W'$, is mapped onto zero by

$\mathrm{id}_{V'} \otimes \lambda$. But since V' is finitely generated, we have $u' = 0$ by assumption. On the other hand, if $\iota: V' \to V$ denotes the inclusion map, then u is the image of u' by the map $\iota \otimes \mathrm{id}_{W'}: V' \otimes_A W' \to V \otimes_A W'$, and hence $u = 0$. Therefore $\mathrm{id}_V \otimes \lambda$ is a monomorphism. ∎

Let S be a commutative R-algebra. Then if A is an R-algebra, $S \otimes_R A$, which is denoted by A^S, is an S-algebra via the natural S-action on it.

In particular, when R is a subring of S with identity in common, A^S is said to be the algebra obtained by the extension of the coefficient ring R to S. For example, if A is R-free with basis $\{x_i; i \in I\}$, then A^S is S-free with basis $\{1_S \otimes x_i; i \in I\}$. Identifying $1_S \otimes x_i$ with x_i, we have that $\{x_i; i \in I\}$ is an S-basis of A^S and $A \subset A^S$.

From Theorems 11.7 and 11.10, we obtain the following, which will be needed in Section 7 of Chapter 4.

Theorem 11.12. *Let R be a principal ideal domain and let S be a commutative R-algebra that is torsion-free as an R-module. For an R-algebra A and V_A, W_A, assume that A and V are finitely generated over R. Then the following isomorphism holds:*

$$\alpha: S \otimes_R \mathrm{Hom}_A(V, W) \xrightarrow{\sim} \mathrm{Hom}_{A^S}(V^S, W^S),$$

where $\alpha(s \otimes f)(t \otimes v) = st \otimes f(v)$.

In particular the following S-algebra isomorphism holds:

$$\mathrm{End}_A(V)^S \simeq \mathrm{End}_{A^S}(V^S).$$

Remark. In the above, V_A is finitely presented by Exercise 11.5.

12. Existence of Injective Hulls

We need some preparation before showing the existence of an injective hull of a given module. Let A be a ring.

Lemma 12.1. *Let I be a right ideal of A and V be an A-module. Then an A-homomorphism $f: I \to V$ can be extended to an A-homomorphism $\tilde{f}: A \to V$ if and only if there exists $v \in V$ such that $f(x) = vx$ for all $x \in I$.*

Proof. If such an extension $\tilde{f}$ exists, then let $v = \tilde{f}(1)$. Conversely, if there exists $v \in V$ satisfying the above, define $\tilde{f}: A \to V$ by $\tilde{f}(a) = va$ for $a \in A$. ■

Theorem 12.2 (Baer). *A necessary and sufficient condition for an A-module V to be injective is that every A-homomorphism $f: I \to V$ from an arbitrary right ideal I of A into V can be extended to an A-homomorphism $\tilde{f}: A \to V$.*

Proof. The necessity is clear from the following diagram:

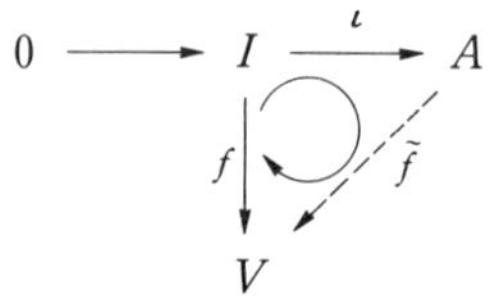

To show the sufficiency, suppose we are given an exact sequence $0 \to U' \xrightarrow{\iota} U$ of A-modules and an A-homomorphism $g: U' \to V$. We assume here for the sake of simplicity that $U' \subset U$ and that ι is the inclusion map.

Let Λ be the set of pairs (W, h) of an A-submodule W of U that contains U' and an A-homomorphism $h: W \to V$ that is an extension of g. Since $(U', g) \in \Lambda$, Λ is not empty. Define a partial order $\le$ in Λ by

$$(W_1, h_1) \le (W_2, h_2) \Leftrightarrow W_1 \subset W_2 \quad \text{and} \quad h_2|_{W_1} = h_1.$$

With this order, Λ becomes an inductively ordered set, and hence Λ has a maximal element, say (W_0, h_0), by Zorn's lemma. We show that $W_0 = U$. If $W_0 \subsetneqq U$, take $u \in U - W_0$ and let $W_1 = W_0 + uA$. The set $I = \{a \in A; ua \in W_0\}$ is a right ideal of A and $f: I \to V(x \mapsto h_0(ux))$ is an A-homomorphism. Thus, by assumption there exists $v \in V$ such that $f(x) = h_0(ux) = vx$ for all $x \in I$. Now define $h_1: W_1 \to V$ by setting $h_1(w_1) = h_0(w_0) + va$ for $w_1 = w_0 + ua \in W_1$ with $w_0 \in W_0$ and $a \in A$. Obviously $h_1(w_1)$ is independent of the above expression of w_1, and it is an A-homomorphism such that $h_1|_{W_0} = h_0$. Thus $(W_0, h_0) \lneqq (W_1, h_1)$, which contradicts the maximality of (W_0, h_0). ■

Let V be a module over an integral domain R. We say that V is *divisible* if for any $v \in V$ and $r(\neq 0) \in R$, there exists $u \in V$ satisfying $ru = v$, equivalently, $rV = V$ for all $r(\neq 0) \in R$.

Lemma 12.3. *Let R be an integral domain and V be an R-module. If V is injective, then V is divisible. The converse holds, provided V is torsion-free.*

Proof. Suppose that V is injective. For any $v \in V$ and $r(\neq 0) \in R$, $f: rR \to V(ra \mapsto va)$ is an R-homomorphism, and hence by Theorem 12.2 there exists $u \in V$ such that $va = f(ra) = u(ra)$ for all $a \in A$. Thus we obtain $v = ru$ by letting $a = 1$.

Conversely, suppose that V is torsion-free and divisible. Let I be an ideal of A and an R-homomorphism $f: I \to V$ be given. Since V is divisible, there exists $v_r \in V$, for each $r \neq 0$, such that $f(r) = v_r r$. Note that such a v_r is unique since V is torsion-free. If r, $s \in I$ are nonzero elements of I, then $f(rs) = f(r)s = v_r rs$. On the other hand, we have $f(rs) = f(sr) = f(s)r = v_s sr$, and hence $v_r = v_s$ for all (nonzero) r, s of I. Let $v = v_r$. Then $f(r) = vr$ for all $r \in I$, which implies that V is injective by Theorem 12.2. ■

Theorem 12.4. *Let R be a principal ideal domain. Then an R-module is injective if and only if it is divisible.*

Proof. It suffices to show that any divisible R-module V is injective. Let $I = aR$ be any ideal of A and $f: I \to V$ be an R-homomorphism. Then since V is divisible, there exists $v \in V$ such that $f(a) = va$ and hence $f(ar) = f(a)r = v(ar)$ for all $r \in R$. Therefore V is injective by Theorem 12.2. ■

Lemma 12.5. *Given a module V over an integral domain R, there is a divisible R-module that contains V.*

Proof. There are a free R-module F and ${}_RW \subset F$ such that $V \simeq F/W$. Let K be the quotient field of R. From the exact sequence $0 \to R \to K$, we get the exact sequence $0 \to R \otimes_R F = F \to K \otimes_R F = F^K$, because F is projective. So we may assume that $F \subset F^K$. Since F^K is divisible as an R-module, so is F^K/W. Therefore $V \simeq F/W$ is contained in the divisible module F^K/W. ■

Using the above results, we first show the following theorem, which can be regarded as a dual version of the fact that any A-module is a homomorphic image of a projective module.

Theorem 12.6. *For any V_A there exists an injective A-module that contains V.*

Proof. Viewing V as a $\mathbf{Z}$-module, there exists a divisible and hence injective $\mathbf{Z}$-module X, which contains V. Note here that $\mathrm{Hom}_{\mathbf{Z}}({}_AA, X)$ is an injective A-module by Corollary 11.2. From the exact sequence $0 \to V \to X$, we get the exact sequence $0 \to \mathrm{Hom}_{\mathbf{Z}}(A, V) \to \mathrm{Hom}_{\mathbf{Z}}(A, X)$, and so we may consider that $\mathrm{Hom}_{\mathbf{Z}}(A, V) \subset \mathrm{Hom}_{\mathbf{Z}}(A, X)$. Then the assertion follows from $V \simeq \mathrm{Hom}_A({}_AA, V) \subset \mathrm{Hom}_{\mathbf{Z}}(A, V) \subset \mathrm{Hom}_{\mathbf{Z}}(A, X)$. ■

Let $U \subset V$ be A-modules. We say that U is an *essential submodule* of V, or V is an *essential extension* of U, if the inclusion map $\iota\colon U \to V$ is essential. This is equivalent to saying that $U \cap W \neq 0$ for any nonzero A-submodule W of V. In that case we write $U \subset_e V$.

Exercise 12.7. $U_1 \subset_e U_2 \subset_e V \Rightarrow U_1 \subset_e V$.

Now we prove the main result of this section.

Theorem 12.8 (Eckman–Schopf). *Any A-module V has an injective hull.*

Proof. There exists an injective A-module, say Q, such that $Q \supset V$. The set of A-submodules of Q that are essential extensions of V is an inductively ordered set w.r.t. inclusion. Therefore it contains a maximal element, say E, by Zorn's lemma. It suffices to show that E is injective. In fact, E is a direct summand of Q, as will be shown below.

The set of A-submodules W of Q such that $E \cap W = 0$ is also an inductively ordered set w.r.t. inclusion. Thus it contains a maximal element, say E'. Let $\lambda\colon E \to Q/E'$ be the monomorphism obtained from the natural isomorphism $E \xrightarrow{\sim} E \oplus E'/E'$.

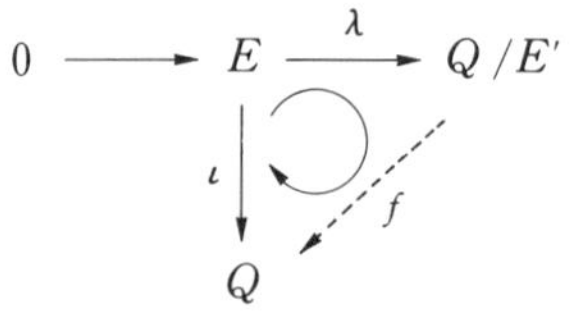

Then by the injectivity of Q, there exists $f: Q/E' \to Q$, which completes the above diagram to a commutative one, where ι denotes the inclusion map. Also, from the maximality of E', it follows that $E \oplus E'/E' \subset_e Q/E'$, and thus $\lambda: E \to Q/E'$ is an essential monomorphism. Hence f is also a monomorphism since $f \circ \lambda = \iota$. Now, it follows from $V \subset_e E = f(E \oplus E'/E') \subset_e f(Q/E')$ that $V \subset_e f(Q/E')$ and hence $E = f(E \oplus E'/E') = f(Q/E')$ because E is maximal. Therefore we have $E \oplus E' = Q$ as desired. This completes the proof of the theorem. ■

13. Discrete Valuation Rings

13.1. Discrete Valuations

Let K be a field. An (exponential) *valuation* of K is a real-valued function v defined on $K^\times = K - \{0\}$ satisfying

(13.1) $$v(ab) = v(a) + v(b),$$

and

(13.2) $$v(a + b) \geq \min\{v(a), v(b)\}.$$

The condition (13.1) means that v is a homomorphism from the multiplicative group $K^\times$ into the additive group $\mathbf{R}^+$ of the real numbers. Thus its image $\Gamma = v(K^\times)$ is a subgroup of $\mathbf{R}^+$, which is called the *value group* of v. We set $v(0) = \infty$, which will be supposed to be larger than any real number. A field K with valuation v, denoted by (K, v), is said to be a *valuation field.*

Remark. The valuation considered above is said to be *non-Archimedian.*

Exercise 13.1. The following holds for a valuation v of K.

(i) $v(\pm 1) = 0$

(ii) $v(a) = v(-a)$

(iii) $v(a^{-1}) = -v(a)$

(iv) $v(b) < v(a) \Rightarrow v(a + b) = v(b)$.

If v is a valuation of a field K, then $R = \{a \in K; v(a) \geq 0\}$ is a subring of K, which is called the *valuation ring* of v. Let $R^{-1} = \{a^{-1}; a(\neq 0) \in R\}$. Then we see that $K = R \cup R^{-1}$. In particular, K is the quotient field of R. The set $P = \{a \in K;\ v(a) > 0\}$ is an ideal of R, which is called the *valuation ideal* of v. The group of units of R is given by $R^\times = \{a \in R;\ v(a) = 0\}$. So the valuation ideal P coincides with the set of nonunits of R. Hence R is a local ring with $J(R) = P$. The field $R^* = R/P$ is said to be the *residue field* of v.

For a positive number λ, we define v' by

$$v'(a) = \lambda v(a) \qquad \text{for} \quad a \in K^\times.$$

Then v' is also a valuation of K, which is said to be an *equivalent* valuation with v, and we write $v \sim v'$. Equivalent valuations give the same valuation ring, valuation ideal, and group of units.

A valuation v of K is said to be *discrete* if its value group is an infinite cyclic group ($\simeq \mathbf{Z}$). The valuation ring of such a valuation is called a *discrete valuation ring*. In case that $v(K^\times) = \mathbf{Z}$, v is said to be a *normalized discrete valuation*.

Before mentioning properties of discrete valuation rings, we show here some elementary facts on integral elements.

Let R be a subring of an integral domain S. An element α of S is said to be *integral* over R if it is a root of some monic polynomial over R, namely, if there exist $a_1, \ldots, a_n \in R$ such that

$$\alpha^n + a_1\alpha^{n-1} + \cdots + a_n = 0. \tag{13.3}$$

Lemma 13.2. *An element α of S is integral over R if and only if there exists a subring T of S satisfying the following two conditions:*

(a) *$T \supset R[\alpha] = \sum_{i \geq 0} R\alpha^i$,*

(b) *T is finitely generated as an R-module.*

Proof. If $\alpha \in S$ is integral over R and satisfies the equation (13.3), then we have $R[\alpha] = \sum_{i=0}^{n-1} R\alpha^i$, and hence $T = R[\alpha]$ satisfies the conditions (a) and (b).

We show the converse. Let $T = R\alpha_1 + \cdots + R\alpha_m$. Since $\alpha\alpha_i \in T$, we may write

$$\alpha\alpha_i = \sum_{j=1}^{m} a_{ij}\alpha_j \qquad (a_{ij} \in R),$$

and hence $\alpha_1, \ldots, \alpha_m$ satisfy the following system of m linear equations:

$$\sum_{j=1}^{m} (\alpha\delta_{ij} - a_{ij})x_j = 0 \qquad (i = 1, 2, \ldots, m), \tag{13.4}$$

where δ_{ij} denotes the Kronecker delta. Therefore, the determinant of the $m \times m$ matrix $(\alpha\delta_{ij} - a_{ij})_{i,j}$ is zero: $\det(\alpha\delta_{ij} - a_{ij})_{i,j} = \alpha^m + c_1\alpha^{m-1} + \cdots + c_m = 0$, with $c_i \in R$ for each i. Thus α is integral over R. ∎

Lemma 13.3. *The set I of integral elements of S over R forms a subring of S containing R.*

Proof. If $\alpha, \beta \in I$, then $R[\alpha] = \sum_{i=0}^{m-1} R\alpha^i$ and $R[\beta] = \sum_{j=0}^{n-1} R\beta^j$ for some m, $n \geq 1$. Thus the subring $R[\alpha, \beta] = \sum_{i,j} R\alpha^i\beta^j$ of S is finitely generated over R and contains $\alpha + \beta$, $\alpha\beta$. Therefore, they are integral over R and hence belong to I. This shows that I is a subring of S. ∎

The subring I in the above lemma is called the *integral closure* of R in S. R is said to be *integrally closed in S* if $I = R$.

Exercise 13.4. With the above notation show that I is integrally closed in S.

We simply say that (an integral domain) R is *integrally closed* if it is integrally closed in its quotient field.

Theorem 13.5. *A principal ideal domain is integrally closed.*

Proof. Let R be a principal ideal domain with quotient field K and let $\alpha \in K$ be integral over R. Since R is a unique factorization domain, we may write $\alpha = b/a$ with $(a, b) = 1$. By assumption, there exist $n \geq 1$ and $c_1, \ldots, c_n \in R$ such that

$$(b/a)^n + c_1(b/a)^{n-1} + \cdots + c_n = 0.$$

By multiplying both sides by a^n, we get

$$b^n + a(c_1b^{n-1} + \cdots + c_na^{n-1}) = 0.$$

So, if a is not a unit of R, then any prime divisor of a is also a divisor of b, which is a contradiction. Thus a is a unit and $\alpha \in R$. ∎

Now returning to discrete valuation rings, we show the following theorem.

Theorem 13.6. *Let v be a discrete valuation of a field K that is normalized, and R and P be its valuation ring and valuation ideal, respectively. Then the following hold.*

(i) *Let $\pi \in R$ be such that $v(\pi) = 1$. Then any $\alpha \in K$ is expressed as $\alpha = \pi^i u$, where $i = v(\alpha)$ and $u \in R^\times$.*

(ii) *Any nonzero ideal of R is of the form $P^i = (\pi^i)$ for some $i \geq 0$. In particular, R is a principal ideal domain.*

(iii) *R is integrally closed.*

Proof.

(i) If $v(\alpha) = i$, then $v(\alpha\pi^{-i}) = 0$, and hence $u = \alpha\pi^{-i}$ is a unit of R.

(ii) Let I be a nonzero ideal of R and let $i = \min\{v(a); a \in I\}$. Then $\pi^i \in I$ by (i). Let $a(\neq 0) \in I$ and let $v(a) = j$. Thus there exists $u \in R^\times$ such that $a = \pi^j u = \pi^i(\pi^{j-i}u) \in (\pi^i)$ and hence $I = (\pi^i)$. In particular, we have $P = (\pi)$ and $I = P^i$.

(iii) This is clear by Theorem 13.5 and (ii). ■

Corollary 13.7. *Two discrete valuations v, v' of K are equivalent if and only if they have the same valuation ideal.*

Proof. It suffices to show the "if" part. Suppose that α and α' have the same valuation ideal, say P. Then the set $R = K - P^{-1}$ is the valuation ring of them, where $P^{-1} = \{\alpha^{-1}; \alpha \in P, \alpha \neq 0\}$. Thus, by Theorem 13.6, two normalized valuations equivalent to v and v', respectively, are identical on R, hence on K. Therefore $v \sim v'$. ■

The next result gives a ring theoretical characterization of discrete valuation rings.

Theorem 13.8. *The following two conditions on a ring R are equivalent.*

(1) *R is a discrete valuation ring.*

(2) *R is a principal ideal domain that is local.*

Proof. The implication "(1) ⇒ (2)" has been already shown. To show the converse, let $P = J(R) = (\pi)$. We first claim that $\bigcap_{i=1}^{\infty} P^i = 0$. If a is a nonzero element of $\bigcap_{i=1}^{\infty} P^i$, then $a = \pi a_1 = \pi^2 a_2 = \cdots$ with $a_1, a_2, \ldots \in R$. Then $a_i = \pi a_{i+1}$ for all i, and we get an ascending series of ideals $(a_1) \subsetneqq (a_2) \subsetneqq \cdots$, which is a contradiction.

Thus for any $a(\neq 0) \in R$, there is a nonnegative integer i such that $a \in P^i$, $a \notin P^{i+1}$. Then a can be expressed as $a = \pi^i u$ with $u \in R - P = R^\times$. Set $v(a) = i$. Letting K be the quotient field of R, define $v: K^\times \to \mathbf{R}$ by $v(b/a) = v(b) - v(a)$. Then it is easy to see that v is a valuation of K with value group $\mathbf{Z}$. ■

13.2. Dedekind Domains and $\mathfrak{p}$*-Adic Valuations*

Let R be an integral domain with quotient field K. A subset $\mathfrak{a}$ of K is said to be a *fractional ideal* if it satisfies the following two conditions:

(13.5) $\mathfrak{a}$ is an R-submodule of K. Namely, if $\alpha, \beta \in \mathfrak{a}$ and $a \in R$, then $\alpha + \beta$, $a\alpha \in \mathfrak{a}$.

(13.6) There exists some $t(\neq 0) \in R$ such that $t\mathfrak{a} \in R$.

If $\mathfrak{a} \subset R$, then the above conditions imply that $\mathfrak{a}$ is an ideal of R, which will be referred to as an *integral ideal.*

For example, for $\alpha \in K$, $\alpha R = \{\alpha r; r \in R\}$ is a fractional ideal. This is denoted by (α) and is said to be a *principal ideal* generated by α.

If $\mathfrak{a}$, $\mathfrak{b}$ are fractional ideals, then

$$\mathfrak{a} + \mathfrak{b} = \{a + b; a \in \mathfrak{a}, b \in \mathfrak{b}\}, \qquad \mathfrak{a}\mathfrak{b} = \left\{\sum_i a_i b_i; a_i \in \mathfrak{a}, b_i \in \mathfrak{b}\right\}$$

are also fractional ideals.

For a nonzero fractional ideal $\mathfrak{a}$, we let $\mathfrak{a}^{-1} = \{\alpha \in K; \alpha\mathfrak{a} \in R\}$. This is clearly an R-submodule of K, even a fractional ideal. Indeed, let $t \in R$ be as in (13.6) and a be any nonzero element of $\mathfrak{a}$. Then $ta \in R$ and $(ta)\mathfrak{a}^{-1} \subset R$ as $ta \in \mathfrak{a}$.

By definition, we see that $\mathfrak{a}^{-1}\mathfrak{a} \subset R$. If the equality $\mathfrak{a}^{-1}\mathfrak{a} = R$ holds, then $\mathfrak{a}$ is said to be *invertible.*

For example, a principal ideal (t) generated by $t \neq 0$ is invertible and $(t)^{-1} = (t^{-1})$. If $\mathfrak{a}$ and $\mathfrak{b}$ are both invertible, then so is $\mathfrak{a}\mathfrak{b}$, and $(\mathfrak{a}\mathfrak{b})^{-1} = \mathfrak{a}^{-1}\mathfrak{b}^{-1}$. In fact, $\mathfrak{a}\mathfrak{b}\mathfrak{a}^{-1}\mathfrak{b}^{-1} = R$ implies that $\mathfrak{a}\mathfrak{b}$ is invertible, and $\mathfrak{a}^{-1}\mathfrak{b}^{-1} = (\mathfrak{a}\mathfrak{b})^{-1}R = (\mathfrak{a}\mathfrak{b})^{-1}$.

An integral domain R is said to be a *Dedekind domain* if it satisfies the following three conditions:

(13.7) R is a Noetherian ring.

(13.8) R is integrally closed.

(13.9) Any nonzero prime ideal of R is maximal.

For example a principal ideal domain is a Dedekind domain. Indeed, (13.7) is trivial and (13.8) has been verified in Theorem 13.5. On the other hand, (13.9) is well known (cf. Hungerford [1], Chapter III).

Theorem 13.9. *Let R be a Dedekind domain. Then the following holds.*

(i) *Every nonzero fractional ideal is invertible.*

(ii) *Any fractional ideal $\mathfrak{a}$ different from R, 0 is expressed uniquely as a product of prime ideals $\mathfrak{p}_i$.*

$$\mathfrak{a} = \mathfrak{p}_1^{e_1} \cdots \mathfrak{p}_n^{e_n} \; (e_i \in \mathbf{N}). \tag{13.10}$$

Proof. We may clearly assume that R is not a field. Thus R has a maximal ideal, which of course is a prime ideal of R. By an ideal, we always mean here a nonzero ideal.

(i) *Step 1.* Any integral ideal contains a product of prime ideals.

If not, then by (13.7) there exists a maximal one, say $\mathfrak{a}$, among the integral ideals that contain no products of prime ideals. Then $\mathfrak{a}$ is not prime, and there exist $b, c \in R - \mathfrak{a}$ such that $bc \in \mathfrak{a}$. Let $\mathfrak{b} = \mathfrak{a} + Rb$, $\mathfrak{c} = \mathfrak{a} + Rc$. Then $\mathfrak{b} \supsetneqq \mathfrak{a}$, $\mathfrak{c} \supsetneqq \mathfrak{a}$, and hence by the maximality of $\mathfrak{a}$, there exist prime ideals $\mathfrak{p}_i$, $\mathfrak{q}_j$ such that $\mathfrak{b} \supset \mathfrak{p}_1 \cdots \mathfrak{p}_r$, $\mathfrak{c} \supset \mathfrak{q}_1 \cdots \mathfrak{q}_s$. Then we have

$$\mathfrak{p}_1 \cdots \mathfrak{p}_r \mathfrak{q}_1 \cdots \mathfrak{q}_s \subset \mathfrak{b}\mathfrak{c} \subset \mathfrak{a}^2 + \mathfrak{a}b + \mathfrak{a}c + Rbc \subset \mathfrak{a},$$

which contradicts the choice of $\mathfrak{a}$.

Step 2. If $\mathfrak{a}$ is an integral ideal such that $\mathfrak{a} \subsetneqq R$, then $R \subsetneqq \mathfrak{a}^{-1}$.

It is clear that $R \subset \mathfrak{a}^{-1}$. Let a be a nonzero element of $\mathfrak{a}$. As is shown above, (a) contains a product of prime ideals. Let r be the smallest integer such that there is a product $\mathfrak{p}_1 \cdots \mathfrak{p}_r$ of prime ideals $\mathfrak{p}_i$ that is contained in (a). Let $\mathfrak{p}$ be a maximal ideal containing $\mathfrak{a}$. Since $\mathfrak{p}_1 \cdots \mathfrak{p}_r \subset \mathfrak{p}$, it follows that $\mathfrak{p}$ contains some $\mathfrak{p}_i$, say $\mathfrak{p}_1$, for instance, $\mathfrak{p} \supset \mathfrak{p}_1$. Then the equality $\mathfrak{p} = \mathfrak{p}_1$ holds by (13.9). Thus we have

$$\mathfrak{p}\mathfrak{p}_2 \cdots \mathfrak{p}_r \subset (a) \subset \mathfrak{a} \subset \mathfrak{p}.$$

Since $\mathfrak{p}_2 \cdots \mathfrak{p}_r \not\subset (a)$ by the choice of r, there exists $b \in \mathfrak{p}_2 \cdots \mathfrak{p}_r$ such that $b \notin (a)$. Let $\lambda = a^{-1}b$. Then $\lambda \notin R$ as $b \notin (a)$, and we have

$$\lambda\mathfrak{a} = a^{-1}b\mathfrak{a} \subset a^{-1}b\mathfrak{p} \subset a^{-1}\mathfrak{p}\mathfrak{p}_2 \cdots \mathfrak{p}_r \subset R.$$

Thus $\lambda \in \mathfrak{a}^{-1}$, and we conclude that $R \subsetneqq \mathfrak{a}^{-1}$.

Step 3. Any integral ideal $\mathfrak{a}$ is invertible.

To see this, let $\mathfrak{b} = \mathfrak{a}^{-1}\mathfrak{a}$. Then $\mathfrak{b}$ is an integral ideal, and $\mathfrak{a}^{-1}\mathfrak{b}^{-1} \subset \mathfrak{a}^{-1}$ since $\mathfrak{a}\mathfrak{a}^{-1}\mathfrak{b}^{-1} = \mathfrak{b}\mathfrak{b}^{-1} \subset R$. Thus, for any $\beta \in \mathfrak{b}^{-1}$, we have $\mathfrak{a}^{-1}\beta \subset \mathfrak{a}^{-1}$, and hence $\mathfrak{a}^{-1}\beta^m \subset \mathfrak{a}^{-1}$ for any natural number m by inductive argument. This implies that $R[\beta] \subset \mathfrak{a}^{-1}[\beta] \subset \mathfrak{a}^{-1}$. Since R is Noetherian, any fractional ideal is finitely generated over R by (13.6). In particular, $\mathfrak{a}^{-1}$, and hence $R[\beta]$, is finitely generated over R. Thus β is integral over R by Lemma 13.2, and $\beta \in R$ by (13.8). Since β is an arbitrary element of $\mathfrak{b}^{-1}$, we conclude that $\mathfrak{b}^{-1} \subset R$, i.e. $\mathfrak{b}^{-1} = R$. As observed in Step 2, this implies that $\mathfrak{b} = \mathfrak{a}\mathfrak{a}^{-1} = R$.

Step 4. If $\mathfrak{a}$ is any fractional ideal, then there exists $t \neq 0$ such that $t\mathfrak{a}$ is an integral ideal. Then $(t\mathfrak{a})(t\mathfrak{a})^{-1} = R$ by the above, and we get $\mathfrak{a}\mathfrak{a}^{-1} = R$, thereby completing the proof of (i).

(ii) Suppose to the contrary that there exists a proper integral ideal not expressed as products of prime ideals and let $\mathfrak{a}$ be a maximal one among those. Take a maximal ideal $\mathfrak{p}$ such that $\mathfrak{p} \supset \mathfrak{a}$. Then $\mathfrak{p} \supsetneqq \mathfrak{a}$ and $\mathfrak{b} = \mathfrak{p}^{-1}\mathfrak{a}$ is an integral ideal. Hence $\mathfrak{p}\mathfrak{b} = \mathfrak{p}\mathfrak{p}^{-1}\mathfrak{a} = R\mathfrak{a} = \mathfrak{a}$ from (i). Since $\mathfrak{a} \subsetneqq \mathfrak{b}$, $\mathfrak{b}$ must be a product of prime ideals, then the same is true for $\mathfrak{a} = \mathfrak{p}\mathfrak{b}$. This is a contradiction.

Next suppose that an integral ideal $\mathfrak{a}$ is expressed as products of prime ideals in two ways: $\mathfrak{a} = \mathfrak{p}_1 \cdots \mathfrak{p}_r = \mathfrak{q}_1 \cdots \mathfrak{q}_s$. Then $\mathfrak{p}_1$ contains some $\mathfrak{q}_1$, for example. But, by (13.9), this implies $\mathfrak{p}_1 = \mathfrak{q}_1$. By multiplying both sides of the above expressions by $\mathfrak{p}_1^{-1}$, we get $\mathfrak{p}_2 \cdots \mathfrak{p}_r = \mathfrak{q}_2 \cdots \mathfrak{q}_s$. Repeating these arguments, we finally obtain $r = s$ and $\mathfrak{p}_i = \mathfrak{q}_i$ for all i after a suitable arrangement of indices. Thus we have proved the assertions for integral ideals.

As for a fractional ideal $\mathfrak{a} \neq R$, take $t \in R$ so that $t\mathfrak{a} = \mathfrak{b}$ is an integral ideal. If we express $\mathfrak{b} = \mathfrak{p}_1 \cdots \mathfrak{p}_r$, $(t) = \mathfrak{q}_1 \cdots \mathfrak{q}_s$ as products of prime ideals, then $\mathfrak{a} = \mathfrak{b}(t)^{-1} = \mathfrak{p}_1 \cdots \mathfrak{p}_r\mathfrak{q}_1^{-1} \cdots \mathfrak{q}_s^{-1}$. The uniqueness will be proved easily by reducing to the case of integral ideals. ■

The expression of $\mathfrak{a}$ as in (13.10) is said to be the *prime ideal decomposition* of $\mathfrak{a}$. We also admit $e_i = 0$ by letting $\mathfrak{p}_i^0 = R$.

Remark. The statement (ii) in the Theorem 13.9 is said to be the *fundamental theorem of ideal theory*. It is also known that either of the conditions (i) and (ii) is equivalent to that R is a Dedekind domain.

We see from Theorem 13.9 that the set of fractional ideals other than 0 forms a group with identity R by defining the multiplication to be the product of ideals. This is called the *ideal group* of K. Note that the ideal group of K is a free abelian group with prime ideals of R as basis, if $R \neq K$.

Exercise 13.10. Let R be a Dedekind domain with quotient field K. Then the following holds for fractional ideals $\mathfrak{a}$, $\mathfrak{b}$ of K.

(i) $\mathfrak{a} \subset \mathfrak{b} \Leftrightarrow \mathfrak{a} = \mathfrak{b}\mathfrak{c}$ for some integral ideal $\mathfrak{c}$.

(ii) If $\mathfrak{a} = \mathfrak{p}_1^{e_1} \cdots \mathfrak{p}_n^{e_n}$, $\mathfrak{b} = \mathfrak{p}_1^{f_1} \cdots \mathfrak{p}_n^{f_n} (e_i, f_i \in \mathbf{Z})$ are the prime ideal decompositions, then

$$\mathfrak{a} + \mathfrak{b} = \mathfrak{p}_1^{\min(e_1, f_1)} \cdots \mathfrak{p}_n^{\min(e_n, f_n)},$$

$$\mathfrak{a} \cap \mathfrak{b} = \mathfrak{p}_1^{\max(e_1, f_1)} \cdots \mathfrak{p}_n^{\max(e_n, f_n)}.$$

[Hint: $\mathfrak{a} + \mathfrak{b}$ is the smallest ideal that contains both $\mathfrak{a}$ and $\mathfrak{b}$, while $\mathfrak{a} \cap \mathfrak{b}$ is the largest ideal that is contained in both $\mathfrak{a}$ and $\mathfrak{b}$].

(iii) Two integral ideals $\mathfrak{a}$ and $\mathfrak{b}$ are said to be *relatively prime* provided $\mathfrak{a} + \mathfrak{b} = R$. Show that this occurs if and only if no prime ideal appears in common in the prime ideal decompositions of $\mathfrak{a}$ and $\mathfrak{b}$.

Lemma 13.11. *Let R be a Dedekind domain and $\mathfrak{a}$, $\mathfrak{b}$ be integral ideals. Then there exists an integral ideal $\mathfrak{c}$ relatively prime to $\mathfrak{b}$ such that $\mathfrak{a}\mathfrak{c}$ is a principal ideal.*

Proof. Let $\mathfrak{a} = \mathfrak{p}_1^{e_1} \cdots \mathfrak{p}_n^{e_n}$, $\mathfrak{b} = \mathfrak{p}_1^{f_1} \cdots \mathfrak{p}_n^{f_n}$ be the prime ideal decompositions, where $e_i, f_i \geq 0$. Choose $a_i \in \mathfrak{p}_i^{e_i} - \mathfrak{p}_i^{e_i+1}$ for each i. Then, since any pair of $\{\mathfrak{p}_1, \ldots, \mathfrak{p}_n\}$ are relatively prime, there exists, by the Chinese remainder theorem (cf. Problem 2 at the end of this chapter), $a \in R$ such that

$$a \equiv a_i (\text{mod } \mathfrak{p}_i^{e_i+1}) \qquad \text{for all } i.$$

In particular, $a \in \mathfrak{p}_i^{e_i} - \mathfrak{p}_i^{e_i+1}$, and hence the prime ideal decomposition of (a) is of the form $(a) = \mathfrak{p}_1^{e_1} \cdots \mathfrak{p}_n^{e_n}\mathfrak{q}_1^{t_1} \cdots \mathfrak{q}_m^{t_m}$. Thus there exists an integral ideal $\mathfrak{c}$

such that $(a) = \mathfrak{a}\mathfrak{c}$. Moreover, we see from Exercise 13.10 that $(a) + \mathfrak{a}\mathfrak{b} = \mathfrak{a}$, i.e., $\mathfrak{a}\mathfrak{c} + \mathfrak{a}\mathfrak{b} = \mathfrak{a}$. By multiplying both sides by $\mathfrak{a}^{-1}$, we obtain $\mathfrak{c} + \mathfrak{b} = R$, namely, $\mathfrak{c}$ and $\mathfrak{b}$ are relatively prime. ■

Let R be a Dedekind domain with quotient field K and assume that $R \neq K$. Fix a nonzero prime ideal $\mathfrak{p}$ of R. For $\alpha(\neq 0) \in K$, let

$$(\alpha) = \mathfrak{p}^e \mathfrak{p}_1^{e_1} \cdots \mathfrak{p}_r^{e_r}$$

be the prime ideal decomposition of (α). We define $v_\mathfrak{p}: K^\times \to \mathbf{Z}$ by $v_\mathfrak{p}(\alpha) = e$. Then it is easy to see that $v_\mathfrak{p}$ is a discrete valuation of K with value group $\mathbf{Z}$. This is called the $\mathfrak{p}$-*adic valuation* of K. We denote the valuation ring by $R_\mathfrak{p}$, i.e., $R_\mathfrak{p} = \{\alpha \in K; v_\mathfrak{p}(\alpha) \geq 0\}$.

Theorem 13.12. *Let R be a Dedekind domain. Then*

$$\bigcap_\mathfrak{p} R_\mathfrak{p} = R,$$

where $\mathfrak{p}$ runs through the prime ideals of R.

Proof. One inclusion "$\supset$" is obvious. If $\alpha(\neq 0) \in \bigcap_\mathfrak{p} R_\mathfrak{p}$ and $(\alpha) = \mathfrak{p}_1^{e_1} \cdots \mathfrak{p}_r^{e_r}$ is the prime ideal decomposition, then $e_i \geq 0$ for all i, and hence (α) is an integral ideal. Therefore, $\alpha \in R$. ■

Theorem 13.13. *Let R be a Dedekind domain and let $\mathfrak{p}$ be a nonzero prime ideal of R. Then*

(i) *$R_\mathfrak{p} = \{b/a; a, b \in R, a \notin \mathfrak{p}\}$.*

(ii) *$\mathfrak{p}R_\mathfrak{p}$ is a maximal ideal of $R_\mathfrak{p}$, and the following isomorphism holds:*

$$R_\mathfrak{p}/\mathfrak{p}R_\mathfrak{p} \simeq R/\mathfrak{p}.$$

Proof. Write v for $v_\mathfrak{p}$.

(i) If $a, b \in R$ and $a \notin \mathfrak{p}$, then $v(b/a) = v(b) - v(a) = v(b) \geq 0$, and hence $b/a \in R_\mathfrak{p}$. On the other hand, let α be a nonzero element of $R_\mathfrak{p}$, and $(\alpha) = \mathfrak{p}_1^{e_1} \cdots \mathfrak{p}_m^{e_m} \mathfrak{q}_1^{-f_1} \cdots \mathfrak{q}_n^{-f_n}$ be the prime ideal decomposition of (α) with e_i, $f_j > 0$. Let $\mathfrak{a} = \mathfrak{q}_1^{f_1} \cdots \mathfrak{q}_n^{f_n}$, $\mathfrak{b} = \mathfrak{p}_1^{e_1} \cdots \mathfrak{p}_m^{e_m}$. Then $\mathfrak{a}$ and $\mathfrak{p}$ are relatively prime since $v(\alpha) \geq 0$. Thus, there exists an integral ideal $\mathfrak{c}$ relatively prime to $\mathfrak{p}$ such that $\mathfrak{a}\mathfrak{c}$ is a principal ideal; $\mathfrak{a}\mathfrak{c} = (a)$. Note that $a \notin \mathfrak{p}$, since (a) is relatively prime to $\mathfrak{p}$. Furthermore, since

$$(\alpha) = \mathfrak{a}^{-1}\mathfrak{b} = (\mathfrak{a}\mathfrak{c})^{-1}\mathfrak{b}\mathfrak{c} = a^{-1}\mathfrak{b}\mathfrak{c},$$

we have $(\alpha a) = \mathfrak{b}\mathfrak{c} \subset R$. Therefore $b = \alpha a \in R$, and $\alpha = b/a$ belongs to the right-hand side of (i).

(ii) Since $\mathfrak{p} = \{a \in R;\ v(a) > 0\}$, it follows that $R \cap \mathfrak{p}R_{\mathfrak{p}} = \mathfrak{p}$. Hence there exists a monomorphism $\varphi\colon R/\mathfrak{p} \to R_{\mathfrak{p}}/\mathfrak{p}R_{\mathfrak{p}}$. To see that φ is an epimorphism, let $\alpha \in R_{\mathfrak{p}}$ and write $\alpha = b/a$, where $a, b \in R$ and $a \notin \mathfrak{p}$. Since $R/\mathfrak{p}$ is a field, there exists $c \in R$ such that $ac \equiv 1 \bmod \mathfrak{p}$. Then $ac = 1 - x$ for some $x \in \mathfrak{p}$, and hence $\alpha - bc = a^{-1}b - bc = a^{-1}b(1 - ac) = \alpha x \in \mathfrak{p}R_{\mathfrak{p}}$. Therefore $bc \equiv \alpha \bmod \mathfrak{p}R_{\mathfrak{p}}$, and hence φ is an epimorphism. ■

As an important example of Dedekind domains there is the ring of algebraic integers in an algebraic number field.

An *algebraic number field* K is a subfield of **C**, which is a finite extension of **Q**. A complex number is said to be an *algebraic integer* if it is integral over **Z**. The set of algebraic integers in a given algebraic number field K is a subring of K, which is called the *ring of integers* of K.

We shall show that the ring of integers of an algebraic number field is a Dedekind domain. First, we prove the following lemma.

Lemma 13.14. *Let K be an algebraic number field of degree n over* **Q** *and let R be the ring of integers of K.*

(i) *Given $\alpha \in K$, there exists $c(\neq 0) \in \mathbf{Z}$ such that $c\alpha \in R$. In particular, K is the quotient field of R.*

(ii) *R is a free* **Z**-*module of rank n.*

(iii) *If $\mathfrak{a}$ is a nonzero ideal of R, then $\operatorname{rank}_{\mathbf{Z}} \mathfrak{a} = n$, and $|R/\mathfrak{a}| < \infty$.*

Proof.

(i) If $\alpha^m + (b_1/a_1)\alpha^{m-1} + \cdots + b_m/a_m = 0$ with $a_i, b_i \in \mathbf{Z}$, then by multiplying both sides by $c = a_1 \ldots a_m$, we get

$$c\alpha^m + c_1\alpha^{m-1} + \cdots + c_m = 0 \qquad (c_i \in \mathbf{Z}).$$

Hence

$$(c\alpha)^m + c_1(c\alpha)^{m-1} + \cdots + c^{m-1}c_m = 0,$$

and thus $c\alpha \in R$.

(ii) Since K is separable over $\mathbf{Q}$, there exists $\theta \in K$ such that

$$K = \mathbf{Q}(\theta) = \mathbf{Q} \oplus \mathbf{Q}\theta \oplus \cdots \oplus \mathbf{Q}\theta^{n-1}.$$

Here, by (i), we may assume that $\theta \in R$. Let $\theta = \theta_0, \theta_1, \ldots, \theta_{n-1}$ be all the conjugates of θ over $\mathbf{Q}$, then each θ_i is an algebraic integer. Let $\Delta = \det(\theta_i^j)_{0 \le i, j \le n-1}$. Then $\Delta \neq 0$ and $\Delta^2 \in \mathbf{Q}$, since Δ^2 is expressed as a symmetric polynomial of $\theta_0, \ldots, \theta_{n-1}$ over $\mathbf{Z}$. But this implies that $\Delta^2 \in \mathbf{Z}$ because $\mathbf{Z}$ is integrally closed. Let $\alpha = \alpha_0 = a_0 + a_1\theta + \cdots + a_{n-1}\theta^{n-1}$ be any element of R, where $a_i \in \mathbf{Q}$, and consider the elements

$$\alpha_i = a_0 + a_1\theta_i + \cdots + a_{n-1}\theta_i^{n-1} \qquad (0 \le i \le n-1).$$

By considering this to be a system of n linear equations on $a_0, a_1, \ldots, a_{n-1}$, we get

$$\text{(13.11)} \qquad a_j = \Delta_j/\Delta \qquad (0 \le j \le n-1),$$

where Δ_j is obtained from Δ by replacing each θ_i^j with α_i. Since θ_i, α_j are algebraic integers, then Δ and Δ_j are also. Hence $b_j = \Delta\Delta_j$ is an algebraic integer, whence it follows that $b_j \in \mathbf{Z}$ since $b_j = a_j\Delta^2 \in \mathbf{Q}$. Now we have $\alpha = b_0(1/\Delta^2) + b_1(\theta/\Delta^2) + \cdots + b_{n-1}(\theta^{n-1}/\Delta^2)$, and hence

$$R \subset \mathbf{Z}(1/\Delta^2) + \mathbf{Z}(\theta/\Delta^2) + \cdots + \mathbf{Z}(\theta^{n-1}/\Delta^2).$$

The right-hand side of the above is a free $\mathbf{Z}$-module of rank n with basis $1/\Delta^2$, $\theta/\Delta^2, \ldots, \theta^{n-1}/\Delta^2$. On the other hand, R contains the free $\mathbf{Z}$-module $\mathbf{Z} \oplus \mathbf{Z}\theta \oplus \cdots \oplus \mathbf{Z}\theta^{n-1}$, which is of rank n. Thus we conclude that R is a free $\mathbf{Z}$-module of rank n.

(iii) If a is a nonzero element of $\mathfrak{a}$, then $aR \subset \mathfrak{a} \subset R$, and $\operatorname{rank}_{\mathbf{Z}} aR = \operatorname{rank}_{\mathbf{Z}} R = n$. Therefore, $\operatorname{rank}_{\mathbf{Z}} \mathfrak{a} = n$ and $|R/\mathfrak{a}| < \infty$ by Theorem 11.8(i). ■

Theorem 13.15. *Let K be an algebraic number field and R be the ring of integers of K. Then R is a Dedekind domain.*

Proof. Since every ideal of R is finitely generated over R by Lemma 13.14(iii), R is Noetherian. R is the integral closure of $\mathbf{Z}$ in K by definition, which implies that R is integrally closed since K is the quotient field of R.

If $\mathfrak{p}$ is a nonzero prime ideal of R, then $\bar{R} = R/\mathfrak{p}$ is an integral domain with only a finite number of elements. Now, for a fixed $\bar{a} \neq 0$ of $\bar{R}$, the map $\bar{R} \to \bar{R}(\bar{x} \mapsto \bar{x}\bar{a})$ is a monomorphism, and then it must also be an isomorphism because $\bar{R}$ is a finite set. Therefore, $\bar{R}$ is a field and $\mathfrak{p}$ is a maximal ideal. ■

Let K be an algebraic number field and R be the ring of integers of K as above. Each prime ideal $\mathfrak{p} \neq 0$ of R provides the $\mathfrak{p}$-adic valuation $v_{\mathfrak{p}}$ of K. The elements of its valuation ring are called the $\mathfrak{p}$-*integers*. The residue field of $v_{\mathfrak{p}}$ is a finite field isomorphic to $R/\mathfrak{p}$. Since $\mathfrak{p} \cap \mathbf{Z}$ is a prime ideal of $\mathbf{Z}$, we have $\mathfrak{p} \cap \mathbf{Z} = (p)$ for some prime number p. Thus $R_{\mathfrak{p}}/\mathfrak{p}R_{\mathfrak{p}}$ is a finite field of characteristic p. If $e = v_{\mathfrak{p}}(p)$, then $v^* = (1/e)v_{\mathfrak{p}}$ is an equivalent valuation with $v_{\mathfrak{p}}$ satisfying $v^*(p) = 1$.

13.3. Completion

Let v be a valuation of a field K. Fix a real number γ such that $0 < \gamma < 1$ and define $\varphi\colon K \to \mathbf{R}$ by the equation

$$\varphi(a) = \gamma^{v(a)}.$$

Then the following holds:

$$\varphi(a) \geq 0; \qquad \varphi(a) = 0 \Leftrightarrow a = 0, \tag{13.12}$$

$$\varphi(ab) = \varphi(a)\varphi(b), \tag{13.13}$$

$$\varphi(a + b) \leq \max(\varphi(a), \varphi(b)). \tag{13.14}$$

Moreover, it holds that $\varphi(a) = \varphi(-a)$ and $\varphi(a^{-1}) = \varphi(a)^{-1}$ by Exercise 13.1. Also the following holds:

$$\varphi(a) < \varphi(b) \Rightarrow \varphi(a + b) = \varphi(b). \tag{13.15}$$

The valuation ring of v is given by $R = \{a \in K; \varphi(a) \leq 1\}$ and the valuation ideal by $P = \{a \in K; \varphi(a) < 1\}$.

Conversely, if we are given a real-valued function $\varphi\colon K \to \mathbf{R}$ satisfying the three conditions from (13.12) to (13.14), then define $v\colon K^\times \to \mathbf{R}$ by the equation $v(a) = \log_\gamma \varphi(a)$, where γ is a real number such that $0 < \gamma < 1$. Then v is an (exponential) valuation of K. In this sense, we call such a φ a *multiplicative valuation* of K. The set $\varphi(K^\times)$ is a multiplicative subgroup of $\mathbf{R}^\times$, which is called the *value group* of φ and is denoted by G_φ. As in the exponential case, φ is said to be discrete if G_φ is an infinite cyclic group.

Two multiplicative valuations φ_1 and φ_2 are said to be equivalent if there exists a positive number r such that $\varphi_2(a) = \varphi_1(a)^r$ for all $a \in K$. This amounts to saying that the two exponential valuations v_1 and v_2 obtained from φ_1 and φ_2 as above are equivalent.

Now, if φ is a multiplicative valuation of K, then by (13.14), it holds that $\varphi(a + b) \leq \varphi(a) + \varphi(b)$. Then the function

$$d(a, b) = \varphi(a - b)$$

defines a metric on K. In particular, K is a Hausdorff space, and the maps $K \times K \to K((x, y) \mapsto x + y)$, $K \times K \to K((x, y) \mapsto xy)$, and $K \times K^\times \to K((x, y) \mapsto xy^{-1})$ are all continuous.

Recall that a sequence $(a_n) = (a_1, a_2, \ldots, a_n, \ldots)$ of elements of K is said to be a *Cauchy sequence* if for any $\varepsilon > 0$ there exists a natural number N such that $d(a_m, a_n) < \varepsilon$ for all $m, n \geq N$. Also (a_n) is said to *converge* on an element a of K if $\lim_{n\to\infty} d(a_n, a) = \lim_{n\to\infty} \varphi(a_n - a) = 0$, in which case we write $\lim_{n\to\infty} a_n = a$. A convergent sequence is always a Cauchy sequence.

If every Cauchy sequence in K converges, K is said to be *complete* or a *complete field*, and the associated valuation ring is said to be a *complete valuation ring*.

Given a valuation field K, let C denote the set of Cauchy sequences in K and define the sum and multiplication in C as follows:

$$(a_n) + (b_n) = (a_n + b_n), \qquad (a_n)(b_n) = (a_n b_n).$$

Then C becomes a commutative ring. If I denotes the subset of C consisting of the sequences converging on 0, then I is an ideal of C, which is a maximal ideal as is easily checked. Hence the factor ring $\tilde{K} = C/I$ is a field.

Exercise 13.16. Prove that the above I is a maximal ideal of C.

If (a_n) is a Cauchy sequence, then $\pm(\varphi(a_m) - \varphi(a_n)) \leq \varphi(a_m - a_n)$ by (13.14). Therefore, the sequence $(\varphi(a_n))$ is a Cauchy sequence in $\mathbf{R}$, and $\lim_{n\to\infty} \varphi(a_n)$ exists. If $(a_n) - (b_n) \in I$, then $\lim_{n\to\infty} \varphi(a_n) = \lim_{n\to\infty} \varphi(b_n)$, and so we can define $\tilde{\varphi}\colon \tilde{K} \to \mathbf{R}$ by $\tilde{\varphi}((\widetilde{a_n})) = \lim_{n\to\infty} \varphi(a_n)$ for $(\widetilde{a_n}) = (a_n) + I$. It is easy to verify that $\tilde{\varphi}$ is a multiplicative valuation of $\tilde{K}$.

For $a \in K$, the sequence $(a) = (a, a, \ldots)$ is a Cauchy sequence, and $(\tilde{a}) \neq (\tilde{b})$ if $a \neq b$. Also, $\tilde{\varphi}((\tilde{a})) = \lim_{n\to\infty} \varphi(a) = \varphi(a)$. Thus, we regard K as a subfield of $\tilde{K}$ by identifying $a \in K$ with $(\tilde{a})$. With this identification, we see that $\tilde{\varphi}$ is an extension of φ, and K is dense in $\tilde{K}$ because $(\widetilde{a_n}) = \lim_{n\to\infty} a_n$.

Exercise 13.17. Show that $\tilde{K}$ is a complete field.

We have thus constructed the complete field $\tilde{K} \supset K$ such that K is dense in $\tilde{K}$. It is known that $\tilde{K}$ is unique up to a K-isomorphism of valuation fields. We call $\tilde{K}$ the *completion* of K.

Theorem 13.18. *Let K be a valuation field with valuation ring R and valuation ideal P. If $\tilde{K}$ is the completion of K with valuation ring $\tilde{R}$ and valuation ideal $\tilde{P}$, then the following field isomorphism holds.*

$$\tilde{R}/\tilde{P} \simeq R/P.$$

Proof. It is clear that $R \subset \tilde{R}$, $R \cap \tilde{P} = P$. Thus it suffices to show that $R + \tilde{P} = \tilde{R}$. Let a be any nonzero element of $\tilde{R}$. Then $0 < \tilde{\varphi}(a) \leq 1$, and there exists $a_n \in K$ such that $\lim_{n\to\infty} a_n = a$. Hence $\lim_{n\to\infty} \tilde{\varphi}(a_n - a) = 0$, and there exists n such that $\tilde{\varphi}(a_n - a) < \tilde{\varphi}(a)$. It then follows from (13.15) that $\varphi(a_n) = \tilde{\varphi}((a_n - a) + a) = \tilde{\varphi}(a) \leq 1$ and $a_n \in R$. Moreover, since $\tilde{\varphi}(a_n - a) < \tilde{\varphi}(a) \leq 1$, we find that $a_n - a \in \tilde{P}$, and hence $a = a_n + (a - a_n) \in R + \tilde{P}$. ■

Exercise 13.19. Let φ be a multiplicative valuation of a field K. Then the following holds:

(i) (a_n) is a Cauchy sequence $\Leftrightarrow \lim_{n\to\infty} \varphi(a_{n+1} - a_n) = 0$.

(ii) If $\lim_{n\to\infty} a_n = a \neq 0$, then $\varphi(a_n) = \varphi(a)$ for a sufficiently large n.

Let v be a normalized discrete valuation of K with valuation ring R and valuation ideal $P = (\pi)$. Then $\{P, P^2, \ldots\}$ is a fundamental neighbourhood system of 0. Therefore, (a_n) is a Cauchy sequence if and only if given a natural number l there exists a natural number N such that $a_m - a_n \in P^l$ whenever m, $n \geq N$. Also $\lim_{n\to\infty} a_n = a$ if and only if, given any l, there exists N such that $a_n - a \in P^l$ whenever $n > N$.

For each $i \in \mathbf{Z}$, take $\pi_i \in K$ so that $v(\pi_i) = i$ (let $\pi_i = \pi^i$, for instance). Let $r \in \mathbf{Z}$ be an integer. For given $c_i \in R (i \geq r)$ and $n \geq r$, let $a_n = \sum_{i=r}^{n} c_i \pi_i$. If $\lim_{n\to\infty} a_n = a$ exists, then we write

$$a = \sum_{i=r}^{\infty} c_i \pi_i. \tag{13.16}$$

As (a_n) is clearly a Cauchy sequence, $\sum_{i=r}^{\infty} c_i \pi_i$ exists if K is complete.

Exercise 13.20. With the notation above, assume that K is complete. If Λ denotes a complete set of representatives of R/P containing 0, then any $a \in K$ is uniquely expressed in the form:

$$a = \sum_{i=r}^{\infty} c_i \pi_i \qquad (c_i \in \Lambda),$$

where $v(a) = r$ if $c_r \neq 0$.

[Hint: If $v(a) = r$, then $a = u\pi_r$ for some $u \in R^\times$. If $u \equiv c_r (\mathrm{mod}\ P)$ with $c_r \in \Lambda$, then $v(a - c_r\pi_r) \geq r + 1$. This determines the next (nonzero) $c_i (i > r)$, and so on.]

If, in particular, K is an algebraic number field with $\mathfrak{p}$-adic valuation, then the completion $\tilde{K}$ is called the $\mathfrak{p}$-*adic number field*. The valuation ring $\tilde{R}$ in it is called the *ring of* $\mathfrak{p}$-*adic integers*, whose elements are said to be $\mathfrak{p}$-*adic integers*. $\tilde{R}$ is a complete discrete valuation ring. If $\tilde{\mathfrak{p}}$ denotes the valuation ideal, then $\tilde{R}/\tilde{\mathfrak{p}} \simeq R/\mathfrak{p}$ is a finite field, where R is the ring of integers of K.

13.4. Extensions of Discrete Valuations

We continue to study discrete valuations.

Lemma 13.21. *Let φ_1, φ_2 be discrete multiplicative valuations. Then the following three conditions are equivalent.*

(1) $\varphi_1 \sim \varphi_2$.

(2) $\{a \in K;\ \varphi_1(a) < 1\} = \{a \in K;\ \varphi_2(a) < 1\}$.

(3) *The topologies on K induced by φ_1 and φ_2, respectively, are the same.*

Proof. Let P_i be the valuation ideal of $\varphi_i (i = 1, 2)$. The condition (2) is the same as that $P_1 = P_2$, thus the equivalence "(1)$\Leftrightarrow$(2)" is clear by Corollary 13.7. Also the implication "(2)$\Rightarrow$(3)" is clear because the set $\{P_i^n;\ n \geq 1\}$ is a fundamental neighbourhood system of 0 in the topology induced by φ_i. To show the implication "(3)$\Rightarrow$(2)", suppose that $\varphi_1(a) < 1$. Then $\lim_{n\to\infty} \varphi_1(a^n) = \lim_{n\to\infty} \varphi_1(a)^n = 0$. Therefore, the assumption implies that $\lim_{n\to\infty} \varphi_2(a^n) = 0$, and hence $\varphi_2(a) < 1$. Similarly, if $\varphi_2(a) < 1$, then $\varphi_1(a) < 1$. ∎

Let φ be a multiplicative valuation of a field K and let V be a vector space over K. A real-valued function $\|v\|$ defined on V is said to be a *norm* on V if it satisfies the following three conditions:

$$\|v\| \geq 0 \quad \text{and} \quad \|v\| = 0 \Leftrightarrow v = 0, \tag{13.17}$$

$$\|av\| = \varphi(a)\|v\|, \quad \text{for all } a \in K \text{ and } v \in V, \tag{13.18}$$

$$\|u + v\| \leq \max\{\|u\|, \|v\|\}. \tag{13.19}$$

If a norm $\| \quad \|$ is given, V becomes a metric space with the metric defined by $d(u, v) = \|u - v\|$, and the maps $V \times V \to V((u, v) \mapsto u + v)$, $K \times V \to V((a, v) \mapsto av)$ are both continuous.

We now assume that V is finite dimensional over K and let $\{u_1, \ldots, u_n\}$ be a K-basis of V.

Exercise 13.22. For $v = a_1 u_1 + \cdots + a_n u_n \in V$, define $\|v\|_0$ by

$$\|v\|_0 = \max\{\varphi(a_i);\ 1 \leq i \leq n\}. \tag{13.20}$$

Then this is a norm on V. Moreover, the following holds concerning the metric induced by this norm.

(i) Let $v_r = a_{r1} u_1 + \cdots + a_{rn} u_n \in V$ be given for $r = 1, 2, \ldots$. Then the sequence (v_r) is a Cauchy sequence if and only if the sequence $(a_{ri})_r$ in K is so for all i. Also $\lim_{r \to \infty} v_r = \sum_{i=1}^{n} a_i u_i$ if and only if $\lim_{r \to \infty} a_{ri} = a_i$ for all i.

(ii) If K is complete, then so is V.

Lemma 13.23. *Suppose that K is complete. Then for any norm $\|v\|$ on V, there exist constants λ, μ such that $\|v\| \leq \lambda \|v\|_0$, $\|v\|_0 \leq \mu \|v\|$ for all $v \in V$, where $\|v\|_0$ is defined by* (13.20). *Thus the two topologies induced by the norms $\|v\|$ and $\|v\|_0$, respectively, are the same.*

Proof. Let $\lambda = \max\{\|u_i\|;\ 1 \leq i \leq n\}$. Then, for $v = \sum_{i=1}^{n} a_i u_i$, we have

$$\|v\| \leq \max\{\varphi(a_i)\|u\|;\ 1 \leq i \leq n\} \leq \lambda \|v\|_0.$$

We proceed by induction on n to show that there exists μ such that $\|v\|_0 \leq \mu \|v\|$. This is trivial when $n = 1$, and therefore we assume that $n > 1$. For $v = a_1 u_1 + \cdots + a_n u_n$, set $\varphi_i(v) = \varphi(a_i)$. We show that, for each i, there exists μ_i such that

$$\varphi_i(v) \leq \mu_i \|v\| \qquad \text{for all } v \in V. \tag{13.21}$$

If this is shown, then we may take $\mu = \max_i \mu_i$.

Without loss of generality, we may assume $i = n$. Let $V' = K u_1 \oplus \cdots \oplus K u_{n-1}$ and consider the restrictions of the norms $\|v\|$ and $\|v\|_0$ to V'. Then by the inductive hypothesis, they induce the same topologies on V'. Since V' is complete in the topology induced by the norm $\|v\|_0$, it is also complete in the topology induced by the norm $\|v\|$. Hence V' is a closed subset of V in the topology induced by the norm $\|v\|$.

Suppose by way of contradiction that there is no μ_n satisfying (13.21) for $i = n$. Then, for any natural number j, there exists $v_j \in V - V'$ such that

$$\varphi_n(v_j) > j\|v_j\|.$$

This holds if v_j is replaced with any nonzero scalar multiple of it. Therefore, we may assume that $v_j = a_{j1}u_1 + \cdots + a_{jn-1}u_{n-1} + u_n$. Then $\|v_j\| < (1/j)\varphi_n(v_j) = 1/j$, whence it follows that $\lim_{j\to\infty} v_j = 0$ in the metric induced by the norm $\|v\|$. Therefore $\lim_{j\to\infty}(v_j - u_n) = -u_n$. But since $v_j - u_n \in V'$ for all j and V' is closed in V, we have that $-u_n \in V'$, a contradiction. ■

Now let K, L be discrete valuation fields with multiplicative valuations φ, Φ, respectively. If $K \subset L$ and Φ is an extension of φ, then we say that (L, Φ) is an *extension* of (K, φ) and write $(K, \varphi) \subset (L, \Phi)$.

Consider L as a vector space over K and let $\|x\| = \Phi(x)$ for $x \in L$. This is a norm on L, i.e., $\|x\|$ satisfies the conditions from (13.17) to (13.19), as is easily verified by using the fact that Φ is an extension of φ. Hence the following theorem holds by Lemmas 13.21 and 13.23.

Theorem 13.24. *Let φ be a complete discrete multiplicative valuation of K and let L be a finite extension of K. If there exists an extension (L, Φ) of (K, φ), then Φ is unique up to equivalence. Furthermore, (L, Φ) is complete.*

Remark. It is known in general that φ always extends to a valuation of L in the above theorem.

Let $(K, \varphi) \subset (L, \Phi)$ be discrete valuation fields. Let R (respectively, S) and $P = (\pi)$ (respectively, $Q = (\Pi)$) be the valuation ring and valuation ideal of K (respectively, L). We see that $P = R \cap Q$, since $R \subset S$, and $R \cap Q$ is a prime ideal of R. Hence $R^* = R/P \simeq (R + Q)/Q$, and we may regard R^* as a subfield of $S^* = S/Q$. We call the extension degree $f = [S^*:R^*]$ to be the *degree* of Φ over φ (possibly $f = \infty$). The *ramification index* of Φ over φ is the integer e such that $\pi S = Q^e$. Then we have $\Pi^e = \pi u$ for some $u \in S^\times$. Therefore, $|G_\Phi : G_\varphi| = e$ and $\{\Phi(1) = 1, \Phi(\Pi), \ldots, \Phi(\Pi^{e-1})\}$ is a complete set of coset representatives of G_φ in G_Φ. If $e = 1$, then we say that Φ is *unramified* over φ.

Theorem 13.25. *Let $(K, \varphi) \subset (L, \Phi)$ be the extension of discrete valuation fields and let $[L:K] = n$. Then, with the notation above, the following holds.*

(i) $ef \leq n$.

(ii) *If K is complete, then the equality holds in the above. To be precise, if $\omega_1, \ldots, \omega_f \in S$ are chosen so that $\{\omega_1^*, \ldots, \omega_f^*\}$ is an R^*-basis of S^*, then $\{\omega_i \Pi^j; 1 \leq i \leq f, 0 \leq j \leq e-1\}$ is an R-basis of S and hence a K-basis of L.*

Proof.

(i) Let $\omega_1, \ldots, \omega_f \in S$ be such that $\{\omega_1^*, \ldots, \omega_f^*\}$ is linearly independent over R^*. First we claim that if $a_1, \ldots, a_r \in K$ and some $a_i \neq 0$, then $x = \sum_{i=1}^r a_i \omega_i$ is not zero and $\Phi(x) = \max\{\varphi(a_i); 1 \leq i \leq r, \varphi(a_i) \neq 0\}$. For this, choose a_k so that $\varphi(a_k) = \max_i\{\varphi(a_i) \neq 0\}$. Then $a_k^{-1} a_i \in R$ and $a_k^{-1} a_k = 1 \notin P$, hence $a_k^{-1} x = \sum_k a_k^{-1} a_i \omega_i \in S - Q$, since $\{\omega_i^*\}$ are linearly independent over R^*. Therefore, $\Phi(a_k^{-1} x) = \varphi(a_k)^{-1} \Phi(x) = 1$, and thus $\Phi(x) = \varphi(a_k) \in G_\varphi$ as claimed. We next show that the er elements $\omega_i \Pi^j (1 \leq i \leq r, 0 \leq j \leq e-1)$ are linearly independent over K. Let $\sum_{i=1}^r \sum_{j=0}^{e-1} a_{ij} \omega_i \Pi^j = 0$ with $a_{ij} \in K$. We want to show that $a_{ij} = 0$ for all i, j. Put $\gamma_j = \sum_{i=1}^r a_{ij} \omega_i$. Then $\sum_{j=0}^{e-1} \gamma_j \Pi^j = 0$. If some a_{ij} is not zero, then $\gamma_j \neq 0$ and $\Phi(\gamma_j) \in G_\varphi$ as claimed above. Now, note that $\Phi(\Pi^j) \not\equiv \Phi(\Pi^k) \bmod G_\varphi$ whenever $j \neq k (0 \leq j,\ k \leq e-1)$, and so $\Phi(\gamma_j \Pi^j) \not\equiv \Phi(\gamma_k \Pi^k) \bmod G_\varphi$ if $\gamma_j \neq 0$, $\gamma_k \neq 0$. Therefore by (13.15), we have

$$\Phi\left(\sum_j \gamma_j \Pi^j\right) = \max\{\Phi(\gamma_j \Pi^j); 0 \leq j \leq e-1, \gamma_j \neq 0\} \neq 0.$$

This contradicts that $\sum_j \gamma_j \Pi^j = 0$. Therefore, the er elements $\{\omega_i \Pi^j\}$ are linearly independent over K. This implies that $f < \infty$ and $ef \leq n = [L:K]$.

(ii) Since K is complete, L is also complete. Apply Exercise 13.20 with K replaced by L, and $\{\pi_i\} = \{\pi^i \Pi^j; i \in \mathbf{Z}, 0 \leq j \leq e-1\}$. If Γ denotes a complete set of representatives of R/P containing 0, then

$$\Lambda = \left\{\sum_{t=1}^{f} c_t \omega_t; c_t \in \Gamma\right\}$$

is a complete set of representatives of S/Q. Therefore, every $x \in L$ is expressed as

$$x = \sum_{i,j}\left(\sum_t c_{ijt} \omega_t\right) \pi^i \Pi^j = \sum_{t=1}^{f} \sum_{j=0}^{e-1}\left(\sum_i c_{ijt} \pi^i\right) \omega_t \Pi^j,$$

where $c_{ijt} \in \Gamma$. Let (r, s) be a minimal member of $\{(i,j); c_{ijt} \neq 0\}$ relative to the lexicographic order. Then, if μ denotes the normalized exponential valuation of L, we have $\mu(x) = re + s(0 \leq s \leq e - 1)$, and hence $\mu(x) \geq 0 \Leftrightarrow r \geq 0$. Consequently, $S = \sum_{t,j} R\omega_t\Pi^j$. Since $\{\omega_t\Pi^j;\ 1 \leq t \leq f,\ 0 \leq j \leq e - 1\}$ is linearly independent over K, this is an R-basis of S and hence a K-basis of L. This completes the proof. ∎

Exercise 13.26. Let $(K, \varphi) \subset (L, \Phi)$ and assume that L is an algebraic extension of K. Then S^* is an algebraic extension of R^*.

The following result will be of use in Section 7 of Chapter 4.

Theorem 13.27. *Let v be a discrete valuation of a field K and let R, P and $F = R/P$ be the valuation ring, valuation ideal, and residue field of v, respectively. Then given an arbitrary algebraic extension F' of F, there exist an algebraic extension K' of K and a discrete valuation v' of K' such that v' is an unramified extension of v with residue field being F-isomorphic to F'.*

Proof. We may assume that v is normalized. We denote by a^* the image of $a \in R$ under the natural map $R \to R/P = F$.

We first deal with the case that F' is a simple extension of F. Therefore, let $F' = F(\zeta)$ and let $f(x) = x^n + a_1^*x^{n-1} + \cdots + a_n^*$ be an irreducible polynomial such that $f(\zeta) = 0$, where $a_i \in R$. Then $g(x) = x^n + a_1x^{n-1} + \cdots + a_n$ is irreducible in $R[x]$ and hence in $K[x]$ (cf. Hungerford [1], p. 163). Then, there exists an extension $K' = K(\alpha)$ of K of degree n with $g(\alpha) = 0$. Any $a \in K'$ is expressed uniquely as $a = a_0 + a_1\alpha + \cdots + a_{n-1}\alpha^{n-1} (a_i \in K)$. Define

$$v'(a) = \min\{v(a_0), \ldots, v(a_{n-1})\}.$$

We shall show that v' is a valuation of K' that gives a desired extension of v.

Let $a = \sum_{i=0}^{n-1} a_i\alpha^i$, $b = \sum_{i=0}^{n-1} b_i\alpha^i$ be arbitrary elements of K'. From the definition of v', it is clear that $v'(a + b) \geq \min\{v'(a), v'(b)\}$. We next prove that $v'(ab) = v'(a) + v'(b)$. This is obvious if either a or b lies in K. Let a_i be such that $v'(a) = v(a_i)$ and set $a' = a_i^{-1}a$. Then $v'(a') = 0$. Similarly, let b_j be such that $v'(b) = v(b_j)$ and $b' = b_j^{-1}b$. Then $v'(b') = 0$. Now it suffices to show that $v'(a'b') = 0$. In fact, if this is shown, then $0 = v'(a_i^{-1}b_j^{-1}ab) = -v(a_i) - v(b_j) + v'(ab)$, which implies the desired equality. Let $R[\alpha] = \sum_{i=0}^{n-1} R\alpha^i$ and $P[\alpha] = \sum_{i=0}^{n-1} P\alpha^i$. Then $P[\alpha]$ is an ideal of $R[\alpha]$ and $R[\alpha]/P[\alpha] = F[\alpha^*]$, where $\alpha^* = \alpha + P[\alpha]$. Since $f(\alpha^*) = 0$, we have that $F[\alpha^*] \simeq F(\zeta)$. Therefore

$F[\alpha^*]$ is a field, and hence $P[\alpha]$ is a prime ideal of $R[\alpha]$. Since $v'(a') = v'(b') = 0$, it follows that $a', b' \in R[\alpha] - P[\alpha]$ and hence $a'b' \notin P[\alpha]$. Consequently, $v'(a'b') = 0$.

Thus v' is a valuation of K'. It is clear from the definition of v' that $R[\alpha]$, $P[\alpha]$, and $R[\alpha]/P[\alpha] \simeq F(\zeta)$ are the valuation ring, valuation ideal, and residue field of v', respectively. The value group of v' is $\mathbf{Z}$, so v' is unramified over v.

Suppose next that F' is an arbitrary algebraic extension of F. Let $\tilde{K}$ be the algebraic closure of K. Consider the triple (K_1, v_1, σ_1), where K_1 is a subfield of $\tilde{K}$ containing K, v_1 is a discrete valuation of K_1, which is an unramified extension of v, and σ_1 is an F-isomorphism from the residue field F_1 of v_1 into F' (regarding F as a subfield of F_1). Let $\mathcal{M}$ denote the set of all such triples.

For (K_1, v_1, σ_1), $(K_2, v_2, \sigma_2) \in \mathcal{M}$, we write $(K_1, v_1, \sigma_1) \leq (K_2, v_2, \sigma_2)$ if $(K_1, v_1) \subset (K_2, v_2)$ and $\sigma_2|_{F_1} = \sigma_1$ (where F_1 is regarded as a subfield of F_2). It is easy to check that $\mathcal{M}$ becomes an inductively ordered set with respect to the above order. Thus it has a maximal element, say (K_0, v_0, σ_0), by Zorn's lemma. Let F_0 be the residue field of v_0 and $\sigma_0(F_0) = F'_0$. If $F'_0 = F'$, then (K_0, v_0) is the desired extension of (K, v). We deduce a contradiction by assuming that $F'_0 \subsetneqq F'$.

Take $\alpha \in F' - F'_0$ and let $F'_1 = F'_0(\alpha)$. Then, as was shown above, there exist an unramified extension $(K_1, v_1) \supset (K_0, v_0)$ and an isomorphism $\sigma_1 \colon F_1 \simeq F'_1$, which extends $\sigma_0 \colon F_0 \simeq F'_0$, with F_1 being the residue field of v_1. This implies that $(K_0, v_0, \sigma_0) < (K_1, v_1, \sigma_1)$, contradicting the maximality of (K_0, v_0, σ_0). ■

14. Algebras over Complete Discrete Valuation Rings

In this section we assume that

R is a complete discrete valuation ring

with valuation v and let $P = (\pi)$ and $F = R/(\pi)$ be the valuation ideal and residue field of v, respectively.

Let V be a finitely generated R-module. Since R is a principal ideal domain, V is a direct sum of cyclic submodules:

$$V = Ru_1 \oplus Ru_2 \oplus \cdots \oplus Ru_r. \tag{14.1}$$

Thus,

$$\pi^n V = R\pi^n u_1 \oplus \cdots \oplus R\pi^n u_r,$$

and we have $\bigcap_{n=0}^{\infty} \pi^n V = 0$ since $\bigcap_{n=0}^{\infty} R\pi^n = 0$. Consequently, for each $v(\neq 0)$ of V, there is $i \geq 0$ such that $v \in \pi^i V - \pi^{i+1} V$. Set $\nu(v) = i$ and $\nu(0) = \infty$. Then the following hold:

(14.2) $$v \in \pi^j V \Rightarrow j \leq \nu(v).$$

(14.3) $$r \in R, v \in V \Rightarrow \nu(rv) \geq \nu(r) + \nu(v).$$

(14.4) $$\nu(u + v) \geq \min\{\nu(u), \nu(v)\}.$$

Fix a real number γ such that $0 < \gamma < 1$, and define real-valued functions $\varphi(r)$, $\|v\|$ by $\varphi(r) = \gamma^{\nu(r)}(r \in R)$ and $\|v\| = \gamma^{\nu(v)}(v \in V)$. Then φ is a multiplicative valuation of R and V is a metric space with the metric defined by $d(u, v) = \|u - v\|$. Note that $\|v\| \geq 0$ and $\|v\| = 0 \Leftrightarrow v = 0$. It also follows from (14.3) and (14.4) that $\|rv\| \leq \varphi(r)\|v\|$ and $\|u + v\| \leq \max\{\|u\|, \|v\|\}$. Therefore, the maps $V \times V \to V((u, v) \mapsto u + v)$ and $R \times V \to V((r, v) \mapsto rv)$ are continuous.

Let $(v_n) = (v_1, v_2, \ldots)$ be a sequence of elements of V. If we write

$$v_n = \sum_{i=1}^{r} r_{ni} u_i$$

according to the decomposition (14.1), then the following holds.

(v_n) is a Cauchy sequence $\Leftrightarrow (r_{ni} u_i)_n$ is a Cauchy sequence for each i.

and

$$\lim_{n \to \infty} v_n = \sum_i r_i u_i \Leftrightarrow \lim_{n \to \infty} r_{ni} u_i = r_i u_i \qquad \text{for each } i.$$

From this, we observe that V is complete in the metric defined above (note that if $\pi^j R u_i = 0$ for some j, then $R u_i$ is discrete and hence complete).

If $V = A$ is a finitely generated R-algebra, the map $A \times A \to A((a, b) \mapsto ab)$ is continuous. Also $A^* = A/\pi A$ is a finite dimensional algebra over F.

Theorem 14.1. *Let A be a finitely generated algebra over R. Then*

(i) $\pi A \subset J(A)$.

(ii) $J(A)^n \subset \pi A$ *for some* $n > 0$.

Proof.

(i) Let V be any irreducible A-module. Then either $\pi V = V$ or $\pi V = 0$. Since $V = vA$ for any $v(\neq 0) \in V$ and A is finitely generated over R, V is finitely generated over R. Thus, if $\pi V = V$, then we have $V = 0$ by the Azumaya–Nakayama lemma, which is a contradiction. Therefore, $\pi V = V(\pi A) = 0$, and thus $\pi A \subset J(A)$.

(ii) From the above, $A^* = A/\pi A$ is a finite dimensional F-algebra and $J(A^*) = J(A)/\pi A$. Since this is nilpotent, the assertion is obvious. ∎

The following theorem on lifting idempotents is important.

Theorem 14.2. *Let A be a finitely generated R-algebra. Let I be an ideal of A contained in $J(A)$ and $\bar{A} = A/I$. Then the following holds.*

(i) *Let $\bar{c}$ be an idempotent of $\bar{A}$ and let*

$$\bar{c} = \bar{c}_1 + \bar{c}_2 + \cdots + \bar{c}_n \tag{14.5}$$

be an idempotent decomposition in $\bar{A}$. Then $\bar{c}$ lifts to an idempotent e of A, and there exist orthogonal idempotents $e_1, e_2, \ldots, e_n$ of A satisfying

$$\bar{e}_i = \bar{c}_i \qquad (i = 1, 2, \ldots, n)$$

$$e = e_1 + e_2 + \cdots + e_n.$$

(ii) *An idempotent e of A is primitive if and only if $\bar{e}$ is primitive in $\bar{A}$.*

(iii) *Let e, f be primitive idempotents of A. Then the following equivalence holds:*

$$eA \simeq fA \Leftrightarrow \bar{e}\bar{A} \simeq \bar{f}\bar{A}.$$

In particular there is a natural one-to-one correspondence between $\widetilde{\text{pi}}(A)$ and $\widetilde{\text{pi}}(\bar{A})$.

Proof. Since $I \subset J(A)$, $I^m \subset \pi A$ for some m, and hence $I^{mr} \subset \pi^r A$ for all $r \geq 0$. Set $\bar{A}^{(r)} = A/I^{mr}$.

(i) Since I/I^m is a nilpotent ideal of $\bar{A}^{(1)} = A/I^m$, $\bar{c}$ lifts to an idempotent of $\bar{A}^{(1)}$ by Theorem 4.9. Namely, there exists $c^{(1)} \in A$ such that

$$(c^{(1)})^2 \equiv c^{(1)} \mod I^m, \qquad c^{(1)} \equiv c \mod I.$$

Next, since I^m/I^{2m} is a nilpotent ideal of $\bar{A}^{(2)}$, there exists $c^{(2)} \in A$ such that

$$(c^{(2)})^2 \equiv c^{(2)} \mod I^{2m}, \qquad c^{(2)} \equiv c^{(1)} \mod I^m.$$

By continuing similar arguments, we find $c^{(r)} \in A$ for each r such that

$$(c^{(r)})^2 \equiv c^{(r)} \mod I^{rm}, \qquad c^{(r)} \equiv c^{(r-1)} \mod I^{(r-1)m},$$

where we set $c^{(0)} = c$ for convenience. Then $(c^{(r)})_r$ is a Cauchy sequence and $e = \lim_{r\to\infty} c^{(r)}$ exists by the completeness of R. This is an idempotent of A that lifts $\bar{c}$ to A.

To show the second, let e be any idempotent of A that lifts $\bar{c}$ to A. Using Theorem 4.10, we see by the same argument as above that there exist $c_i^{(r)} \in A (i = 1, 2, \ldots, n)$ for each r such that

$$c_i^{(r)} \equiv c_i^{(r-1)} \mod I^{(r-1)m},$$

and

$$e \equiv c_1^{(r)} + \cdots + c_n^{(r)} \mod I^{rm}$$

is an idempotent decomposition in $\bar{A}^{(r)}$. Each $(c_i^{(r)})_r$ is a Cauchy sequence, and $e_i = \lim_{r\to\infty} c_i^{(r)}$ exists. Then $e_1, e_2, \ldots, e_n$ satisfy our requirement.

(ii) Since $J(A)$ has no idempotent, an argument similar to the proof of Theorem 4.11 works by making use of the above (i).

(iii) One implication, ($\Rightarrow$), is trivial. The other one, ($\Leftarrow$), follows because eA and fA are, by Theorem 10.14, projective covers of $\bar{e}\bar{A}$ and $\bar{f}\bar{A}$, respectively. ■

As an immediate consequence of Theorems 14.1 and 14.2, we have

Theorem 14.3. *Every finitely generated algebra over a complete discrete valuation ring R is semiperfect.*

Before proceeding, we show the following lemma.

Lemma 14.4. *Let A be a semiperfect ring. Then*

$$A \text{ is a local ring} \Leftrightarrow 1_A \text{ is a unique idempotent of } A.$$

Proof. ($\Rightarrow$). This has been shown in Theorem 5.9. Let us show the converse. By assumption, $\bar{1}_A$ is a unique idempotent of $\bar{A} = A/J(A)$, hence $\bar{A}_A$ is indecomposable. But since $\bar{A}$ is Artinian, this implies that $\bar{A}$ is a division ring. ■

For the rest, we assume that *A is a finitely generated R-algebra.*

Lemma 14.5. *If V_A is finitely generated over A, then* $\operatorname{End}_A(V)$ *is finitely generated over R and hence a semiperfect ring.*

Proof. Since R is Noetherian and $\operatorname{End}_R(V) \supset \operatorname{End}_A(V)$, it suffices to show that $\operatorname{End}_R(V)$ is finitely generated over R. By assumption, V is finitely generated over R; $V = Rv_1 + \cdots + Rv_n$. Let $X = Rx_1 \oplus \cdots \oplus Rx_n$ be a free R-module and consider an R-homomorphism $\varphi: X \to V(x_i \mapsto v_i)$. Note that $E(X) = \operatorname{End}_R(X) \simeq M_n(R)$ is finitely generated over R. Consequently, $E_0(X) = \{\sigma \in E(X); \sigma(\operatorname{Ker} \varphi) \subset \operatorname{Ker} \varphi\}$ is also finitely generated over R. We show that $\operatorname{End}_R(V)$ is a homomorphic image of $E_0(X)$. To show this, let $\bar{X} = X/\operatorname{Ker} \varphi(\simeq V)$. Each $\sigma \in E_0(X)$ induces $\bar{\sigma}: \bar{X} \to \bar{X}$ such that $\bar{\sigma}(\bar{x}) = \overline{\sigma(x)}$. Hence we have a homomorphism $E_0(X) \to \operatorname{End}_R(\bar{X})$ that sends each σ onto $\bar{\sigma}$. We claim that this is an epimorphism. In fact, for any $\rho \in \operatorname{End}_R(\bar{X})$, take $y_i \in X$ so that $\bar{y}_i = \rho(\bar{x}_i)$ for each i. Then there exists a homomorphism $\lambda: X \to X$ such that $\lambda(x_i) = y_i$. It is clear that $\bar{\lambda} = \rho$, which proves our claim. Thus $\operatorname{End}_R(V) \simeq \operatorname{End}_R(\bar{X})$ is a homomorphic image of $E_0(X)$ and hence finitely generated over R. ■

Now we have the following theorem from Theorem 5.5 and the above two lemmas.

Theorem 14.6. *Let V be a finitely generated A-module. Then*

$$V \text{ is indecomposable} \quad \Leftrightarrow \quad \operatorname{End}_A(V) \text{ is a local ring.}$$

The K–S–A theorem applies to indecomposable decompositions by the above theorem. Namely, we have

Theorem 14.7. *If V_A is finitely generated, then V is a finite direct sum of indecomposable A-submodules. Furthermore, if*

$$V = V_1 \oplus V_2 \oplus \cdots \oplus V_n$$

and

$$V = W_1 \oplus W_2 \oplus \cdots \oplus W_m$$

are both indecomposable decompositions of V, then it holds that $n = m$ and $V_i \simeq W_i$ for all i, after a suitable arrangement of indices.

Proof. Since V is Noetherian by Theorem 2.5, the first statement follows at once from Theorem 6.2(i). The second one follows from the K–S–A theorem.

Problems

In the following A denotes a ring.

1. (Principle of idealization) Let U be an (A, A)-bimodule and let $B = A \oplus U$. Then B becomes a ring with the multiplication defined by

$$(a_1, u_1)(a_2, u_2) = (a_1 a_2, a_1 u_2 + u_1 a_2),$$

and U is an ideal of B.

2. Let $I_1, \ldots, I_r$ be ideals of A, and suppose that $I_i + I_j = A$ whenever $i \neq j$. Then the following holds.

(i) $I_i + \prod_{j \neq i} I_j = A$.

(ii) **(Chinese remainder theorem)** The natural homomorphism

$$\varphi: A \to \bigoplus_{j=1}^{r} A/I_j \qquad (a \mapsto (a + I_j)_j)$$

is an epimorphism. Consequently, the following isomorphism holds:

$$A/\bigcap_{j=1}^{r} I_j \simeq \bigoplus_{j=1}^{r} A/I_j.$$

3. Let V_A be a Noetherian module. Then any epimorphism $\varphi: V \to V$ is necessarily an isomorphism.

4. Let $U = V \oplus W$, and let π_V and π_W be projections on V and W respectively. Then the following holds.

(i) For a submodule L of U, $\pi_V(L)/L \cap V \simeq \pi_W(L)/L \cap W$.

(ii) If U has a composition series, the following two conditions are equivalent.

(1) Every submodule L of U is expressed in the form:

$$L = L_1 \oplus L_2 \qquad \text{with} \quad L_1 \subset V \quad \text{and} \quad L_2 \subset W.$$

(2) V and W have no irreducible constituent in common.

5. If $\varphi: A \to A'$ is an epimorphism of rings, then $\varphi(J(A)) \subset J(A')$.

6. Suppose that $\bar{A} = A/J(A)$ is Artinian, and let V be a finitely generated A-module. Then $VJ(A)$ coincides with the intersection of all maximal A-submodules of V.

7. Suppose that the identity of A is expressed as orthogonal sums of n idempotents in two ways: $1 = \sum_{i=1}^{n} e_i = \sum_{i=1}^{n} f_i$. If $e_i A \simeq f_i A$ for all i, then there exists a unit u of A such that $u^{-1} e_i u = f_i$ for all i.

8. Let I be an ideal of A contained in $J(A)$ and let $\bar{A} = A/I$. If two idempotents e, f of A satisfy $\bar{e} = \bar{f}$, then there exists c of I such that $(1 + c)^{-1} e(1 + c) = f$.

9. The following three conditions on V_A with composition series are equivalent.

(1) V is completely reducible.

(2) Every irreducible submodule of V is a direct summand of V.

(3) Every maximal submodule of V is a direct summand of V.

10. $\operatorname{soc}(U \oplus V) = \operatorname{soc}(U) \oplus \operatorname{soc}(V)$.

11. Let A be a semisimple Artinian ring and let $A = A_1 \oplus \cdots \oplus A_n$ be the decomposition into the direct sum of simple components A_i. Then the following holds.

(i) Every ideal of A is a direct sum of some A_i's.

(ii) If A' is a homomorphic image of A, then A' is isomorphic to a direct sum of some of the A_i as rings.

12. Let A be an Artinian ring. Then the following holds.

(i) There are only a finite number of maximal ideals, and the intersection of them coincides with $J(A)$.

(ii) Let I be a right ideal of A. If I contains no idempotent, then $I \subset J(A)$.

13. Let A be Noetherian and let $x, y \in A$. If $xy = 1$, then $yx = 1$.

14. Let P, Q be finitely generated projective A-modules. Then

$$P/PJ(A) \simeq Q/QJ(A) \Leftrightarrow P \simeq Q.$$

15. Let e be an idempotent of A such that $AeA = A$. Let $\Gamma = eAe$ and set $\mathfrak{F}(V) = Ve$ for V_A, $\mathfrak{G}(W) = W \otimes_\Gamma eA$ for W_Γ. Note that $\mathfrak{F}(V)$ and $\mathfrak{G}(W)$ are a Γ- and A-module under the natural actions, respectively. Then the following holds.

(i) $\mathfrak{G}(\mathfrak{F}(V)) \simeq V$, $\mathfrak{F}(\mathfrak{G}(W)) \simeq W$.

(ii) The restriction map induces an isomorphism:

$$\operatorname{Hom}_A(U, V) \simeq \operatorname{Hom}_\Gamma(Ue, Ve).$$

In particular, there is a ring isomorphism $A \stackrel{\sim}{\to} \mathrm{End}_\Gamma(Ae)(a \mapsto \varphi_a)$, where $\varphi_a(xe) = axe$.

(iii) $\mathfrak{F}$ and $\mathfrak{G}$ preserve direct sum, inclusion, and projectivity.

16. Suppose that $1 = \sum_{i=1}^{k} \sum_{\lambda=1}^{n_i} e_{i\lambda}$ is a primitive idempotent decomposition of 1 such that $e_{i\lambda}A \simeq e_{j\mu}A \Leftrightarrow i = j$. If we set $e_{i1} = e_i$ and $e = e_1 + \cdots + e_k$, then $AeA = A$. Moreover, if we let $\Gamma = eAe$ as in the preceding problem, then $\Gamma = \bigoplus_{i=1}^{k} e_i\Gamma$, and Γ-isomorphism $e_i\Gamma \simeq e_j\Gamma$ holds only if $i = j$; (Γ is called the *basic ring* of A).

17. Let P be a finitely generated projective A-module and let $E = \mathrm{End}_A(P)$. Then the following holds.

(i) $J(E) = \{\varphi \in E; \mathrm{Im}\ \varphi \subset PJ(A)\}$.

(ii) $E/J(E) \simeq \mathrm{End}_A(P/PJ(A))$.

18. The following condition on U_A is a necessary and sufficient condition for U to be a finitely generated projective A-module:

There exist $u_1, \ldots, u_n \in U$ and $\varphi_1, \ldots, \varphi_n \in \mathrm{Hom}_A(U, A)$ such that $u = \sum_i u_i\varphi_i(u)$ for all $u \in U$.

19. Let I be an ideal of A. If $Z(A)$ has no idempotent other than the identity, then the same is true for $Z(A/I)$, provided I satisfies one of the following conditions:

(1) $I = NA$, where N is a nilpotent ideal of $Z(A)$.

(2) $I \subset J(A)^2$, where $J(A)$ is assumed to be nilpotent.

20. Given two exact sequences

$$V'_A \xrightarrow{\lambda} V_A \xrightarrow{\lambda'} V''_A \to 0,$$

$$_AW' \xrightarrow{\mu} {}_AW \xrightarrow{\mu'} {}_AW'' \to 0,$$

the following sequence is also exact:

$$(V' \otimes_A W) \oplus (V \otimes_A W') \xrightarrow{f} V \otimes_A W \xrightarrow{\lambda' \otimes \mu'} V'' \otimes_A W'' \to 0,$$

where $f = \lambda \otimes \mathrm{id}_W + \mathrm{id}_V \otimes \mu$.

21. Let V_A be A-flat and let W be a left A-module. Then the following equality holds in $V \otimes_A W$ for two A-submodules W_1, W_2 of W:

$$V \otimes_A W_1 \cap V \otimes_A W_2 = V \otimes_A (W_1 \cap W_2).$$

22. Let R be a commutative ring and I, L be ideals of R. Then the following isomorphism holds:

$$R/I \otimes_R R/L \simeq R/(I + L).$$

2 Algebras and Their Representations

The first three sections of this chapter give some fundamental facts about representations of algebras, which will be followed by theories of simple algebras, Schur indices, crossed products, and Frobenius algebras.

1. Fundamental Concepts of Representations

1.1. Representations and Representation Modules

Let R be a commutative ring. Let A, B be R-algebras. A ring homomorphism $f: A \to B$ is said to be an R-algebra homomorphism if it is an R-homomorphism as well.

An (R-)*representation* of an R-algebra A is an R-algebra homomorphism from A into a full matrix ring over R:

$$\mathbf{X}: A \to M_n(R) \qquad (a \mapsto X(a)). \tag{1.1}$$

The integer n is called the *degree* of $\mathbf{X}$. We assume $X(1) = I$ as before, where I denotes the identity matrix. The representation $\mathbf{X}$ is said to be *faithful* if $\mathbf{X}$ is a monomorphism. This is always the case if A is a simple ring, for instance.

If $R^\times$ denotes the group of units of R, then we have

$$\mathrm{GL}_n(R) = \{M \in M_n(R);\ \det M \in R^\times\}.$$

Let $\mathbf{Y}: A \to M_n(R)$ be another representation of A. We say $\mathbf{X}$ and $\mathbf{Y}$ are (R-)*equivalent* and write $\mathbf{X} \sim \mathbf{Y}$ or $\mathbf{X} \sim_R \mathbf{Y}$, provided there exists $P \in \mathrm{GL}_n(R)$ such that

$$Y(a) = P^{-1}X(a)P \qquad \text{for all } a \in A.$$

The representation $\mathbf{X}$ gives rise to an R-homomorphism

$$\chi_{\mathbf{X}}: A \to R \qquad (a \mapsto \operatorname{tr} X(a)),$$

which is called the (R-)*character* defined by the representation $\mathbf{X}$. It is clear that

$$\mathbf{X} \sim \mathbf{Y} \Rightarrow \chi_{\mathbf{X}} = \chi_{\mathbf{Y}}.$$

Now let V be an A-module that is R-free with basis $\{v_1, \ldots, v_n\}$;

$$V = Rv_1 \oplus Rv_2 \oplus \cdots \oplus Rv_n.$$

For $a \in A$, write

(1.2) $$v_i a = \sum_{j=1}^{n} \alpha_{ij}(a) v_j \qquad (\alpha_{ij}(a) \in R,\ i = 1, 2, \ldots, n)$$

and let $X(a) = (\alpha_{ij}(a))_{i,j} \in M_n(R)$. Then the map

(1.3) $$\mathbf{X}: A \to M_n(R) \qquad (a \mapsto X(a))$$

is clearly a representation of A, which is called the *representation of A defined by the A-module V (relative to the R-basis $\{v_i\}$)*.

Remark. The expression (1.2) can be written as

$$\begin{pmatrix} v_1 \\ \vdots \\ v_n \end{pmatrix} a = \begin{pmatrix} v_1 a \\ \vdots \\ v_n a \end{pmatrix} = X(a) \begin{pmatrix} v_1 \\ \vdots \\ v_n \end{pmatrix}.$$

Exercise 1.1

(i) Show that the map $\mathbf{X}$ given in (1.3) is indeed a representation of A.

(ii) Let $\mathbf{Y}$ be the representation of A defined by V relative to another basis $\{u_1, \ldots, u_n\}$. Show that $\mathbf{X} \sim \mathbf{Y}$.

[Hint: There exists a non-singular matrix P such that

$$\begin{pmatrix} v_1 \\ \vdots \\ v_n \end{pmatrix} = P \begin{pmatrix} u_1 \\ \vdots \\ u_n \end{pmatrix}.$$

Then $P^{-1}X(a)P = Y(a)$ for all $a \in A$.]

Given a representation $\mathbf{X}: A \to M_n(R)(a \mapsto X(a))$, we consider a free R-module $V = Rv_1 \oplus \cdots \oplus Rv_n$, and define the action of $a \in A$ on each v_i by (1.2) and extend it to the whole V linearly. This makes V into a right A-module, and the representation of A defined by V relative to the basis $\{v_i\}$ coincides with $\mathbf{X}$. The A-module V obtained in this way is said to be a (right) *representation module* for $\mathbf{X}$.

In general, an A-module that is finitely generated and free as an R-module will be sometimes referred to as a *representation module* of A.

Exercise 1.2. Let $\mathbf{X}$ and $\mathbf{Y}$ be two representations of A with representation modules V and W, respectively. Show that

$$\mathbf{X} \sim \mathbf{Y} \Leftrightarrow V \simeq W \qquad (A\text{-isomorphic}).$$

Thus the equivalence classes of representations of A correspond bijectively to the isomorphism classes of representation modules of A.

Remark. If V is a left A-module that is finitely generated and free as an R-module with R-basis $\{v_i\}_i$, we set, for $a \in A$,

$$av_j = \sum_i \alpha_{ij}(a)v_i. \tag{1.4}$$

Then we have a representation $\mathbf{X}$ of A by assigning $X(a) = (\alpha_{ij}(a))$ to each a of A. We say that ${}_AV$ is a left representation module for $\mathbf{X}$. The above expression (1.4) is written as

$$a(v_1, \ldots, v_n) = (av_1, \ldots, av_n) = (v_1, \ldots, v_n)X(a).$$

Given representations $\mathbf{X}_1, \ldots, \mathbf{X}_m$ of A, the representation defined by

$$\mathbf{X}: a \mapsto \begin{pmatrix} X_1(a) & & 0 \\ & \ddots & \\ 0 & & X_m(a) \end{pmatrix}$$

is called the *direct sum* of $\mathbf{X}_1, \ldots, \mathbf{X}_m$ and is denoted by $\mathbf{X} = \mathbf{X}_1 \oplus \cdots \oplus \mathbf{X}_m$. If V_i is a representation module for $\mathbf{X}_i$, then $V = V_1 \oplus \cdots \oplus V_m$ gives a representation module for $\mathbf{X}$. If, in particular, $\mathbf{X}_1 = \cdots = \mathbf{X}_m$, we write $\mathbf{X} = m\mathbf{X}_1$ and $V = mV_1$ for brevity.

A representation $\mathbf{X}$ of A is said to be *indecomposable* if $\mathbf{X}$ is not equivalent to a direct sum of two representations of A. This is equivalent to a representation module for $\mathbf{X}$ being indecomposable as an A-module.

If $\mathbf{X} \sim \mathbf{X}_1 \oplus \cdots \oplus \mathbf{X}_m$ and each $\mathbf{X}_i$ is indecomposable, then this is called an *indecomposable decomposition* of $\mathbf{X}$ and each $\mathbf{X}_i$ an *indecomposable component* of $\mathbf{X}$.

By an R-*lattice* we mean a finitely generated R-module that is torsion-free. We assume for the remainder of this section that

(1.5) *R is a principal ideal domain.*

In this case, an R-lattice is just a finitely generated free R-module.

Let $W \subset V$ be R-lattices. We say that W is *pure* in V if $\bar{V} = V/W$ is torsion-free. So, if W is pure in V, then $\bar{V}$ is R-free, and hence W is a direct summand of V:

$$V = W \oplus W', \qquad W' \simeq \bar{V}.$$

Now assume that the above V is an A-module and $W \neq 0$ is an A-submodule of it. If $\{v_1, \ldots, v_r\}$ and $\{v_{r+1}, \ldots, v_n\}$ are R-bases of W and W', respectively, then $\{v_1, \ldots, v_n\}$ forms an R-basis of V. Then the representation $\mathbf{X}$ of A defined by V is given by

$$\text{(1.6)} \qquad \mathbf{X}\colon a \mapsto X(a) = \begin{pmatrix} Y(a) & 0 \\ * & Y'(a) \end{pmatrix},$$

where $\mathbf{Y}\colon a \mapsto Y(a)$ and $\mathbf{Y}'\colon a \mapsto Y'(a)$ are defined by W and $\bar{V}$ relative to the bases $\{v_1, \ldots, v_r\}$ and $\{\bar{v}_{r+1}, \ldots, \bar{v}_n\}$, respectively.

Exercise 1.3. Show that the above $\mathbf{X}$ is equivalent to the following representation:

$$\mathbf{X}'\colon a \mapsto X'(a) = \begin{pmatrix} Y'(a) & * \\ 0 & Y(a) \end{pmatrix}.$$

An R-representation of A is said to be *irreducible* if it is not equivalent to a representation $\mathbf{X}$ of the form (1.6). This is equivalent to that a representation module for it has no nontrivial pure submodule.

In particular, if R is a field, then every proper A-submodule of V is pure. Therefore the following holds.

Theorem 1.4. *Let A be an algebra over a field K. Then a representation of A is irreducible if and only if a representation module for it is an irreducible A-module.*

Let A be an algebra over a field K and $\mathbf{X}$ be a representation of A with representation module V_A. Then V has a composition series

$$0 = V_0 \subset V_1 \subset \cdots \subset V_m = V.$$

Take a K-basis $\{v_1, \ldots, v_{n_1}\}$ of V_1 and extend it to a K-basis $\{v_1, \ldots, v_{n_2}\}$ of V_2. By continuing this, we obtain a K-basis $\{v_i\}_i$ of V. The representation $\mathbf{Y}$ of A relative to this basis is of the form

$$\mathbf{Y}: a \mapsto Y(a) = \begin{pmatrix} Y_1(a) & & & 0 \\ & Y_2(a) & & \\ & & \ddots & \\ & * & & Y_m(a) \end{pmatrix}, \tag{1.7}$$

where each $\mathbf{Y}_i: a \mapsto Y_i(a)$ is the representation defined by the irreducible module $\bar{V}_i = V_i/V_{i-1}$ relative to the basis $\{\bar{v}_{n_{i-1}+1}, \ldots, \bar{v}_{n_i}\}$.

Note that the set $\{\mathbf{Y}_1, \ldots, \mathbf{Y}_m\}$ is unique up to equivalence by the Jordan–Hölder theorem. We call each $\mathbf{Y}_i$ an *irreducible constituent* of $\mathbf{X}$.

If the above V is completely reducible, then $\mathbf{X} \sim \mathbf{Y}_1 \oplus \cdots \oplus \mathbf{Y}_m$, in which case $\mathbf{X}$ is said to be *completely reducible.*

Let $\mathbf{X}$ and $\mathbf{Y}$ be two K-representations of A with representation modules V and W, respectively. We write

$$V \leftrightarrow W, \qquad \mathbf{X} \leftrightarrow \mathbf{Y},$$

provided the irreducible constituents of V and W are the same up to the order of occurrence and isomorphism.

If this is so, we have $\chi_{\mathbf{X}} = \chi_{\mathbf{Y}}$. We also have $\mathbf{X} \leftrightarrow \mathbf{Y}$ if $\mathbf{X} \sim \mathbf{Y}$. The converse is true if both $\mathbf{X}$ and $\mathbf{Y}$ are completely reducible.

1.2. R-forms

Let $S \supset R$ be rings having identity in common, and let A be an R-algebra. An R-algebra homomorphism $\mathbf{X}: A \to M_n(S)$ is said to be an *S-representation* of A.

This extends to an S-algebra homomorphism $\mathbf{X}^S: A^S = S \otimes_R A \to M_n(s)(s \otimes a \mapsto sX(a))$, which gives an S-representation of A^S. Conversely, any

S-representation of A^S induces an S-representation of A via the homomorphism $A \to A^S (a \mapsto 1_S \otimes a)$. Thus the study of the S-representations of A is the same as the study of the representations of A^S.

Note that any R-representation $\mathbf{X}\colon A \to M_n(R) (\subset M_n(S))$ can be regarded as an S-representation of A, hence we obtain the S-representation $\mathbf{X}^S$ of A^S. If V is a representation module for $\mathbf{X}$, then $V^S = S \otimes_R V$ gives a representation module for $\mathbf{X}^S$.

Let U be any A^S-module. We say that U is *realizable* in R if there exists an A-module V such that

$$V^S \simeq U. \tag{1.8}$$

We call such V an *R-form* of U.

If the above V is R-free and defines an R-representation $\mathbf{X}$ of A of degree n, then U is S-free and, for any S-representation $\mathbf{Y}$ defined by U, there exists $P \in \mathrm{GL}_n(S)$ such that

$$P^{-1} Y(a) P = X(a) \qquad \text{for all} \quad a \in A. \tag{1.9}$$

In this case, $\mathbf{Y}$ is said to be *S-equivalent* to $\mathbf{X}$.

In general, an S-representation $\mathbf{Y}$ of A is said to be *realizable* in R if there exist an R-representation $\mathbf{X}$ of A and a nonsingular matrix $P \in \mathrm{GL}_n(S)$ satisfying (1.9).

In the case that S is the quotient field of R, the following holds.

Lemma 1.5. *Let K be the quotient field of R, and suppose that A is a finitely generated R-algebra. Let M be a finitely generated A^K-module and V be a finitely generated A-submodule of M. If V contains a K-basis of M, then V is an R-form of M.*

Proof. Clearly, V is R-torsion-free, and a subset of V is R-free if and only if it is linearly independent over K. Therefore our assumption implies that $\operatorname{rank}_R V = \dim_K M$, and if $\{v_1, \ldots, v_n\}$ is an R-basis of V, then this is also a K-basis of M. Thus we have

$$M = \bigoplus_{i=1}^{n} K v_i \simeq V^K = K \otimes_R V = \bigoplus_{i=1}^{n} K(1 \otimes v_i),$$

namely, V is an R-form of M. ∎

Theorem 1.6. *Let R, K, A be as above. Then the following holds.*

(i) *Any representation module M of A^K has an R-form.*

(ii) *Any K-representation of A is K-equivalent to some R-representation of A.*

Proof. It suffices to show (i). Let $A = \sum_{i=1}^{n} Ra_i$ and let $\{m_1, \ldots, m_n\}$ be a K-basis of M. Then $V = \sum_{i,j} R(m_i a_j)$ is an A-submodule of M that contains a K-basis of M. Therefore V is an R-form of M. ■

We now assume for the remainder of this section that

(1.10) *R is a discrete valuation ring with maximal ideal* (π).

Moreover, we let K and $F = R/(\pi)$ denote the quotient field and the residue field of R, respectively. The image of $a \in R$ under the natural map $R \to F$ is denoted by a^*. More generally, for an R-module V, V^* denotes the F-module $V/\pi V$, and v^* indicates the image of $v \in V$ under the natural map $V \to V^*$.

Exercise 1.7. Let A be an R-algebra and M be an A^K-module. If two A-submodules V and W of M are both R-forms of M, then there exists a natural number m such that $\pi^m V \subset W$.

If V_R is R-free, then $V = R \otimes_R V \to V^K = K \otimes_R V (v \mapsto 1 \otimes v)$ is a monomorphism (cf. Chapter 1, 10.3), and we understand that $V \subset V^K$ by identifying v with $1 \otimes v$. Also if W is an R-submodule of V, we denote by KW the K-submodule of V generated by W over K.

Exercise 1.8. Let $W \subset V$ be R-lattices. Then the following five conditions are equivalent.

(1) W is pure in V.
(2) $W = KW \cap V$.
(3) $W = U \cap V$ for some K-submodule U of V^K.
(4) $\alpha W = W \cap \alpha V$ for all $\alpha \in R$.
(5) $\pi W = W \cap \pi V$.

Now, let A be a finitely generated R-algebra and let $\mathbf{M}$ be a K-representation of A with representation module M_{A^K}. Take an R-form, say V, of M and consider the R-representation $\mathbf{X}$ of A defined by V relative to some R-basis $\{v_i\}$ of V. So $\mathbf{M} \sim_K \mathbf{X}$. By denoting by $X(a)^*$ the F-matrix obtained from $X(a)$ by reducing each entry mod(π), we get an F-representation

$$\mathbf{X}^*\colon A^* \to M_n(F) \qquad (a \mapsto X(a)^*)$$

of A^*, which is defined by the A^*-module V^* relative to the basis $\{v_i^*\}$.

Thus we have constructed an F-representation $\mathbf{X}^*$ of A^* from a K-representation $\mathbf{M}$ of A^K via a certain R-representation $\mathbf{X}$ that is K-equivalent to $\mathbf{M}$. But an R-form of M is not unique in general, not even up to isomorphism. Therefore, different choices of R-forms of M may induce nonequivalent F-representations in general (cf. Exercise 1.10 below). However, the set of irreducible constituents of $\mathbf{X}^*$ is uniquely determined by $\mathbf{M}$, as will be shown in the following theorem.

Theorem 1.9. *Let A be as above and let* $\mathbf{M}$, $\mathbf{N}$ *be K-representations of A with representation modules M_{A^K}, N_{A^K}, respectively. Let V_A, W_A, respectively, be R-forms of M, N which define R-representations* $\mathbf{X}$, $\mathbf{Y}$ *of A. Then the following holds.*

(i) $M \simeq N \Rightarrow V^* \leftrightarrow W^*$.

(ii) $\mathbf{X} \sim_K \mathbf{Y} \Rightarrow \mathbf{X}^* \leftrightarrow \mathbf{Y}^*$.

Proof. (Serre[1]). It suffices to show the first assertion. We may assume that $M = N$. Then there exists an integer m such that $\pi^m W \subset V$. But since $\pi^m W$ is A-isomorphic with W, we may also assume that $W \subset V$ by replacing $\pi^m W$ with W. Take an integer n such that $\pi^n V \subset W$. We shall prove that $V^* \leftrightarrow W^*$ by induction on n.

If $n = 1$, $\pi W \subset \pi V \subset W \subset V$, and we have the following two exact sequences:

$$0 \to W/\pi V \to V/\pi V \to V/W \to 0,$$

$$0 \to \pi V/\pi W \to W/\pi W \to W/\pi V \to 0.$$

From this, $V^* \leftrightarrow W^*$, since $V/W \simeq \pi V/\pi W$.

We next suppose that $n > 1$ and set $U = \pi^{n-1} V + W$. Then U is also an R-form of M, which satisfies that $\pi^{n-1} V \subset U \subset V$ and $\pi U \subset W \subset U$. Hence $V^* \leftrightarrow U^* \leftrightarrow W^*$ by the inductive hypothesis. ■

Exercise 1.10. Suppose that K is the 2-adic field with R as its valuation ring. Let $A = RG$ be the group ring of the group $G = \langle\sigma\rangle$ of order 2. There are two representations **X**, **Y** of degree 2 such that

$$X(\sigma) = \begin{pmatrix} 1 & 0 \\ 0 & -1 \end{pmatrix}, \qquad Y(\sigma) = \begin{pmatrix} 1 & 1 \\ 0 & -1 \end{pmatrix}.$$

Then the following holds.

(i) **X** and **Y** are K-equivalent but not R-equivalent.

(ii) **X*** and **Y*** are not F-equivalent.

2. Algebras over Fields

Throughout the remainder of this chapter, we assume, unless otherwise specified explicitly, that

All algebras and their modules are finitely generated over the coefficient rings under consideration.

Let K be a field and A be a K-algebra. Since A is Artinian, it has a direct sum decomposition

$$A = \bigoplus_{i=1}^{k} \bigoplus_{\mu=1}^{n_i} e_{i\mu} A, \tag{2.1}$$

where $1 = \sum_{i,\mu} e_{i\mu}$ is a primitive idempotent decomposition of 1, and it is assumed that $e_{i\mu}A \simeq e_{j\nu}A \Leftrightarrow i = j$ (see Chapter 1, Section 8.1).

Corresponding to the above, there exists an irreducible decomposition of $\bar{A} = A/J(A)$ such that

$$\bar{A} = \bigoplus_{i=1}^{k} \bigoplus_{\mu=1}^{n_i} \bar{e}_{i\mu} \bar{A}, \tag{2.2}$$

with $\bar{e}_{i\mu}\bar{A} \simeq \bar{e}_{j\nu}\bar{A}$, if and only if $i = j$. Set $e_i = e_{i1}$ for brevity and let IRR(A) be a complete set of representatives for the isomorphism classes of irreducible A-modules. Then we have

$$\mathrm{IRR}(A) = \{\bar{e}_1\bar{A}, \ldots, \bar{e}_k\bar{A}\}.$$

Thus, if we denote by

$$\mathbf{F}_1, \ldots, \mathbf{F}_k$$

the K-representations of A defined by $\bar{e}_1\bar{A}, \ldots, \bar{e}_k\bar{A}$, respectively, then they are representatives for the equivalence classes of irreducible representations of A.

In view of Chapter 1, 8.1, every representation of A is completely reducible if and only if A is semisimple. It also holds that a K-representation $\mathbf{X}$ of A is completely reducible if and only if $X(J(A)) = 0$ (cf. Chapter 1, 8.2). Thus a completely reducible representation of A is actually considered as that of $\bar{A}$.

With the notation of (2.2), let $\bar{A}_i = \bigoplus_{\mu=1}^{n_i} \bar{e}_{i\mu}\bar{A}$. Then each $\bar{A}_i$ is a simple algebra, and

$$\bar{A} = \bar{A}_1 \oplus \bar{A}_2 \oplus \cdots \oplus \bar{A}_k \tag{2.3}$$

is the direct sum of the simple components $\bar{A}_i$ of $\bar{A}$. $D_i = \mathrm{End}_A(\bar{e}_i\bar{A})$ is a division ring such that

$$\bar{A}_i \simeq M_{n_i}(D_i),$$

which are often identified with each other. If f_i is the degree of $\mathbf{F}_i$, then

$$f_i = n_i(\dim_K D_i). \tag{2.4}$$

A K-algebra that is a division ring such as D_i above is called a *division algebra* (over K).

The following result is obvious.

Theorem 2.1. *The following three conditions are equivalent for an irreducible representation* $\mathbf{F}$ *of* A *(respectively, an irreducible* A*-module* V*).*

(1) $\mathbf{F} \sim \mathbf{F}_i$ ($V \simeq \bar{e}_i\bar{A}$, respectively).
(2) $F(\bar{A}_i) \neq 0$ ($V\bar{A}_i \neq 0$, respectively).
(3) If $j \neq i$, then $F(\bar{A}_j) = 0$ ($V\bar{A}_j = 0$, respectively).

As is seen in the above theorem, every irreducible representation $\mathbf{F}$ of A is actually regarded as that of some simple component, say $\bar{A}_i$, of $\bar{A}$, in which case $\mathbf{F}$ is said to belong to $\bar{A}_i$.

Given a representation $\mathbf{X}$ of A, let us consider the irreducible constituents of it and write

$$\mathbf{X} \leftrightarrow \bigoplus_{i=1}^{k} m_i \mathbf{F}_i.$$

Therefore, if χ_i denotes the character defined by $\mathbf{F}_i$, then

$$\chi_{\mathbf{X}} = \sum_{i=1}^{k} m_i \chi_i.$$

Each m_i is referred to as the *multiplicity* of $\mathbf{F}_i$ (or respectively, χ_i) in $\mathbf{X}$ (or respectively, in $\chi_{\mathbf{X}}$). Of course $\mathbf{F}_i$ is an irreducible constituent of $\mathbf{X}$ if and only if $m_i > 0$. The module theoretic analogue of this notion is defined quite similarly.

A character defined by an irreducible representation is called an *irreducible character.*

Theorem 2.2. *If* Char $K = 0$, *then the following holds for two representations* $\mathbf{X}$, $\mathbf{Y}$ *of* A:

$$\mathbf{X} \leftrightarrow \mathbf{Y} \Leftrightarrow \chi_{\mathbf{X}} = \chi_{\mathbf{Y}}.$$

In particular, if $\mathbf{X}$ *and* $\mathbf{Y}$ *are completely reducible, then each of the above conditions is equivalent to* $\mathbf{X} \sim \mathbf{Y}$.

Proof. The implication ($\Rightarrow$) is clear, and so we show the converse. Let $\bar{c}_i$ be the identity of $\bar{A}_i$, and let χ_i be the character defined by $\mathbf{F}_i$. Then $F_i(c_j) = \delta_{ij}I$. Thus, if $\chi_{\mathbf{X}} = \sum_i m_i \chi_i$, then we have $\chi_{\mathbf{X}}(c_i) = m_i f_i$. Therefore the multiplicity m_i is determined by the character $\chi_{\mathbf{X}}$ and the assertion follows. The second half is obvious. ∎

The representation defined by A_A is called the right *regular representation* and is denoted by $\mathbf{\Gamma}$, and the representation defined by ${}_AA$ is called the left regular representation, denoted by $\mathbf{\Gamma}'$.

Theorem 2.3. *The representation* $\mathbf{F}_i'$ *of* A *defined by* $\bar{A}\bar{e}_i$ *is equivalent to* $\mathbf{F}_i$.

Proof. By using the same notation as in the proof of the above theorem, we have $F_i'(c_i) = I$, whence it follows that $\mathbf{F}_i' \sim \mathbf{F}_i$. ∎

Each $\mathbf{F}_i$ is an irreducible constituent of the representations of A defined by $(A/J(A))_A$ and by ${}_A(A/J(A))$. This yields the following result.

Theorem 2.4. *Any irreducible representation of* A *is an irreducible constituent of the right and left regular representations of* A.

If, in particular, A is semisimple, we obtain the following from Theorem 2.3.

Theorem 2.5. *If A is semisimple, then the right and left regular representations are equivalent.*

Principal indecomposable representations of A are the representations defined by principal indecomposable A-modules. Thus, if $\mathbf{P}_i$ is the representation defined by $e_i A$, then

$$\{\mathbf{P}_1, \mathbf{P}_2, \ldots, \mathbf{P}_k\}$$

gives a complete set of representatives for equivalence classes of principal indecomposable representations of A.

Remark. In general $\mathbf{\Gamma}$ and $\mathbf{\Gamma}'$ are not equivalent. If they are equivalent, A is called a Frobenius algebra, which will be studied in Section 8 of this chapter.

3. Absolutely Irreducible Representations

Let A be an algebra over a field K and let L be an extension field of K. We first consider the relationship between the K-equivalence and the L-equivalence of given two representations of A. An L-representation of A extends naturally to that of A^L. Especially, if $\mathbf{X}$ is a K-representation of A with representation module V_A, then we have the L-representation $\mathbf{X}^L$ of A^L which is defined by V^L.

Theorem 3.1. *Let $\mathbf{X}$, $\mathbf{Y}$ be K-representations of A with representation modules V, W, respectively. If $\mathbf{X}$ and $\mathbf{Y}$ are L-equivalent, then they must be K-equivalent. In other words, we have the following*:

$$(V^L)_{A^L} \simeq (W^L)_{A^L} \Rightarrow V_A \simeq W_A.$$

Proof. Let n be the degree of $\mathbf{X}$ and $\{a_1, \ldots, a_s\}$ be a K-basis of A. By assumption, there exists $P = (\xi_{ij}) \in M_n(L)$, satisfying the following two conditions:

(3.1) $\qquad X(a_i)P = PY(a_i) \qquad (1 \leq i \leq s),$

(3.2) $\qquad \det P \neq 0.$

Case 1. L is an algebraic extension of K: Let L_1 be the subfield of L generated by $\{\xi_{ij}\}$ over K. Then L_1 is a finite extension of K, and **X** and **Y** are L_1-equivalent, thus $V^{L_1} \simeq W^{L_1}$. Let $\{\omega_1, \ldots, \omega_m\}$ be a K-basis of L_1. Then

$$V^{L_1} = L_1 \otimes_K V = \bigoplus_{i=1}^{m} \omega_i \otimes_K V.$$

holds. But since $\omega_i \otimes_K V \simeq V$ as A-modules, we find that $V^{L_1} \simeq mV$ as A-modules. Likewise, we have $W^{L_1} \simeq mW$ as A-modules. Then by the K–S–A theorem, we conclude that $V_A \simeq W_A$.

Case 2. $|K| = \infty$: The condition (3.1) can be interpreted as a system of linear equation on the n^2 indeterminates ξ_{ij} with coefficients in K. Since it has a nontrivial solution in L, it must have one in K, as is well known. Let $\{P_1, \ldots, P_r\}$ be a system of fundamental solutions of it with $P_i \in M_n(K)$ and let $x_1, \ldots, x_r$ be indeterminates. Then $\det(x_1P_1 + \cdots + x_rP_r) = f(x_1, \ldots, x_r)$ is a polynomial over K. This is a nonzero polynomial, since if we write $P = \sum_i \lambda_i P_i$ with $\lambda_i \in L$, then $f(\lambda_1, \ldots, \lambda_r) \neq 0$ by (3.2). But since K is infinite, we can find $\mu_1, \ldots, \mu_r \in K$ such that $f(\mu_1, \ldots, \mu_r) \neq 0$, using induction on r. Now by letting $Q = \sum_i \mu_i P_i$, we have $\det Q \neq 0$ and $Q^{-1}X(a)Q = Y(a)$ for all $a \in A$, and hence $\mathbf{X} \sim_K \mathbf{Y}$.

Case 3. Conclusion: Let $\tilde{K}$ be the algebraic closure of K in that of L. Then $\tilde{K}$ is an infinite field.

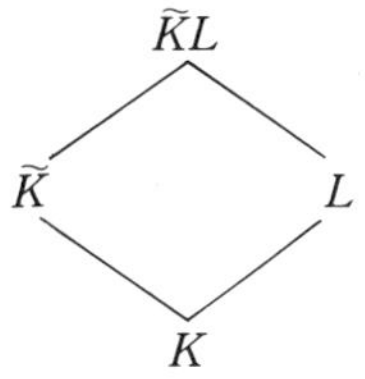

By assumption, **X** and **Y** are $\tilde{K}L$-equivalent and hence $\tilde{K}$-equivalent by the above step. Therefore they are K-equivalent as shown in Case 1. ∎

We continue to assume that A is an algebra over a field K and keep the notation of the preceding section.

Let **F** be an irreducible representation of A, and let V be a representation module for it. In general, **F** does not remain irreducible under field extensions. We say that **F** or V is *absolutely irreducible* if, for any extension field L of K, $\mathbf{F}^L$ remains irreducible, that is, V^L is irreducible as an A^L-module.

To give some conditions equivalent to the absolute irreducibility, we need a lemma.

Lemma 3.2. *Let D be a division algebra over K.*

(i) *If $K \subsetneqq D$, then there exists a finite extension L of K such that D^L is not a division ring.*

(ii) *If K is algebraically closed, then $D = K$.*

Proof.

(i) Let $a \in D - K$ and $L = K[a]$, so L is a finite extension field of K by 1.7 in Chapter 1. If $D^L = L \otimes_K D$ is a division ring, then the subring $L \otimes_K L$ is a field by the same reason as above. However the map $f: L \otimes_K L \to L(\sum_i a_i \otimes b_i \mapsto \sum a_i b_i)$ is an epimorphism as K-algebras, which fails to be an isomorphism by comparing the dimensions of both sides. This is a contradiction.

(ii) This is clear by the above, since no finite extension of K exists besides K. ■

Theorem 3.3. *Let $\mathbf{F}_i$ be an irreducible representation of A of degree n_i with representation module $\bar{e}_i\bar{A}$, where $\bar{A} = A/J(A)$ and $e_i \in \mathrm{pi}(A)$. Then the following five conditions are equivalent.*

(1) *$\mathbf{F}_i$ is absolutely irreducible.*

(2) *$\mathbf{F}_i^{\tilde{K}}$ is irreducible, where $\tilde{K}$ is any algebraically closed field containing K.*

(3) *$\mathrm{End}_A(\bar{e}_i\bar{A}) = K$.*

(4) *$\bar{A}_i \simeq M_{n_i}(K)$, where $\bar{A}_i$ is the simple component of $\bar{A}$ to which $\mathbf{F}_i$ belongs.*

(5) *$\mathbf{F}_i: A \to M_{n_i}(K)$ is an epimorphism.*

Proof.

(1) ⇒ (2). Clear from the definition.

(2) ⇒ (3). From the assumption, $\mathrm{End}_{A^{\tilde{K}}}((\bar{e}_i\bar{A})^{\tilde{K}})$ is a division algebra over the algebraically closed field $\tilde{K}$, and so it coincides with $\tilde{K}$. But since $\mathrm{End}_{A^{\tilde{K}}}((\bar{e}_i\bar{A})^{\tilde{K}}) \simeq \tilde{K} \otimes_K \mathrm{End}_A(\bar{e}_i\bar{A})$ (cf. Chapter 1, 11.12), we have $\mathrm{End}_A(\bar{e}_i\bar{A}) = K$ by comparing the dimensions over K.

(3) ⇒ (4). This is clear.

(4) ⇒ (5). As $\bar{A}_i$ is simple, $\mathbf{F}_i: \bar{A}_i \to M_{n_i}(K)$ is a monomorphism, therefore it must be an isomorphism by comparing the dimensions over K.

(5) ⇒ (1). Clearly $\mathbf{F}_i$ induces an isomorphism $\bar{A}_i \xrightarrow{\sim} M_{n_i}(K)$. Thus we have $\bar{A}_i^L \simeq M_{n_i}(L)$ for any extension field L of K and $\bar{A}_i^L \simeq n_i W$ with irreducible A^L-module W. On the other hand, since $\bar{A}_i \simeq n_i(\bar{e}_i\bar{A})$, it follows that $\bar{A}_i^L \simeq n_i(\bar{e}_i\bar{A})^L$. Thus $(\bar{e}_i\bar{A})^L$ is isomorphic to W and hence irreducible. ■

As an immediate consequence of the above theorem we have

Corollary 3.4. *If K is algebraically closed, then every irreducible representation of a K-algebra A is absolutely irreducible.*

The following result is also referred to as *Schur's lemma.*

Lemma 3.5 (Schur). *Let $\mathbf{F}$ be an absolutely irreducible representation of A of degree n.*

(i) *Let $P \in M_n(K)$. If $PF(a) = F(a)P$ for all $a \in A$, then $P = \lambda I$, where $\lambda \in K$ and I denotes the identity matrix.*

(ii) *If z is an element of the center $Z(A)$ of A, then $F(z) = \zeta(z)I$. Thus $\zeta: Z(A) \to K(z \mapsto \zeta(z))$ is a one-dimensional representation of $Z(A)$.*

In particular, if A is commutative, then any absolutely irreducible representation of A is one-dimensional.

Proof. It holds that $F(A) = M_n(K)$ by Theorem 3.3(5), whence we have (i). Part (ii) is immediate from (i). ■

Lemma 3.6. *Suppose that K is algebraically closed.*

(i) *Given a K-representation $\mathbf{X}$ of A and $a \in A$, there exists a representation $\mathbf{X}'$ equivalent to $\mathbf{X}$ such that $X'(a)$ is a lower triangular matrix:*

$$X'(a) = \begin{pmatrix} \lambda_1 & & & 0 \\ & \lambda_2 & & \\ & & \ddots & \\ * & & & \lambda_n \end{pmatrix}. \tag{3.3}$$

In particular, if a is nilpotent, then $\chi_{\mathbf{X}}(a) = 0$.

(ii) *Let $\mathbf{F}$ be an irreducible representation of A and let $e \in \mathrm{pi}(A)$ be such that $F(e) \neq 0$. Then there exists a representation $\mathbf{F}'$ equivalent to $\mathbf{F}$ such that*

$$F'(e) = \begin{pmatrix} 1 & & & 0 \\ & 0 & & \\ & & \ddots & \\ 0 & & & 0 \end{pmatrix}. \tag{3.4}$$

In particular, we have $\chi_{\mathbf{F}}(e) = 1$.

Proof.

(i) Let $B = K[a]$. Then B is a commutative subalgebra of A. Thus there exists a nonsingular matrix P such that $P^{-1}X(a)P$ is a matrix of the form written in (3.3), since the irreducible representations of B are all one-dimensional. If a is nilpotent, then all λ_i must be zero.

(ii) Let $\bar{A} = A/J(A)$, then $\bar{e}\bar{A}$ affords $\mathbf{F}$, and we have $\bar{e}\bar{A} = \bar{e}\bar{A}\bar{e} \oplus \bar{e}\bar{A}(1 - \bar{e})$ as $K[e]$-modules. Since $\bar{e}\bar{A}\bar{e} = K\bar{e}$ is one-dimensional, we can choose a basis $\{\bar{a}_1, \ldots, \bar{a}_n\}$ of $\bar{e}\bar{A}$ so that $\bar{a}_1 = \bar{e}$ and $\bar{a}_i \in \bar{e}\bar{A}(1 - \bar{e})$ for all $i \geq 2$. If $\mathbf{F}'$ is the representation relative to this basis, then $F'(e)$ is of the form given in (3.4). ∎

Using the above lemma, we prove the following theorem.

Theorem 3.7. *Let I be a right ideal of A. If I has a K-basis consisting of nilpotent elements, then I is nilpotent.*

Proof. Let $\tilde{K}$ be the algebraic closure of K. It suffices to show that $I^{\tilde{K}}$, for which our assumption is clearly retained, is nilpotent. Therefore we may assume that $K = \tilde{K}$. If $\bar{I} = (I + J(A))/J(A) = 0$, then $I \subset J(A)$. Thus, we need only to show that $I = 0$, assuming $J(A) = 0$.

If $I \neq 0$, then I contains a primitive idempotent, say e, of A. Consider a representation $\mathbf{F}$ of A afforded by eA. Then we have $\chi_{\mathbf{F}}(I) = 0$ by (i) of the above lemma, while $\chi_{\mathbf{F}}(e) = 1$ by the second part of it, which is a contradiction. ∎

Concerning the multiplicity of an absolutely irreducible A-modules, we have the following.

Theorem 3.8. *Let $e \in \mathrm{pi}(A)$ and assume that $\bar{e}\bar{A}$ is absolutely irreducible. Then the multiplicity of $\bar{e}\bar{A}$ in a given A-module V is equal to $\dim_K Ve = \dim_K \mathrm{Hom}_A(eA, V)$.*

Proof. Remember that the multiplicity of $\bar{e}\bar{A}$ in V is equal to the composition length of the eAe-module Ve (see Chapter 1, 8.11). Now, we see that

$$eAe/J(eAe) = eAe/eAe \cap J(A) \simeq \bar{e}\bar{A}\bar{e} = K;$$

hence the unique irreducible eAe-module is one-dimensional. Therefore the composition length of $(Ve)_{eAe}$ equals $\dim_K Ve$. ■

Lemma 3.9. *Let L be an extension field of a field K.*

(i) *If M is an irreducible A^L-module, then M is an irreducible constituent of some $(\bar{e}_i\bar{A})^L$, and such i is unique* $(1 \leq i \leq k)$.

(ii) *Let U, V be irreducible A-modules. Then U^L and V^L have no irreducible constituent in common unless $U \simeq V$. In particular, it holds that* $|\mathrm{IRR}(A)| \leq |\mathrm{IRR}(A^L)|$.

Proof.

(i) If $A \leftrightarrow \bigoplus_i m_i(\bar{e}_i\bar{A})$, then $A^L \leftrightarrow \bigoplus_i m_i(\bar{e}_i\bar{A})^L$, and hence M is an irreducible constituent of some $(\bar{e}_i\bar{A})^L$. Moreover if we choose $c_j \in A$ so that $\bar{c}_j$ is the identity of $\bar{A}_j$, then $Mc_i \neq 0$ and $Mc_j = 0$ for all $j \neq i$. This proves the uniqueness of the above integer i.

(ii) This is clear from the above. ■

An extension field L of K is said to be a *splitting field* for A, provided the semisimple L-algebra $A^L/J(A^L)$ splits into a direct sum of full matrix rings over L. This is equivalent to any irreducible L-representation of A being absolutely irreducible.

Exercise 3.10.

(i) If L is a splitting field for A, then so is any extension field of L.

(ii) If L contains the algebraic closure of K, then L is a splitting field for A.

Theorem 3.11. *There exists a finite extension of K that is a splitting field for A.*

Proof. Let $\tilde{K}$ be the algebraic closure of K and let $\mathbf{M}_1, \ldots, \mathbf{M}_l$ be representatives for the equivalence classes of irreducible representations of $A^{\tilde{K}}$. Let furthermore $A = \bigoplus_{\nu=1}^{m} Ka_\nu$ and $M_p(a) = (\alpha_{ij}^{(p)}(a))$. If L denotes the field generated by $\{\alpha_{ij}^{(p)}(a_\nu); i, j, p, \nu\}$ over K, then L is a finite extension of K. We show that L is a splitting field for A. To do so, let $\mathbf{M}'_p$ be the restriction of $\mathbf{M}_p$

to A^L. Since $(\mathbf{M}'_p)^{\tilde{K}} = \mathbf{M}_p$, each $\mathbf{M}'_p$ is an absolutely irreducible L-representation of A, and $\mathbf{M}'_p$ and $\mathbf{M}'_q$ are equivalent only if $p = q$. Therefore by Lemma 3.9(ii), $\{\mathbf{M}'_1, \ldots, \mathbf{M}'_l\}$ gives a complete set of representatives for the equivalence classes of irreducible representations of A^L. ■

We also show the following fact.

Lemma 3.12. *Let L be an extension field of K and let $M_n(A)$ be the full matrix ring over A. Then*

$$L \text{ is a splitting field for } A \Leftrightarrow L \text{ is a splitting field for } M_n(A).$$

Proof. Note that $M_n(A)^L \simeq M_n(A^L)$, and $J(M_n(A^L)) = M_n(J(A^L))$ by Chapter 1, 3.11. Thus, if $A^L/J(A^L) \simeq \bigoplus_i M_{n_i}(D_i)$ with division algebras D_i, then

$$M_n(A)^L/J(M_n(A)^L) \simeq M_n(A^L/J(A^L)) \simeq \bigoplus_i M_{nn_i}(D_i),$$

whence the assertion follows easily. ■

The following fact holds for the characters of absolutely irreducible representations.

Theorem 3.13. *Suppose that $\mathbf{F}_1, \ldots, \mathbf{F}_r$ are absolutely irreducible representations of A, any two of which are not equivalent to each other, and let χ_i be the character defined by $\mathbf{F}_i$. Then $\chi_1, \ldots, \chi_r$ are linearly independent over K.*

Proof. We may assume that each $\mathbf{F}_i$ is defined by $\bar{e}_i\bar{A}$, where $e_i \in \mathrm{pi}(A)$. Then we have $\chi_i(e_j) = \delta_{ij}$ by Lemma 3.6(ii), whence the assertion follows easily. ■

If K is a splitting field for the K-algebra A, then A is called a *split K-algebra* or *a* K-algebra of the *split type*. The next result is concerned with the number of the irreducible representations of such an algebra.

Theorem 3.14. *Let A be a split K-algebra that is semisimple. Then the number of nonequivalent (absolutely) irreducible representations of A is equal to the dimension of $Z(A)$ over K.*

Proof. By assumption, each simple component of A is a full matrix ring over K:

$$A = A_1 \oplus \cdots \oplus A_k, \qquad A_i \simeq M_{n_i}(K).$$

It then follows that $Z(A) = Z(A_1) \oplus \cdots \oplus Z(A_k)$ and $Z(A_i) \simeq K$. Therefore, $k = \dim_K Z(A)$. ■

Remark. In the nonsemisimple case, the following result due to R. Brauer is known. Namely, let A be a split K-algebra. We let $S(A) = \sum_{a,b\in A} K[a,b]$, where $[a,b] = ab - ba$, and when Char $K = p > 0$, let $T(A) = \{a \in A;\ a^{p^n} \in S(A)$ for some $n\}$. Then the number k of nonequivalent irreducible representations of A is given as follows:

(i) $k = \dim_K A/(S(A) + J(A))$.

(ii) If Char $K = p > 0$, it holds that $T(A) = S(A) + J(A)$, and hence $k = \dim_K A - \dim_K T(A)$.

(See Curtis and Reiner[1], Section 83B for the proof.)

We close this section by studying irreducible representations of the tensor product of two split K-algebras.

Let A, B be K-algebras. We keep the notations of Section 2 for A, and for B, we let the decompositions corresponding to (2.1) and (2.2) be

$$B = \bigoplus_{i=1}^{l} \bigoplus_{\mu=1}^{m_i} f_{i\mu} B,$$

and

$$\bar{B} = B/J(B) = \bar{B}_1 \oplus \cdots \oplus \bar{B}_l, \qquad \bar{B}_i = \bigoplus_{\mu=1}^{m_i} \bar{f}_{i\mu} \bar{B},$$

respectively. Also we abbreviate f_{i1} and $\otimes_K$ to f_i and $\otimes$, respectively.

Note that $J(A) \otimes B + A \otimes J(B)$ is a nilpotent ideal of $A \otimes B$, hence $J(A) \otimes B + A \otimes J(B) \subset J(A \otimes B)$. And we have

$$A \otimes B/(J(A) \otimes B + A \otimes J(B)) \simeq \bar{A} \otimes \bar{B} = \bigoplus_{i,j} \bar{A}_i \otimes \bar{B}_j,$$

which is not semisimple in general. However, if both A and B are split K-algebras, then $\bar{A}_i \otimes \bar{B}_j = M_{n_i}(K) \otimes M_{m_j}(K) = M_{n_i m_j}(K)$ is a simple algebra and hence $A \otimes B$ is semisimple. In particular, it holds that $J(A \otimes B) = J(A) \otimes B + A \otimes J(B)$, and $A \otimes B$ is also a split K-algebra.

Theorem 3.15. *Let A, B be split K-algebras. If V_A, W_B are irreducible, then $V \otimes W$ is an (absolutely) irreducible $A \otimes B$-module. Conversely, every irreducible $A \otimes B$-module can be expressed as $V \otimes W$ for suitable irreducible modules V_A and W_B. Namely, the following holds.*

$$\mathrm{IRR}(A \otimes B) = \{V \otimes W;\ V \in \mathrm{IRR}(A),\ W \in \mathrm{IRR}(B)\},$$
$$|\mathrm{IRR}(A \otimes B)| = |\mathrm{IRR}(A)||\mathrm{IRR}(B)|.$$

Proof. Using the same notation as above, we may assume that $V = \bar{e}_i \bar{A}$ and $W = \bar{f}_j \bar{B}$. Then $\bar{A}_i \otimes \bar{B}_j \simeq n_i m_j V \otimes W$ and $\bar{A}_i \otimes \bar{B}_j \simeq M_{n_i m_j}(K)$, which is a direct sum of $n_i m_j$ irreducible $A \otimes B$-modules. Hence, $V \otimes W$ must be irreducible. Moreover, since

$$\overline{A \otimes B} = \bar{A} \otimes \bar{B} = \bigoplus_{i,\mu} \bigoplus_{j,\nu} \bar{e}_{i\mu} \bar{A} \otimes \bar{f}_{j\nu} \bar{B},$$

any irreducible $A \otimes B$-module is isomorphic to some $\bar{e}_i \bar{A} \otimes \bar{f}_j \bar{B}$. The remaining assertions are easy. ■

4. Simple Algebras

4.1. Central Simple Algebras

If A is a subring of a ring B, we denote by $C_B(A)$ the subset of B consisting of the elements that commute with every element of A:

$$C_B(A) = \{b \in B;\ ab = ba \text{ for all } a \in A\}.$$

This is clearly a subring of B, which is called the *centralizer* of A (in B).

Now let K be a field and A, B be K-algebras. We denote the tensor product of A and B over K by $A \otimes B$ as before. We understand that $A, B \subset A \otimes B$ by identifying $a \in A$ and $b \in B$ with $a \otimes 1_B$ and $1_A \otimes b$, respectively. Then every element of A and of B commute with each other, and it holds that $A \otimes B = AB$, where we set $AB = \{\sum_i a_i b_i;\ a_i \in A,\ b_i \in B\}$.

Lemma 4.1. *Let A, B be K-algebras, possibly infinite-dimensional over K. Let A_1 and B_1 be subrings of A and B, respectively, with identities in common. Then the following holds.*

(i) $C_{A\otimes B}(A_1) = C_A(A_1) \otimes B$. *In particular* $C_{A\otimes B}(A) = Z(A) \otimes B$.

(ii) $C_{A\otimes B}(A_1 \otimes B_1) = C_A(A_1) \otimes C_B(B_1)$.

(iii) $Z(A \otimes B) = Z(A) \otimes Z(B)$.

Proof.

(i) Let $\{b_i: i \in I\}$ be a K-basis of B, then every $x \in A \otimes B$ is written as a finite sum $x = \sum_i a_i \otimes b_i$ with uniquely determined $a_i \in A$. If $a \in A$, then $ax = \sum_i (aa_i) \otimes b_i$ and $xa = \sum_i (a_i a) \otimes b_i$, and hence

$$x \in C_{A\otimes B}(A_1) \Leftrightarrow a_i \in C_A(A_1) \text{ for all } i.$$

This yields our assertion.

(ii) Since $1_A \in A_1$ and $1_B \in B_1$, we see that $A_1, B_1 \subset A_1B_1$ and $A_1 \otimes B_1 = A_1B_1$. Now using (i), we have

$$\begin{aligned} C_{A\otimes B}(A_1 \otimes B_1) &= C_{A\otimes B}(A_1) \cap C_{A\otimes B}(B_1) \\ &= (C_A(A_1) \otimes B) \cap (A \otimes C_B(B_1)) \\ &= C_A(A_1) \otimes C_B(B_1). \end{aligned}$$

(iii) This follows at once from (ii), with $A_1 = A$ and $B_1 = B$. ■

Now, let $A = M_n(D)$ be a simple K-algebra with division algebra D belonging to it. We understand that $D \subset A$ by identifying $d \in D$ with dI, where I denotes the identity matrix. Then we see that $Z(A) = Z(D)$, which is a field. If $Z(A) = K$, then A is said to be a *central simple algebra* over K. A division algebra with center K is said to be a *central division algebra* over K. In general, any simple K-algebra A is considered as an algebra over its center $Z = Z(A)$, that is, a central simple algebra over the field Z.

Theorem 4.2. *Let A be a central simple algebra over K and let B be any K-algebra. Then the following holds. (We do not assume here that* $\dim_K B$ *is finite except for* (iii).)

(i) *Every ideal of* $A \otimes B$ *is of the form* $A \otimes I$ *with ideal I of B.*

(ii) *If B is simple, then* $A \otimes B$ *is simple. In particular,* A^L *is central simple over L for any extension field L of K, and if B is central simple over K, then so is* $A \otimes B$.

(iii) $J(A \otimes B) = A \otimes J(B)$.

Proof.

(i) Let $\{b_i; i \in I\}$ be a K-basis of B. Then $A \otimes B = \bigoplus_i A \otimes b_i$. Let us consider $A \otimes B$ as an (A, A)-bimodule via the natural action. Since ${}_AA_A$ is irreducible, it follows from the above that $A \otimes B$ is a direct sum of simple modules isomorphic to A. Thus, if M is an ideal of $A \otimes B$, then as an (A, A)-bimodule we have $M = \bigoplus_j M_j$ with ${}_A(M_j)_A \simeq {}_AA_A$. If $c_j \in M_j$ corresponds to 1_A under the isomorphism above, then $ac_j = c_ja$ holds for all $a \in A$, that is, $c_j \in C_{A \otimes B}(A) = Z(A) \otimes B = B$. Therefore, $M = \bigoplus_j A \otimes c_j = A \otimes I$, where $I = \bigoplus_j Kc_j$. Since $BM = MB = M$, I is an ideal of B.

(ii) The first assertion is clear from (i), and the rest follows from Lemma 4.1(iii).

(iii) Since $A \otimes J(B)$ is a nilpotent ideal of $A \otimes B$, the inclusion $J(A \otimes B) \supset A \otimes J(B)$ is clear. On the other hand, we see from the above that $J(A \otimes B) = A \otimes I$ for some ideal I of B. Then, as I must be nilpotent, it follows that $I \subset J(B)$, and hence $J(A \otimes B) \subset A \otimes J(B)$. ■

Our next objective is to study the centralizer of a simple subalgebra of a given central simple algebra over K. To this end, we need some preliminary lemmas.

Lemma 4.3. *Suppose that A is a simple K-algebra and let $A = M_r(D)$, where D is a division algebra. Let U be an irreducible A-module. Then, for an A-module V and $E = \mathrm{End}_A(V)$, the following holds.*

(i) *If $V \simeq sU$, then $E \simeq M_s(D)$. Also ${}_EV \simeq r({}_EM)$, where M denotes an irreducible left E-module.*

(ii) $(\dim_K A)(\dim_K E) = (\dim_K V)^2$.

(iii) *For $a \in A$ define $\varphi_a \colon V \to V$ by $\varphi_a(v) = va$. Then φ_a is an E-homomorphism, and we have the following isomorphism:*

$$\varphi \colon A \xrightarrow{\sim} \mathrm{End}_E(V) \qquad (a \mapsto \varphi_a).$$

Proof. Since $\mathrm{End}_A(U) \simeq D$, we have, by Chapter 1, 5.2, that $E \simeq M_s(D)$, which is a simple algebra. Also, $\dim_K V = rs \dim_K D$ as $\dim_K U = r \dim_K D$, whence part (ii) follows, and the second half of part (i) holds because $\dim_K M = s \dim_K D$. In (iii) it is clear that $\varphi_a \in \mathrm{End}_E(V)$ and φ is a monomorphism because A is a simple ring. Now we apply (i) to ${}_EV$ in place of V_A and have $\mathrm{End}_E(V) \simeq M_r(D) = A$. Thus by comparing the dimensions, we conclude that φ is an isomorphism. ■

If V is a faithful A-module, the representation $\mathbf{X}: A \to M_n(K)$ defined by V is a monomorphism. So we understand that $A \subset M_n(K)$ by identifying $a \in A$ with $X(a)$. Let $E = \mathrm{End}_A(V)$ and let E° be the opposite ring of it. Then V is a faithful (right) E°-module, and we understand that $E^\circ \subset M_n(K)$ via the representation of E° relative to the same basis of V as is used for $\mathbf{X}$. In this case, E° coincides with the centralizer of A in $M_n(K)$, and therefore we obtain the following from Lemma 4.3.

Lemma 4.4. *Let A be a simple subalgebra of $B = M_n(K)$ and let $A = M_r(D)$ with division algebra D. Then the following holds.*

(i) $C_B(A) \simeq M_s(D^\circ)$ *for some s.*

(ii) $(\dim_K A)(\dim_K C_B(A)) = \dim_K B = n^2$.

For a K-algebra A, let A° denote the opposite ring of A. Then the K-algebra $A^\circ \otimes A$, denoted by A^e, is called the *enveloping algebra* of A. A is an A^e-module via the action $x(a^\circ \otimes b) = axb$ for $a, b, x \in A$.

Lemma 4.5. *Let A be a central simple K-algebra of dimension n. Then the following isomorphism holds:*

$$A^e = A^\circ \otimes A \simeq M_n(K).$$

Proof. By Theorem 4.2(ii) A^e is simple. In particular, A is a faithful A^e-module. Hence there is a monomorphism $\varphi: A^e \to M_n(K)$, and we see that φ is an isomorphism by comparing the dimensions. ■

We are now ready to prove the following fundamental theorem on simple algebras.

Theorem 4.6. *Let B be a central simple K-algebra and A be a simple subalgebra of B with common identity. Then the following statements are valid.*

(i) *$C_B(A)$ is a simple algebra and the division algebra belonging to it is isomorphic to the division algebra belonging to $B \otimes A^\circ$.*

(ii) $(\dim_K A)(\dim_K C_B(A)) = \dim_K B$.

(iii) $C_B(C_B(A)) = A$.

(iv) $Z(A) = A \cap C_B(A) = Z(C_B(A))$.

(v) *If A is also central simple over K, then so is $C_B(A)$, and moreover, $A \otimes C_B(A) \simeq B$.*

(vi) (**Skolem–Noether**) *Every K-algebra homomorphism $\sigma\colon A \to B$ extends to an inner automorphism of B, i.e., there exists $b \in B^\times$ such that $\sigma(a) = b^{-1}ab$ for all $a \in A$.*

Proof. We have already proved (i) and (ii) for the case $B = M_n(K)$ in Lemma 4.4. To prove them for the general case, we may assume that $B^e = B^\circ \otimes B = M_n(K) \supset B^\circ \otimes A$. Then $B^\circ \otimes A$ is simple since B° is central simple, and by Lemma 4.1 we have $C_{B^e}(B^\circ \otimes A) = Z(B^\circ) \otimes C_B(A) = C_B(A)$, whence (i) and (ii) follow immediately from Lemma 4.4.

(iii) Since $A \subset C_B(C_B(A))$, it suffices to show that they have the same dimension over K. But this is easily deduced from the equality (ii), with A replaced by $C_B(A)$.

(iv) The first equality is obvious, and the second one follows from (iii).

(v) The first assertion is clear by (iv). Consider the homomorphism: $A \otimes C_B(A) \to B(\sum_i a_i \otimes c_i \mapsto \sum a_i c_i)$. Since $A \otimes C_B(A)$ is simple, we see that this is in fact an isomorphism by the comparison of their dimensions.

(vi) If V_B is an irreducible B-module, $D = \mathrm{End}_B(V)$ is a central division algebra such that $\mathrm{End}_D(V) = B$ by Lemma 4.3. Hence $D^\circ \otimes A$ is simple, and V is a $D^\circ \otimes A$-module; $v(d^\circ \otimes a) = (dv)a(v \in V,\ d \in D,\ a \in A)$.

On the other hand, we let $V^{(\sigma)}$ be the $D^\circ \otimes A$-module obtained from the same K-module V with the action defined by

$$v \cdot (d^\circ \otimes a) = (dv)\sigma(a).$$

Then, since $D^\circ \otimes A$ is simple, there exists a $D^\circ \otimes A$-isomorphism $\varphi\colon V \xrightarrow{\sim} V^{(\sigma)}$. This implies that $\varphi(dv) = \varphi(v(d^\circ \otimes 1_A)) = \varphi(v) \cdot (d^\circ \otimes 1_A) = d\varphi(v)$, i.e. $\varphi \in \mathrm{End}_D(V) = B$. Thus, there exists $b \in B$ such that $\varphi(v) = vb$ for all $v \in V$. Note that b is a unit of B, since φ is an isomorphism. Now we have, for $v \in V$, $a \in A$,

$$vab = \varphi(va) = \varphi(v(1^\circ \otimes a)) = \varphi(v) \cdot (1^\circ \otimes a) = \varphi(v)\sigma(a) = vb\sigma(a),$$

whence it follows that $ab = b\sigma(a)$ and hence $\sigma(a) = b^{-1}ab$. ■

One obtains, as a direct consequence of (vi) of Theorem 4.6, that an algebra endomorphism of the central simple K-algebra B must be an inner automorphism. This is also true for a matrix ring over a commutative local ring, as will be shown below.

Let R be a commutative local ring with maximal ideal $\mathfrak{m}$ and residue field $F = R/\mathfrak{m}$. For an R-module V, we let $V^* = V/\mathfrak{m}V$ and v^* denote the image of $v \in V$ under the natural map $V \to V^*$. If $\varphi: V \to W$ is an R-homomorphism, φ^* denotes the induced F-homomorphism $V^* \to W^*$.

Lemma 4.7. *Let R be as above and let $A = M_n(R)$. The set of $1 \times n$ matrices over R, denoted by R^n, is naturally considered as a right A-module. If an A-module V is R-free of rank n, then V is A-isomorphic to R^n.*

Proof. Since $A^* = A/\mathfrak{m}A = M_n(F)$ is a simple algebra over F, there exists an A^*-isomorphism $\lambda: F^n \xrightarrow{\sim} V^*$.

$$\begin{array}{ccc} R^n & \xrightarrow{*} & F^n \\ \varphi \downarrow & & \downarrow \lambda \\ V & \xrightarrow{*} & V^* \longrightarrow 0 \end{array}$$

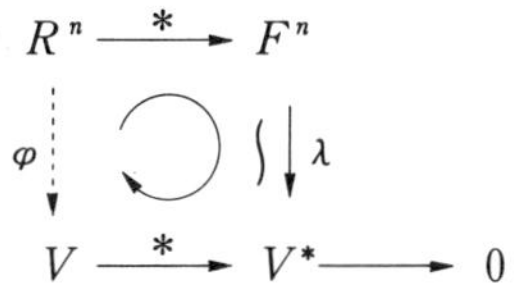

Then, since R^n is A-projective, there exists an A-homomorphism $\varphi: R^n \to V$, which completes the above diagram to a commutative one. But φ must be an isomorphism because R^n and V are projective covers of V^* as R-modules (Chapter 1, 10.14). ∎

Theorem 4.8. *Let R be a commutative local ring, and $A = M_n(R)$. Then any R-algebra endomorphism σ of A is an inner automorphism.*

Proof. Let R^n be the right A-module as above. Now, let us set $V = R^n$ and view it as a right A-module via $v \cdot a = v\sigma(a)$. Then by the above lemma there exists an A-isomorphism $\varphi: R^n \xrightarrow{\sim} V$. Since $\varphi \in \mathrm{End}_R(R^n) = M_n(R) = A$, there is $a \in A$ such that $\varphi(v) = va$ for all $v \in R^n$, where a must be a unit of A because φ is an isomorphism. Observe that, for $v \in R^n$ and $x \in A$,

$$vxa = \varphi(vx) = \varphi(v) \cdot x = \varphi(v)\sigma(x) = va\sigma(x),$$

whence it follows that $xa = a\sigma(x)$ and thus $\sigma(x) = a^{-1}xa$. ∎

4.2. Brauer Groups

Two central simple algebras A and B over K are said to be *similar*, provided the division algebras belonging to them respectively are isomorphic. In that case we write $A \sim B$. This defines an equivalence relation. Each equivalence class is called an *algebra class* and is denoted by (A) if it contains A.

Lemma 4.9. *The following holds for central simple K-algebras.*
(i) $A \sim A' \Leftrightarrow M_m(A) \simeq M_n(A')$ *for some* m, n.
(ii) *If* $A \sim A'$ *and* $B \sim B'$, *then* $A \otimes B \sim A' \otimes B'$.

Proof.
(i) $A = M_r(D)$, $A' = M_s(D')$, where D, D' are division algebras.
($\Rightarrow$). If $D \simeq D'$, then $M_s(A) \simeq M_{rs}(D) \simeq M_{rs}(D') \simeq M_r(A')$.
($\Leftarrow$). If $M_{mr}(D) \simeq M_{ns}(D')$, then $D \simeq D'$ by Chapter 1, 8.7.
(ii) If $M_m(A) \simeq M_n(A')$ and $M_r(B) \simeq M_s(B')$, then $M_m(A) \otimes M_r(B) \simeq M_n(A') \otimes M_s(B')$. Therefore we have $M_{mr}(A \otimes B) \simeq M_{ns}(A' \otimes B')$. ■

Define the multiplication of (A) and (B) by $(A) \times (B) = (A \otimes B)$. This is well defined by virtue of the above lemma. Moreover, it holds that

$$(A) \times (K) = (K) \times (A) = (A), \qquad (A^\circ) \times (A) = (A) \times (A^\circ) = (K).$$

Thus the set of algebra classes forms a group, with (K) as identity, and the inverse of (A) is given by (A°). This is called the *Brauer group* (over K) and is denoted by $\mathscr{B}(K)$.

Exercise 4.10. Let A, B be central simple K-algebras and let L be an extension field of K. Then the following hold for central simple L-algebras A^L, B^L (cf. 4.2(ii)).
(i) If $A \sim B$, then $A^L \sim B^L$.
(ii) L is a splitting field for A if and only if $A^L \sim L$.

By the above exercise, L can be said to be a splitting field for the algebra class (A) if it is a splitting field for the central simple K-algebra A. If so, then $A^L \simeq M_n(L)$ for some n. Since $\dim_K A = \dim_L A^L = n^2$, we obtain the following.

Theorem 4.11. *The dimension of a central simple K-algebra over K is a square.*

On the other hand, the following holds concerning a splitting field for a central simple algebra.

Theorem 4.12. *Let A be a central simple K-algebra. Then the following holds.*

(i) *Any maximal commutative subfield of A is a splitting field for A.*

(ii) *Let L be a finite extension of K, which is a splitting field for A. Then there exists a central simple K-algebra B such that $B \supset L$, $B \sim A$ and $C_B(L) = L$. In particular, L is a maximal commutative subfield of B.*

Proof.

(i) Let L be a maximal commutative subfield of A. Then by Theorem 4.6 $C_A(L)$ is a simple algebra with center $Z(C_A(L)) = L$, and the division algebra D belonging to it is isomorphic with the one belonging to A^L. We may assume that $L \subset D$. But if $D \neq L$, then $L[a]$, where $a \in D - L$, is a subfield of $D(\subset A)$ properly containing L (Chapter 1, 1.7). This contradicts the maximality of L, and hence we have $D = L$. Therefore A^L is a full matrix ring over L.

(ii) By assumption $A^L \simeq M_n(L)$. If V is an irreducible A^L-module, then $\dim_L V = n$ and $\dim_K V = ln$, where $l = \dim_K L$. Let us consider the K-endomorphism ring E of V: $E = \mathrm{End}_K(V) \simeq M_{ln}(K)$. This contains both A° and L in the sense that any element of them induces a natural K-endomorphism of V, and it holds that $C_E(A^\circ) \supset L$ in E. Let $B = C_E(A^\circ)$, then B is a central simple K-algebra similar to A and $\dim_K B = l^2$ by Lemma 4.4. And by Theorem 4.6, we have $(\dim_K L)(\dim_K C_B(L)) = \dim_K B = l^2$; hence, $\dim_K C_B(L) = l$. This proves that $C_B(L) = L$, since $C_B(L) \supset L$. ∎

We conclude this section with the following theorem.

Theorem 4.13. *Let $A = M_n(D)$ be a central simple K-algebra, where D is a division algebra. Then the following holds.*

(i) *There exists a maximal commutative subfield L that is separable over K.*

(ii) *There exists a finite Galois extension E of K that is a splitting field for A.*

Proof. It suffices to show the existence of L satisfying (i). Indeed, a minimal Galois extension containing such L will satisfy (ii). We may assume Char $K = p > 0$.

Let L be a maximal one among the commutative subfields of D that are separable over K. Let $C = C_D(L)$. If $C = L$, then L must be a maximal subfield of D. We deduce a contradiction by assuming that $C \supsetneqq L$. Let $c \in C - L$. Then $L[c]$ is a subfield of D and c^{p^e} is separable over L for some $e \geq 0$ (see Hungerford [1], p. 283). Hence $c^{p^e} \in L$ by the maximality of L and $e > 0$ as $c \notin L$. Take the smallest integer $e \geq 1$ such that $c^{p^e} \in L$ and let $d = c^{p^{e-1}}$. Then $d \notin L$ and $d^p \in L$. Let σ be the inner automorphism of C defined by d, i.e., $\sigma(x) = d^{-1}xd$ for $x \in C$. Then $\sigma^p = 1$ and hence $(\sigma - 1)^p = 0$. Note that $\sigma - 1 \neq 0$, since $L = Z(C)$ by Theorem 4.6(iv). Let k be the largest integer such that $(\sigma - 1)^k \neq 0$ and choose $u \in C$ so that $(\sigma - 1)^k u \neq 0$. If we set $v = (\sigma - 1)^k u$ and $w = (\sigma - 1)^{k-1}u$, then $(\sigma - 1)v = 0$ from the choice of k, and $\sigma(w) = v + w$ because $v = (\sigma - 1)w$. Hence by letting $t = v^{-1}w$, we have

$$\sigma(t) = \sigma(v)^{-1}\sigma(w) = v^{-1}(v + w) = 1 + t.$$

Thus σ induces a nontrivial L-automorphism of $L[t]$. This implies that the separability degree of $L[t]$ over L is larger than one. Thus there exists a subfield of $L[t]$ different from L that is separable over L (see Hungerford [1], Chapter V, Section 6). This contradicts the maximality of L. ■

5. Separable Algebras

Let A be a K-algebra and let L be an extension field of K. Since $J(A)^L$ is a nilpotent ideal of A^L, it holds that $J(A)^L \subset J(A^L)$. However, they do not necessarily coincide.

A semisimple K-algebra A is said to be *separable* over K, provided A^L remains semisimple for any extension field L of K.

Lemma 5.1. *Let A be a K-algebra.*

(i) *If A is a split K-algebra that is semisimple, then A is separable over K.*

(ii) *$\bar{A} = A/J(A)$ is separable if and only if $J(A^L) = J(A)^L$ for any extension field L of K. If this is the case, then, for a completely reducible A-module V, V^L is also completely reducible as an A^L-module.*

Proof.

(i) If $A \simeq \bigoplus_i M_{n_i}(K)$, then $A^L \simeq \bigoplus_i M_{n_i}(L)$ for any field $L \supset K$. Thus A^L is semisimple.

(ii) We know that $J(A)^L \subset J(A^L)$, and the equality holds if and only if $\bar{A}^L = A^L/J(A)^L$ is semisimple. Thus, if $J(A)^L = J(A^L)$, then $V^L J(A^L) = V^L J(A)^L = 0$, since $VJ(A) = 0$ by assumption. Therefore V^L is completely reducible. ■

Lemma 5.2. *Let A be a K-algebra and let $\tilde{K}$ be the algebraic closure of K. If $A^{\tilde{K}}$ is semisimple, then A must be separable.*

Proof. Let L be any field containing K and $\tilde{L}$ be the algebraic closure of L. We may assume that $\tilde{L} \supset \tilde{K}$. By assumption, $A^{\tilde{K}}$ is separable over $\tilde{K}$, hence $A^{\tilde{L}}$ is semisimple and A^L must be semisimple. ■

If A is a field, then the separability over K given above coincides with the so-called separability of field extensions, i.e., we have the following theorem.

Theorem 5.3. *Let Z be a finite extension field of K. Then Z is separable as a K-algebra if and only if Z is a separable extension of K, that is, there exist exactly $[Z:K]$ distinct K-monomorphisms $Z \to \tilde{K}$, where $\tilde{K}$ denotes the algebraic closure of K.*

Proof. Every irreducible $\tilde{K}$-representation of Z has degree one, namely, it is just a K-monomorphism from Z into $\tilde{K}$. Thus, the number of K-monomorphisms from Z into $\tilde{K}$ is precisely $\dim_{\tilde{K}} Z^{\tilde{K}}/J(Z^{\tilde{K}})$, and this is equal to $[Z:K]$ if and only if $J(Z^{\tilde{K}}) = 0$, because $[Z:K] = \dim_K Z = \dim_{\tilde{K}} Z^{\tilde{K}}$. ■

If A is semisimple and $A = \bigoplus_{i=1}^k A_i$ is the direct sum of its simple components A_i, then A is separable if and only if each A_i is separable. By using this, we prove the following theorem.

Theorem 5.4. *A semisimple K-algebra A is separable if and only if the center $Z = Z(A)$ is separable over K.*

Proof. We may assume that A is simple. For an extension field L of K we have

$$A^L = L \otimes_K A = L \otimes_K (Z \otimes_Z A) = (L \otimes_K Z) \otimes_Z A,$$

and since A is central simple over Z, $J(A^L) = J(L \otimes_K Z) \otimes_Z A$ by Theorem 4.2(iii). Therefore $J(A^L) = 0 \Leftrightarrow J(L \otimes_K Z) = 0$, that is, A is separable if and only if Z is separable. ■

If K is a perfect field (e.g., if Char $K = 0$ or K is finite), then every K-algebra is separable by the above theorem.

The separability condition of an algebra is closely related to the structure of its enveloping algebra. To see this, we need two preliminary lemmas.

Lemma 5.5. *Let A be a separable K-algebra. Then, for a semisimple K-algebra B, $A \otimes_K B$ is also semisimple.*

Proof. We may assume that B is simple. Let $Z = Z(B)$. Then

$$A \otimes_K B = A \otimes_K (Z \otimes_Z B) = (A \otimes_K Z) \otimes_Z B.$$

Since $A \otimes_K Z$ is semisimple and B is central simple over Z, we conclude from Theorem 4.2(ii) that $A \otimes_K B$ is semisimple. ■

Lemma 5.6. *Let A be a K-algebra. Then A is projective as a* (*right*) *A^e-module if and only if there exist $a_1, \ldots, a_n,\ b_1, \ldots, b_n \in A$ satisfying the following conditions* (a) *and* (b):

(a) $\sum_{i=1}^n a_i^\circ \otimes (b_i x) = \sum_{i=1}^n (xa_i)^\circ \otimes b_i$ *for all* $x \in A$,

(b) $\sum_{i=1}^n a_i b_i = 1$.

Proof. Define $\lambda\colon A^e \to A$ by $\lambda(\sum_i x_i^\circ \otimes y_i) = \sum_i x_i y_i$. It is easy to see that λ is an A^e-epimorphism. Thus, A is A^e-projective if and only if there exists an A^e-homomorphism $\mu\colon A \to A^e$ such that $\lambda \circ \mu = \mathrm{id}_A$.

Assuming such μ exists, we set $\mu(1) = \sum_{i=1}^{n} a_i^\circ \otimes b_i$. Then (b) holds from $(\lambda \circ \mu)(1) = 1$. We also observe that, for $x \in A$,

$$\mu(x) = \mu(1(1^\circ \otimes x)) = \mu(1)(1^\circ \otimes x) = \sum_i a_i^\circ \otimes (b_i x).$$

On the other hand,

$$\mu(x) = \mu(1(x^\circ \otimes 1)) = \mu(1)(x^\circ \otimes 1) = \sum_i (a_i^\circ x_i^\circ) \otimes b_i = \sum_i (xa_i)^\circ \otimes b_i,$$

whence (a) follows.

To show the converse, define $\mu\colon A \to A^e$ by

$$\mu(x) = \sum_i a_i^\circ \otimes (b_i x)\left(= \sum_i (xa_i)^\circ \otimes b_i\right).$$

Then it is easy to see that μ is an A^e-homomorphism and that $\lambda \circ \mu = \mathrm{id}_A$. ■

We now show the following theorem.

Theorem 5.7. *The following three conditions on the K-algebra A are equivalent.*

(1) *A is separable.*
(2) *A^e is semisimple.*
(3) *A is A^e-projective.*

Proof. We have already shown "(1) ⇒ (2)" in Lemma 5.5, while the implication "(2) ⇒ (3)" is clear from Chapter 1, 10.4. Therefore, it remains only to show that "(3) ⇒ (1)". Let L be any extension field of K. We shall prove that every A^L-module V is completely reducible. By taking any A^L-submodule W of V, we shall show that $W \mid V$. Since

$$(A^L)^e = (A^L)^\circ \otimes A^L = (A^\circ \otimes_K L) \otimes_L (L \otimes_K A) = L \otimes_K A^e,$$

A^L is $(A^L)^e$-projective by our assumption. Therefore there exist $a_1, \ldots, a_n, b_1, \ldots, b_n \in A^L$, satisfying the two conditions in Lemma 5.6.

Now, since W is a direct summand of V as an L-module, there exists an L-homomorphism $f: V \to W$ such that $f(w) = w$ for all $w \in W$. Using this, define $\mu(f): V \to W$ by

$$\mu(f)(v) = \sum_{i=1}^{n} f(va_i)b_i.$$

Then by condition (b) of Lemma 5.6, we see that $\mu(f)(w) = w$ for all $w \in W$. Thus, in order to complete the proof, it is sufficient to show that $\mu(f)$ is an A^L-homomorphism. Given $v \in V$, define an L-homomorphism $g: (A^L)^e = (A^L)^\circ \otimes_L A^L \to V$ by $g(a^\circ \otimes b) = f(va)b$. Then we have by the condition (a) of Lemma 5.6 that $g(\sum_i a_i^\circ \otimes (b_i x)) = g(\sum_i (xa_i)^\circ \otimes b_i)$ for all $x \in A^L$. However by the definition of g, the left-hand side of this equality is $\mu(f)(v)x$, and the right-hand side is $\mu(f)(vx)$. Thus $\mu(f)$ is an A^L-homomorphism. ∎

6. The Schur Index

Let $A = M_n(D)$ be a simple K-algebra, where D is a division algebra, and let $Z = Z(A)$. Since D is central simple over Z, we know that $\dim_Z D$ is a square (Theorem 4.11). If $\dim_Z D = m^2$, we call m the *Schur index* of the simple K-algebra A.

In this section, we shall study the Schur index from the representation-theoretic point of view. First of all, the following holds for central simple algebras.

Theorem 6.1. *Let $A = M_n(D)$ be a central simple K-algebra with division algebra D, and let m be the Schur index of A; $m^2 = \dim_K D$.*

(i) *If L is any maximal commutative subfield of D, then $[L:K] = m$ and L is a splitting field for A.*

(ii) *If a finite extension E of K is a splitting field for A, then $m \mid [E:K]$.*

(iii) *Let V be an irreducible A-module. Let E be any splitting field for A and $\tilde{V}$ be an (absolutely) irreducible A^E-module. Then*

(a) $V^E \simeq m\tilde{V}$ *(A^E-isomorphism),*

(b) *For a positive integer r, it holds that*

$$r\tilde{V} \text{ is realizable in } K \Leftrightarrow m \mid r.$$

Proof.

(i) Since $C_D(L) = L$ by the maximality of L, it follows from Theorem 4.6(ii) that $(\dim_K L)^2 = \dim_K D = m^2$, namely, $\dim_K L = m$. L is a splitting field for A by Lemma 3.12 and Theorem 4.12.

(ii) E is also a splitting field for D, and we have $D^E \simeq M_m(E)$ (note that $\dim_E D^E = \dim_K D = m^2$). In particular, $D^E \simeq mW$, where W is an irreducible D^E-module. Thus, by considering that $D \subset D^E$, we find

$$\dim_K E = \dim_D D^E = m \dim_D W.$$

This proves (ii).

(iii) We see that $\dim_K V = nm^2$ and $\dim_E \tilde{V} = nm$ since $A^E \simeq M_{nm}(E)$. Thus, as V^E is completely reducible, part (a) follows at once by comparing the dimensions of both sides.

(b) ($\Rightarrow$). Suppose that $r\tilde{V} \simeq W^E$ for some W_A and $W \simeq sV$. Then $r\tilde{V} \simeq sV^E \simeq sm\tilde{V}$, and thus $r = sm$.

($\Leftarrow$). If $r = sm$, then $(sV)^E \simeq r\tilde{V}$. ■

Now returning to a general K-algebra A, let E be a splitting field for A and $\tilde{V}$ be an (absolutely) irreducible A^E-module. If $\bar{A} = A/J(A) = \bigoplus_i \bar{A}_i$ is the decomposition into the direct sum of simple components $\bar{A}_i$, then there exists a unique i such that $\tilde{V}\bar{A}_i = 0$. This means that $\tilde{V}$ is actually considered as an absolutely irreducible module of the simple algebra $\bar{A}_i$. We denote the Schur index of $\bar{A}_i$ by $m_K(\tilde{V})$ and call it the *Schur index* of $\tilde{V}$ over K. If $\tilde{\zeta}$ denotes the character of A afforded by $\tilde{V}$, we write $m_K(\tilde{\zeta})$ for $m_K(\tilde{V})$, which is also called the Schur index of $\tilde{\zeta}$.

Before proceeding, we introduce here some notations. Given any extension field E of K, each $\sigma \in \mathrm{Aut}(E/K)$ gives rise to a K-algebra automorphism of $A^E = E \otimes_K A$, denoted also by σ, such that $(\lambda \otimes a)^\sigma = \lambda^\sigma \otimes a (\lambda \in E, a \in A)$. For an E-representation $\mathbf{X}: A \to M_n(E)$ of A, we denote by $X(a)^\sigma$ the matrix obtained from $X(a)$ by replacing the entries with their images under σ. Then $\mathbf{X}^\sigma: A \to M_n(E)(a \mapsto X(a)^\sigma)$ is also an E-representation of A with the character $\chi_{\mathbf{X}}^\sigma$ given by $\chi_{\mathbf{X}}^\sigma(a) = (\chi(a))^\sigma$ for $a \in A$. We call $\mathbf{X}^\sigma$ and $\chi_{\mathbf{X}}^\sigma$ *algebraic conjugates* of $\mathbf{X}$ and $\chi_{\mathbf{X}}$, respectively. If $\mathbf{X}$ is defined by an A-module U, we let U^σ denote a representation module for $\mathbf{X}^\sigma$, which is also called an algebraic conjugate of U.

Remark 1. If, in the above, U is an A^E-submodule of A^E, then the image U^σ of U under the automorphism σ of A^E gives a representation module for $\mathbf{X}^\sigma$.

Now we assume, for the rest of this section, that $A = M_n(D)$ is a simple algebra which is separable over K, where D is a division algebra.

Remark 2. For a group ring KG, $KG/J(KG)$ is always separable, as will be shown in Chapter 3, 1.28.

As is shown in Theorem 4.13, there exists a finite Galois extension E of K that is a splitting field for A. Let $\mathfrak{G} = \mathrm{Gal}(E/K)$ be the Galois group.

Let V be an irreducible A-module, $\tilde{V}$ an (absolutely) irreducible A^E-module, and ζ, $\tilde{\zeta}$ be characters defined by V, $\tilde{V}$, respectively. Here we understand that $\tilde{\zeta}$ is a function defined on A but not on $\tilde{A}$. Let $K(\tilde{\zeta})$ be the field generated by $\{\tilde{\zeta}(a); a \in A\}$ over K. We also denote by $\mathfrak{G}_{\tilde{\zeta}}$ the set of all $\sigma \in \mathfrak{G}$ such that $\tilde{\zeta}^\sigma = \tilde{\zeta}$, i.e., $\tilde{\zeta}(a)^\sigma = \tilde{\zeta}(a)$ for all $a \in A$. Thus $\mathfrak{G}_{\tilde{\zeta}} = \{\sigma \in \mathfrak{G}; \mu^\sigma = \mu$ for all $\mu \in K(\tilde{\zeta})\}$. Then $K(\tilde{\zeta}) = \{\lambda \in E; \lambda^\sigma = \lambda$ for all $\sigma \in \mathfrak{G}_{\tilde{\zeta}}\}$ by the fundamental theorem of Galois theory. Note also that $\mathfrak{G}_{\tilde{\zeta}} = \{\sigma \in \mathfrak{G};\ \tilde{V}^\sigma \simeq \tilde{V}$ (A^E-isomorphic)$\}$ (cf. Theorem 3.13).

We now prove the following result.

Theorem 6.2. *Let $Z = Z(A)$ and $t = \dim_K Z$. Then the following holds.*

(i) $|\mathfrak{G} : \mathfrak{G}_{\tilde{\zeta}}| = t$.

(ii) $Z \simeq K(\tilde{\zeta})$ *(K-isomorphic).*

(iii) *If $\mathfrak{G} = \sum_{i=1}^{t} \mathfrak{G}_{\tilde{\zeta}}\sigma_i$ with $\sigma_1 = 1$, then we have the following irreducible decomposition of V^E:*

$$V^E \simeq m_K(\tilde{V})(\tilde{V}^{\sigma_1} \oplus \cdots \oplus \tilde{V}^{\sigma_t}). \tag{6.1}$$

In other words, $m_K(\tilde{V})$ coincides with the multiplicity of $\tilde{V}$ *in* V^E.

Proof. Since A is separable, A^E is semisimple. Let

$$A^E = \bigoplus_i B_i, \qquad B_i = M_{s_i}(E) \tag{6.2}$$

be the direct sum of the simple components B_i of A^E. Then the center $Z(A^E) = Z(A)^E = Z^E$ is written as

$$Z^E = \bigoplus_i Ec_i, \tag{6.3}$$

where we assume that $1 = \sum_i c_i$ is the central primitive idempotent decomposition of 1 with $B_i = A^E c_i$. Note that the subscript i runs over $\{1, \ldots, t\}$ in (6.2) and (6.3), since $\dim_E Z^E = \dim_K Z = t$.

Now, considering that $Z \subset Z^E$, we express $z \in Z$ as

$$(6.4) \qquad z = \sum_{i=1}^{t} \omega_i(z) c_i \qquad (\omega_i(z) \in E),$$

according to the decomposition (6.3). Then each $\omega_i\colon Z \to E(z \mapsto \omega_i(z))$ is an irreducible E-representation of the K-algebra Z, and $Z_i = \operatorname{Im} \omega_i$ is a subfield of E that is K-isomorphic to Z. Moreover, $\{\omega_i; 1 \le i \le t\}$ is the set of all the irreducible E-representations of Z.

Since each $\sigma \in \mathfrak{G}$ induces a K-automorphism of A^E, it induces a permutation on $\{B_1, \ldots, B_t\}$, hence on $\{c_1, \ldots, c_t\}$ and $\{\omega_1, \ldots, \omega_t\}$. We note that $\mathfrak{G}$ acts transitively on $\{c_1, \ldots, c_t\}$. In fact, if $\{c_1, \ldots, c_r\}$ is one of $\mathfrak{G}$-orbits, then $\sum_{i=1}^{r} c_i$ is $\mathfrak{G}$-stable and hence a central idempotent of A. But since A is simple, it must be the identity of A, hence $r = t$.

From the above, there exists $\sigma_i \in \mathfrak{G}$ for each i such that $B_1^{\sigma_i} = B_i$. Thus we have $s_1 = \cdots = s_t$, and if we set $\mathfrak{G}_1 = \{\sigma \in \mathfrak{G}; B_1^\sigma = B_1\}$, then $\mathfrak{G} = \sum_{i=1}^{t} \mathfrak{G}_1 \sigma_i$, where we assume $\sigma_1 = 1$.

We may assume that $\tilde{V}$ is a minimal right ideal of B_1, and hence $\tilde{V}^\sigma \subset B_1^\sigma$ (Remark 1). Thus, we see that $\mathfrak{G}_1 = \{\sigma \in \mathfrak{G}; \tilde{V}^\sigma \simeq \tilde{V}(A^E\text{-isomorphic})\} = \mathfrak{G}_{\tilde{\zeta}}$, and thus $|\mathfrak{G}:\mathfrak{G}_{\tilde{\zeta}}| = t$.

To show (ii), let $z \in Z$. Then

$$(6.5) \qquad z = z^\sigma = \sum_{i=1}^{t} \omega_i(z)^\sigma c_i^\sigma.$$

Thus, if $\sigma \in \mathfrak{G}_1$, then $c_1^\sigma = c_1$, and we find that $\omega_1(z)^\sigma = \omega_1(z)$ for all $z \in Z$, i.e., $\mathfrak{G}_{\tilde{\zeta}} = \mathfrak{G}_1 \subset \mathfrak{G}_{Z_1}$. But since $|\mathfrak{G}:\mathfrak{G}_{Z_1}| = \dim_K Z_1 = t = |\mathfrak{G}:\mathfrak{G}_{\tilde{\zeta}}|$, we have $\mathfrak{G}_{\tilde{\zeta}} = \mathfrak{G}_{Z_1}$, and by taking the fixed field of $\mathfrak{G}_{Z_1}$, we obtain $Z_1 = K(\tilde{\zeta})$, which is K-isomorphic to Z.

Finally we show (6.1). Let $m = m_K(\tilde{V})$. Then $\dim_K A = n^2 m^2 t$, while we see from (6.2) that $A^E \simeq s_1(\bigoplus_{i=1}^{t} \tilde{V}^{\sigma_i})$. Since $\dim_E \tilde{V} = s_1$, we have, by evaluating the dimension of A^E over E, that $n^2 m^2 t = (s_1)^2 t$ or $s_1 = nm$. From this and $A^E \simeq nV^E$, we get (6.1). ■

Next we treat the special case where the above $K(\tilde{\zeta})$ is a Galois extension of K, which is always guaranteed for group rings, as we shall see later (Chapter 3, Section 1.5).

Theorem 6.3. *In addition to the assumption of Theorem 6.2, suppose that $K(\tilde{\zeta})$ is a Galois extension of K. Let $Z_1 = K(\tilde{\zeta})$ and U be an irreducible A^{Z_1}-module such that $\tilde{V} \mid U^E$. Then*

(i) $V^{Z_1} \simeq U^{\sigma_1} \oplus \cdots \oplus U^{\sigma_t}$.

(ii) $U^E \simeq m_K(\tilde{V})\tilde{V}$, *and hence* $m_K(\tilde{V}) = m_{K(\tilde{\zeta})}(\tilde{V})$.

(iii) *If $\tilde{V}$ is realizable in a field M with $K \subset M \subset E$, then it holds that $Z_1 \subset M$ and*

$$m_K(\tilde{V}) \mid [M:Z_1].$$

Moreover, there exists an extension field L of Z_1 with $[L:Z_1] = m_K(\tilde{V})$ such that $\tilde{V}$ is realizable in L.

(iv) *For a positive integer r, we have*

$$r\tilde{V} \text{ is realizable in } K(\tilde{\zeta}) \Leftrightarrow m_K(\tilde{V}) \mid r.$$

Proof. We use the same notation as in the proof of Theorem 6.2. Also, we understand that $A \subset A^{Z_1} \subset A^E$ (via the natural identifications).

By the assumption that Z_1 is a Galois extension, we have $Z_1^{\sigma_i} = Z_1$, so that $\omega_1, \ldots, \omega_t$ are the distinct irreducible Z_1-representations of the K-algebra Z. Consequently, Z^{Z_1} is isomorphic to the direct sum of t copies of Z_1, and hence 1 is the sum of t primitive idempotents in Z^{Z_1} that are orthogonal to each other. This implies that the primitive idempotent decomposition $1 = \sum_{i=1}^t c_i$ in Z^E has been actually realized in $Z^{Z_1} (\subset Z^E)$, namely

$$Z^{Z_1} = \bigoplus_{i=1}^{t} Z_1 c_i.$$

Therefore,

$$A^{Z_1} = Z_1 \otimes_K Z \otimes_Z A = \bigoplus_{i=1}^{t} Z_1 c_i \otimes_Z A$$

gives the simple decomposition of A^{Z_1}. Let $A_i = Z_1 c_i \otimes_Z A$. Then $A_i = A^{Z_1} c_i = A_1^{\sigma_i}$. Also, we have $A_i = \{c_i \otimes a;\, a \in A\}$, since $\omega_i(z)c_i \otimes a = c_i z \otimes a = c_i \otimes (za)$. Therefore, the map $f_i\colon A \to A_i (a \mapsto c_i \otimes a)$ is a K-algebra isomorphism, and $A_i \simeq M_n(D_i)$ with $D_i \simeq D$ follows.

Now we have $A^E = (A^{Z_1})^E = \bigoplus_{i=1}^t A_i^E$ and $A_i^E = A^E c_i = B_i$. Since we have assumed that $\tilde{V} \subset B_1$, we may take U as a minimal right ideal of A_1. Then

$$A^{Z_1} \simeq n(U^{\sigma_1} \oplus \cdots \oplus U^{\sigma_t}) \qquad (U^{\sigma_1} = U).$$

This proves (i), since $A^{Z_1} \simeq nV^{Z_1}$. Also this yields that

$$A^E \simeq n((U^{\sigma_1})^E \oplus \cdots \oplus (U^{\sigma_t})^E)$$
$$\simeq nm(\tilde{V}^{\sigma_1} \oplus \cdots \oplus \tilde{V}^{\sigma_t}),$$

where $m = m_K(\tilde{V})$. Therefore $U^E \simeq m\tilde{V}$, proving (ii).

Finally, by considering $\tilde{V}$ as an absolutely irreducible module over the central simple Z_1-algebra A_1, we obtain (iii) and (iv) from Theorem 6.1. ■

As another consequence of Theorem 6.2, we have

Theorem 6.4. *Let A be a K-algebra and suppose that $A/J(A)$ is separable. Let L be an extension field of K, and let $\tilde{V}$ be an absolutely irreducible A^L-module. Then, for an A-module W, the multiplicity of $\tilde{V}$ in W^L is a multiple of $m_K(\tilde{V})$.*

Proof. For an A^L-module M, let $c(M, \tilde{V})$ denote the multiplicity of $\tilde{V}$ in M. If $W \leftrightarrow \bigoplus_i r_i V_i$, where $\{V_1, \ldots, V_l\} = \mathrm{IRR}(A)$, then $c(W^L, \tilde{V}) = \sum_i r_i c(V_i^L, \tilde{V})$, and there exists a unique i such that $c(V_i^L, \tilde{V}) \neq 0$. Thus, we may assume that A is simple (and separable), and it suffices to show that $c(V^L, \tilde{V}) = m_K(\tilde{V})$ for an irreducible A-module V. Let m be the Schur index of A, then $m_K(\tilde{V}) = m$.

Let $\tilde{L}$ be the algebraic closure of L, and $E(\subset \tilde{L})$ be a finite Galois extension of K, which is a splitting field for A. If $\tilde{V}_1$ is an (absolutely) irreducible A^E-module such that $\tilde{V}_1^{\tilde{L}} = \tilde{V}^{\tilde{L}}$, then $c(V^E, \tilde{V}_1) = m_K(\tilde{V}_1) = m$ by Theorem 6.2(iii). On the other hand, if U is an irreducible constituent of V^L that is not isomorphic to $\tilde{V}$, then $c(U^{\tilde{L}}, \tilde{V}^{\tilde{L}}) = 0$. Consequently, we have

$$c(V^L, \tilde{V}) = c(V^{\tilde{L}}, \tilde{V}^{\tilde{L}}) = c(V^E, \tilde{V}_1) = m. \quad ■$$

7. Crossed Products

7.1. Cohomology Groups

Let G be a group. By a *G-module* V, we mean the module V on which G acts from the right in the following way:

(1) $(u + v)a = ua + va$ (2) $(va)b = v(ab)$ (3) $v1 = v$

for $u, v \in V$ and $a, b, 1 \in G$.

Let V be a G-module and n be a positive integer. By an *n-cochain* f, we mean a map

$$f: G \times G \times \cdots \times G \to V \qquad ((x_1, \ldots, x_n) \mapsto f(x_1, \ldots, x_n)).$$

We denote by $C^n(G, V)$ the set of all n-cochains. This is an additive group with the addition defined by

$$(f + g)(x_1, \ldots, x_n) = f(x_1, \ldots, x_n) + g(x_1, \ldots, x_n).$$

We set $C^0(G, V) = V$ for convenience.

Now, define δ_n: $C^n(G, V) \to C^{n+1}(G, V)$ as follows:

$$(7.1) \quad \begin{cases} \delta_0 v(x) = v - vx, \\ \delta_n f(x_1, \ldots, x_{n+1}) = f(x_2, \ldots, x_{n+1}) \\ \qquad + \sum_{i=1}^{n} (-1)^i f(x_1, \ldots, x_i x_{i+1}, \ldots, x_{n+1}) \\ \qquad + (-1)^{n+1} f(x_1, \ldots, x_n) x_{n+1} \ (n \geq 1). \end{cases}$$

Then the following holds.

$$(7.2) \qquad \delta_n(f + g) = \delta_n f + \delta_n g,$$

$$(7.3) \qquad \delta_{n+1}(\delta_n f) = 0.$$

Exercise 7.1. Prove (7.3) above.

Write $Z^n(G, V)$ for Ker δ_n and $B^n(G, V)$ for Im δ_{n-1}. The elements of $Z^n(G, V)$ and $B^n(G, V)$ are said to be *n-cocycles* and *n-coboundaries*, respectively. We set $B^0(G, V) = 0$. We see from (7.3) that $B^n(G, V) \subset Z^n(G, V)$, and we denote the factor group $Z^n(G, V)/B^n(G, V)$ by $H^n(G, V)$, which is called the *n-th cohomology group* of G with coefficients in V.

If two n-cocycles f and g are congruent mod $B^n(G, V)$, we say that they are *cohomologous* and write $f \sim g$.

Example 7.2.

(1) $H^0(G, V) = Z^0(G, V) = \{v \in V;\ vx = v \text{ for all } x \in G\}$.

(2) If $f \in C^1(G, V)$, then

$$\delta_1 f(x, y) = f(y) - f(xy) + f(x)y.$$

Consequently, we have the following:

$$f \in Z^1(G, V) \Leftrightarrow f(xy) = f(x)y + f(y),$$
$$f \in B^1(G, V) \Leftrightarrow \text{there exists } v \in V \text{ such that } f(x) = v - vx.$$

(3) If $f \in C^2(G, V)$, then

$$\delta_2 f(x, y, z) = f(y, z) - f(xy, z) + f(x, yz) - f(x, y)z.$$

Consequently, we have the following:

$$(7.4) \quad \begin{cases} f \in Z^2(G, V) \Leftrightarrow f(xy, z) + f(x, y)z = f(y, z) + f(x, yz), \\ f \in B^2(G, V) \Leftrightarrow f(x, y) = g(y) - g(xy) + g(x)y \text{ for some } g \in C^1(G, V). \end{cases}$$

A 2-cocycle of G is also called a *factor set* of G.

Theorem 7.3. *Let V be a G-module and $n \geq 1$. Then the following holds.*

(i) *The order of each element of $H^n(G, V)$ divides $|G|$.*

(ii) *Let $f \in Z^n(G, V)$. Suppose that the map*

$$\theta: V \to V \qquad (v \mapsto |G|v)$$

is a monomorphism and that

$$\sum_{x \in G} f(x, x_1, \ldots, x_{n-1}) \in \operatorname{Im} \theta$$

for arbitrary $x_1, \ldots, x_{n-1} \in G$. Then $f \sim 0$.

In particular, we have $H^n(G, V) = 0$ if θ is an isomorphism.

Proof. For $f \in Z^n(G, V)$, let $g(x_1, \ldots, x_{n-1}) = \sum_{x \in G} f(x, x_1, \ldots, x_{n-1})$. Then it follows from $\delta_n f(x, x_1, \ldots, x_n) = 0$ that

$$f(x_1, \ldots, x_n) - f(xx_1, x_2, \ldots, x_n) + \sum_{i=1}^{n-1} (-1)^{i+1} f(x, \ldots, x_i x_{i+1}, \ldots, x_n)$$
$$+ (-1)^{n+1} f(x, x_1, \ldots, x_{n-1})x_n = 0$$

Summing up the above equality with x ranging over G, we get

$$|G| f(x_1, \ldots, x_n) = g(x_2, \ldots, x_n) + \sum_{i=1}^{n-1} (-1)^i g(x_1, \ldots, x_i x_{i+1}, \ldots, x_n)$$
$$+ (-1)^n g(x_1, \ldots, x_{n-1})x_n = \delta_{n-1} g(x_1, \ldots, x_n) \cdots (*).$$

Therefore $|G| f \sim 0$. This proves (i).

Now there exists, under the assumption of (ii), $h \in C^{n-1}(G, V)$ such that $g(x_1, \ldots, x_{n-1}) = |G| h(x_1, \ldots, x_{n-1})$. Hence it follows from (∗) above that $|G| f(x_1, \ldots, x_n) = |G| \delta_{n-1} h(x_1, \ldots, x_n)$, whence $f = \delta_{n-1} h$, since θ is a monomorphism. The latter half of (ii) is obvious. ■

7.2. *Twisted Group Rings*

We say that G acts on a ring S if there is a homomorphism $\varphi: G \to \mathrm{Aut}(S)$, where $\mathrm{Aut}(S)$ denotes the group of ring automorphisms of S. For $x \in G$ and $\alpha \in S$, the image of α by $\varphi(x)$ is denoted by α^x as usual. We also denote by S^G the set of elements of S, fixed by all elements of G. This is a subring of S, which is called the *fixed ring* of G in S.

Now, we assume henceforth that S is a commutative ring. If G acts on S, then G acts on the abelian group $S^\times$. Thus, cocycles and coboundaries will be considered, and they are written multiplicatively. For example,

$$\begin{aligned} &f: G \times G \times \cdots \times G \to S^\times \quad \text{is a cocycle} \\ &\Leftrightarrow \delta_n f(x_1, \ldots, x_{n+1}) \\ &= f(x_2, \ldots, x_{n+1}) \cdot \prod_{i=1}^{n} f(x_1, \ldots, x_i x_{i+1}, \ldots, x_{n+1})^{(-1)^i} \\ &\cdot f(x_1, \ldots, x_n)^{(-1)^{n+1} x_{n+1}} = 1. \end{aligned}$$

From Theorem 7.3, we obtain the following.

Theorem 7.4. *Let K be a perfect field of positive characteristic p and P be a p-group acting on K. Then $H^n(P, K^\times) = 1$ for all $n \geq 1$.*

Proof. By assumption, the map $\theta: K^\times \to K^\times (\alpha \mapsto \alpha^{|P|})$ is an isomorphism (Hungerford [1], p. 289). Hence the result is a direct consequence of Theorem 7.3. ■

Recall that $\alpha: G \times G \to S^\times$ is a factor set (i.e., a 2-cocycle) if and only if it satisfies the following:

(7.5) $\qquad \alpha(xy, z)\alpha(x, y)^z = \alpha(y, z)\alpha(x, yz) \qquad$ for all $\quad x, y, z \in G$.

And two factor sets α, β are cohomologous if and only if there exists a map $\gamma: G \to S^\times (x \mapsto \gamma(x))$ such that

$$\alpha(x, y) = \beta(x, y)\gamma(y)\gamma(xy)^{-1}\gamma(x)^y. \tag{7.6}$$

For α, $\beta \in Z^2(G, S)$, define $\alpha\beta$ by

$$\alpha\beta: (x, y) \mapsto \alpha(x, y)\beta(x, y).$$

Then $Z^2(G, S^\times)$ is an abelian group with identity $1: (x, y) \to 1$ (for all x, $y \in G$).

Exercise 7.5. The following holds for a factor set α.

(i) $\alpha(1, z) = \alpha(x, 1)^z$.

(ii) $\alpha(1, 1) = \alpha(x, 1)$.

[Hint: Part (i) follows from (7.5) by letting $y = 1$, and part (ii) by letting $z = 1$.]

Given $\alpha \in Z^2(G, S^\times)$, let us consider a set of symbols $\{u_x; x \in G\}$ and the free S-module

$$A = \bigoplus_{x \in G} u_x S.$$

We define the multiplication in A by

$$u_x u_y = u_{xy}\alpha(x, y), \qquad \gamma u_x = u_x \gamma^x \qquad (\gamma \in S)$$

and extend it by linearity to all elements of A:

$$\left(\sum_x u_x \beta_x\right)\left(\sum_y u_y \gamma_y\right) = \sum_{x, y} u_{xy}\alpha(x, y)\beta_x^y \gamma_y.$$

Then A becomes a ring. In fact, the associative law of multiplication follows from (7.5), and we see from Exercise 7.5 that $u_1\alpha(1, 1)^{-1}$ is the identity of A. We denote the ring A by (S, G, α) and call it a *twisted group ring* of G over S. If R is a subring of S^G with 1_S, then (S, G, α) is an algebra over R.

Exercise 7.6. Prove that each u_x in the above is a unit of A, and $u_x^{-1} = u_{x^{-1}}\alpha(1, 1)^{-1}\alpha(x, x^{-1})^{-1}$.

Lemma 7.7. *Let $A = (S, G, \alpha)$ and $B = (S, G, \beta)$ be twisted group rings. Then there exists a ring isomorphism $\varphi: A \cong B$ such that $\varphi(u_x S) = v_x S$ with $\varphi(s) = s$ for all $s \in S$ if and only if $\alpha \sim \beta$.*

Proof. If such φ exists, then $\varphi(u_x) = v_x\gamma(x)$ for some $\gamma(x) \in S^\times$. And we have

$$\varphi(u_x u_y) = \varphi(u_{xy}\alpha(x, y)) = \varphi(u_{xy})\alpha(x, y) = v_{xy}\gamma(xy)\alpha(x, y).$$

On the other hand, we also have

$$\varphi(u_x u_y) = \varphi(u_x)\varphi(u_y) = v_x\gamma(x)v_y\gamma(y) = v_{xy}\beta(x, y)\gamma(x)^y\gamma(y).$$

Thus,

$$\alpha(x, y) = \beta(x, y)\gamma(y)\gamma(xy)^{-1}\gamma(x)^y,$$

and, therefore, $\alpha \sim \beta$.

If, conversely, there exists a map $\gamma: G \to S^\times$ satisfying the above equality, define $\varphi: A \to B$ by $\varphi(\sum_x u_x s_x) = \sum_x v_x\gamma(x)s_x$, where $s_x \in S$. Then this gives a desired isomorphism: $A \simeq B$. ■

Let $A = (S, G, \alpha) = \bigoplus_x u_x S$ be the twisted group ring. If we set $v_1 = u_1\alpha(1, 1)^{-1}$ and $v_x = u_x (x \neq 1)$, then $A = \bigoplus_x v_x S$. This yields a new factor set β such that $\beta \sim \alpha$ and $v_x v_y = v_{xy}\beta(x, y)$. Since v_1 is the identity of A,

$$\beta(x, 1) = \beta(1, x) = 1 \qquad \text{for all } x \in G.$$

holds. Any factor set β satisfying the above equality is said to be a *normalized factor set*.

Exercise 7.8. If β is a normalized factor set, then the following holds:

$$\alpha(x, x^{-1})^x = \alpha(x^{-1}, x).$$

[Hint: Let $y = x^{-1}$, $z = x$ in (7.5).]

7.3. Crossed Products

Let L be a Galois extension of a field K with Galois group G and let $\alpha \in Z^2(G, L^\times)$. Then the twisted group ring (G, L, α) is said to be the *crossed product* of L and G with factor set α, which is denoted by $(L/K, \alpha)$. Note that $(L/K, \alpha)$ is a K-algebra.

Theorem 7.9. *Using the same notation as above, let $n = [L:K]$. Then*

(i) *$(L/K, \alpha)$ is a central simple K-algebra, which contains L as a maximal commutative subfield. Hence L is a splitting field for it.*

(ii) *If $\alpha \sim 1$, then $(L/K, \alpha) \simeq M_n(K)$.*

(iii) *For $\beta \in Z^2(G, L^\times)$, it holds that*

$$(L/K, \alpha) \simeq (L/K, \beta) \quad \text{as } K\text{-algebras} \quad \Leftrightarrow \quad \alpha \sim \beta.$$

Proof. Set $A = (L/K, \alpha) = \bigoplus_x u_x L$ and let $u_x u_y = u_{xy}\alpha(x, y)$, $\gamma u_x = u_x \gamma^x$. We may assume that u_1 is the identity of A and that $L \subset A$.

(i) If $Z(A) \ni z = \sum_x u_x \zeta_x$, then for any $\gamma \in L$, we have

$$\sum_x u_x \zeta_x \gamma = \sum_x \gamma u_x \zeta_x = \sum_x u_x \gamma^x \zeta_x,$$

whence $\zeta_x(\gamma - \gamma^x) = 0$ for all $x \in G$. If $x \neq 1$, then there is $\gamma \in L$ such that $\gamma^x \neq \gamma$, and hence $\zeta_x = 0$, $z = u_1 \zeta_1 = \zeta_1$. Also from $zu_x = u_x z$ it follows that $\zeta_1^x = \zeta_1$ for all $x \in G$. Therefore, $\zeta_1 \in K$ and $Z(A) = K$.

Next let $I \neq 0$ be any ideal of A, and let $I \ni a = \sum_x u_x \alpha_x \neq 0$ ($\alpha_x \in L$) be chosen so that it has the least number of nonzero coefficients among the elements in I. Suppose $I \neq A$. Then, since any nonzero $u_x \alpha_x$ is a unit, there are at least two α_x's that are different from zero. Multiplying a by a suitable $(u_x \alpha_x)^{-1}$ if necessary, we may assume that $a = 1 + \sum_{x \neq 1} u_x \alpha_x$. Then, for any $\gamma \in L$, we have

$$I \ni a\gamma - \gamma a = \sum_{x \neq 1} u_x(\alpha_x \gamma - \gamma^x \alpha_x).$$

But, according to the choice of a, this holds only if $\alpha_x(\gamma - \gamma^x) = 0$ for all $x \neq 1$ and $\gamma \in L$. Consequently, $\alpha_x = 0$ for all $x \neq 1$, i.e., $a = 1$, a contradiction. Therefore A must be simple.

By Theorem 4.6(ii), we have $(\dim_K L)(\dim_K C_A(L)) = \dim_K A$. Then we find that $\dim_K C_A(L) = n$ as $\dim_K L = n$ and $\dim_K A = n^2$. Therefore $C_A(L) = L$ and L must be a maximal commutative subfield.

(ii) We may assume that $\alpha = 1$. Define $\varphi: A \to \mathrm{End}_K(L)$ via $\varphi(\sum_x u_x \alpha_x): \gamma \mapsto \sum_x (\alpha_x \gamma)^{x^{-1}}$ for $\gamma \in L$. It is easy to see that φ is a K-algebra homomorphism. Since A is simple, this must be a monomorphism, and we find that φ is an isomorphism by comparing the dimensions.

(iii) Let $B = (L/K, \beta) = \bigoplus_x v_x L$ with $v_x v_y = v_{xy}\beta(x, y)$.

($\Leftarrow$). Clear by Lemma 7.7.

($\Rightarrow$). Suppose that we are given a K-isomorphism $f: A \xrightarrow{\sim} B$. Then the isomorphism $L \simeq f(L) \subset B$ extends to an inner automorphism of B (Theorem 4.6(vi)). So we may assume $f(\gamma) = \gamma$ for all $\gamma \in L$. Let $u'_x = f(u_x)$. Then $u'_x u'_y = u'_{xy}\alpha(x, y)$ and $\gamma u'_x = u'_x \gamma^x$. But since $\gamma v_x = v_x \gamma^x$, we find that $\gamma u'_x v_x^{-1} = u'_x v_x^{-1} \gamma$ for all $\gamma \in L$, thus $u'_x v_x^{-1} \in C_B(L) = L$. Therefore, $u'_x = v_x \gamma(x)$ with $\gamma(x) \in L^\times$, and we have $\alpha(x, y) = \beta(x, y)\gamma(y)\gamma(xy)^{-1}\gamma(x)^y$, i.e., $\alpha \sim \beta$. ■

As was shown above, the crossed product is a central simple algebra over K. We next show that every central simple algebra is similar to a crossed product.

Theorem 7.10. *Let A be a central simple K-algebra and E be a Galois extension of K, which is a splitting field for A. Then there exists a central simple K-algebra B such that $B \supset E$, $C_B(E) = E$, $B \sim A$, and $B \simeq (E/K, \alpha)$ for some factor set α.*

Proof. As was shown in Theorems 4.12 and 4.13 there exist E, B satisfying the first three conditions in the above. We want to show that B satisfies the last condition. To see this, let $n = [E:K]$ and $G = \mathrm{Gal}(E/K)$. We know that $\dim_K B = n^2$ by Theorem 4.6. On the other hand, each $x \in G$ is extended to an inner automorphism of B; there exists $u_x \in B^\times$ such that $\gamma^x = u_x^{-1}\gamma u_x$ for all $\gamma \in E$. We claim that $\{u_x; x \in G\}$ are E-free. Suppose the contrary and let $\sum_x u_x\gamma_x = 0$ be a nontrivial linear relation ($\gamma_x \in E$), which is assumed to be chosen so that $\#\{x \in G; \gamma_x \neq 0\}$ is minimal. Since u_x is a unit of B, there must be at least two nonzero coefficients, say γ_y, γ_z. Let $\lambda \in E$ be such that $\lambda^y \neq \lambda^z$. Then in the linear relation

$$0 = \lambda\left(\sum_x u_x\gamma_x\right) - \left(\sum_x u_x\gamma_x\right)\lambda^y = \sum_x u_x(\lambda^x\gamma_x - \gamma_x\lambda^y),$$

the coefficient of u_y is zero, whereas that of u_z is not. This contradicts the choice of the linear relation above. Therefore $\{u_x; x \in G\}$ is E-free as claimed. Since $\dim_E B = n$, this forms an E-basis of B:

$$B = \bigoplus_{x \in G} u_x E, \qquad u_x^{-1}\gamma u_x = \gamma^x \quad (\gamma \in E).$$

Moreover, since $(u_x u_y)^{-1}\gamma(u_x u_y) = \gamma^{xy}$ and $u_{xy}^{-1}\gamma u_{xy} = \gamma^{xy}$, we get $u_{xy}^{-1}u_x u_y \in C_B(E) = E$. Thus, if we set $\alpha(x, y) = u_{xy}^{-1}u_x u_y$, then from the associative law $(u_x u_y)u_z = u_x(u_y u_z)$, we find that $\alpha: G \times G \to E^\times$ is a factor set and thus $B = (E/K, \alpha)$. ∎

Remark. For an extension field E of K, the set of algebra classes that have E as a splitting field forms a subgroup of $\mathscr{B}(K)$. This is denoted by $\mathscr{B}(E/K)$. If E is a Galois extension with $G = \mathrm{Gal}(E/K)$, then every algebra class of $\mathscr{B}(E/K)$ is represented by some crossed product $(E/K, \alpha)$, and we know that

$$(E/K, \alpha) \sim (E/K, \beta) \Leftrightarrow (E/K, \alpha) \simeq (E/K, \beta) \Leftrightarrow \alpha \sim \beta.$$

Moreover, it is known that $(E/K, \alpha) \otimes_K (E/K, \beta) \sim (E/K, \alpha\beta)$ (cf. Reiner[1]). Thus we get an isomorphism $\mathscr{B}(E/K) \simeq H^2(G, E^\times)$, which assigns $\alpha B^2(G, E^\times)$ to each $(E/K, \alpha)$.

Recently Fein, Kantor, and Schacher [1] showed the remarkable result that if $E \supsetneq K$ are both number fields, then $\mathscr{B}(E/K)$ is an infinite group. The proof of the result is based on the following fact on permutation groups.

Theorem. *If G is a transitive permutation group on a set X, where $|X| > 1$, then there is a prime p such that some p-element of G fixes no point of X.*

However, the proof of this theorem uses the classification of the finite simple groups, and no elementary proof is known at present.

8. Frobenius Algebras and Symmetric Algebras

8.1. Dual Modules

Let R be a commutative ring and A be an R-algebra. If V is a right A-module, then $\mathrm{Hom}_R(V, R)$, denoted by $V^\wedge$, is a left A-module. This is called the *dual module* of V. For $v \in V$, $\sigma \in V^\wedge$, we set $\langle v, \sigma \rangle = \sigma(v)$. Then by the definition of the A-action on $V^\wedge$, the following holds for $a \in A$:

$$\langle va, \sigma \rangle = \langle v, a\sigma \rangle.$$

If U is a left A-module, then $U^\wedge$ is a right A-module. Our arguments below will be mainly concerned with the duals of right A-modules, but the corresponding results will be valid for left A-modules, too.

Given $v \in V$, we have an R-homomorphism $\varphi(v)\colon V^\wedge \to R(\sigma \mapsto \langle v, \sigma \rangle)$, which yields a homomorphism $\varphi\colon V \to V^{\wedge\wedge} = \mathrm{Hom}_R(V^\wedge, R)$.

Exercise 8.1. Prove that the above φ is an A-homomorphism.

For a subset W of V, define $W^\perp$ by

$$W^\perp = \{\sigma \in V^\wedge; \langle w, \sigma \rangle = 0 \qquad \text{for all } w \in W\},$$

which is an R-submodule of $V^\wedge$, called the *orthogonal complement* of W in $V^\wedge$. If W is an A-submodule of V, then $W^\perp$ is also an A-submodule of $V^\wedge$. If $W_1 \subset W_2$, then $W_1^\perp \supset W_2^\perp$. Also, $V^\perp = 0$, $0^\perp = V^\wedge$.

Similarly we define, for $M \subset V^\wedge$, the orthogonal complement of M in V by $M^\perp = \{v \in V;\ \langle v, \sigma \rangle = 0 \text{ for all } \sigma \in M\}$. If $W \subset V$, then $W \subset (W^\perp)^\perp$.

Given an R-homomorphism $f\colon U \to V$, the *transposed map* ${}^t f$ of f is the R-homomorphism defined by ${}^t f\colon V^\wedge \to U^\wedge (\sigma \mapsto \sigma \circ f)$.

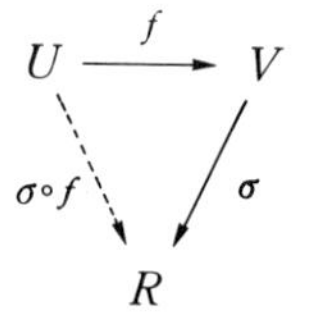

The following equality is obvious by definition:

$$\langle u, {}^t f(\sigma) \rangle = \langle f(u), \sigma \rangle, \qquad \text{for} \quad u \in U, \sigma \in V^\wedge.$$

Exercise 8.2.
(i) If $f \in \mathrm{Hom}_A(U, V)$, then ${}^t f \in \mathrm{Hom}_A(V^\wedge, U^\wedge)$.
(ii) If $f \in \mathrm{Hom}_R(U, V)$, $g \in \mathrm{Hom}_R(V, W)$, then ${}^t(g \circ f) = {}^t f \circ {}^t g$.

If

$$U \xrightarrow{f} V \xrightarrow{g} W \to 0$$

is exact, then the following sequence is also exact (cf. Chapter 1, 9.4):

$$0 \to W^\wedge \xrightarrow{{}^t g} V^\wedge \xrightarrow{{}^t f} U^\wedge.$$

In particular, if $g\colon V \to W$ is an epimorphism, then ${}^t g\colon W^\wedge \to V^\wedge$ is a monomorphism, and if g is an isomorphism, so is ${}^t g$.

Exercise 8.3. If $W_A \subset V_A$, then the following A-isomorphism holds:

$$(V/W)^\wedge \simeq W^\perp.$$

[Hint: If $f\colon V \to V/W$ is the natural map, then $\mathrm{Im}\ {}^t f = W^\perp$.]

We assume henceforth that

(8.1) *all A-modules are finitely generated and free over R.*

Let V be an A-module with R-basis $\{v_i;\ 1 \le i \le n\}$; $V = Rv_1 \oplus \cdots \oplus Rv_n$. For each i there exists a (unique) $\sigma_i \in V^\wedge$ such that $\langle v_j, \sigma_i \rangle = \delta_{ij}$. Then, as will be shown below, the set $\{\sigma_i;\ 1 \le i \le n\}$ gives an R-basis of $V^\wedge$, which is called the *dual basis* to $\{v_i\}$.

Theorem 8.4. *With the notation above, the following holds.*

(i) $\{\sigma_i;\ 1 \le i \le n\}$ *is an R-basis of* $V^\wedge$, *and each* $\sigma \in V^\wedge$ *is expressed uniquely in the form*:

$$\sigma = \sum_{i=1}^{n} \langle v_i, \sigma \rangle \sigma_i. \tag{8.2}$$

(ii) *The R-representation of A defined by V relative to* $\{v_i\}$ *coincides with that of A defined by* ${}_A(V^\wedge)$ *relative to* $\{\sigma_i\}$.

(iii) $\varphi\colon V \to V^{\wedge\wedge}$ $(\varphi(v)\colon \sigma \mapsto \langle v, \sigma \rangle)$ is an A-isomorphism.

Proof.

(i) We have $\langle v_j, \sum_i \langle v_i, \sigma \rangle \sigma_i \rangle = \sum_i \langle v_i, \sigma \rangle \langle v_j, \sigma_i \rangle = \langle v_j, \sigma \rangle$ for all j, hence (8.2) holds. If $\sum_i \alpha_i \sigma_i = 0$, where $\alpha_i \in R$, then $0 = \langle v_j, \sum_i \alpha_i \sigma_i \rangle = \alpha_j$; therefore $\{\sigma_i\}$ is R-free.

(ii) For $a \in A$, write

$$v_i a = \sum_{j=1}^{n} \alpha_{ij}(a) v_j \qquad (\alpha_{ij}(a) \in R).$$

Then we see that $\langle v_i, a\sigma_j \rangle = \langle v_i a, \sigma_j \rangle = \alpha_{ij}(a)$ and, hence, by (8.2)

$$a\sigma_j = \sum_{i=1}^{n} \alpha_{ij}(a) \sigma_i,$$

which proves our assertion (cf. (1.4)).

(iii) If $v = \sum_i \alpha_i v_i \ne 0$, then $\alpha_i \ne 0$ for some i. Thus, $\varphi(v)(\sigma_i) = \langle v, \sigma_i \rangle = \alpha_i \ne 0$, namely, φ is a monomorphism. On the other hand, $\{\varphi(v_i)\}$ gives the dual basis to the basis $\{\sigma_i\}$ of $V^\wedge$, and hence φ is an epimorphism as well. ■

In view of (iii) of the above theorem, we may identify V with $V^{\wedge\wedge}$ via the isomorphism φ.

Lemma 8.5. *Let* $f\colon U \to V$ *be an R-homomorphism. Then, with the above identification, we have* ${}^t({}^tf) = f$.

Proof. For $u \in U$, $\sigma \in V^\wedge$ and ${}^t({}^tf)\colon U = U^{\wedge\wedge} \to V^{\wedge\wedge} = V$, we observe

$$\langle {}^t({}^tf)(u), \sigma \rangle = \langle u, {}^tf(\sigma) \rangle = \langle f(u), \sigma \rangle.$$

Therefore ${}^t({}^tf)(u) = f(u)$ for all $u \in U$. ■

Lemma 8.6. *For U_A, V_A satisfying* (8.1), *the following isomorphisms hold.*

(i) $$(U \oplus V)^\wedge \simeq U^\wedge \oplus V^\wedge \text{ (}A\text{-isomorphism).}$$

In particular $_A(V^\wedge)$ is indecomposable if and only if V is indecomposable.

(ii) $$\mathrm{Hom}_A(U, V) \simeq \mathrm{Hom}_A(V^\wedge, U^\wedge).$$

Proof.

(i) There is a natural isomorphism (cf. Chapter 1, 9.6),

$$\mathrm{Hom}_R(U \oplus V, R) \simeq \mathrm{Hom}_R(U, R) \oplus \mathrm{Hom}_R(V, R),$$

which is also an A-isomorphism as is easily seen. The second half is clear from $V^{\wedge\wedge} = V$.

(ii) This follows easily from Lemma 8.5. In fact, the inverse of the R-homomorphism $t\colon \mathrm{Hom}_A(U, V) \to \mathrm{Hom}_A(V^\wedge, U^\wedge)(f \mapsto {}^t f)$ is given by

$$t'\colon \mathrm{Hom}_A(V^\wedge, U^\wedge) \to \mathrm{Hom}_A(U, V)\,(g \mapsto {}^t g). \qquad \blacksquare$$

Finally, we specialize the situation as follows:

(8.3) *A is an algebra over a field K and all A-modules are assumed to be finitely generated over K.*

Lemma 8.7. *With the assumption* (8.3), *let W be a K-submodule of V_A. Then the following holds.*

(i) $\dim_K W + \dim_K W^\perp = \dim_K V$.

(ii) $(W^\perp)^\perp = W$.

(iii) *The correspondence $\perp\colon W \mapsto W^\perp$ gives rise to a bijection between the set of A-submodules of V and that of $V^\wedge$, which inverts inclusion.*

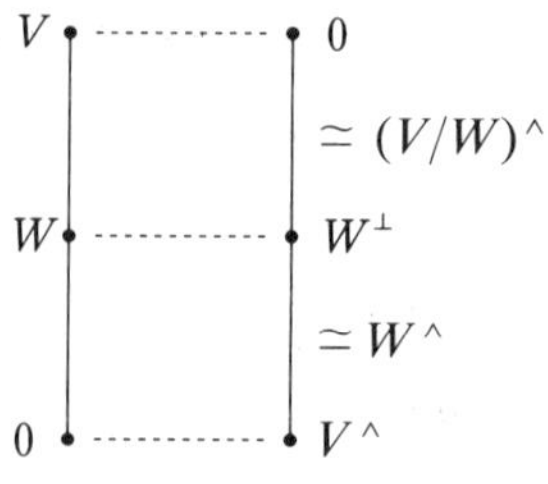

Moreover the following A-isomorphisms hold:

$$_A(V/W)^\wedge \simeq {}_A(W^\perp), \qquad {}_A(W^\wedge) \simeq {}_A(V^\wedge/W^\perp).$$

Proof. Let $\{v_1, \ldots, v_r\}$ be a K-basis of W. We extend it to a K-basis $\{v_1, \ldots, v_n\}$ of V. Let $\{\sigma_i; 1 \leq i \leq n\}$ be the dual basis to $\{v_i\}$.

(i) We readily find that $W^\perp = K\sigma_{r+1} \oplus \cdots \oplus K\sigma_n$. Hence (i) holds.

(ii) Clearly, $(W^\perp)^\perp \supset W$ holds, whence we have the equality by comparing their dimensions.

(iii) The correspondence $\perp$ is injective by (ii). It is surjective because $(U^\perp)^\perp = U$ for any $_AU \subset V^\wedge$.

For the rest, we remark that the first isomorphism has been given in Exercise 8.3. To show the second isomorphism, let $f: V^\wedge \to W^\wedge (\sigma \mapsto \sigma|_W)$ be the restriction map. By definition Ker $f = W^\perp$, hence we have a natural monomorphism $\bar{f}: V^\wedge/W^\perp \to W^\wedge$. But since both $V^\wedge/W^\perp$ and $W^\wedge$ have the same dimension over K, $\bar{f}$ must be an isomorphism, which is an A-isomorphism as is easily verified. ■

Exercise 8.8. Let $f \in \mathrm{Hom}_K(U, V)$. Then the following holds.

(i) $(\mathrm{Im}\ f)^\perp = \mathrm{Ker}\ {}^tf$,

(ii) $(\mathrm{Ker}\ f)^\perp = \mathrm{Im}\ {}^tf$.

[Remark: Part (i) holds for a general commutative ring K.]

Theorem 8.9. *With the assumption* (8.3) *the following holds.*

(i) P_A *is projective* $\Leftrightarrow$ $_A(P^\wedge)$ *is injective.*

(ii) Q_A *is injective* $\Leftrightarrow$ $_A(Q^\wedge)$ *is projective.*

Proof.

(i) ($\Leftarrow$). Given the diagram (a) below with concrete arrows, we obtain the diagram (b) below with concrete arrows:

(a)

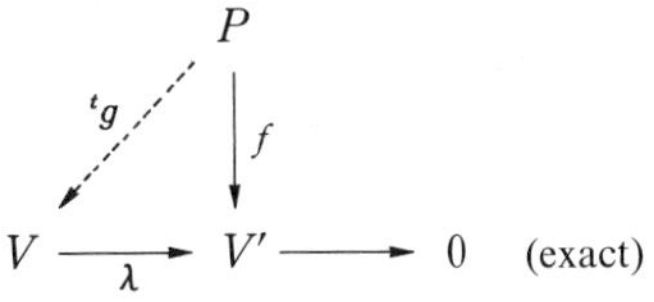

(b)

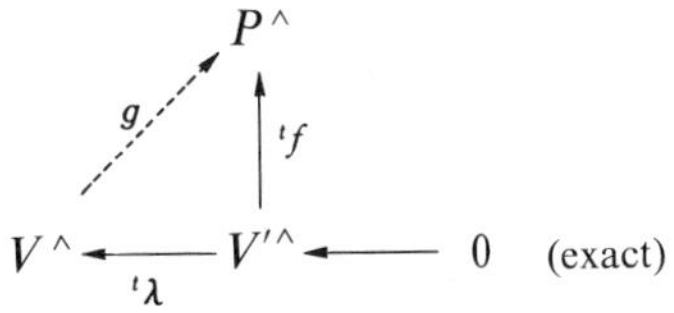

If ${}_A(P^\wedge)$ is injective, there exists $g: V^\wedge \to P^\wedge$, which completes the diagram (b) to a commutative one. Then ${}^t g: P = P^{\wedge\wedge} \to V = V^{\wedge\wedge}$ completes (a) to a commutative diagram. Therefore P_A is projective.

The rest will be proved similarly. ■

8.2. *Frobenius and Symmetric Algebras*

Assumption (8.3) will be kept throughout.

A K-algebra A is said to be a *Frobenius algebra*, provided its right and left regular representations are equivalent. In view of Theorem 8.4 (ii), this amounts to saying that

(8.4) A is a Frobenius algebra $\Leftrightarrow A_A \simeq (A^\wedge)_A$ (A-isomorphic).

Let, as in Section 2,

$$A = \bigoplus_{i=1}^{k} \bigoplus_{\mu=1}^{n_i} e_{i\mu} A \qquad (e_{i\mu} A \simeq e_{j\nu} A \Leftrightarrow i = j)$$

be an indecomposable decomposition of A_A and set $e_i = e_{i1}$. Then

$$A_A \simeq \bigoplus_{i=1}^{k} n_i e_i A. \tag{8.5}$$

In the same way, we have

$${}_A A \simeq \bigoplus_{i=1}^{k} n_i A e_i,$$

and this gives an indecomposable decomposition of $(A^\wedge)_A$ as follows:

$$(A^\wedge)_A \simeq \bigoplus_{i=1}^{k} n_i (A e_i)^\wedge. \tag{8.6}$$

Consequently, if A is a Frobenius algebra, then by applying the K–S–A theorem to the decompositions (8.5) and (8.6), we find a permutation π on $\{1, \ldots, k\}$ such that

$$n_i = n_{\pi(i)}, \qquad e_i A \simeq (A e_{\pi(i)})^\wedge \qquad (i = 1, \ldots, k). \tag{8.7}$$

Now, let $J = J(A)$. Then $Ae_{\pi(i)}$ has a unique maximal A-submodule $Je_{\pi(i)}$, hence $(Ae_{\pi(i)})^\wedge$ has a unique minimal A-submodule $(Je_{\pi(i)})^\perp$, which is isomorphic to $(\bar{A}\bar{e}_{\pi(i)})^\wedge$. But since $\bar{e}_{\pi(i)}\bar{A}$ and $(\bar{A}\bar{e}_{\pi(i)})^\wedge$ belong to the same simple component of $\bar{A}$, they must be isomorphic.

Summarizing the above, we obtain the following theorem.

Theorem 8.10. *Let A be a Frobenius algebra over K and let* (8.5) *be an indecomposable decomposition of A_A. Then there exists a permutation π on $\{1, \ldots, k\}$ such that* (8.7) *holds. Moreover, we have*

$$\operatorname{soc}(e_i A) \simeq \bar{e}_{\pi(i)} \bar{A} \qquad (i = 1, \ldots, k)$$

The next fact is one of the main properties of Frobenius algebras.

Theorem 8.11. *Let A be a Frobenius algebra and V be an A-module. Then V is projective if and only if it is injective.*

Proof. First of all note that A_A, $(A^\wedge)_A$, which are A-isomorphic, are both projective and injective by Theorem 8.9. Now, if V is projective, there is an integer r such that $V | rA$, and hence V is injective. If, conversely, V is injective, then $V^\wedge$ is projective and $V^\wedge | s({}_AA)$ for some integer s. Then, it follows that $V = V^{\wedge\wedge} | s(A^\wedge)_A$, and thus V is projective. ■

Remark. In general a K-algebra A is said to be a *quasi-Frobenius algebra* if A_A is injective.

Before going into the further study of Frobenius algebras, we recall here some elementary facts on bilinear forms.

Suppose that we are given a (K-)bilinear form f on a K-algebra A:

$$f: A \times A \to K. \tag{8.8}$$

Then given $a \in A$, there is a K-linear map

$$\theta_a: A \to K \qquad (x \mapsto f(a, x)).$$

Consequently, we obtain a K-linear map

$$\theta: A \to A^\wedge (a \mapsto \theta_a). \tag{8.9}$$

Conversely, starting with the K-linear map (8.9), we obtain the bilinear form (8.8) by setting $f(a, b) = \theta_a(b)$.

A bilinear form $f: A \times A \to K$ is said to be *nonsingular*, provided

$$f(a, A) = 0 \Rightarrow a = 0$$

holds. This is equivalent to saying that the linear map θ given in (8.9) is an isomorphism.

Exercise 8.12. Let f be as above and let $\{a_1, \ldots, a_n\}$ be a K-basis of A. Then the following conditions are equivalent.

(1) f is nonsingular.

(2) $\det(f(a_i, a_j)) \neq 0$.

(3) $f(A, b) = 0 \Rightarrow b = 0$.

For a K-submodule W of A, we define

$$W^{\perp(l)} = \{a \in A;\ f(a, W) = 0\}, \qquad W^{\perp(r)} = \{a \in A;\ f(W, a) = 0\}. \tag{8.10}$$

If f is nonsingular, then the following holds.

$$\dim_K W + \dim_K W^{\perp(l)} = \dim_K A, \tag{8.11}$$

$$\dim_K W + \dim_K W^{\perp(r)} = \dim_K A, \tag{8.12}$$

$$(W^{\perp(l)})^{\perp(r)} = W, \qquad (W^{\perp(r)})^{\perp(l)} = W. \tag{8.13}$$

We say the bilinear form $f\colon A \times A \to K$ is *associative* if

$$f(ab, c) = f(a, bc) \qquad \text{for all} \quad a, b, c \in A.$$

and *symmetric* if

$$f(a, b) = f(b, a) \qquad \text{for all} \quad a, b \in A.$$

If A has a nonsingular associative bilinear form $f\colon A \times A \to K$, then the associated linear map $\theta\colon A \to A^\wedge$ of (8.9) is an isomorphism as right A-modules, and it follows that A is a Frobenius algebra. In fact, we observe that

$$\theta_{ab}(x) = f(ab, x) = f(a, bx) = \theta_a(bx) = (\theta_a b)(x),$$

and thus $\theta_{ab} = \theta_a b$, which means that θ_a is an A-homomorphism.

Conversely, if A is a Frobenius algebra and $\theta\colon A_A \to (A^\wedge)_A (a \mapsto \theta_a)$ is an A-isomorphism, we obtain a nonsingular associative bilinear form $f\colon A \times A \to K$ by setting $f(a, b) = \theta_a(b)$.

Thus we have shown the equivalence of (1)$\Leftrightarrow$(2) in the following theorem.

Theorem 8.13. *The following three conditions on a K-algebra A are equivalent.*

(1) *A is a Frobenius algebra.*

(2) *A has a nonsingular associative bilinear form $f\colon A \times A \to K$.*

(3) *There exists a K-linear map $\lambda\colon A \to K$ such that* Ker λ *contains no right ideal of A different from zero.*

Proof.

(2) $\Rightarrow$ (3). Define λ by $\lambda(a) = f(a, 1)$. If $\lambda(aA) = 0$, then $f(aA, 1) = f(a, A) = 0$, whence we have $a = 0$. Therefore Ker λ contains no right ideal other than 0.

(3) $\Rightarrow$ (2). Define $f: A \times A \to K$ by $f(a, b) = \lambda(ab)$. Then f is clearly associative. If $f(a, A) = 0$, then $\lambda(aA) = 0$, and it follows that $a = 0$. ∎

An element λ of $A^\wedge$ is said to be *regular* if it satisfies condition (3) of the above theorem. It is said to be *symmetric* if the following holds for all $a, b \in A$:

$$\lambda(ab) = \lambda(ba).$$

If there exists a regular symmetric linear function $\lambda \in A^\wedge$, then A is called a *symmetric algebra.*

The next theorem can be proved quite similarly to Theorem 8.13; therefore, we omit the detail.

Theorem 8.14. *The following conditions on a K-algebra A are equivalent.*

(1) *A is symmetric.*

(2) *A has a nonsingular bilinear form $f: A \times A \to K$, which is associative and symmetric.*

Exercise 8.15. An element $\lambda \in A^\wedge$ is regular if and only if Ker λ contains no left ideal different from zero. [Hint: Use Exercise 8.12.]

Let us show a couple of examples of symmetric algebras.

Theorem 8.16. *If A is semisimple, then A is symmetric.*

Proof. We may assume that A is simple. First, we shall show that A is symmetric as a Z-algebra, where $Z = Z(A)$. Because A is a central simple Z-algebra, $A^\circ \otimes_Z A$ is also central simple over Z, and it follows that A is a unique irreducible right $A^\circ \otimes_Z A$-module. On the other hand, A can be viewed as a left $A^\circ \otimes_Z A$-module by $(a^\circ \otimes b)x = bxa$. Then $A^\wedge = \mathrm{Hom}_Z(A, Z)$ is an irreducible right $A^\circ \otimes_Z A$-module. Consequently, there is an isomorphism

$$g: A \simeq A^\wedge$$

as $A^\circ \otimes_Z A$-modules. Let $\lambda = g(1_A) \in A^\wedge$. We claim that $\lambda: A \to Z$ is regular and symmetric. In fact, we have $[\lambda \cdot (a^\circ \otimes b)](x) = \lambda(bxa)$ for a, b, $x \in A$, because g is an $A^\circ \otimes_Z A$-isomorphism. Since $1_A(a^\circ \otimes 1) = a = 1_A(1^\circ \otimes a)$, we have $[\lambda \cdot (a^\circ \otimes 1)](b) = [\lambda \cdot (1^\circ \otimes a)](b)$, namely, $\lambda(ba) = \lambda(ab)$ and λ is symmetric. If $a \neq 0$, then $0 \neq g(a) = \lambda \cdot (1^\circ \otimes a)$. Therefore $[\lambda \cdot (1^\circ \otimes a)](A) = \lambda(aA) = Z$ and λ is regular, as claimed.

Now, let $\mu: Z \to K$ be any K-linear map such that $\mu(1) = 1$. Then $\mu \circ \lambda: A \to K$ is a symmetric K-linear map. Moreover if $A \ni a \neq 0$, then $\mu(\lambda(aA)) = \mu(Z) = K$, and hence $\mu \circ \lambda$ is regular. Therefore A is a symmetric K-algebra as asserted. ■

Let G be a group acting on a field L. If $\alpha: G \times G \to L^\times$ is a factor set, then the twisted group ring

$$(L, G, \alpha) = \bigoplus_{x \in G} u_x L \qquad (u_x u_y = u_{xy}\alpha(x, y),\ \gamma u_x = u_x \gamma^x\ (\gamma \in L))$$

is an algebra over the fixed field $K = L^G$. If in particular $K = L$, that is, G acts trivially on L, then (K, G, α) is called the *generalized group ring* of G over K with factor set α, which is denoted by $K^{(\alpha)}G$.

Theorem 8.17. *With the above notation, the following holds.*

(i) *The twisted group ring is a symmetric algebra over the fixed field $K = L^G$ of G.*

(ii) *If* Char $L \nmid |G|$, *then (L, G, α) is semisimple.*

Proof. Let $A = (L, G, \alpha)$. We may assume that $u_1 = 1$. In particular it holds that $\alpha(x, x^{-1})^x = \alpha(x^{-1}, x)$ (Exercise 7.8).

(i) Let H be the kernel of the action of G on L, i.e., $H = \{x \in G;\ \gamma^x = \gamma$ for all $\gamma \in L.\}$. Then, as is well known, L is a Galois extension of K with Galois group $\bar{G} = G/H$. Furthermore, the trace map $T_{L/K}: L \to K(\gamma \mapsto \sum_{\bar{x} \in \bar{G}} \gamma^{\bar{x}})$ is an epimorphism; $T_{L/K}(L) = K$ (cf. Hungerford [1], Chapter V).

Now, define $\lambda: A \to L$ by $\lambda(\sum_x u_x\alpha_x) = \alpha_1$. We show the composite map $\mu = T_{L/K} \circ \lambda: A \to K$ is regular and symmetric. To show that μ is regular, let $a = \sum_x u_x\alpha_x$ be any nonzero element of A. Then there exists x such that $\alpha_x \neq 0$, and hence $\lambda(a(u_x\alpha_x)^{-1}) = 1$. This yields that $\lambda(aA) = L$ and $\mu(aA) = T_{L/K}(L) = K$, i.e., μ is regular. For $a = \sum_x u_x\alpha_x$, $b = \sum_y u_y\beta_y$, we have

$$\lambda(ab) = \sum_x \alpha(x, x^{-1})(\alpha_x)^{x^{-1}}\beta_{x^{-1}}, \qquad \lambda(ba) = \sum_x \alpha(x^{-1}, x)(\beta_{x^{-1}})^x\alpha_x.$$

Here we see that $(\alpha(x, x^{-1})(\alpha_x)^{x^{-1}}\beta_{x^{-1}})^x = \alpha(x^{-1}, x)(\beta_{x^{-1}})^x\alpha_x$, then taking the trace of both sides, we obtain the same element of K. Therefore $\mu(ab) = \mu(ba)$ and μ is symmetric.

(ii) For $a \in A$, let

$$au_y = \sum_{x \in G} u_x\gamma_{x,y}(a) \qquad (\gamma_{x,y}(a) \in L).$$

Then we obtain a ring homomorphism $\mathbf{X}: A \to M_n(L) \quad (a \mapsto X(a) = (\gamma_{x,y}(a))$, where $n = |G|$.

We deduce a contradiction assuming that $J(A) \neq 0$. As is seen in the proof of (i), $J(A)$ contains an element a of the form $a = 1 + \sum_{x \neq 1} u_x\beta_x$. Since a is nilpotent, $X(a)$ is also nilpotent and hence $\operatorname{tr} X(a) = 0$. On the other hand, if $x \neq 1$, then the diagonal entries of $X(u_x\beta_x)$ are zero and $\operatorname{tr} X(u_x\beta_x) = 0$. This means that $\operatorname{tr} X(a) = |G| \cdot 1 \neq 0$, a contradiction. Therefore $J(A) = 0$. ■

Returning to a general K-algebra A, we let, for a subset S of A,

$$l(S) = \{a \in A; aS = 0\} \qquad \text{and} \qquad r(S) = \{a \in A; Sa = 0\},$$

which are called the left and the right *annihilator ideal* of S, respectively.

Theorem 8.18. *If A is a Frobenius algebra, then, for any right ideal I and any left ideal L of A, the following holds.*

(i) $r(l(I)) = I$, $\dim_K I + \dim_K l(I) = \dim_K A$.

(ii) $l(r(L)) = L$, $\dim_K L + \dim_K r(L) = \dim_K A$.

Proof. By assumption, there is a nonsingular associative bilinear form $f: A \times A \to K$. We know that the K-linear map $\lambda: A \to K(\lambda(a) = f(a, 1))$ is regular. Thus, for any $x \in A$,

$$xI = 0 \Leftrightarrow \lambda(xI) = 0 \Leftrightarrow f(x, I) = 0.$$

This implies that $l(I) = I^{\perp(l)}$. Similarly we have $r(L) = L^{\perp(r)}$. Our assertion now follows from the facts (8.11)–(8.13). ■

Remark. It is known that the converse of the above theorem is also true (Nakayama [2]). (See Curtis and Reiner [1].)

Next we study a special automorphism of a Frobenius algebra.

Let A be a Frobenius algebra with nonsingular associative bilinear form $f: A \times A \to K$ and let $\lambda \in A^\wedge$ be the K-linear map defined by $\lambda(a) = f(a, 1)$. Using the notation of (8.9), we have an isomorphism $\theta: A \xrightarrow{\sim} A^\wedge$.

Similarly, we have an isomorphism $\theta': A \xrightarrow{\sim} A^\wedge (\theta \mapsto \theta'_a)$ such that $\theta'_a(x) = f(x, a)$ for $a, x \in A$. Consequently, for each $a \in A$, there exists a unique $b \in A$ such that $\theta_a = \theta'_b$. If we write $b = a^\sigma$, then we get a bijection $\sigma: A \to A (a \mapsto a^\sigma)$, which satisfies

$$(8.14) \qquad f(a, x) = f(x, a^\sigma), \qquad \text{equivalently,} \qquad \lambda(ax) = \lambda(xa^\sigma).$$

Theorem 8.19. *Let A be a Frobenius algebra and let $\sigma: A \to A$ be a map satisfying* (8.14). *Then*

(i) *σ is a K-algebra automorphism of A.*

(ii) *For a right ideal I of A, there are A-isomorphisms*

$$I \simeq (A/l(I)^\sigma)^\wedge, \qquad A/I \simeq (l(I)^\sigma)^\wedge.$$

(iii) *If π is the permutation given in* (8.7), *then*

$$Ae_{\pi(i)} \simeq Ae_i^\sigma.$$

Proof.

(i) Clearly σ is a K-automorphism. Also

$$\lambda(x(ab)^\sigma) = \lambda(abx) = \lambda(bxa^\sigma) = \lambda(xa^\sigma b^\sigma),$$

whence we have $(ab)^\sigma = a^\sigma b^\sigma$ from the regularity of λ.

(ii) We know that $\theta: A \xrightarrow{\sim} A^\wedge (a \mapsto \theta_a)$ is an A-isomorphism. By definition, the orthogonal complement of $\theta_I = \{\theta_a; a \in I\}$ is $I^{\perp(r)}$.

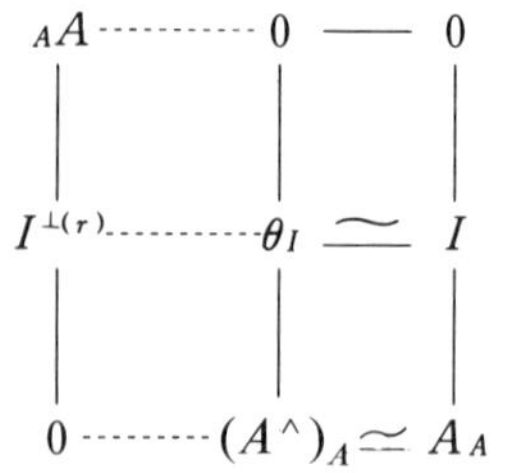

Hence we have the following A-isomorphisms:

$$I \simeq \theta_I \simeq (A/I^{\perp(r)})^{\wedge},$$
$$A/I \simeq A^{\wedge}/\theta_I \simeq (I^{\perp(r)})^{\wedge}.$$

Thus, it remains to show that $I^{\perp(r)} = l(I)^{\sigma}$. But, noting that $l(I)^{\sigma} = l(I^{\sigma})$ and xI^{σ} is a right ideal of A, we have

$$x \in l(I^{\sigma}) \Leftrightarrow \lambda(xI^{\sigma}) = 0 \Leftrightarrow \lambda(Ix) = f(I, x) = 0 \Leftrightarrow x \in I^{\perp(r)},$$

as required.

(iii) Apply the above (ii) to $I = e_i A$. Then $l(I) = A(1 - e_i)$, and $A/l(I)^{\sigma} \simeq Ae_i^{\sigma}$. Therefore $(e_i A)_A \simeq ((Ae_i^{\sigma})^{\wedge})_A$, and we have $Ae_i^{\sigma} \simeq Ae_{\pi(i)}$ by (8.7). ■

The above σ is called the *Nakayama automorphism* of the Frobenius algebra A.

If in particular A is a symmetric algebra, then $\sigma = \mathrm{id}_A$ by (8.14), and it follows from the above theorem that $\pi(i) = i$ for all $i (1 \leq i \leq k)$. So we have the following theorem from (8.7) and Theorem 8.10.

Theorem 8.20. *Let A be a symmetric algebra. Then, for any $e \in \mathrm{pi}(A)$, the following A-isomorphisms hold.*

$$(eA)_A \simeq (Ae)_A^{\wedge}, \qquad \mathrm{soc}(eA) \simeq \mathrm{hd}(eA) = \bar{e}\bar{A}.$$

In particular, the representation of A defined by eA is equivalent to that of A defined by Ae.

Theorem 8.21. *Let A be a symmetric algebra over K and suppose that K is a splitting field for A. Then*

(i) *Given P_A and V_A with P projective, we have*

$$\dim_K \mathrm{Hom}_A(P, V) = \dim_K \mathrm{Hom}_A(V, P).$$

(ii) *The Cartan matrix $C = (c_{ij})$ is symmetric; $c_{ij} = c_{ji}$. Also we have $c_{ii} \geq 2$, unless $e_i A$ is irreducible.*

Proof.

(i) We may assume that $P = e_i A$ for some $e_i \in \mathrm{pi}(A)$. We proceed by induction on the dimension of V. If V is an irreducible module such that

$V \simeq \bar{e}_j\bar{A}$, then the left-hand side of the above equality is $\dim_K \bar{e}_j\bar{A}\bar{e}_i = \delta_{ij}$, which is equal to the right-hand side because $\mathrm{soc}(e_iA) \simeq \bar{e}_i\bar{A}$.

Assume next that V has a proper submodule $W \neq 0$, and consider the exact sequence

$$0 \to W \to V \to V/W \to 0.$$

Since P is also injective, the following sequence is exact:

$$0 \to \mathrm{Hom}_A(V/W, P) \to \mathrm{Hom}_A(V, P) \to \mathrm{Hom}_A(W, P) \to 0.$$

Consequently, we have

$$\dim_K \mathrm{Hom}_A(V, P) = \dim_K \mathrm{Hom}_A(W, P) + \dim_K \mathrm{Hom}_A(V/W, P).$$

By applying the inductive hypothesis to W and V/W, we get the assertion from the above.

(ii) $c_{ij} = \dim_K \mathrm{Hom}_A(e_jA, e_iA) = \dim_K \mathrm{Hom}_A(e_iA, e_jA) = c_{ji}$. The second half is immediate from Theorem 8.20. ■

Note that the following holds for the annihilator ideals of the radical of a general K-algebra A:

$$l(J(A)) = \{a \in A;\ aJ(A) = 0\} = \mathrm{soc}(A_A),$$
$$r(J(A)) = \{a \in A;\ J(A)a = 0\} = \mathrm{soc}({}_AA),$$

and both of them are ideals of A.

Theorem 8.22. *If A is a Frobenius algebra, then*

$$l(J(A)) = r(J(A))$$

holds, and hence $\mathrm{soc}(A_A) = \mathrm{soc}({}_AA)$.

Proof. Let $J = J(A)$. Let σ be the Nakayama automorphism of A given as in (8.14). Then $J^\sigma = J$, and we have the following:

$$x \in r(J) \Leftrightarrow \lambda(Jx) = 0 \Leftrightarrow \lambda(xJ^\sigma) = \lambda(xJ) = 0 \Leftrightarrow x \in l(J). \qquad ■$$

Finally, we study factor rings of symmetric algebras. Suppose that A is a symmetric algebra and let $\lambda \in A^\wedge$ be a regular and symmetric function. For $c \in A$, we define the K-linear map

$$\lambda_c\colon A \to K \qquad (x \mapsto \lambda(cx)).$$

Since λ is regular, the map $A \to A^\wedge (c \mapsto \lambda_c)$ must be a K-isomorphism. This implies that $A^\wedge = \{\lambda_c; c \in A\}$.

Theorem 8.23. *Let A be a symmetric algebra. Then, for every ideal I of A, we have the following.*

(i) $r(I) = l(I)$.

(ii) *The following two conditions are equivalent.*

(1) *A/I is a symmetric algebra.*

(2) *$r(I) = l(I) = cA$ for some $c \in Z(A)$.*

(iii) *$r(J(A)) = l(J(A)) = cA$ for some $c \in Z(A)$.*

Proof. Let $\lambda \in A^\wedge$ be regular and symmetric, and let $\bar{A} = A/I$.

(i) We see that

$$x \in r(I) \Leftrightarrow \lambda(Ix) = 0 \Leftrightarrow \lambda(xI) = 0 \Leftrightarrow x \in l(I),$$

proving (i).

(ii) (1) $\Rightarrow$ (2). There is a K-linear map $\bar{\mu}: \bar{A} \to K$, which is regular and symmetric. Then the induced K-linear map $\mu: A \to K (x \mapsto \bar{\mu}(\bar{x}))$ is written, as remarked above, in the form $\mu = \lambda_c$ for some $c \in A$. Since μ is symmetric, we have $\lambda(cxy) = \lambda(cyx)$ for all $x, y \in A$. Hence

$$\lambda(cxy) = \lambda(cyx) = \lambda(xcy),$$

and it follows that $cx = xc$, namely, $c \in Z(A)$.

Now, since $\lambda((cA)I) = \lambda(cI) = \mu(I) = 0$, we have $(cA)I = 0$ and $cA \subset l(I)$. On the other hand, if $x \in r(cA)$, then $\lambda(cAx) = 0$, and it follows that $\mu(Ax) = \bar{\mu}(\bar{A}\bar{x}) = 0$. Since $\bar{\mu}$ is regular, this forces that $\bar{x} = 0$, i.e., $x \in I$. Thus we have shown that $r(cA) \subset I$. Using this, we have $cA = Ac = l(r(Ac)) \supset l(I)$, whence the equality $cA = l(I)$ follows.

(2) $\Rightarrow$ (1). The K-linear map $\mu = \lambda_c: A \to K$ is symmetric and $\mu(I) = \lambda(cI) = 0$. Therefore, we obtain a K-linear map $\bar{\mu}: \bar{A} \to K (\bar{a} \mapsto \mu(a))$. We shall show that $\bar{\mu}$ is regular. In fact, if $\bar{\mu}(\bar{a}\bar{A}) = \mu(aA) = 0$, then $\lambda(caA) = 0$, and we have $0 = caA = a(cA)$. Therefore, $a \in l(cA) = l(r(I)) = I$, namely, $\bar{a} = 0$.

(iii) This is clear by (ii), since $A/J(A)$ is a symmetric algebra (Theorem 8.16). ■

8.3. Heller Operators and Frobenius Algebras

We continue to assume (8.3).

Let A be a K-algebra and V be an A-module. We denote by $P(V)$ and $I(V)$ a projective cover and an injective hull of V, respectively. Thus, there are exact sequences

$$0 \to \Omega(V) \xrightarrow{\iota} P(V) \xrightarrow{f} V \to 0, \tag{8.15}$$

$$0 \to V \xrightarrow{h} I(V) \xrightarrow{g} \Omega^{-1}(V) \to 0. \tag{8.16}$$

Lemma 8.24. *With the notation above, the following hold.*

(i) *$P(V)$, $I(V)$ are both finite-dimensional over K, and there are the A-isomorphisms:*

$$P(V)^\wedge \simeq I(V^\wedge), \qquad \Omega(V)^\wedge \simeq \Omega^{-1}(V^\wedge),$$

$$I(V)^\wedge \simeq P(V^\wedge), \qquad \Omega^{-1}(V)^\wedge \simeq \Omega(V^\wedge).$$

(ii) *If $\lambda: P(V) \to V$ is an epimorphism, then* $\operatorname{Ker} \lambda \simeq \Omega(V)$. *Likewise if $\mu: V \to I(V)$ is a monomorphism, then* $\operatorname{Coker} \mu \simeq \Omega^{-1}(V)$.

Proof.

(i) There exist a finitely generated projective A-module P and an A-epimorphism $P \to V$. In particular, P is finite dimensional over K, and we have $P \simeq P(V) \oplus P'$ by Chapter 1, 10.18. Therefore $P(V)$ is also finite-dimensional over K. From the sequence (8.15), we get an exact sequence

$$0 \to V^\wedge \xrightarrow{{}^t f} P(V)^\wedge \to \Omega(V)^\wedge \to 0.$$

Using the fact that f is essential, we easily check that ${}^t f$ is also essential. This yields the upper two isomorphisms in (i). In particular $I(V) \simeq P(V^\wedge)^\wedge$, and it follows that $I(V)$ is finite dimensional over K.

The lower two isomorphisms will be obtained by dualizing the exact sequence (8.16).

(ii) The first statement follows from Chapter 1, 10.18 and the second from Chapter 1, 10.19. ■

Exercise 8.25. Prove the following two isomorphisms:

$$\Omega(V) \simeq \Omega^{-1}(V^\wedge)^\wedge, \Omega^{-1}(V) \simeq \Omega(V^\wedge)^\wedge.$$

Throughout the remainder of this subsection, we assume that A is a Frobenius algebra over K. Remember that every projective A-module is injective and vice versa. In particular, if $V = U \oplus P$ and P is projective, then $\Omega(V) = \Omega(U)$ and $\Omega^{-1}(V) = \Omega^{-1}(U)$ (cf. Chapter 1, 10.17 and the remark preceding it). Hence, as far as the Heller operators are concerned, every module under consideration may be assumed to have no projective module as a direct summand. Such an A-module is said to be *projective-free.*

Theorem 8.26. *If V is projective-free, then the following hold.*

(i) *$\Omega(V)$ and $\Omega^{-1}(V)$ are also projective-free.*

(ii) $\Omega(\Omega^{-1}(V)) \simeq V \simeq \Omega^{-1}(\Omega(V))$.

(iii) *If V is indecomposable, so are $\Omega(V)$ and $\Omega^{-1}(V)$.*

Proof.

(i) We may assume that the map ι in (8.15) is the inclusion map. Suppose that $\Omega(V) = W \oplus P_1$ with nonzero projective module P_1. Since $P_1(\subset P(V))$ is injective, we have $P(V) = P_1 \oplus P_2$, and it follows that $f|_{P_2}: P_2 \to V$ is an epimorphism. This contradicts that $f: P(V) \to V$ is essential.

The proof of the statement for $\Omega^{-1}(V)$ is dual to the above.

(ii) Since $I(V)$ is projective in the exact sequence (8.16), we see from Chapter 1, 10.18 that

$$I(V) \simeq P(\Omega^{-1}(V)) \oplus P', \qquad V \simeq \operatorname{Ker} g \simeq \Omega(\Omega^{-1}(V)) \oplus P'$$

for some P'. However, since P' is projective and V is projective-free, it follows that $P' = 0$, and the first isomorphism holds.

The proof of the second isomorphism is dual to the above.

(iii) This is clear by (i) and (ii). ■

Exercise 8.27. Prove the assertion concerning $\Omega^{-1}(V)$ in (i) and the second isomorphism in (ii).

The following result is useful.

Corollary 8.28. *Let*

$$0 \to U \overset{\mu}{\to} P \overset{\lambda}{\to} V \to 0$$

be an exact sequence of A-modules, where P is projective and V is projective-free. Then the following two conditions are equivalent.

(1) *P is a projective cover of V.*

(2) *P is an injective hull of U.*

Proof.

(1) ⇒ (2). If $P = P(V)$, then by Lemma 8.24(ii), we have $U \simeq \Omega(V)$, and so $\Omega^{-1}(U) \simeq \Omega^{-1}(\Omega(V)) \simeq V$. On the other hand, since P is injective, there exists Q_A such that $P \simeq I(U) \oplus Q$, $V \simeq \operatorname{Coker} \mu \simeq \Omega^{-1}(U) \oplus Q$ by Chapter 1, 10.19. Therefore, $Q = 0$ and $P \simeq I(U)$.

(2) ⇒ (1). The proof is dual to the above. ■

Also, we have the following theorem.

Theorem 8.29. *Let V_A be projective-free. Then the following holds.*

(i) $\operatorname{soc}(\Omega(V)) = \operatorname{soc}(P(V))$.

(ii) *Suppose that A is symmetric and let $J = J(A)$. Then*

$$\operatorname{soc}(\Omega(V)) \simeq V/VJ.$$

Proof.

(i) Let $P = P(V)$, and let $P = \bigoplus_i P_i$ be a decomposition into the direct sum of principal indecomposable modules P_i. Note that $\operatorname{soc}(P_i)$ is irreducible. Since $P/\Omega(V) \simeq V$ is projective-free, we see easily that $P_i \cap \Omega(V) \neq 0$, and hence $\operatorname{soc}(P_i) \subset \Omega(V)$ for all i. Therefore $\operatorname{soc}(P) \subset \Omega(V)$, and we have $\operatorname{soc}(P) = \operatorname{soc}(\Omega(V))$.

(ii) Since A is symmetric, $\operatorname{soc}(P) \simeq P/PJ$. On the other hand, we know that $\Omega(V) \subset PJ$ (Chapter 1, 10.15(ii)). Hence the isomorphism $V \simeq P/\Omega(V)$ induces $V/VJ \simeq P/PJ$. By combining this with (i), we get the assertion (ii). ■

Problems

The hypothesis (8.3) is assumed throughout. We let $J = J(A)$.

1. Let $A = \{a = (\alpha_{ij}) \in M_3(K);\ \alpha_{31} = \alpha_{32} = 0\}$. Compute J and A/J, and describe the irreducible A-modules.

2. Let $A \subset M_n(K)$ and suppose that A contains both $e_{ii}M_n(K)$ and $M_n(K)e_{ii}$ for some matrix unit e_{ii} of $M_n(K)$. Then $e_{ii}A$ is injective.

3. Let $A \subset M_n(\mathbf{C})$ and suppose that $a = {}^t(\overline{\alpha_{ij}}) \in A$ for every $a = (\alpha_{ij}) \in A$. Then A is semisimple. Here $\bar{\alpha}$ denotes the complex conjugate of $\alpha \in \mathbf{C}$.

4. Let ζ_d be a primitive dth root of unity in $\mathbf{C}$ and let G be a cyclic group of order m. Then the following $\mathbf{Q}$-algebra isomorphism holds:

$$\mathbf{Q}G \simeq \bigoplus_{d|m} \mathbf{Q}[\zeta_d].$$

5. If B is a finite dimensional K-algebra without identity, then B contains a nonzero nilpotent ideal. Prove this in the following way.

(i) $A = K \oplus B$ is a K-algebra with the multiplication:

$$(\lambda, a)(\lambda', a') = (\lambda\lambda', \lambda a' + \lambda' a + aa'),$$

which has the identity $(1, 0)$ and contains B as an ideal.

(ii) $J(A) \neq 0$.

(iii) $J(A) \cap B \neq 0$.

6. Suppose that A is commutative and K is algebraically closed. Let B be a K-subalgebra of A not necessarily having identity. Then every (non-zero) one dimensional K-representation $\varphi: B \to K$ of B extends to a representation $\hat{\varphi}: A \to K$ of A such that $\hat{\varphi}(1_A) = 1$.

7. Let D be a non-commutative division algebra over $\mathbf{R}$. Prove the following.

(i) $D \supset \mathbf{C}$ and $\dim_{\mathbf{R}} D = 4$.

(ii) There exists $j \in D$ such that

$$j^{-1}ij = -i, \qquad j^2 = -1 \quad (i = \sqrt{-1}).$$

(Hint: The complex conjugation in $\mathbf{C}$ extends to an inner automorphism of D.)

(iii)

$$D \simeq Q_{\mathbf{R}} = \left\{\begin{pmatrix} \alpha & \beta \\ -\bar{\beta} & \bar{\alpha} \end{pmatrix};\ \alpha, \beta \in \mathbf{C}\right\}$$

(the *division ring of quaternions*).

8. Let L be a separable extension field of K. Then
(i) $J(A^L) = J(A)^L$.
(ii) For any A^L-module V, there exists an A-module W such that $V \mid W^L$.

9. With the assumption of Theorem 6.3 let L be an intermediate field of K and E. Then
(i) $m_L(\tilde{V}) \mid m_K(\tilde{V})$.
(ii) $m_K(\tilde{V}) \mid m_L(\tilde{V}) \dim_K L$.

10. Let V be a G-module and $T \leq G$. For $f \in Z^2(G, V)$, f_T denotes the restriction of f to $T \times T$. Prove that

$$f_T \sim 0 \Rightarrow |G:T| f \sim 0$$

in the following way. By assumption there exists $g: T \to V$ such that

$$f(x, y) = g(y) - g(xy) + g(x)y, \qquad \text{for all} \quad x, y \in T.$$

Fixing a set of coset representatives $T \backslash G$, we define $d: G \to V$ by

$$d(cx) = f(c, x) - g(c)x - g(1) \qquad (c \in T, x \in T \backslash G)$$

and $h: G \times G \to V$ by

$$h(a, b) = f(a, b) + d(b) - d(ab) + d(a)b.$$

Then $h \sim f$ and the following holds.
(i) $h(a, b) = 0 \qquad (a \in T, b \in G)$.
(ii) $h(ax, b) = h(x, b) \qquad (a \in T, x \in T \backslash G, b \in G)$.
(iii) Set $\rho(y) = \sum_{x \in T \backslash G} h(x, y)$. Then from the equation

$$h(xy, z) + h(x, y)z = h(y, z) + h(x, yz),$$

we have

$$|G:T| h(y, z) = \rho(z) - \rho(yz) + \rho(y)z.$$

11. Let $f \in Z^2(G, V)$. Then the following holds:

$$f \sim 0 \Leftrightarrow f_S \sim 0 \text{ for all prime } p \text{ dividing } |G| \text{ and } S \in \mathrm{Syl}_p(G).$$

12. Let $G = \langle x, y; x^5 = y^{11} = 1, x^{-1}yx = y^3 \rangle$, $H = \langle x \rangle$, and let $L = \mathbf{Q}[\zeta_{11}]$. Regarding H as a subgroup of $(\mathbf{Z}/(11))^\times = \mathrm{Gal}(L/\mathbf{Q})$, take the crossed product $A = (L/K, 1) = \bigoplus_{x \in H} u_x L \simeq M_5(K)$, where $K = L^H$. Then the following holds.
(i) $\varphi: \mathbf{Q}G \to A$ ($\varphi(x) = u_x$, $\varphi(y) = \zeta_{11}$) is an epimorphism.
(ii) $\mathbf{Q}G \simeq \mathbf{Q} \oplus \mathbf{Q}[\zeta_5] \oplus M_5(K)$.

13. If a generalized group ring $K^{(\alpha)}G$ has a one-dimensional K-representation, then it is isomorphic to the group ring KG.

14. For an A-module V the following hold ($i \geq 0$).

(i) $(V/S_i(V))^{\wedge} = J^i V^{\wedge}$.

(ii) $(S_{i+1}(V)/S_i(V))^{\wedge} \simeq J^i V^{\wedge}/J^{i+1}V^{\wedge}$ (A-isomorphic).

15. (i) Let e be an idempotent of A. Then there is an A-isomorphism:

$$l(J)e \simeq \mathrm{Hom}_A(eA/eJ, {}_AA_A).$$

(ii) $r(l(J)) = J \Leftrightarrow$ Every irreducible right A-module is isomorphic to a right ideal of A.

16. Consider $B = A \oplus {}_A(A^{\wedge})_A$ as a K-algebra via the principle of idealization (Chapter 1, Problem 1) and define $\lambda\colon B \to K$ by $\lambda(a, \varphi) = \varphi(1)(a \in A, \varphi \in A^{\wedge})$. Then λ is regular and symmetric, and hence B is a symmetric algebra.

17. Let A be a Frobenius algebra and $\bar{A} = A/J$. If K is a splitting field for A, then, for any $e \in \mathrm{pi}(A)$, the multiplicity of $\bar{e}\bar{A}$ in A_A is equal to $\dim_K eA$.

18. A is a symmetric algebra $\Leftrightarrow {}_AA_A \simeq {}_A(A^{\wedge})_A$.

19. Let A be a symmetric algebra.

(i) For any V_A, an A-isomorphism ${}_AV^{\wedge} \simeq \mathrm{Hom}_A(V, {}_AA_A)$ holds.

(ii) Give an alternative proof to the fact that $\mathrm{hd}(eA) \simeq \mathrm{soc}(eA)$ by making use of the fact that $l(J) = Ac$ for some $c \in Z(A)$.

20. (Wedderburn). If A/J is separable, then there exists a subring B of A such that $A = B \oplus J$. Prove this in the following way.

(i) Let Γ be a basic ring of A (cf. Chapter 1, Problem 16). Then $\bar{\Gamma} = \Gamma/J(\Gamma)$ is also separable. If the assertion holds for Γ, then it holds for A.

(ii) We may assume $J^2 = 0$ by induction on $\dim_K A$.

(iii) Let L be any extension field of K. If our assertion holds for A^L, then it holds for A; Suppose that $A^L = A' \oplus J^L$ for some subalgebra A'. If $\{1, x_\alpha; \alpha \in I\}$ denotes a K-basis of L, then $W = \bigoplus_{\alpha \in I} x_\alpha \otimes J$ is an (A, A)-submodule of J^L such that $J^L = J \oplus W$. Then $B = (A' \oplus W) \cap A$ is a subalgebra of A, and $A = B \oplus J$ holds.

(iv) From the above we may assume that $A = \Gamma$ and K is algebraically closed.

21. Let Λ, Γ be algebras over a principal ideal domain R. Then for V_Λ and W_Γ, which are assumed to be finitely generated and free over R, the following isomorphism holds:

$$\mathrm{End}_\Lambda(V) \otimes \mathrm{End}_\Gamma(W) \simeq \mathrm{End}_{\Lambda \otimes \Gamma}(V \otimes W),$$

where $\otimes = \otimes_R$.

3 Representations of Groups

After the preparatory Chapters 1 and 2, we shall establish here the fundamental theory of representations of finite groups, including modular representations.

The first section explains basic methods. This is followed by classical theory of ordinary representations in Section 2; a complete description of Clifford theory in Section 3 and Section 5; Brauer's characterization of characters in Section 4 (following the method of Issacs[2]); Schur's theory of projective representations in Section 5; and an introductory exposition of the theory of modular representations in Section 6, with emphasis on Brauer characters. Module-theoretical treatments of modular representations and Brauer's main theorems will be given in Chapter 5.

We have provided new approaches at several stages; for instance, new proofs are presented for the orthogonality relations (Theorems 2.2 and 2.5) and for the results on the degree of irreducible representations, ordinary or projective (Theorems 2.4 and 5.6).

1. Representations of Groups and Group Rings

1.1. Representations of Groups

Let G be a group and let R be a commutative ring. By an *R-representation* or a representation of G over R, we mean a group homomorphism

$$\mathbf{X}: G \to \mathrm{GL}_n(R) \qquad (x \mapsto X(x)). \tag{1.1}$$

The integer n is called the *degree* of $\mathbf{X}$. An *R-character* of G means the character defined by some R-representation of G.

By extending (1.1) linearly to the group ring RG, we obtain the representation

$$\mathbf{X}: RG \to M_n(R) \qquad \left(\sum_x \alpha_x x \mapsto \sum_x \alpha_x X(x)\right) \tag{1.2}$$

of RG. The character defined by this representation is just the linear extension to RG of the R-character of G given above.

Conversely, if a representation $\mathbf{X}$ of RG is given as in (1.2), then $X(1) = I$ (the identity matrix), and it follows that $X(x) \in \mathrm{GL}_n(R)$ for every $x \in G$. Thus $\mathbf{X}$ gives rise to a representation of G. Thus the study of representations of G is the same as that of representations of RG. If V is a representation module for the representation (1.2) of RG, it is also referred to as a representation module for the representation (1.1) of G, and the character defined by $\mathbf{X}$ is denoted by $\chi_{\mathbf{X}}$.

When we say that G acts on an R-module V, the action of G is supposed to be R-linear. In other words, we are given a group homomorphism $\varphi: G \to \mathrm{GL}(V)$, where $\mathrm{GL}(V)$ denotes the group of units of $\mathrm{End}_R(V)$. Then the action of G can be extended by linearity to that of RG, making V into an RG-module.

For a G-module V, define

$$\mathrm{Ker}_G\, V = \{x \in G;\, vx = v \quad \text{for all} \quad v \in V\}.$$

This is precisely the kernel of the homomorphism $\varphi: G \to \mathrm{GL}(V)$, and hence it is a normal subgroup of G. If $\mathrm{Ker}_G\, V = 1$, then V is said to be a *faithful G-module.* If $\mathbf{X}: G \to \mathrm{GL}_n(R)$ is a representation defined by V, then $\mathrm{Ker}_G\, V$ coincides with $\mathrm{Ker}\, \mathbf{X}$. We call $\mathbf{X}$ a *faithful representation* of G if $\mathrm{Ker}\, \mathbf{X} = 1$.

We denote by $\mathrm{Inv}_G\, V$ the set of G-invariant elements of V, i.e.,

$$\mathrm{Inv}_G\, V = \{v \in V;\, vx = v \quad \text{for all} \quad x \in G\},$$

which is clearly an R-submodule of V.

Remark. The notion of faithfulness defined above is different from that in a representation of RG. Namely, if the representation (1.2) is faithful, then so is the representation (1.1), but the converse is not true. For example, if $G = \langle x \rangle$ is a cyclic group of prime order p, and ζ denotes a primitive pth root of unity in $\mathbf{C}$, then $\mathbf{X}: G \to \mathbf{C}^\times (x^i \mapsto \zeta^i)$ is a faithful representation of G. But the corresponding representation $\mathbf{X}: \mathbf{C}G \to \mathbf{C}$ is not faithful, as is easily seen by comparing dimensions.

Example 1.1. The *trivial representation*: By assigning 1_R to every element of G, we get a one-dimensional representation

$$\mathbf{1}_G: G \to R^\times.$$

This is defined by the *trivial G-module R_G*, namely, R is considered as a G-module via $\alpha x = \alpha$ for $\alpha \in R$ and $x \in G$. In general, any one-dimensional representation of G is naturally identified with the character defined by it, thus it is also called a *linear character* of G. In particular, $\mathbf{1}_G$ is called the *trivial character* of G.

1.2. Induced Modules

Let $H \leq G$, and let V be an RG-module. We denote by V_H the RH-module obtained by restricting the action of RG on V to RH. If $\mathbf{X}$ is a representation of G defined by a representation module V, then the restriction $\mathbf{X}_H$ of $\mathbf{X}$ to H is defined by V_H.

Let W be an RH-module. Then

$$W^G = W \otimes_{RH} RG \tag{1.3}$$

is an RG-module and is called the *induced module* of W. Similarly, an induced module of a left RH-module can be defined.

If $G = \sum_{i=1}^r Ht_i$ is a right coset decomposition of G relative to H, then

$$RG = \bigoplus_{i=1}^{r} (RH)t_i. \tag{1.4}$$

Thus, RG is RH-free as a left RH-module (and similarly RG is free as a right RH-module).

From (1.3) and (1.4), it follows that

$$W^G = \bigoplus_{i=1}^{r} W \otimes t_i.$$

The action of each $x \in G$ is described as follows: for each t_i, there exist a unique $h \in H$ and $t_{i'}$ such that

$$t_i x = h t_{i'}, \tag{1.5}$$

and it holds, for $w \in W$, that

$$(w \otimes t_i)x = w \otimes (t_i x) = w \otimes (h t_{i'}) = (wh) \otimes t_{i'}. \tag{1.6}$$

Now we assume that $W = \bigoplus_{i=1}^{m} R w_i$ is R-free and let $\mathbf{Y}$ be the representation of H defined by W relative to the basis $\{w_1, \ldots, w_m\}$. Then

$$\{w_1 \otimes t_1, \ldots, w_m \otimes t_1, \ldots, w_1 \otimes t_r, \ldots, w_m \otimes t_r\}$$

forms an R-basis of W^G, and thus W^G is R-free. We also denote by $\mathbf{Y}^G$ the representation of G defined by W^G and call it the *induced representation* (obtained from $\mathbf{Y}$).

We see from (1.5) that $h = t_i x t_{i'}^{-1} \in H$, while $t_i x t_j^{-1} \notin H$ if $j \neq i'$. Thus, in view of (1.6), the matrix presentation of $\mathbf{Y}^G$ relative to the basis of W^G given above is expressed as

$$\mathbf{Y}^G \colon x \mapsto Y^G(x) = \begin{pmatrix} Y(t_1 x t_1^{-1}) & \cdots & Y(t_1 x t_r^{-1}) \\ \vdots & & \vdots \\ Y(t_r x t_1^{-1}) & \cdots & Y(t_r x t_r^{-1}) \end{pmatrix}, \tag{1.7}$$

where we let $Y(t_i x t_j^{-1}) = 0$, the 0-matrix of degree m, unless $t_i x t_j^{-1} \in H$. Thus the character $\chi_{\mathbf{Y}^G}$ defined by $\mathbf{Y}^G$ is given by

$$\chi_{\mathbf{Y}^G}(x) = \sum_{i=1}^{r} \chi_{\mathbf{Y}}(t_i x t_i^{-1}), \tag{1.8}$$

where we set $\chi_{\mathbf{Y}}(t_i x t_i^{-1}) = 0$ unless $t_i x t_i^{-1} \in H$.

Lemma 1.2. *Let $H \leq G$. Then*

$$\operatorname{Ker}_G W^G = \bigcap_{a \in G} a^{-1}(\operatorname{Ker}_H W)a.$$

Proof. If $x \in \operatorname{Ker}_G W^G$, then, with the notation of (1.5), we find from (1.6) that $i' = i$ and $h \in \operatorname{Ker}_H W$. Namely, $x \in \bigcap_i t_i^{-1}(\operatorname{Ker}_H W)t_i$, and, conversely, this implies that $x \in \operatorname{Ker}_G W^G$. Therefore we have $\operatorname{Ker}_G W^G = \bigcap_i t_i^{-1}(\operatorname{Ker}_H W)t_i$, and this coincides with $\bigcap_{a \in G} a^{-1}(\operatorname{Ker}_H W)a$ since $\operatorname{Ker}_H W \lhd H$. ■

Example 1.3. Let G be a permutation group on a finite set Ω. For $\omega \in \Omega$ and $x \in G$, we denote by ωx the image of ω under the action of x. If $R\Omega = \bigoplus_{\omega \in \Omega} R\omega$ is the free R-module with Ω as basis, then this is a G-module with the action defined by $(\sum_\omega \alpha_\omega \omega)x = \sum_\omega \alpha_\omega(\omega x)$. We call $R\Omega$ a *permutation module* of G (over R).

With the above notation, let $\Omega = \{\omega_1, \ldots, \omega_n\}$ and set $\omega_i x = \omega_{i(x)}$ for $x \in G$. If $\mathbf{P}: x \to P(x) = (\rho_{ij}(x))$ denotes the representation of G defined by $R\Omega$ relative to the basis Ω, then $\rho_{ij}(x) = \delta_{i(x),j}$ (the Kronecker delta); namely, each row and column of $P(x)$ have 1 at exactly one position and 0 at all the rest. We call $\mathbf{P}$ a *permutation representation* of G, which is said to be *transitive* if G acts on Ω transitively.

If $\Omega/G = \{\Omega_1, \ldots, \Omega_t\}$ denotes the set of G-orbits on Ω, then we have a decomposition

$$R\Omega = R\Omega_1 \oplus \cdots \oplus R\Omega_t$$

as RG-modules. Consequently $\mathbf{P}$ splits into a direct sum of transitive permutation representations.

Exercise 1.4. In the situation of Example 1.3, the following holds.

(i) $\chi_{\mathbf{P}}(x) = \#\{\omega \in \Omega;\ \omega x = \omega\} \cdot 1_R$.

(ii) $\mathrm{Inv}_G(R\Omega) = \bigoplus_{i=1}^t R\hat{\Omega}_i$, where $\hat{\Omega}_i = \sum_{\omega \in \Omega_i} \omega$.

Suppose that G acts transitively on Ω. The *stabilizer* G_ω of $\omega \in \Omega$ is the subgroup of G defined by $G_\omega = \{x \in G;\ \omega x = \omega\}$. If $G = \sum_{i=1}^r G_\omega t_i$, then we have $\Omega = \{\omega t_1, \ldots, \omega t_r\}$.

Exercise 1.5. With the same notation and assumption as above, let $H = G_\omega$. Prove that the permutation representation $\mathbf{P}$ on Ω is equivalent to the induced representation $(\mathbf{1}_H)^G$ of $\mathbf{1}_H$. Hence, the following holds.

(i) $\chi_{\mathbf{P}}(x) = \#\{Ht \in H\backslash G;\ Htx = Ht\} \cdot 1_R$.

(ii) $\mathrm{Ker}_G\,\mathbf{P} = \bigcap_{a \in G} a^{-1}Ha$.

Example 1.6. The representation of G defined by RG relative to the basis $G = \{x_1, \ldots, x_g\}$ is a permutation representation, which is called the *regular* *representation* of G and denoted by $\boldsymbol{\Gamma}_G$. If $\{1\}$ denotes the identity subgroup of G, then $(RG)_{RG} \simeq (R_{\{1\}})^G$, and

$$\boldsymbol{\Gamma}_G = (\mathbf{1}_{\{1\}})^G : x \mapsto (\delta_{x_i x, x_j})_{i,j},$$

with the character γ_G given by

$$\gamma_G(x) = \begin{cases} |G| \cdot 1_R & \text{if } x = 1, \\ 0 & \text{otherwise.} \end{cases} \tag{1.9}$$

This formula will be needed later.

Example 1.7. G acts on itself by *conjugation*; $a^x = x^{-1}ax$ for $a, x \in G$. The G-orbits are the conjugate classes of G, and the set of them is denoted by $\mathrm{Cl}(G)$. With respect to this action $\mathrm{Inv}_G(RG)$ coincides with $Z(RG)$, the center of RG. Consequently, if $\mathrm{Cl}(G) = \{C_1, \ldots, C_l\}$, then

$$Z(RG) = \bigoplus_{i=1}^{l} R\hat{C}_i.$$

Now returning to general induced modules, let $H \leq K$ be subgroups of G, and let W be an RH-module. If $K = \sum_i Hs_i$, $G = \sum_j Kt_j$, then $G = \sum_{i,j} Hs_i t_j$. Thus, we have the following RG-isomorphism:

$$(W^K)^G \simeq W^G.$$

For $x \in G$, we define

$$W^x = \{w^x;\ w \in W\},$$

which is isomorphic to W as R-modules via the natural map $w \mapsto w^x$. And this is endowed with the H^x-module structure defined by

$$w^x h^x = (wh)^x \qquad (h \in H),$$

where $H^x = x^{-1}Hx$. We call W^x a *G-conjugate* of W.

Lemma 1.8. *Let $H \leq G$ and let W be an RH-module. Then the following RG-isomorphism holds for $x \in G$.*

$$(W^x)^G \simeq W^G.$$

In particular if V is an RG-module, then $V^x \simeq V$ as RG-modules.

Proof. If $G = \sum_i Ht_i$, then $G = \sum_i H^x t_i^x = \sum_i H^x(x^{-1}t_i)$. Thus, there exists an R-isomorphism $f: W^G = \bigoplus_i W \otimes t_i \to (W^x)^G = \bigoplus_i W^x \otimes (x^{-1}t_i)$ such that $f(w \otimes t_i) = w^x \otimes (x^{-1}t_i)$, which is in fact an RG-isomorphism, as is easily seen.

To show the second half, let $H = G$ in the above. Then $f: V \to V^x (v \mapsto v^x x^{-1})$ gives an RG-isomorphism. ■

The following theorem, which is called the *Mackey decomposition theorem* is useful.

Theorem 1.9 (Mackey). *Let H, K be subgroups of G and let $G = \sum_{t \in H\backslash G/K} HtK$ be a double coset decomposition relative to H, K. Given an RH-module W, we have the following decomposition of the RK-module $(W^G)_K$:*

$$(1.10) \qquad (W^G)_K \simeq \bigoplus_{t \in H\backslash G/K} (W^t_{H^t \cap K})^K \qquad (RK\text{-isomorphism}),$$

where $\bigoplus_{t \in H\backslash G/K}$ indicates the direct sum with t running through a complete set of representatives for double cosets of G relative to H, K.

Proof. Let $K = \sum_j (H^t \cap K)s_{tj}$. Then $G = \sum_{t,j} Hts_{tj}$, and it follows that

$$W^G = \bigoplus_t \left(\bigoplus_j W \otimes ts_{tj} \right).$$

Set $W_t = \bigoplus_j W \otimes ts_{tj}$. There is an R-isomorphism

$$f: W_t \to (W^t_{H^t \cap K})^K = \bigoplus_j W^t \otimes s_{tj}$$

such that $f(w \otimes ts_{tj}) = w^t \otimes s_{tj}$. To show that f is an RK-isomorphism, let $x \in K$ and write $s_{tj}x = h^t s_{tj'}$ with $h^t \in H^t \cap K$. Then we see that

$$(w \otimes ts_{tj})x = w \otimes th^t s_{tj'} = (wh) \otimes ts_{tj'},$$

$$(w^t \otimes s_{tj})x = w^t \otimes h^t s_{tj'} = (wh)^t \otimes s_{tj'}.$$

Therefore f is an RK-isomorphism and (1.10) follows. ■

Remark. In the following, $\bigoplus_{t \in H\backslash G}$ indicates as above the direct sum, with t running through a complete set of representatives for right cosets of H in G.

If we let $H = K$ in the above theorem, then we obtain the following theorem.

Theorem 1.10. *For an RH-module W, we have*

$$(W^G)_H \simeq W \oplus \left(\bigoplus_{\substack{t \in H\backslash G/H \\ t \notin H}} (W^t_{H^t \cap H})^H \right).$$

In particular, we have $W | (W^G)_H$.

1.3. Hom *and* $\otimes_R$

Let U, V be RG-modules. For $f \in \mathrm{Hom}_R(U, V)$ and $x \in G$, we define $f^x: U \to V$ by $f^x(u) = f(ux^{-1})x$ $(u \in V)$. With this action of G, $\mathrm{Hom}_R(U, V)$ becomes a G-module, and we have

$$\mathrm{Inv}_G(\mathrm{Hom}_R(U, V)) = \mathrm{Hom}_{RG}(U, V).$$

In particular the dual module $V^\wedge = \mathrm{Hom}_R(V, R)$ of V is a G-module in the above sense, where R is regarded as the trivial G-module. This is called the *contragredient module* of V. We denote the action of $x \in G$ on $\sigma \in V^\wedge$ by σx, thus

$$\langle v, \sigma x \rangle = \langle v x^{-1}, \sigma \rangle.$$

Remark. In the general context of Chapter 2, $V^\wedge$ is a left RG-module such that $\langle v, x\sigma \rangle = \langle vx, \sigma \rangle$. Define the right RG-module structure on $V^\wedge$ by $\sigma x = x^{-1}\sigma$. Then this makes $V^\wedge$ into the contragredient module of V.

Exercise 1.11. Let $\mathbf{X}: G \to \mathrm{GL}_n(R)$ be a representation of G defined by a representation module V relative to a basis $\{v_i\}$. Then the representation of G defined by the contragredient module $V^\wedge$ relative to the basis dual to $\{v_i\}$ is given by $\hat{\mathbf{X}}: x \mapsto {}^tX(x^{-1})$, where ${}^tX(x^{-1})$ denotes the transpose of $X(x^{-1})$.

The representation $\hat{\mathbf{X}}$ defined above is called the *contragredient representation* of $\mathbf{X}$.

An RG-module V is said to be *self-contragredient* if $V^\wedge \simeq V$.

Lemma 1.12.

(i) $(RG)_{RG}$ *is self-contragredient.*

(ii) *If V is a finitely generated projective RG-module, then the contragredient module $V^\wedge$ is also projective.*

(iii) *Let K be a field and let P be a projective cover of K_G. Then P is self-contragredient.*

Proof.

(i) Let Γ be the regular representation of G. Then we have $\Gamma(x)^t\Gamma(x) = I$ for all $x \in G$, hence $\hat{\Gamma}(x) = {}^t\Gamma(x^{-1}) = \Gamma(x)$. Therefore, $RG \simeq (RG)^\wedge$.

(ii) By assumption, $V | n(RG)$ for some $n > 0$, hence $V^\wedge | n(RG)^\wedge \simeq n(RG)$.

(iii) Since $\mathrm{hd}(P) \simeq K_G$, it follows that $\mathrm{soc}(P^\wedge) \simeq K_G$, hence $P \simeq P^\wedge$ because P is projective. ■

Next, let us consider the tensor product $U \otimes_R V$, where U, V are RG-modules as above. This is a G-module with the action

$$(u \otimes v)x = ux \otimes vx \qquad (x \in G).$$

Exercise 1.13. For $A = (\alpha_{ij}) \in M_m(R)$ and $B = (\beta_{ij}) \in M_n(R)$, we define the tensor product $A \otimes B$ as the following square matrix of degree mn:

$$A \otimes B = \begin{pmatrix} \alpha_{11}B & \cdots & \alpha_{1m}B \\ \vdots & & \vdots \\ \alpha_{m1}B & \cdots & \alpha_{mm}B \end{pmatrix}.$$

(i) There is an isomorphism:

$$M_m(R) \otimes_R M_n(R) \xrightarrow{\sim} M_{mn}(R)\left(\sum_i A_i \otimes_R B_i \mapsto \sum_i A_i \otimes B_i\right).$$

[Hint: If $\{e_{ij}\}$, $\{f_{kl}\}$ are the matrix units of $M_m(R)$, $M_n(R)$, respectively, then $\{e_{ij} \otimes_R f_{kl}\}$ are a basis of $M_m(R) \otimes_R M_n(R)$ and $\{e_{ij} \otimes f_{kl}\}$ are the matrix units of $M_{mn}(R)$.]

(ii) Let U, V be representation modules. Let $\mathbf{X}$ and $\mathbf{Y}$ be representations of G defined by U and V, respectively, relative to the bases $\{u_i\}$ and $\{v_k\}$, respectively. Then the representation of G defined by $U \otimes_R V$ relative to the basis $\{u_i \otimes v_k\}$ of $U \otimes_R V$ arranged in lexicographic order with respect to (i, k) is given by

$$\mathbf{X} \otimes \mathbf{Y}\colon x \mapsto X(x) \otimes Y(x) \qquad (x \in G).$$

The representation $\mathbf{X} \otimes \mathbf{Y}$ of G defined above is called the *tensor product* of the two representations $\mathbf{X}$ and $\mathbf{Y}$. It is clear that

$$\chi_{\mathbf{X}\otimes\mathbf{Y}}(x) = \chi_{\mathbf{X}}(x)\chi_{\mathbf{Y}}(x) \qquad (x \in G).$$

Theorem 1.14. *Let U, V, W be RG-modules.*

(i) *If both U and V are representation modules of G, then*

$$(U \otimes_R V)^\wedge \simeq U^\wedge \otimes_R V^\wedge \qquad (RG\text{-isomorphic}).$$

(ii) *If U is a representation module of G, then we have the following RG-isomorphism:*

$$f\colon U^\wedge \otimes V \simeq \operatorname{Hom}_R(U, V)$$

where $f(\sigma \otimes v)(u) = \langle u, \sigma\rangle v$ for $\sigma \in U^\wedge$, $v \in V$, and $u \in U$.

(iii) *If both U and V are representation modules of G, then*

$$\operatorname{Hom}_{RG}(U \otimes_R V, W) \simeq \operatorname{Hom}_{RG}(U, V^\wedge \otimes_R W).$$

Proof.

(i) Define $\varphi\colon U^\wedge \otimes_R V^\wedge \to (U \otimes_R V)^\wedge$ by $\varphi(\sigma \otimes \rho)(u \otimes v) = \langle u, \sigma\rangle\langle v, \rho\rangle$. It is easy to see that φ is an RG-homomorphism. Let $\{u_i\}$, $\{v_j\}$ be bases of U, V, respectively. If $\{\sigma_i\}$, $\{\rho_j\}$ denote the dual bases to $\{u_i\}$, $\{v_j\}$, respectively, then $\{\varphi(\sigma_i \otimes \rho_j)\}$ is the dual basis to the R-basis $\{u_i \otimes v_j\}$ of $U \otimes_R V$. Therefore φ is an isomorphism.

(ii) For $x \in G$, we have

$$f(\sigma x \otimes vx) = \langle u, \sigma x\rangle vx,$$

and

$$f(\sigma \otimes v)^x(u) = [f(\sigma \otimes v)(ux^{-1})]x = \langle ux^{-1}, \sigma\rangle vx = \langle u, \sigma x\rangle vx,$$

which means that f is an RG-homomorphism.

To see that f is an isomorphism, let $\{u_i\}$ be an R-basis of U and $\{\sigma_i\}$ be the dual basis to it. If $f(\sum_i \sigma_i \otimes v_i) = 0$, then, for any j, we have

$$0 = f\left(\sum_i \sigma_i \otimes v_i\right)(u_j) = \sum_i \langle u_j, \sigma_i\rangle v_i = v_j.$$

Thus f is a monomorphism. Also, for any $\varphi \in \operatorname{Hom}_R(U, V)$, we have

$$f\left(\sum_i \sigma_i \otimes \varphi(u_i)\right)(u_j) = \sum_i \langle u_j, \sigma_i\rangle\varphi(u_i) = \varphi(u_j) \qquad (\text{for all } j).$$

Therefore, $\varphi = f(\sum_i \sigma_i \otimes \varphi(u_i))$, and f is an epimorphism.

(iii) From (i) and (ii), we have the following RG-isomorphisms:

$$\begin{aligned}\mathrm{Hom}_R(U \otimes_R V, W) &\simeq (U \otimes_R V)^\wedge \otimes_R W \simeq U^\wedge \otimes_R V^\wedge \otimes_R W \\ &\simeq \mathrm{Hom}_R(U, V^\wedge \otimes_R W).\end{aligned}$$

Then by taking the sets of G-invariant elements of the first and the last of the above G-modules, we get the assertion. ■

Lemma 1.15. *Let $H \leq G$. The following RG-isomorphisms hold for V_{RG} and W_{RH}:*

(i) $(W^\wedge)^G \simeq (W^G)^\wedge$.

(ii) $V \otimes_R W^G \simeq (V_H \otimes_R W)^G$.

Proof. Let $G = \sum_i Ht_i$ with $t_1 = 1$.

(i) Define

$$\varphi\colon \{\mathrm{Hom}_R(W, R)\}^G \to \mathrm{Hom}_R(W^G, R)$$

by

$$\varphi\Big(\sum_i f_i \otimes t_i\Big)\Big(\sum_i w_i \otimes t_i\Big) = \textstyle\sum_i f_i(w_i).$$

We see easily that φ is an R-isomorphism. To see that φ is an RG-homomorphism, let $x \in G$ and $t_i x = h_i t_{i'} (h_i \in H)$. Then

$$\varphi\Big(\Big(\sum_i f_i \otimes t_i\Big)x\Big)\Big(\sum_i w_i \otimes t_i\Big) = \varphi\Big(\sum_i f_i^{h_i} \otimes t_{i'}\Big)\Big(\sum_i w_i \otimes t_i\Big) = \sum_i f_i(w_{i'} h_i^{-1}),$$

$$\begin{aligned}\varphi\Big(\sum_i f_i \otimes t_i\Big)^x\Big(\sum_i w_i \otimes t_i\Big) &= \varphi\Big(\sum_i f_i \otimes t_i\Big)\Big(\sum_i w_{i'} \otimes t_{i'} x^{-1}\Big) \\ &= \varphi\Big(\sum_i f_i \otimes t_i\Big)\Big(\sum_i w_{i'} h_i^{-1} \otimes t_i\Big) \\ &= \sum_i f_i(w_{i'} h_i^{-1}).\end{aligned}$$

Therefore φ is an RG-homomorphism.

(ii) Define

$$\psi\colon V \otimes_R W^G \to (V_H \otimes_R W)^G$$

by

$$\psi(v \otimes (w \otimes t_i)) = (vt_i^{-1} \otimes w) \otimes t_i.$$

We see easily that ψ is an R-isomorphism. To see that ψ is an RG-homomorphism, let $x \in G$ and $t_i x = ht_{i'} (h \in H)$. Then

$$\begin{aligned}\psi(vx \otimes (w \otimes t_i)x) &= \psi(vx \otimes (wh \otimes t_{i'})) = (vxt_{i'}^{-1} \otimes wh) \otimes t_{i'}\\ &= (vt_i^{-1}h \otimes wh) \otimes t_{i'} = ((vt_i^{-1} \otimes w) \otimes t_i)x.\end{aligned}$$

Therefore ψ is an RG-homomorphism. ■

Corollary 1.16. *If P is RG-projective, then, for any representation module V of G, $V \otimes_R P$ is RG-projective.*

Proof. If $P | n(RG)$ for some $n > 0$, then $V \otimes P | V \otimes n(RG) \simeq n(V \otimes_R RG)$. Thus, it suffices to show that $V \otimes_R RG$ is RG-free.

Note that $RG \simeq R^G = (R_{\{1\}})^G$, and so we have

$$V \otimes_R RG \simeq V \otimes_R R^G \simeq (V_{\{1\}} \otimes_R R)^G \simeq (V_{\{1\}})^G.$$

However, since V is R-free, $(V_{\{1\}})^G$ is RG-free. ■

Lemma 1.15(ii) yields the following, which is called the *Mackey tensor product theorem.*

Theorem 1.17 (Mackey). *Let $H, K \leq G$. Given U_{RH} and W_{RK}, we have the following RG-isomorphism:*

$$U^G \otimes_R W^G \simeq \bigoplus_{t \in H\backslash G/K} (U^t_{H^t \cap K} \otimes_R W_{H^t \cap K})^G.$$

Proof. Write $\otimes$ for $\otimes_R$. We know that $U^G \otimes W^G \simeq ((U^G)_K \otimes W)^G$. And it follows from the Mackey decomposition theorem that

$$(U^G)_K \otimes W \simeq \bigoplus_{t \in H\backslash G/K} (U^t_{H^t \cap K})^K \otimes W \simeq \bigoplus_t (U^t_{H^t \cap K} \otimes W_{H^t \cap K})^K.$$

Thus, the assertion follows by taking the induced modules of both ends. ■

Exercise 1.18. Let P be RG-projective and V be a representation module of G. Then $\mathrm{Hom}_R(P, V)$ and $\mathrm{Hom}_R(V, P)$ are both RG-projective. [Hint: Use Theorem 1.14(ii).]

Theorem 1.19. *Let $H \leq G$. Then the following isomorphisms hold for V_{RG} and W_{RH}.*

(i) $\mathrm{Hom}_{RH}(W, V_H) \simeq \mathrm{Hom}_{RG}(W^G, V)$.

(ii) $\mathrm{Hom}_{RH}(V_H, W) \simeq \mathrm{Hom}_{RG}(V, W^G)$.

Proof.

(i) This follows directly from Chapter 1, 11.3(i) by letting $A = RG$, $B = RH$. In fact, the above isomorphism sends each $g \in \mathrm{Hom}_{RH}(W, V_H)$ onto $g' \in \mathrm{Hom}_{RG}(W^G, V)$, which is defined by

$$g'\left(\sum_{t \in H\backslash G} w_t \otimes t\right) = \sum_t g(w_t)t.$$

(ii) Given $f \in \mathrm{Hom}_{RH}(V_H, W)$, we define $\tilde{f}: V \to W^G$ by

$$\tilde{f}(v) = \sum_{t \in H\backslash G} f(vt^{-1}) \otimes t.$$

Clearly, $\tilde{f} \in \mathrm{Hom}_{RG}(V, W^G)$, and the map $f \to \tilde{f}$ is a monomorphism. To see that this is an epimorphism, take an arbitrary $h \in \mathrm{Hom}_{RG}(V, W^G)$ and set $h(v) = \sum_{t \in H\backslash G} h_t(v) \otimes t$ $(h_t(v) \in W)$. Since h is an RG-homomorphism, it follows that $h_1 \in \mathrm{Hom}_{RH}(V_H, W)$ and $h_t(v) = h_1(vt^{-1})$ for all $t \in H\backslash G$. Thus we have $h = \tilde{h}_1$. ∎

Lemma 1.20. *Let $H \leq G$ and $G = \sum_i Ht_i$ with $t_1 = 1$. Then the following isomorphism holds for W_{RH}:*

$$\mathrm{Tr}: \mathrm{Inv}_H(W) \xrightarrow{\sim} \mathrm{Inv}_G(W^G) \qquad \left(w \mapsto \sum_i w \otimes t_i\right).$$

Proof. It is clear that $\mathrm{Tr}(w)$ is G-invariant and Tr is a monomorphism. To see that this is an epimorphism, let $\mathrm{Inv}_G(W^G) \ni u = \sum_i w_i \otimes t_i$. Thus $ut_i = w_1 \otimes t_i + \cdots = u$, whence we have $w_i = w_1$ for all i. Therefore $u = \mathrm{Tr}(w_1)$. Furthermore, since $uh = u$ for all $h \in H$, it follows that $w_1 \in \mathrm{Inv}_H(W)$. ∎

Exercise 1.21. Let H, $K \leq G$. Given U_{RH} and W_{RK}, the following isomorphism holds:

$$\mathrm{Hom}_{RG}(U^G, W^G) \simeq \bigoplus_{t \in H\backslash G/K} \mathrm{Hom}_{R[H^t \cap K]}(U^t, W).$$

1.4. Group Rings over Fields

K denotes a field throughout the remainder of this section.

Theorem 1.22 (Maschke). *Let KG be the group ring of G over K. Then*

$$KG \text{ is semisimple} \Leftrightarrow \mathrm{Char}\, K \nmid |G|.$$

Proof. The implication ($\Leftarrow$) has been already shown in Chapter 2, 8.17.

($\Rightarrow$) If Char $K = p$ is a divisor of $|G|$, then $z = \sum_{x \in G} x$ is a central element of KG satisfying $z^2 = |G|z = 0$. Thus $z(KG)$ is a nilpotent ideal of KG and hence $J(KG) \neq 0$. ■

We say that K is a splitting field for G if it is a splitting field for KG.

Theorem 1.23. *If* Char $K \nmid |G|$, *then the number of nonequivalent absolutely irreducible representations of G is equal to the number of the conjugacy classes of G.*

Proof. Assume that K is a splitting field for G. The assertion follows from Chapter 2, 3.14 since

$$Z(KG) = \bigoplus_{i=1}^{l} K\hat{C}_i,$$

where $\mathrm{Cl}(G) = \{C_1, \ldots, C_l\}$. ■

As for the group ring of a p-group, the following holds.

Theorem 1.24. *Let* Char $K = p > 0$ *and let* G *be a p-group. Then*

(i) $J(KG) = \bigoplus_{x \in G} K(x - 1) = \{\sum_x \alpha_x x; \sum_x \alpha_x = 0\}$.

(ii) *The trivial representation is a unique irreducible* K*-representation of* G.

(iii) KG *is a local ring.*

(iv) *Every finitely generated projective* KG*-module* V *is free.*

Proof. Let $A = KG$.

(i) Let $I = \bigoplus_{x \in G} K(x - 1)$. It is clear that I is an ideal of A and $A/I \simeq K$, thus $I \supset J(A)$. On the other hand, if $|G| = p^n$, then $(x - 1)^{p^n} = x^{p^n} - 1 = 0$. Therefore I must be nilpotent by Chapter 2, 3.7 and hence $I \subset J(A)$.

Parts (ii) and (iii) are immediate from $A/J(A) \simeq K$.

(iv) Since A_A is indecomposable, $V \simeq nA$ for some $n > 0$. ■

Remark. In the above theorem, part (iv) is true without assuming that V is finitely generated.

Recall that $A = KG$ is a symmetric algebra (Chapter 2, 8.17). Thus, if $e \in \text{pi}(A)$, then

$$\text{hd}(eA) \simeq \text{soc}(eA) \simeq \bar{e}\bar{A},$$

where $\bar{A} = A/J(A)$. Also, the representation of G defined by Ae is equivalent to that of G defined by eA. If K is a splitting field for G, then the Cartan invariants satisfy $c_{ij} = c_{ji}$.

Exercise 1.25. $H \lhd G \Rightarrow J(KH)KG \subset J(KG)$. [Hint: Note that $xJ(KH) = J(KH)x$ for $x \in G$. So, $J(KH)KG$ is nilpotent.]

If P is a principal indecomposable KG-module, then $P_H | (KG)_H$, and as $(KG)_H$ is KH-free, it follows that P_H is KH-projective, which is a direct sum of principal indecomposable KH-modules.

On the other hand, if I is a right ideal of KH, then $I \cdot KG = \bigoplus_{t \in H \backslash G} It$, which is KG-isomorphic to the induced module I^G. In particular if $I | (KH)_{KH}$, it follows that $I^G | (KG)_{KG}$. Thus, if Q is a principal indecomposable KH-module, then $Q^G | KG$, that is, Q^G is KG-projective.

Theorem 1.26. *Suppose that* Char $K = p > 0$ *and let* $|G| = p^a h$ *with* $(p, h) = 1$. *If P is a finitely generated projective KG-module, then*

$$\dim_K P \equiv 0 \qquad (\text{mod } p^a).$$

Proof. Let H be a Sylow p-subgroup of G. As observed above, P_H is projective, hence free by Theorem 1.24. This implies the assertion. ∎

Now assume that K *is a splitting field for* G, H and let $\{P\}_A$ (respectively, $\{Q\}_Q$) denote a complete set of representatives for nonisomorphic principal indecomposable KG-modules (respectively, KH-modules).

Theorem 1.27 (Frobenius–Nakayama reciprocity). *With the notation and the assumption above, let* $\bar{P} = \text{hd}(P)$, $\bar{Q} = \text{hd}(Q)$. *Then*

(i) $P_H \simeq \bigoplus_Q r_{PQ} Q \Leftrightarrow \bar{Q}^G \leftrightarrow \sum_P r_{PQ} \bar{P}$.

(ii) $Q^G \simeq \bigoplus_P s_{PQ} P \Leftrightarrow \bar{P}_H \leftrightarrow \sum_Q s_{PQ} \bar{Q}$.

(iii) $P_H \leftrightarrow \sum_Q t_{PQ} \bar{Q} \Leftrightarrow Q^G \leftrightarrow \sum_P t_{PQ} \bar{P}$.

Proof. Let $A = KG$, $B = KH$. If P, $P' \in \{P\}$, then $\dim_K \text{Hom}_A(P, \bar{P}') = \delta_{P,P'}$. In fact, the left-hand side gives the multiplicity of $\bar{P}$ in $\bar{P}'$, which is equal to $\delta_{P,P'}$. Similarly, $\dim_K \text{Hom}_B(Q, \bar{Q}') = \delta_{Q,Q'}$ if Q, $Q' \in \{Q\}$.

(i) ($\Rightarrow$). From the isomorphism given in the assumption, it follows that

$$\dim_K \text{Hom}_B(P_H, \bar{Q}) = \sum_{Q'} r_{PQ'} \dim_K[\text{Hom}_B(Q', \bar{Q})] = r_{PQ}.$$

But since $\text{Hom}_B(P_H, \bar{Q}) \simeq \text{Hom}_A(P, \bar{Q}^G)$ by Theorem 1.19 (ii), r_{PQ} must be the multiplicity of $\bar{P}$ in $\bar{Q}^G$.

(ii) ($\Rightarrow$). From the isomorphism given in the assumption, it follows that

$$\dim_K \text{Hom}_A(Q^G, \bar{P}) = \sum_{P'} s_{P'Q} \dim_K[\text{Hom}_A(P', \bar{P})] = s_{PQ}.$$

But since $\text{Hom}_A(Q^G, \bar{P}) \simeq \text{Hom}_B(Q, \bar{P}_H)$ by Theorem 1.19 (i), s_{PQ} must be the multiplicity of $\bar{Q}$ in $\bar{P}_H$.

(iii) ($\Rightarrow$). From the assumption, we have, by using Chapter 2, 8.21(i), that

$$t_{PQ} = \dim_K \text{Hom}_B(Q, P_H) = \dim_K \text{Hom}_B(P_H, Q),$$

which is equal to $\dim_K \text{Hom}_A(P, Q^G)$, and this is just the multiplicity of $\bar{P}$ in Q^G. ∎

1.5. Schur Indices

Let $k \subset K$ be fields. Then there is a natural isomorphism $K \otimes_k kG \simeq KG$ of K-algebras such that $\alpha \otimes x \mapsto \alpha x$, and we identify them with each other below.

Thus the group ring KG is just the scalar extension of kG for every subfield k of K, which is quite useful in the study of group rings. By using this, we prove the following result.

Lemma 1.28. *$KG/J(KG)$ is separable over K.*

Proof. It suffices to show that $J(LG) = J(KG)^L$ for any extension field L of K (Chapter 2, 5.1). Let k be the prime field of K. Then, since k is a perfect field, $kG/J(kG)$ is separable, and it follows that $J(KG) = J(kG)^K$ and $J(LG) = J(kG)^L$. But since $J(kG)^L = (J(kG)^K)^L = J(KG)^L$, we conclude that $J(LG) = J(KG)^L$. ■

The above lemma, together with Chapter 2, 5.1, yield the following.

Corollary 1.29. *If V_{KG} is completely reducible, then, for any extension field L of K, $(V^L)_{LG}$ is completely reducible.*

In order to study the Schur indices of absolutely irreducible G-modules, let E be, as in Chapter 2, Section 6, a splitting field for G, which is a finite Galois extension of K and let $\mathfrak{G} = \mathrm{Gal}(E/K)$. Let furthermore $\tilde{V}$ be an (absolutely) irreducible EG-module, $\tilde{\zeta}$ the character of G defined by $\tilde{V}$, and $K(\tilde{\zeta})$ be the field generated by $\{\tilde{\zeta}(x);\ x \in G\}$ over K.

Let n be the *exponent* of G, namely, the smallest positive integer n such that $x^n = 1$ for all $x \in G$. It is clear that $n \,|\, |G|$.

Note that $K(\tilde{\zeta})$ is an abelian extension of K. In fact, if U_n denotes the group of nth roots of unity in the algebraic closure of K, then $K(U_n)$ is an abelian extension of K (Hungerford [1], Chapter V, Theorem 8.1) and we have $K(\tilde{\zeta}) \subset K(U_n)$ because $\tilde{\zeta}(x)$ is a sum of elements of U_n for $x \in G$. Thus $K(\tilde{\zeta})$ is an abelian extension of K.

Therefore the results 6.2, 6.3 and 6.4 of Chapter 2 hold in our case. We summarize them in the following theorem for convenience.

Theorem 1.30. *With the above notation let $Z_1 = K(\tilde{\zeta})$ and $m_K(\tilde{V})$ be the Schur index of $\tilde{V}$ over K. Take the irreducible modules V_{KG}, U_{Z_1G} such that $\tilde{V} | V^E$, $\tilde{V} | U^E$. Let furthermore $\mathfrak{G} = \sum_{i=1}^{t} \mathfrak{G}_{\zeta}\sigma_i$ with $\sigma_1 = 1$. Then the following holds.*

(i) $V^{Z_1} \simeq U^{\sigma_1} \oplus \cdots \oplus U^{\sigma_t}$, $\quad \dim_K Z_1 = t$.

(ii) $U^E \simeq m_K(\tilde{V})\tilde{V}$, $\quad V^E \simeq m_K(\tilde{V})(\tilde{V}^{\sigma_1} \oplus \cdots \oplus \tilde{V}^{\sigma_t})$.

(iii) *$\tilde{V}$ is realizable in some extension field L of Z_1 of degree $m_K(\tilde{V})$. If $\tilde{V}$ is realizable in a field M such that $K \subset M \subset E$, then $Z_1 \subset M$ and $\dim_{Z_1} M$ is a multiple of $m_K(\tilde{V})$.*

(iv) *For an integer r, we have*

$$r\tilde{V} \text{ is realizable in } K(\tilde{\zeta}) \Leftrightarrow m_K(\tilde{V}) | r.$$

(v) *For a KG-module W, the multiplicity of $\tilde{V}$ in W^E is a multiple of $m_K(\tilde{V})$.*

The following theorem of Wedderburn is essential when we work with fields of prime characteristic. See e.g., Hungerford [1], Chapter IX for the proof.

Theorem 1.31. *A finite division ring is a field.*

Using this, we show the following theorem.

Theorem 1.32. *Suppose that* Char $K = p > 0$ *and let E be a splitting field for KG. Then the following holds.*

(i) *We have $m_K(\tilde{V}) = 1$ for any absolutely irreducible EG-module $\tilde{V}$. Consequently, if $\tilde{\zeta}$ denotes the character of a representation defined by $\tilde{V}$, then $\tilde{V}$ is realizable in $K(\tilde{\zeta})$.*

(ii) *If n is the exponent of G, $K(U_n)$ is a splitting field for G.*

(iii) *If V_{KG} is irreducible, we have, for any extension field $L \supset K$,*

$$V^L \simeq W_1 \oplus \cdots \oplus W_r \text{ (an irreducible decomposition)}$$

where $W_i \not\simeq W_j$ if $i \neq j$.

(iv) *For an irreducible KG-module V, there exist a finite extension field L and an absolutely irreducible LG-module U such that*

$$V \simeq U$$

as KG-modules.

Proof.

(i) Let k be the prime subfield of K. We have the decomposition

$$kG/J(kG) \simeq \bigoplus_i M_{n_i}(k_i)$$

into the direct sum of simple components, where each k_i is a finite field by Wedderburn's theorem. By tensoring the above with K, we get

$$KG/J(KG) \simeq (kG/J(kG))^K \simeq \bigoplus_i M_{n_i}(K \otimes_k k_i).$$

Since k_i is separable over k, $K \otimes_k k_i$ is semisimple, and it follows that $K \otimes_k k_i \simeq \bigoplus_j L_{ij}$, where each L_{ij} is a finite extension of K. Thus

$$KG/J(KG) \simeq \bigoplus_{i,j} M_{n_i}(L_{ij}),$$

and hence the Schur index of each $M_{n_i}(L_{ij})$ is 1. This proves (i).

(ii) This is clear from (i).

(iii) If $\tilde{L}$ is the algebraic closure of L, then every (absolutely) irreducible constituent of $V^{\tilde{L}}$ has multiplicity 1, and the same is true for V^L, since $V^{\tilde{L}} = (V^L)^{\tilde{L}}$.

(iv) Let, as in Theorem 1.30, E be a finite Galois extension of K and $\tilde{V}$ be an absolutely irreducible EG-module such that $\tilde{V} \mid V^E$. Let $\tilde{\zeta}$ be the character of G defined by $\tilde{V}$, $L = K(\tilde{\zeta})$ and $t = \dim_K L$. Then by Theorem 1.30(i), we have an irreducible decomposition

$$V^L \simeq U_1 \oplus \cdots \oplus U_t.$$

Since $V^L \simeq tV$ as KG-modules, each U_i is isomorphic to V as KG-modules. Also, since $m_K(\tilde{V}) = 1$, U_i is an absolutely irreducible LG-module. ■

2. Ordinary Representations

By an *ordinary representation* of G we mean a representation over a field K of characteristic zero, and the character defined by it is called an *ordinary character*. Such a representation is always completely reducible and determined uniquely up to equivalence by the character. Thus, we shall concentrate here on characters.

All representations considered in this section will be taken over the complex field **C**. In particular, every irreducible representation is absolutely irreducible.

2.1. Orthogonality Relations of Characters

The complex group ring $\mathbf{C}G$ is isomorphic to a direct sum of full matrix rings over $\mathbf{C}$:

(2.1) $$\mathbf{C}G = \bigoplus_{i=1}^{k} A_i, \qquad A_i \simeq M_{n_i}(\mathbf{C})$$

Let $\mathbf{F}_i$ be an irreducible representation of G arising from the simple component A_i. Let χ_i denote the character defined by $\mathbf{F}_i$ and let $\mathrm{Irr}(G)$ be the set of them:

$$\mathrm{Irr}(G) = \{\chi_1, \ldots, \chi_k\}.$$

We know that $|\mathrm{Irr}(G)| = |\mathrm{Cl}(G)|$ and $\chi_i(1) = n_i$.

The regular representation $\mathbf{\Gamma}_G$ of G decomposes as follows:

$$\mathbf{\Gamma}_G \sim \bigoplus_{i=1}^{k} \chi_i(1)\mathbf{F}_i.$$

Therefore the character γ_G defined by $\mathbf{\Gamma}_G$ is given by

(2.2) $$\gamma_G = \sum_{i=1}^{k} \chi_i(1)\chi_i.$$

This, together with (1.9), yields the following.

Theorem 2.1.

(i) $\sum_{i=1}^{k} \chi_i(1)^2 = |G|$.

(ii) $\sum_{i=1}^{k} \chi_i(1)\chi_i(x) = 0$, if $x \neq 1$.

For a commutative ring R, let $\mathbf{F}_R(G)$ denote the set of functions on G with values in R; $\mathbf{F}_R(G) = \{f\colon G \to R\}$.

Define the sum $f + g$, the multiplication fg, and the scalar multiple αf for $f, g \in \mathbf{F}_R(G)$ and $\alpha \in R$ by

$$(f + g)(x) = f(x) + g(x), \qquad (fg)(x) = f(x)g(x), \qquad (\alpha f)(x) = \alpha f(x).$$

Then $\mathbf{F}_R(G)$ is an R-algebra. Given $a \in G$, we denote by φ_a the function such that $\varphi_a(a) = 1$, $\varphi_a(b) = 0 (b \neq a)$. Then $\{\varphi_a; a \in G\}$ forms an R-basis of $\mathbf{F}_R(G)$, in particular we have $\mathrm{rank}_R \mathbf{F}_R(G) = |G|$.

When $R = \mathbf{C}$, we abbreviate $\mathbf{F}_R(G)$ to $\mathbf{F}(G)$.

We define the *inner product* $(f, g)_G$ in $\mathbf{F}(G)$ by

$$(f, g)_G = \frac{1}{|G|} \sum_{x \in G} f(x)g(x^{-1}),$$

which is a symmetric, nonsingular bilinear form over $\mathbf{C}$.

We say f and g are *orthogonal* if $(f, g)_G = 0$.

We sometimes abbreviate $(f, g)_G$ to (f, g).

Now we set

$$F_i(x) = (\alpha_{\mu\nu}^{(i)}(x))_{\mu, \nu}.$$

Then we have the following orthogonality relations between the coordinate functions $\alpha_{\mu\nu}^{(i)}$.

Theorem 2.2. $(\alpha_{\mu\nu}^{(i)}, \alpha_{\rho\sigma}^{(j)})_G = \delta_{ij}\delta_{\mu\sigma}\delta_{\nu\rho}\dfrac{1}{\chi_i(1)}$.

Proof (Nagao[1]). The set of triples $\{(i, \mu, \nu); 1 \le i \le k, 1 \le \mu, \nu \le \chi_i(1)\}$ consists of $|G|$ elements. Define the square matrices A, B of degree $|G|$ as follows:

$$A = (\alpha_{\mu\nu}^{(i)}(x))_{(i, \mu, \nu), x}, \qquad B = \left(\frac{\chi_i(1)}{|G|}\alpha_{\nu\mu}^{(i)}(x^{-1})\right)_{(i, \mu, \nu), x},$$

where the triples (i, μ, ν) are used as the row indices arranged in lexicographic order, and x runs through G as the column indices.

Then the (x, y)-entry of ${}^tAB$ is given by

$$\sum_{i, \mu, \nu} \alpha_{\mu\nu}^{(i)}(x)\frac{\chi_i(1)}{|G|}\alpha_{\nu\mu}^{(i)}(y^{-1}) = \sum_{i, \mu} \frac{\chi_i(1)}{|G|}\alpha_{\mu\mu}^{(i)}(xy^{-1}) = \frac{1}{|G|}\sum_i \chi_i(1)\chi_i(xy^{-1}) = \delta_{xy}.$$

Namely, we have ${}^tAB = I$, and it follows that

$$B{}^tA = I.$$

Comparing the $((i, \mu, \nu), (j, \rho, \sigma))$-entry of both sides, we get

$$\frac{\chi_i(1)}{|G|}\sum_{x \in G} \alpha_{\nu\mu}^{(i)}(x^{-1})\alpha_{\rho\sigma}^{(j)}(x) = \delta_{(i, \mu, \nu)(j, \rho, \sigma)} = \delta_{ij}\delta_{\mu\rho}\delta_{\nu\sigma},$$

which implies the result. ■

Exercise 2.3. Show that $(\alpha_{\mu\nu}^{(i)}; 1 \le i \le k, 1 \le \mu, \nu \le \chi_i(1)\}$ is linearly independent and therefore a $\mathbf{C}$-basis of $\mathbf{F}(G)$.

From Theorem 2.2, we deduce the following important theorem.

Theorem 2.4. $\chi_i(1) \,|\, |G|$, *for all i.*

Proof. Let K be an algebraic number field that is a splitting field for G. Let R be the ring of integers in K and $\mathfrak{p}$ be an arbitrary prime ideal of R. Then $R_{\mathfrak{p}}$, the set of $\mathfrak{p}$-integers in K, is a principal ideal domain with quotient field K. We may assume that each $\mathbf{F}_i$ is a K-representation and therefore an $R_{\mathfrak{p}}$-representation by Chapter 2, 1.6. From Theorem 2.2, we get

$$\sum_{x \in G} \alpha_{\mu\nu}^{(i)}(x)\alpha_{\nu\mu}^{(i)}(x^{-1}) = \frac{|G|}{\chi_i(1)}.$$

Consequently, $|G|/\chi_i(1) \in R_{\mathfrak{p}}$ and thus $|G|/\chi_i(1) \in \bigcap_{\mathfrak{p}} R_{\mathfrak{p}} = R$. Since $|G|/\chi_i(1) \in \mathbf{Q}$, we conclude that $|G|/\chi_i(1) \in \mathbf{Z}$, as desired. ∎

Since $\chi_i(x) = \sum_\mu \alpha_{\mu\mu}^{(i)}(x)$, we get the following orthogonality relation from Theorem 2.2.

Theorem 2.5 (The first orthogonality relation of characters).

$$(\chi_i, \chi_j)_G = \frac{1}{|G|} \sum_{x \in G} \chi_i(x)\chi_j(x^{-1}) = \delta_{ij}.$$

Proof.

$$(\chi_i, \chi_j) = \sum_{\mu, \nu} (\alpha_{\mu\mu}^{(i)}, \alpha_{\nu\nu}^{(j)}) = \delta_{ij} \sum_{\mu} (\alpha_{\mu\mu}^{(i)}, \alpha_{\mu\mu}^{(i)}) = \delta_{ij}\chi_i(1)\frac{1}{\chi_i(1)} = \delta_{ij}.$$ ∎

Exercise 2.6. (Generalized orthogonality relations). Show that

$$\frac{1}{|G|} \sum_{x \in G} \chi_i(xy)\chi_j(x^{-1}) = \delta_{ij}\frac{\chi_i(y)}{\chi_i(1)}.$$

[Hint: Use $\alpha_{\mu\mu}^{(i)}(xy) = \sum_\rho \alpha_{\mu\rho}^{(i)}(x)\alpha_{\rho\mu}^{(i)}(y)$.]

A function $f \in \mathbf{F}_R(G)$ is said to be a *class function* if

$$f(x) = f(a^{-1}xa), \qquad \text{for all } a, x \in G.$$

The set of class functions in $\mathbf{F}_R(G)$ is denoted by $\mathbf{CF}_R(G)$. They are regarded as functions defined on $\mathrm{Cl}(G)$. In particular, $\mathbf{CF}_R(G)$ is a free R-module of rank $k(=|\mathrm{Cl}(G)|)$.

Example 2.7. Every R-character of G is a class function; in fact

$$\operatorname{tr} X(a^{-1}xa) = \operatorname{tr}(X(a)^{-1}X(x)X(a)) = \operatorname{tr} X(x),$$

and thus $\chi_{\mathbf{X}}(a^{-1}xa) = \chi_{\mathbf{X}}(x)$ holds.

When $R = \mathbf{C}$, we abbreviate $\mathbf{CF}_R(G)$ to $\mathbf{CF}(G)$. We see from Theorem 2.5 that $\mathrm{Irr}(G) = \{\chi_1, \ldots, \chi_k\}$ is a basis of $\mathbf{CF}(G)$, which is orthonormal w.r.t. the inner product (,). Thus, if $f \in \mathbf{CF}(G)$, then

$$f = \sum_{i=1}^{k} (f, \chi_i)\chi_i.$$

In particular, for any $\mathbf{C}$-character χ of G, the multiplicity of χ_i in χ is equal to (χ, χ_i).

If a representation $\mathbf{X}$ of G is restricted to a cyclic subgroup $\langle x \rangle$, then it is a direct sum of one-dimensional representations. Thus, there exists $P \in \mathrm{GL}_n(C)$ such that

$$P^{-1}X(x)P = \begin{pmatrix} \varepsilon_1 & & & 0 \\ & \varepsilon_2 & & \\ & & \ddots & \\ 0 & & & \varepsilon_n \end{pmatrix},$$

and it follows that

(2.5) $$\chi_{\mathbf{X}}(x) = \varepsilon_1 + \cdots + \varepsilon_n.$$

If $x^m = 1$, then $\varepsilon_i^m = 1$, namely, each ε_i is a root of unity. In particular, $\chi_{\mathbf{X}}(x)$ is an algebraic integer.

Also, since $\varepsilon_i^{-1} = \bar{\varepsilon}_i$, the complex conjugate of ε_i, we have $\chi_{\mathbf{X}}(x^{-1}) = \sum_i \varepsilon_i^{-1} = \overline{\chi_{\mathbf{X}}(X)}$. Therefore $\overline{\chi_{\mathbf{X}}(x)}$ gives the character defined by the contragredient representation $\hat{\mathbf{X}}\colon x \mapsto {}^t X(x^{-1})$ of $\mathbf{X}$.

From the above remark, the first orthogonality relation of characters can be expressed as

$$(2.6)\qquad \sum_{x\in G}\chi_i(x)\overline{\chi_j(x)}=\delta_{ij}|G|.$$

Also, from (2.5) we have

$$|\chi_{\mathbf{X}}(x)|\le n=\chi_{\mathbf{X}}(1),$$

where $|\alpha|$ is the absolute value of $\alpha\in\mathbf{C}$.

Exercise 2.8.

(i) $|\chi_{\mathbf{X}}(x)|=n\Leftrightarrow X(x)=\varepsilon I$ (a scalar matrix).

[Hint: If $\varepsilon_i\neq\varepsilon_j$ for some i, j in (2.5), then $|\chi_{\mathbf{X}}(x)|<n$.]

(ii) Let $\bar{G}=G/\text{Ker }\mathbf{X}$ and let $Z_G(\bar{G})$ denote the inverse image of $Z(\bar{G})$ under the natural map $G\to\bar{G}: Z_G(\bar{G})=\{x\in G;\ \bar{x}\in Z(\bar{G})\}$. Then, if $\mathbf{X}$ is an irreducible representation, we have

$$|\chi_{\mathbf{X}}(x)|=n\Leftrightarrow x\in Z_G(\bar{G}).$$

(iii) $\chi_{\mathbf{X}}(x)=n\Leftrightarrow x\in\text{Ker }\mathbf{X}$.

Remark. We sometimes write Ker $\chi_{\mathbf{X}}$ for Ker $\mathbf{X}$. In view of the above exercise, we may write

$$\text{Ker }\chi_{\mathbf{X}}=\{x\in G;\ \chi_{\mathbf{X}}(x)=\chi_{\mathbf{X}}(1)\}.$$

Let x_ρ be a representative of $C_\rho\in\text{Cl}(G)$. Then $|C_\rho|=|G|/|C_G(x_\rho)|$, where $C_G(x_\rho)$ is the centralizer of x_ρ in G. We see from (2.6) that

$$(2.7)\qquad \sum_{\rho=1}^{k}\frac{1}{|C_G(x_\rho)|}\chi_i(x_\rho)\overline{\chi_j(x_\rho)}=\delta_{ij}.$$

The following square matrix of degree k

$$X=(\chi_i(x_j))_{i,j}$$

is called the *character table* of G. We set

$$Y=\left(\frac{1}{|C_G(x_j)|}\overline{\chi_i(x_j)}\right)_{i,j}.$$

Then $X^tY=I$ (the identity matrix) by (2.7). Thus X is nonsingular, and it follows that ${}^tYX=I$. Therefore we get the following theorem.

Theorem 2.9 **(The second orthogonality relation of characters).**

$$\sum_{i=1}^{k} \overline{\chi_i(x_\rho)}\chi_i(x_\sigma) = \delta_{\rho\sigma}|C_G(x_\rho)|.$$

2.2. Character Rings, Restriction Maps, and Induction Maps

It is clear that **CF**(G) is a subring of **F**(G). Also,

$$X(G) = \mathbf{Z}\chi_1 \oplus \cdots \oplus \mathbf{Z}\chi_k$$

is a subring of **CF**(G), which is called the *character ring* of G. Each element $\chi = \sum_i m_i\chi_i$ of $X(G)$ is said to be a *generalized character*. If all m_i are nonnegative, then χ is a character of G (unless $\chi = 0$).

Lemma 2.10. *A generalized character χ of G is an irreducible character of G if and only if the following two conditions hold*:

$$(1)\quad (\chi, \chi)_G = 1, \qquad (2)\quad \chi(1) > 0.$$

Proof. The "only if" part is clear. To show the converse, let $\chi = \sum_i m_i\chi_i$. Then we have $\sum_i m_i^2 = 1$ by (1), and it follows that $\chi = \pm\chi_i$ for some i, whence we have $\chi = \chi_i$ by (2). ■

Exercise 2.11. If we set $\bar{\chi}_i = \chi_{i'}$, then

$$\chi_i\chi_j = \sum_\nu m_{ij\nu}\chi_\nu \Leftrightarrow \chi_{i'}\chi_\nu = \sum_j m_{ij\nu}\chi_j.$$

Now, let $G \geq H$ and $\varphi \in$ **CF**(G). Then the restriction φ_H of φ to H is also a class function of H, and thus we obtain the *restriction map*

$$\mathrm{Res}_H^G\colon \mathbf{CF}(G) \to \mathbf{CF}(H) \qquad (\varphi \mapsto \varphi_H).$$

Given $\theta \in$ **CF**(H), we define the function θ^G of G by

$$\theta^G(x) = \sum_{t\in H\backslash G} \theta(txt^{-1}), \tag{2.8}$$

where we let $\theta(txt^{-1}) = 0$ if $txt^{-1} \notin H$.

Because θ is a class function, (2.8) is also expressed as

$$\theta^G(x) = \frac{1}{|H|}\sum_{t\in G} \theta(txt^{-1}). \tag{2.9}$$

This implies that θ^G is a class function of G, and we obtain a $\mathbf{C}$-linear map

$$\mathrm{Ind}_H^G\colon \mathbf{CF}(H) \to \mathbf{CF}(G) \qquad (\theta \mapsto \theta^G),$$

which is said to be the *induction map* from H to G. As is clear from the definition, $\theta^G(x) = 0$ if x is not G-conjugate to any element of H.

If φ, θ are the characters defined by representations $\mathbf{X}$ and $\mathbf{Y}$ of G and H, respectively, then φ_H is the character defined by $\mathbf{X}_H$, while θ^G is the character defined by the induced representation $\mathbf{Y}^G$ of G by (1.8). Thus the restriction and the induction maps, respectively, give rise to the maps

$$\mathrm{Res}_H^G\colon X(G) \to X(H), \qquad \mathrm{Ind}_H^G\colon X(H) \to X(G).$$

Exercise 2.12. If $H \le K \le G$ and $\varphi \in \mathbf{CF}(G)$, $\theta \in \mathbf{CF}(H)$, then

$$(\varphi_K)_H = \varphi_H, \qquad (\theta^K)^G = \theta^G.$$

With the notation of Theorem 1.27, we have $P = \bar{P}$, $Q = \bar{Q}$ as $\mathbf{C}G$ is semisimple. Thus, the following theorem holds.

Theorem 2.13 (Frobenius reciprocity). *Let $G \ge H$.*

(i) *If $\chi \in \mathrm{Irr}(G)$, $\zeta \in \mathrm{Irr}(H)$, then*

$$\chi_H = \sum_{\theta \in \mathrm{Irr}(H)} r_{\chi\theta}\theta \Leftrightarrow \zeta^G = \sum_{\chi \in \mathrm{Irr}(G)} r_{\chi\zeta}\chi.$$

(ii) *If $\varphi \in \mathbf{CF}(G)$, $\theta \in \mathbf{CF}(H)$, then*

(2.10) $$(\varphi_H, \theta)_H = (\varphi, \theta^G)_G.$$

Proof.

(i) This is clear by Theorem 1.27.

(ii) If $\chi \in \mathrm{Irr}(G)$, $\zeta \in \mathrm{Irr}(H)$, then part (i) asserts that $(\chi_H, \zeta)_H = (\chi, \zeta^G)_G$. But since $\varphi = \sum_\chi m_\chi \chi$, $\theta = \sum_\zeta n_\zeta \zeta$ with $m_\chi, n_\zeta \in \mathbf{C}$, (2.10) is true in general. ∎

Theorem 2.14. *Let $\varphi \in \mathbf{CF}(G)$, $\theta \in \mathbf{CF}(H)$. Then the following holds.*

(i) $(\varphi_H\theta)^G = \varphi\theta^G$.

(ii) $(\varphi_H)^G = \varphi(\mathbf{1}_H)^G$.

(iii) $\mathrm{Ind}_H^G(\mathbf{CF}(H))$ *and* $\mathrm{Ind}_H^G(X(H))$ *are ideals of* $\mathbf{CF}(G)$ *and* $X(G)$, *respectively.*

Proof.

(i) Since φ and θ are **C**-linear combinations of characters of G and H, respectively, it suffices to show the assertion in case that both are characters. Then the result is clear from Lemma 1.15(ii).

(ii) This follows immediately from (i) by letting $\theta = \mathbf{1}_H$.

(iii) This is clear from (i). ■

For $t \in G$ and $\theta \in \mathbf{CF}(H)$, define $\theta^t \in \mathbf{CF}(H^t)$ by

$$\theta^t(x) = \theta(txt^{-1}) \qquad (x \in H^t = t^{-1}Ht).$$

If θ is the character defined by a representation module W of H, then θ^t is the character defined by H^t-module W^t.

From the Mackey decomposition theorem, we have the following.

Theorem 2.15. *Let $G \geq H$, K and let $\theta \in \mathbf{CF}(H)$, $\eta \in \mathbf{CF}(K)$. Then*

(i)
$$(\theta^G)_K = \sum_{t \in H\backslash G/K} (\theta^t_{H^t \cap K})^K.$$

(ii)
$$(\theta^G, \eta^G)_G = \sum_{t \in H\backslash G/K} (\theta^t_{H^t \cap K}, \eta_{H^t \cap K})_{H^t \cap K}.$$

Proof. We may assume that θ, η are characters of H, K respectively.

(i) This is clear from Theorem 1.9.

(ii) Using the Frobenius reciprocity twice, we get

$$(\theta^G, \eta^G)_G = ((\theta^G)_K, \eta)_K = \sum_{t \in H\backslash G/K} ((\theta^t_{H^t \cap K})^K, \eta)_K$$
$$= \sum_t (\theta^t_{H^t \cap K}, \eta_{H^t \cap K})_{H^t \cap K}. \quad ■$$

Exercise 2.16. Let $\zeta \in \mathrm{Irr}(H)$. Then $\zeta^G \in \mathrm{Irr}(G)$ if and only if, for any $t \in G - H$, $\zeta_{H^t \cap H}$ and $\zeta^t_{H^t \cap H}$ have no irreducible constituent in common. [Hint: Use Theorem 2.15(ii).]

Suppose that G acts on a finite set Ω, and let ρ_Ω be the character of G defined by the permutation module $\mathbf{C}\Omega$. This is said to be a *permutation character* of G. We know that the value $\rho_\Omega(x)(x \in G)$ equals the number of fixed points of x (Exercise 1.4). Let $\omega \in \Omega$ and let G_ω be the stabilizer of ω in G. If G is transitive, then the number r of the G_ω-orbits on Ω is called the *rank* of G. If $r = 2$, then G is said to be *2-transitive*.

Remark. Since $\{\omega\}$ is one of the G_ω-orbits on Ω, it holds that $r \geq 2$ whenever $|\Omega| > 1$. And G is 2-transitive if and only if G_ω is transitive on $\Omega - \{\omega\}$.

Theorem 2.17. *With the above notation, the following holds:*

(i) $$(\rho_\Omega, \mathbf{1}_G)_G = \frac{1}{|G|} \sum_{x \in G} \rho_\Omega(x) = \#\{G\text{-orbits on } \Omega\}.$$

(Namely, the mean value of the numbers of fixed points of the elements of G is equal to the number of the G-orbits.) In particular,

$$G \text{ is transitive on } \Omega \Leftrightarrow (\rho_\Omega, \mathbf{1}_G)_G = 1.$$

(ii) *Suppose that G is transitive and let* $H = G_\omega$ $(\omega \in \Omega)$. *Then*

$$r = |H \backslash G/H| = (\rho_\Omega, \rho_\Omega)_G.$$

(Namely, the mean value of the square of the numbers of fixed points of the elements of G is equal to the rank of G.)

In particular, G is 2-transitive on Ω *if and only if* $\rho_\Omega = \mathbf{1}_G + \chi$ *with some irreducible character* $\chi \neq \mathbf{1}_G$.

Proof.

(i) Let $\{\Omega_1, \ldots, \Omega_t\}$ be the set of G-orbits on Ω, $\omega_i \in \Omega_i$, and $H_i = G_{\omega_i}$. Then $\rho_\Omega = \sum_i \rho_{\Omega_i}$. Since $\rho_{\Omega_i} = (\mathbf{1}_{H_i})^G$, we get

$$(\rho_{\Omega_i}, \mathbf{1}_G)_G = ((\mathbf{1}_{H_i})^G, \mathbf{1}_G)_G = (\mathbf{1}_{H_i}, \mathbf{1}_{H_i})_{H_i} = 1,$$

whence it follows that $(\rho_\Omega, \mathbf{1}_G) = \sum_{i=1}^t (\rho_{\Omega_i}, \mathbf{1}_G)_G = t$.

(ii) $(\rho_\Omega, \rho_\Omega)_G = (\rho_\Omega, (\mathbf{1}_H)^G)_G = ((\rho_\Omega)_H, \mathbf{1}_H)_H = r$. Also, we have

$$(\rho_\Omega, \rho_\Omega)_G = ((\mathbf{1}_H)^G, (\mathbf{1}_H)^G)_G = \sum_{t \in H\backslash G/H} (\mathbf{1}_{H^t \cap H}, \mathbf{1}_{H^t \cap H})_{H^t \cap H} = |H \backslash G/H|.$$

To show the latter half, we may write, as G is transitive, $\rho_\Omega = \mathbf{1}_G + \sum_{\chi \neq \mathbf{1}_G} m_\chi \chi$. Then $r = 1 + \sum_\chi m_\chi^2$, whence the assertion is clear. ∎

2.3. Brauer's Permutation Lemma

We show here a useful lemma due to Brauer and a couple of applications of it.

For a finite set Ω, we let S^{Ω} denote the symmetric group on Ω. Then, a permutation representation of G on Ω is interpreted as a homomorphism $\mu: G \to S^{\Omega}$. Let $\Omega = \{1, 2, \ldots, n\}$ and let

$$\mu(x) = \begin{pmatrix} i \\ i^{\mu(x)} \end{pmatrix}.$$

The corresponding permutation representation of G on Ω is given by

$$\mathbf{P}_\mu: x \mapsto P_\mu(x) = (p_{ij}^{(\mu)}(x)),$$

where

$$p_{ij}^{(\mu)}(x) = \begin{cases} 1, & \text{if } j = i^{\mu(x)} \\ 0, & \text{otherwise.} \end{cases}$$

If ρ_μ denotes the character defined by $\mathbf{P}_\mu$, then

$$\rho_\mu(x) = \#\{i \in \Omega;\ i^{\mu(x)} = i\}.$$

Lemma 2.18 (Brauer). *Let $\Omega = \{1, 2, \ldots, n\}$ and $\mu: G \to S^{\Omega}$, $v: G \to S^{\Omega}$ be homomorphisms. If there exists $B = (\beta_{ij}) \in \mathrm{GL}_n(K)$ such that $\mu(x)$ permutes the rows of B and $v(x)$ permutes the columns of B satisfying*

$$\beta_{i^{\mu(x)} j^{v(x)}} = \beta_{ij}$$

for all i, j and $x \in G$, then the following holds.

(i) For any $x \in G$, the number of rows fixed by $\mu(x)$ equals that of the columns fixed by $v(x)$.

(ii) The number of G-orbits on the set of rows of B equals that of the G-orbits on the set of colums of B.

Proof. It suffices to show that $\rho_\mu(x) = \rho_v(x)$ for all $x \in G$ (cf. Theorem 2.17). We have $P_\mu(x) B P_v(x^{-1}) = (\beta_{i^{\mu(x)} j^{v(x)}})_{i,j}$, which is equal to B by assumption. Thus, $B^{-1} P_\mu(x) B = P_v(x)$ and hence $\rho_\mu(x) = \rho_v(x)$, as required. ■

Remark. The conclusion of the above theorem remains valid if we replace the assumption with the following

$$\beta_{i^{\mu(x)} j} = \beta_{i j^{v(x)}}$$

for all i, j and $x \in G$. In fact, this implies that $P_\mu(x)B = BP_\nu(x^{-1})$. Hence, $B^{-1}P_\mu(x)B = P_\nu(x^{-1})$, and we get $\rho_\mu(x) = \rho_\nu(x^{-1}) = \rho_\nu(x)$.

The above lemma is called *Brauer's permutation lemma*. Using this lemma, we show the following fact.

Lemma 2.19. *Let a group A act on the sets* $\mathrm{Irr}(G) = \{\chi_1, \ldots, \chi_k\}$ *and* $\mathrm{Cl}(G) = \{C_1, \ldots, C_k\}$, *and assume that*

$$\chi_i^a(x_j^a) = \chi_i(x_j), \textit{ for all } a \in A \textit{ and all } i, j,$$

where $x_j \in C_j$, and x_j^a denotes a representative of C_j^a. Then the following holds.

(i) *For any $a \in A$, the number of irreducible characters fixed by a equals that of conjugate classes fixed by a.*

(ii) *The number of A-orbits on* $\mathrm{Irr}(G)$ *equals that of A-orbits on* $\mathrm{Cl}(G)$.

Proof. The character table $X = (\chi_i(x_j))$ is nonsingular. A acts on rows and columns of X, and the condition of Lemma 2.18 is satisfied by assumption. Therefore the result follows. ■

For example, if a group A acts on G as automorphisms, then A acts on $\mathrm{Irr}(G)$ and $\mathrm{Cl}(G)$ naturally, and the conclusions (i) and (ii) in the above lemma hold, where we set $\chi_i^a(x) = \chi_i(x^{a^{-1}})$.

A character χ of G is said to be a *real character* if $\bar{\chi} = \chi$. An element x of G is said to be real if it is conjugate to x^{-1}, and a conjugate class of G is a *real conjugate class* if it consists of real elements. Now the following holds.

Corollary 2.20. *The number of irreducible real characters of G is equal to that of the real conjugate classes.*

Proof. Let $A = \langle a \rangle$ be a group of order 2 and set $\chi_i^a = \bar{\chi}_i$, $C_j^a = \{x^{-1};\ x \in C_j\}$, then A acts on $\mathrm{Irr}(G)$ and $\mathrm{Cl}(G)$. Since $\bar{\chi}_i(x_j^{-1}) = \chi_i(x_j)$, the assertion follows immediately from Lemma 2.19. ■

2.4. Central Idempotents of Group Rings

For $\chi \in \mathrm{Irr}(G)$, let A_χ be the simple component of $\mathbf{C}G$, which provides an irreducible module defining χ. If e_χ denotes the identity of A_χ, then

$$\mathbf{C}G = \bigoplus_{\chi \in \mathrm{Irr}(G)} A_\chi, \qquad A_\chi \simeq M_{\chi(1)}(\mathbf{C}), \tag{2.11}$$

$$Z(\mathbf{C}G) = \bigoplus_{\chi} \mathbf{C}e_\chi, \tag{2.12}$$

and we have $\{e_\chi; \chi \in \mathrm{Irr}(G)\} = \mathrm{pi}(Z(\mathbf{C}G))$.

In order to obtain the explicit description of the central primitive idempotent e_χ, we first prove the following.

Lemma 2.21 (Fourier's inversion formula). *If $u = \sum_{x \in G} \alpha_x x \in CG$, then*

$$\alpha_x = \frac{1}{|G|} \sum_{\chi \in \mathrm{Irr}(G)} \chi(1)\chi(ux^{-1}).$$

(*Here χ is considered as a character of* $\mathbf{C}G$.)

Proof. We have

$$ux^{-1} = \alpha_x + \sum_{y \neq x} \alpha_y yx^{-1}.$$

By taking the values of the regular character of both sides, we get from (2.2) and Theorem 2.1,

$$\sum_{\chi} \chi(1)\chi(ux^{-1}) = \alpha_x |G|. \qquad \blacksquare$$

Theorem 2.22.

$$e_\chi = \frac{\chi(1)}{|G|} \sum_{x \in G} \overline{\chi(x)}x. \tag{2.13}$$

Proof. Let $e_\chi = \sum_{x \in G} \alpha_x x$. From the above lemma, we have

$$\alpha_x = \frac{1}{|G|} \sum_{\chi' \in \mathrm{Irr}(G)} \chi'(1)\chi'(e_\chi x^{-1}),$$

whence the assertion follows since $\chi'(e_\chi x^{-1}) = \delta_{\chi\chi'}\chi(x^{-1}) = \delta_{\chi\chi'}\overline{\chi(x)}$. $\blacksquare$

2.5. Representations of $Z(\mathbf{C}G)$

Let $z \in Z(\mathbf{C}G)$. According to (2.12), we can express z as

$$z = \sum_{\chi} \omega_\chi(z) e_\chi \qquad (\omega_\chi(z) \in \mathbf{C}).$$

Thus $\omega_\chi\colon Z(\mathbf{C}G) \to \mathbf{C}$ is a one-dimensional representation of $Z(\mathbf{C}G)$ and as χ runs through $\mathrm{Irr}(G)$, ω_χ covers all the irreducible representation of $Z(\mathbf{C}G)$. Each ω_χ is said to be the linear character of $Z(\mathbf{C}G)$ corresponding to χ.

Given $\chi \in \mathrm{Irr}(G)$, let $\mathbf{F}_\chi$ be a representation of $\mathbf{C}G$ defining χ. If $z \in Z(\mathbf{C}G)$, then

$$F_\chi(z) = F_\chi(\omega_\chi(z) e_\chi) = \omega_\chi(z) I$$

By taking the trace of both sides, we get

$$\chi(z) = \chi(1)\omega_\chi(z). \tag{2.14}$$

In particular, if x_i is a representative of $C_i \in \mathrm{Cl}(G)$, then

$$\chi(\hat{C}_i) = |C_i|\chi(x_i) = \frac{|G|}{|C_G(x_i)|}\chi(x_i),$$

and hence, from (2.14), we obtain the following result.

Theorem 2.23.

$$\omega_\chi(\hat{C}_i) = \frac{\chi(\hat{C}_i)}{\chi(1)} = \frac{|G|\chi(x_i)}{|C_G(x_i)|\chi(1)}.$$

Let us write

$$\hat{C}_i \hat{C}_j = \sum_{l} t_{ijl} \hat{C}_l. \tag{2.15}$$

Comparing the coefficients of x_l on both sides, we find that

$$t_{ijl} = \#\{(a, b) \in C_i \times C_j;\ ab = x_l\}, \tag{2.16}$$

which is a nonnegative integer.

Since ω_χ is a one-dimensional representation of $Z(\mathbf{C}G)$, we obtain from (2.15) that

$$\omega_\chi(\hat{C}_i)\omega_\chi(\hat{C}_j) = \sum_{l} t_{ijl}\omega_\chi(\hat{C}_l), \tag{2.17}$$

which yields the following theorem.

Theorem 2.24.

$$t_{ijl} = \frac{|C_i||C_j|}{|G|} \sum_{\chi \in \mathrm{Irr}(G)} \frac{\chi(x_i)\chi(x_j)\overline{\chi(x_l)}}{\chi(1)}.$$

Proof. It follows from (2.17) and Theorem 2.23 that

$$\frac{|C_i|\chi(x_i)}{\chi(1)} \frac{|C_j|\chi(x_j)}{\chi(1)} = \sum_l t_{ijl} \frac{|C_l|\chi(x_l)}{\chi(1)}.$$

Therefore,

$$|C_i||C_j| \frac{\chi(x_i)\chi(x_j)}{\chi(1)} = \sum_l t_{ijl}|C_l|\chi(x_l).$$

Multiplying both sides of the above by $\overline{\chi(x_l)}$ and summing up with χ running through Irr(G), we get, from the second orthogonality relation,

$$|C_i||C_j| \sum_\chi \frac{\chi(x_i)\chi(x_j)\overline{\chi(x_l)}}{\chi(1)} = t_{ijl}|C_l| \frac{|G|}{|C_l|},$$

whence the result follows. ■

Remark. The integer t_{ijl} is determined by group-theoretical properties, as observed in (2.16). Thus the above theorem gives the connection between the characters and the structure of the group G.

We obtain the following important result from (2.17).

Theorem 2.25. *$\omega_\chi(\hat{C}_i)$ is an algebraic integer.*

Proof. $S = \sum_x \mathbf{Z}\omega_\chi(\hat{C}_i)$ is a subring of $\mathbf{C}$ by (2.17), which is finitely generated over $\mathbf{Z}$. Hence, each element of S is an algebraic integer (cf. Chapter 1, 13.2). ■

Below, we give an alternative proof to Theorem 2.4 by making use of the above theorem.

An alternative proof to Theorem 2.4. If $\chi \in \mathrm{Irr}(G)$, then

$$\sum_i \omega_\chi(\hat{C}_i)\overline{\chi(x_i)} = \sum_i \frac{|C_i|\chi(x_i)\overline{\chi(x_i)}}{\chi(1)}$$

$$= \frac{1}{\chi(1)} \sum_{x \in G} \chi(x)\overline{\chi(x)} = \frac{|G|}{\chi(1)}.$$

Since $\omega_\chi(\hat{C}_i)$ and $\overline{\chi(x_i)}$ are algebraic integers, $|G|/\chi(1)$ is also an algebraic integer that lies in $\mathbf{Q}$. Therefore, $|G|/\chi(1) \in \mathbf{Z}$. ■

2.6. Burnside's Theorem

As an application of the theory of characters, we shall show here a famous result known as Burnside's $p^a q^b$ theorem.

Lemma 2.26. *Let $\chi \in \mathrm{Irr}(G)$ and $\mathbf{F}$ be an irreducible representation that defines χ. Furthermore, let $C \in \mathrm{Cl}(G)$ and $x \in C$. If $(\chi(1), |C|) = 1$, then either $F(x) = \varepsilon I$ (a scalar matrix) or $\chi(x) = 0$.*

(***Remark.*** According to Exercise 2.8, $F(x) = \varepsilon I \Leftrightarrow x \in Z_G(G/\operatorname{Ker}\chi)$.)

Proof. By assumption, there exist $r, s \in \mathbf{Z}$ such that $r|C| + s\chi(1) = 1$. Thus,

$$r\frac{|C|\chi(x)}{\chi(1)} + s\chi(x) = \frac{\chi(x)}{\chi(1)}.$$

Since $|C|\chi(x)/\chi(1)$ and $\chi(x)$ are both algebraic integers, this implies that $\xi = \chi(x)/\chi(1)$ is also an algebraic integer.

If $F(x)$ is not a scalar matrix, then $|\xi| < 1$ by Exercise 2.8. Let $m = o(x)$, E the field generated by a primitive mth root of unity over $\mathbf{Q}$ and $\mathfrak{G} = \mathrm{Gal}(E/\mathbf{Q})$. Since $\chi(x)$ is a sum of mth roots of unity, so is $\chi(x)^\sigma$ for $\sigma \in \mathfrak{G}$, and it follows that $|\xi^\sigma| = |\chi(x)^\sigma|/\chi(1) \le 1$. Hence

$$\left|\prod_{\sigma \in \mathfrak{G}} \xi^\sigma\right| < 1.$$

On the other hand, we see that $\eta = \prod_{\sigma \in \mathfrak{G}} \xi^\sigma \in \mathbf{Z}$ as it is clearly an algebraic integer lying in $\mathbf{Q}$. But since $|\eta| < 1$, it follows that $\eta = 0$ and $\chi(x) = 0$. ■

Lemma 2.27. *If G is a non-abelian simple group, then there is no $C \in \mathrm{Cl}(G)$ other than $\{1\}$ such that $|C|$ is a power of a prime.*

Proof. Suppose to the contrary that there is $C(\neq\{1\}) \in \mathrm{Cl}(G)$ such that $|C| = p^r$ for some prime p. Let γ_G be the character defined by the regular representation of G and $x \in C$. Then

$$0 = \gamma_G(x) = 1 + \sum_{\chi} \chi(1)\chi(x), \tag{2.18}$$

where χ runs through $\mathrm{Irr}(G) - \{\mathbf{1}_G\}$. By assumption we have

$$Z_G(G/\operatorname{Ker}\chi) = 1$$

for any $\chi \neq \mathbf{1}_G$. Thus, Lemma 2.26 implies that $\chi(x) = 0$ if $p \nmid \chi(1)$, since then $(|C|, \chi(1)) = 1$. Hence $\alpha = \sum_{\chi \neq \mathbf{1}_G} (\chi(1)/p)\chi(x)$ is an algebraic integer, which is equal to $-1/p$ by (2.18), a contradiction. ■

We are now ready to prove the following theorem.

Theorem 2.28 (Burnside). *If $|G| = p^a q^b$ where p, q are primes, then G is solvable.*

Proof. We may assume that G is non-abelian and that a Sylow p-subgroup, say P, of G is nontrivial. If it is shown that G is not simple, then the assertion follows by induction on the order of G. Now suppose to the contrary that G is simple. Let $x(\neq 1) \in Z(P)$ and C be a conjugate class of G containing x. Then, since $C_G(x) \supset P$, it follows that $|C| = |G\colon C_G(x)|$ is a power of q. But this contradicts Lemma 2.27. ■

Remark. It had been a long-standing problem to prove Theorem 2.28 by a purely group-theoretic method. But in 1970 Goldschmidt[1] succeeded in it when $|G|$ is odd, while the even-order case was solved by Matsuyama[1] in 1973.

3. The Clifford Theory

Let $G \trianglerighteq H$. We first show Clifford's theorems over arbitrary fields that describe relations between irreducible G-modules and irreducible H-modules.

Let K be a field.

If W is a KH-module, then, for $t \in G$, W^t is an $H(=H^t)$-module, and

$$T(W) = \{t \in G;\ W^t \simeq W\ (KH\text{-isomorphic})\}$$

is a subgroup of G containing H. This is called the *inertial group* of W (in G). Note that the number of the distinct G-conjugates of W is equal to $|G:T(W)|$.

In the following, we denote by $\mathrm{IRR}(KG)$ a complete set of representatives for the isomorphism classes of irreducible KG-modules. But we mean by $V \in \mathrm{IRR}(KG)$ that V is an irreducible KG-module for convenience.

Theorem 3.1 (Clifford). *Let $G \trianglerighteq H$. Let $V \in \mathrm{IRR}(KG)$, and let W be an irreducible KH-submodule of V_H. Set $T = T(W)$. Then the following holds.*

(i) *There is an integer e such that*

$$V_H \simeq e\Bigl(\bigoplus_{a \in T\backslash G} W^a\Bigr). \tag{3.1}$$

In particular, V_H is completely reducible.

(ii) *There exists $U \in \mathrm{IRR}(KT)$ such that $U_H \simeq eW$ and $U^G \simeq V$.*

Proof. If $x \in G$, then Wx is an irreducible KH-submodule of V, which is isomorphic to W^x. Since $\sum_{x \in G} Wx$ is a G-submodule of V and V is irreducible, it follows that $V = \sum_x Wx$. Hence, V_H is completely reducible. If $V_H = U_1 \oplus \cdots \oplus U_r$ is the homogenous decomposition, then $r = |G:T|$. Choose a right coset decomposition $G = \sum_{i=1}^r Ta_i$ so that $Wa_i \subset U_i (1 \le i \le r)$, where $a_1 = 1$. Then it turns out that $U_i = U_1 a_i$, and therefore, if $U_1 \simeq eW$, then $U_i \simeq eW^{a_i}$ and (3.1) follows. Furthermore, we see that U_1 is a KT-module and that $U_1^G \simeq V$. ■

The integer e in (3.1) is said to be the *ramification index* of V relative to H. For $W \in \mathrm{IRR}(KH)$, let

$$\mathrm{IRR}(KG|W) = \{V \in \mathrm{IRR}(KG);\ W | V_H\}.$$

Theorem 3.2. *Let $G \rhd H$, $\mathrm{IRR}(KH) \ni W$ and $T = T(W)$. Then we have the following.*

(i) *If $U \in \mathrm{IRR}(KT|W)$, then U^G is irreduciible.*

(ii) *The map Ind_T^G: $\mathrm{IRR}(KT|W) \to \mathrm{IRR}(KG|W)$ $(U \mapsto U^G)$ is a bijection, and the ramification index of $U \in \mathrm{IRR}(KT|W)$ and that of U^G relative to H are the same.*

Proof.

(i) Since $G \rhd H$, we have, from the Mackey decomposition theorem,

$$(U^G)_H \simeq U_H \oplus \left(\bigoplus_{\substack{a \in T\backslash G \\ a \notin T}} U_H^a \right),$$

where $U_H^a \simeq eW^a$ if $U_H \simeq eW$. Thus $(U^G)_H$ is completely reducible.

Now let V be an irreducible G-submodule of U^G. Since $V_H | (U^G)_H \simeq e(\bigoplus_{a \in T\backslash G} W^a)$, we have that $V \in \mathrm{IRR}(KG|W)$ and $V = U_1^G$ with $U_1 \in \mathrm{IRR}(KT|W)$ by Theorem 3.1(ii). We shall show that $U_1 \simeq U$. We know that $\mathrm{Hom}_{KG}(V, U^G) \neq 0$. On the other hand, by Exercise 1.21, it holds that

$$\mathrm{Hom}_{KG}(U_1^G, U^G) \simeq \mathrm{Hom}_{KT}(U_1, U) \oplus \left(\bigoplus_{\substack{a \in T\backslash G/T \\ a \notin T}} \mathrm{Hom}_{K[T^a \cap T]}(U_1^a, U) \right). \tag{3.2}$$

Note that $(U_1)_H \simeq fW$ implies that $(U_1^a)_H \simeq fW^a$. Thus,

$$\mathrm{Hom}_{KH}((U_1^a)_H, U_H) = 0,$$

whenever $a \notin T$, and hence the right-hand side of (3.2) reduces to $\mathrm{Hom}_{KT}(U_1, U)$, which therefore is not zero. Therefore the irreducibility implies that $U_1 \simeq U$, and thus $U^G \simeq V \in \mathrm{IRR}(KG|W)$.

(ii) Ind_T^G is surjective by Theorem 3.1. The injectiveness has been shown in the proof of (i). The assertion concerning the ramification indices is obvious. ■

As a special case of the above theorem, we obtain the following.

Corollary 3.3. *Let $G \rhd H$ and let $W \in \mathrm{IRR}(KH)$. Then*

$$T(W) = H \Rightarrow W^G \textit{ is irreducible.}$$

Exercise 3.4. With the assumption of Theorem 3.2, let $T(W) \leq X \leq G$. Then the map

$$\mathrm{Ind}_X^G\colon \mathrm{IRR}(KX\,|\,W) \to \mathrm{IRR}(KG\,|\,W) \qquad (U \mapsto U^G)$$

is a bijection, and the ramification index of U and that of U^G relative to H are the same.

A KG-module V is said to be *imprimitive* if there exists a proper subgroup H of G and a KH-module W such that $V \simeq W^G$, and *primitive* if otherwise. The same terms will be used for representations of G.

The following result is useful.

Theorem 3.5. *Let A be an abelian normal subgroup of G that is not contained in the center $Z(G)$ of G. Assume that K is algebraically closed and let V be an irreducible KG-module that is faithful as a G-module, i.e., $\mathrm{Ker}_G V = 1$. If W is an irreducible A-submodule of V_A, then $T(W) < G$ and V is imprimitive.*

Proof. Suppose that $T(W) = G$, then $V_A \simeq eW$. Since K is algebraically closed, we have $\dim_K W = 1$, and hence V_A defines a representation $\mathbf{X}$ such that $X(a)$ is a scalar matrix for all $a \in A$ (relative to a suitable basis of V). Thus $X(A) \subset Z(X(G))$, which implies that $A \subset Z(G)$ because X is faithful. This contradicts the assumption. Therefore $T(W) \neq G$, and it follows that $U^G \simeq V$ for some $U \in \mathrm{IRR}(KT(W)|\,W)$. ■

A KG-module V is said to be a *monomial module* if there exists a subgroup H of G and a KH-module W with $\dim_K W = 1$ such that $V \simeq W^G$. A K-representation of G defined by a monomial module is said to be a *monomial representation.*

In order to give an application of Theorem 3.5 to nilpotent groups, we prove the following lemma.

Lemma 3.6. *Let G be a nilpotent group. If G is not abelian, then G has an abelian normal subgroup A such that $A \not\leq Z(G)$.*

Proof. Let $Z_2(G)$ be the second center of G: $Z_2(G)/Z(G) = Z(G/Z(G))$. Our assumption implies that $Z_2(G) \neq Z(G)$. Take any $a \in Z_2(G) - Z(G)$. Then $A = \langle Z(G), a\rangle$ will clearly meet our requirement. ■

Theorem 3.7. *Let G be a nilpotent group, and suppose that K is an algebraically closed field. Then every irreducible KG-module V is monomial.*

Proof. We shall proceed by induction on the order of G. Let $\bar{G} = G/\operatorname{Ker}_G V$, then V is faithful as a $\bar{G}$-module. If $\bar{G}$ is abelian, then the assertion is obvious as $\dim_K V = 1$. Thus, we may assume that $\bar{G}$ is non-abelian. Then, by Theorem 3.5 and Lemma 3.6, there exist $\bar{T} < \bar{G}$ and a $K\bar{T}$-module U such that $V \simeq U^{\bar{G}}$, whence we have $V \simeq U^G$ by naturally considering U as a T-module. We apply now the inductive hypothesis to U_{KT} and obtain the result. ■

Now, as in Section 2, let $\operatorname{Irr}(G)$ be the set of irreducible $\mathbf{C}$-characters of G. Also, assuming that $G \trianglerighteq H$, we let $T(\theta) = \{t \in G; \theta^t = \theta\}$ for $\theta \in \mathbf{CF}(H)$, where $\theta^t(h) = \theta(tht^{-1})$. This is called the *inertial group* of θ.

Given $\zeta \in \operatorname{Irr}(H)$, we set

$$\operatorname{Irr}(G|\zeta) = \{\chi \in \operatorname{Irr}(G); (\chi_H, \zeta)_H \neq 0, \quad \text{i.e.} \quad (\chi, \zeta^G)_G \neq 0\}.$$

The next result is a direct consequence of Theorems 3.1, 3.2, and Corollary 3.3.

Theorem 3.8 (Clifford). *Let $G \trianglerighteq H$, $\operatorname{Irr}(H) \ni \zeta$ and $T = T(\zeta)$.*

(i) *If $\chi \in \operatorname{Irr}(G|\zeta)$, then*

$$\chi_H = e\left(\sum_{a \in T\backslash G} \zeta^a\right).$$

(*The above integer e is also said to be the ramification index of χ relative to H.*)

(ii) *Given $\chi \in \operatorname{Irr}(G|\zeta)$, there exists a unique $\eta \in \operatorname{Irr}(T|\zeta)$ such that*

$$\eta^G = \chi, \qquad \eta_H = e\zeta.$$

Moreover, $\operatorname{Ind}_T^G\colon \operatorname{Irr}(T|\zeta) \to \operatorname{Irr}(G|\zeta)$ $(\eta \mapsto \eta^G)$ is a bijection.

(iii) *If $T \leq X \leq G$, then $\operatorname{Ind}_X^G\colon \operatorname{Irr}(X|\zeta) \to \operatorname{Irr}(G|\zeta)$ $(\eta \mapsto \eta^G)$ is a bijection.*

(iv) *If $T(\zeta) = H$, then $\zeta^G \in \operatorname{Irr}(G)$.*

A character defined by a monomial module is said to be a *monomial character*. Let us call G an *M-group* if every $\chi \in \operatorname{Irr}(G)$ is monomial, that is, for any $\chi \in \operatorname{Irr}(G)$, there always exist $H \leq G$ and a linear character λ of H such that $\chi = \lambda^G$. Now, Theorem 3.7 yields the following.

Theorem 3.9. *Every nilpotent group is an M-group.*

Remark. The above fact will be needed in the next section. Besides it, there are various facts known about M-groups (cf. Issacs[2]). For example, every M-group is solvable (K. Taketa). For a group-theoretical characterization of an M-group, see Parks[1].

4. Some Brauer Theorems

In this section, we shall be only concerned with **C**-representations and **C**-characters. We begin with a definition.

A group E is said to be a (p-) *elementary group* if it is a direct product of a p-group P and a cyclic group C:

$$E = P \times C, \tag{4.1}$$

thus E is nilpotent, and it may be assumed that the group C in (4.1) is a p'-group. In fact, if $C = C_p \times C_{p'}$, then replace P and C with $P \times C_p$ and $C_{p'}$, respectively, in (4.1).

Exercise 4.1. If $G = P \times C$ with $(|P|, |C|) = 1$, then any subgroup of G is a direct product of a subgroup of P and that of C. In particular, if G is p-elementary, so are its subgroups and factor groups.

Let $\mathscr{E}(G)$ be the set of elementary subgroups of G, and let $X(G)$ be, as in Section 2, the character ring of G, namely, $X(G)$ consists of the generalized characters of G. Now the fundamental theorem of Brauer on characters can be stated as follows.

Theorem 4.2 (Brauer).

(i) (A characterization of generalized characters) Let $\varphi \in \mathbf{CF}(G)$. *Then the following holds.*

$$\varphi \in X(G) \Leftrightarrow \varphi_E \in X(E) \textit{ for all } E \in \mathscr{E}(X).$$

(ii) (Induction theorem) *Every* $\chi \in \mathrm{Irr}(G)$ *can be expressed as*

$$\chi = \sum_{i=1}^{r} a_i \lambda_i^G,$$

where $a_i \in \mathbf{Z}$, *and* λ_i *is a character of* $E_i \in \mathscr{E}(G)$ *of degree* 1.

For the proof of the above theorem, we need a number of preliminary results. Set $\mathscr{E} = \mathscr{E}(G)$ for brevity and let

$$R(G, \mathscr{E}) = \{\varphi \in \mathbf{CF}(G);\ \varphi_E \in X(E) \text{ for all } E \in \mathscr{E}\},$$
$$I(G, \mathscr{E}) = \{\varphi \in \mathbf{CF}(G);\ \varphi = \sum_i a_i \zeta_i^G, \text{ where } a_i \in \mathbf{Z}, \zeta_i \in \mathrm{Irr}(E_i) \text{ and } E_i \in \mathscr{E}\}.$$

It is clear that

$$I(G, \mathscr{E}) \subset X(G) \subset R(G, \mathscr{E}),$$

and $R(G, \mathscr{E})$ is a subring of $\mathbf{CF}(G)$ with the identity $\mathbf{1}_G$ in common.

Lemma 4.3.

(i) $I(G, \mathscr{E})$ *is an ideal of* $R(G, \mathscr{E})$.

(ii) $I(G, \mathscr{E})$ *consists of all class functions* φ *of* G *such that* $\varphi = \sum_i a_i \lambda_i^G$, *where* $a_i \in \mathbf{Z}$ *and* λ_i *is a linear character of* $E_i \in \mathscr{E}$.

Proof.

(i) We need only to show that $\zeta^G\theta = (\zeta\theta_E)^G \in I(G, \mathscr{E})$ for all $\zeta \in \mathrm{Irr}(E)$, $\theta \in R(G, \mathscr{E})$, where $E \in \mathscr{E}$. But, since by assumption $\theta_E \in X(E)$, this is clear.

(ii) If $\zeta_i \in \mathrm{Irr}(E_i)$, then, since $E_i \in \mathscr{E}$ is nilpotent, there exist $H_i \leq E_i$ and a linear character λ_i of H_i such that $\zeta_i = \lambda_i^{E_i}$. Thus $\zeta_i^G = \lambda_i^G$, and the assertion holds. ■

Now, the two assertions of Theorem 4.2 follow simultaneously once the following is shown.

Theorem 4.4 (Brauer). $I(G, \mathscr{E}) \ni \mathbf{1}_G$.

In fact this theorem, together with Lemma 4.3(i), yields that $I(G, \mathscr{E}) = X(G) = R(G, \mathscr{E})$. Then the first equality implies the induction theorem by Lemma 4.3(ii), while the second equality implies the characterization of generalized characters.

As for permutation characters, there has been an analogous result to Theorem 4.2 given by Solomon. We shall first prove Solomon's theorem, which is of interest in its own right, and then Brauer's theorem by making use of it, following Issacs[2].

Remark. There are several proofs to Brauer's theorem besides the one we shall give here; for instance, those due to Brauer and Tate[1] and Asano[1], which are essentially the same although obtained independently. Also, there is a proof due to Tachikawa[1] that makes use of a result of Roquette[1].

Before stating Solomon's theorem, we introduce here further definitions and notations.

Let p be a prime. A subgroup Q of G is said to be a *p-complement* of G if it is a p'-group, and $|G:Q|$ is a power of p. We remark that if Q is a p-complement, then $G = PQ$, where $P \in \mathrm{Syl}_p(G)$. If in addition, Q is normal in G, then it consists of all the p'-elements of G, which is therefore a characteristic subgroup of G.

A group H is called a *p-quasi-elementary group* if it has a normal p-complement C that is cyclic. Thus, if H is p-quasi-elementary, we have

$$H = PC \rhd C, \qquad P \cap C = 1, \tag{4.2}$$

where $P \in \mathrm{Syl}_p(H)$, and C is a cyclic p'-group. It is clear that every subgroup of a p-quasi-elementary group is also p-quasi-elementary. A group is said to be *quasi-elementary* if it is p-quasi-elementary for some prime p. For a group G, let us denote by $\mathscr{E}'(G)$ and $\mathscr{E}'_p(G)$ the sets of quasi-elementary subgroups and p-quasi-elementary subgroups of G, respectively.

Remark. If in general a group H admits a factorization as in (4.2), we say that H is the *semidirect product* of P and C and write $H = P \ltimes C$ or $H = C \rtimes P$.

For a group G, we denote by $P(G)$ the submodule of the character ring $X(G)$ generated by the set $\{(\mathbf{1}_H)^G; H \leq G\}$ over $\mathbf{Z}$.

The following holds by the Mackey tensor product theorem.

Lemma 4.5. *Let $G \geq H, K$. Then*

$$(\mathbf{1}_H)^G(\mathbf{1}_K)^G = \sum_{t \in H\backslash G/K} (\mathbf{1}_{H^t \cap K})^G. \tag{4.3}$$

From the above lemma, it turns out that $P(G)$ is a subring of $X(G)$ with the common identity $\mathbf{1}_G$.

If $\mathscr{H}$ is a set of subgroups of G, we define

$$P(G, \mathscr{H}) = \sum_{H \in \mathscr{H}} \mathbf{Z}(\mathbf{1}_H)^G.$$

We set $\mathscr{E}' = \mathscr{E}'(G)$, $\mathscr{E}'_p = \mathscr{E}'_p(G)$ for brevity.

Lemma 4.6. *Both $P(G, \mathscr{E}'_p)$ and $P(G, \mathscr{E}')$ are ideals of $P(G)$.*

Proof. If $H \in \mathscr{E}'_p$, then $H^t \cap K \in \mathscr{E}'_p$ for any $K \leq G$. Thus the assertion is immediate from (4.3). ∎

Theorem 4.7 (Solomon). *With the above notation we have the following.*

(i) $\mathbf{1}_G \in P(G, \mathscr{E}')$, *i.e.*, $P(G, \mathscr{E}') = P(G)$.

(ii) *There exists a natural number m relatively prime to p such that $m\mathbf{1}_G \in P(G, \mathscr{E}'_p)$. Hence $P(G)/P(G, \mathscr{E}'_p)$ is a finite p'-group.*

We need two lemmas for the proof of this theorem.

Lemma 4.8 (Banaschewski). *Let $X(\neq \varnothing)$ be a finite set and let $\mathbf{F}_\mathbf{Z}(X)$ be the ring of all $\mathbf{Z}$-valued functions on X (with $\mathbf{1}_X(: x \mapsto 1)$ as the identity). If a subring R of $\mathbf{F}_\mathbf{Z}(X)$ does not contain $\mathbf{1}_X$, then there exist $x_0 \in X$ and a prime p such that $p \mid f(x_0)$ for all $f \in R$.*

Proof. For $x \in X$, the set $I_x = \{f(x); f \in R\}$ is an ideal of Z. If $I_y \neq \mathbf{Z}$ for some $y \in X$, then there exists a prime p such that $I_y \subset (p)$ and the assertion holds. Thus, assume that $I_x = \mathbf{Z}$ for all $x \in X$. Then, for each $x \in X$, there exists $f_x \in R$ such that $f_x(x) = 1$, and thus

$$\prod_{x \in X} (f_x - \mathbf{1}_X) = 0$$

holds. Expanding this product, we find that $\mathbf{1}_X$ can be expressed as a **Z**-linear combination of products of the functions f_x, and thus $\mathbf{1}_X \in R$, which is a contradiction. ■

Lemma 4.9. *Given $a \in G$, there exists $H \in \mathscr{E}'_p(G)$ satisfying*

(4.4) $$(\mathbf{1}_H)^G(a) \not\equiv 0 \pmod{p}.$$

Proof. Let a_p, $a_{p'}$ be the p-part, p'-part of a, respectively, and set $C = \langle a_{p'} \rangle$, $N = N_G(C)$, then $a \in N$ and $\bar{a}$ is a p-element of $\bar{N} = N/C$. Choose $H \leq N$ so that $\bar{H} \in \mathrm{Syl}_p(\bar{N})$ with $\bar{a} \in \bar{H}$. Then C is a normal p-complement of H and $H \in \mathscr{E}'_p(G)$. We shall show that H satisfies (4.4).

Recall that $(\mathbf{1}_H)^G(a) = \#\{t \in H\backslash G; Hta = Ht\}$. If $Hta = Ht$, then $tat^{-1} \in H$ and it follows that $ta_{p'}t^{-1} \in H$. But since C is a normal p-complement of H, this forces $tCt^{-1} = C$, thus $t \in N$, and we have

$$(\mathbf{1}_H)^G(a) = \#\{t \in H\backslash N; Hta = Ht\}.$$

Note that $Hta_{p'} = Ht$ if $t \in N$. Consequently, we have that $Hta = Ht \Leftrightarrow Hta_p = Ht$ and thus $(\mathbf{1}_H)^G(a) = (\mathbf{1}_H)^N(a_p)$. Now, since the length of each orbit of the p-group $\langle a_p \rangle$ on $H\backslash N$ is either 1 or a power of p, we have

$$|N:H| \equiv (\mathbf{1}_H)^N(a_p) \equiv (\mathbf{1}_H)^G(a) \pmod{p},$$

whence the result follows, since $|N:H|$ is prime to p. ■

We are now ready to prove Solomon's theorem.

Proof of Theorem 4.7.

(i) If $\mathbf{1}_G \notin P(G, \mathscr{E}')$, then, as $P(G, \mathscr{E}')$ is a subring of $\mathbf{F}_\mathbf{Z}(G)$, there exist $a \in G$ and a prime p such that $p \mid \varphi(a)$ for all $\varphi \in P(G, \mathscr{E}')$ by Lemma 4.8. But this contradicts Lemma 4.9.

(ii) Note that $R = P(G, \mathscr{E}'_p) + \mathbf{Z}(p\mathbf{1}_G)$ is a subring of $\mathbf{F}_\mathbf{Z}(G)$. Thus, if $\mathbf{1}_G \notin R$, then there exist $a \in G$ and a prime q such that $q \mid \varphi(a)$ for all $\varphi \in R$. But then it must be $q = p$ as $p\mathbf{1}_G(a) = p$, which is impossible by Lemma 4.9. Thus $\mathbf{1}_G \in R$, and there exists $n \in \mathbf{Z}$ such that $(1 - np)\mathbf{1}_G \in P(G, \mathscr{E}'_p)$. Therefore, it suffices to let m be the positive one of $\pm(1 - np)$. The latter half is obvious, since $P(G)$ is finitely generated over $\mathbf{Z}$ and $mP(G) = (m\mathbf{1}_G)P(G) \subset P(G, \mathscr{E}'_p)$. ∎

Now Brauer's theorem (Theorem 4.4) will be yielded from Theorem 4.7(i) as follows.

Proof of Theorem 4.4. We proceed by induction on $|G|$.

Case 1. G is not quasi-elementary: Since $|H| < |G|$ for any $H \in \mathscr{E}'$, we have that $\mathbf{1}_H \in I(H, \mathscr{E}(H))$, and therefore $(\mathbf{1}_H)^G \in I(G, \mathscr{E})$. Thus by Theorem 4.7, we get

$$\mathbf{1}_G \in P(G, \mathscr{E}') = \sum_{H \in \mathscr{E}'} Z(\mathbf{1}_H)^G \subset I(G, \mathscr{E}).$$

Case 2. G is quasi-elementary: We have $G = P \rtimes C$, where $P \in \mathrm{Syl}_p(G)$ and C is a cyclic p'-group which is normal in G. Let D be the centralizer of P in C, namely, $D = \{c \in C;\ ac = ca \text{ for all } a \in P\}$ and let $PD = E$. Then $E = P \times D \in \mathscr{E}$ and $D \leq Z(G)$. The assertion is trivial if $G = E$. Thus, assume that $E < G$ and let

$$(\mathbf{1}_E)^G = \mathbf{1}_G + \varphi. \tag{4.5}$$

Then it suffices to show that any irreducible constituent χ of φ lies in $I(G, \mathscr{E})$.

Since $G = EC$ with $E \cap C = D$, it follows from (4.5) that

$$\mathbf{1}_C + \varphi_C = ((\mathbf{1}_E)^G)_C = (\mathbf{1}_D)^C,$$

whence we have $(\chi_C, \mathbf{1}_C) = 0$ as $((\mathbf{1}_D)^C, \mathbf{1}_C)_C = 1$. Take any irreducible constituent λ of χ_C and let T be the inertial group of λ in G. Then there exists $\eta \in \mathrm{Irr}(T|\lambda)$ such that $\eta^G = \chi$. If $T < G$, we apply the inductive hypothesis to obtain that $\eta \in I(T, \mathscr{E}(T))$. Thus $\eta^G = \chi \in I(G, \mathscr{E})$.

Let us deduce a contradiction assuming that $T = G$. If $T = G$, then $K = \mathrm{Ker}\ \lambda$ is a normal subgroup of G that is properly contained in C as $\lambda \neq \mathbf{1}_C$. If $x \in C$, then $\lambda(x) = \lambda^{a^{-1}}(x) = \lambda(x^a)$ for all $a \in P$. Therefore $x^{-1}x^a \in \mathrm{Ker}\ \lambda = K$, and it follows that $Kx = Kx^a = (Kx)^a$, namely, P permutes the elements of Kx. Since $|Kx| = |K|$ is relatively prime to p, there exists a P-orbit of length 1, i.e., $Kx \cap D \neq \varnothing$, which yields $C = KD$. On the other hand, since $D \leq Z(G)$, D is contained in the kernel of $((\mathbf{1}_E)^G)_C$, and hence $D \leq \mathrm{Ker}\ \lambda = K$, since λ is an irreducible constituent of $((\mathbf{1}_E)^G)_C$. Thus we have $K = C$, a contradiction. ∎

The fundamental theorem of Brauer has a number of important applications. Here we show a result concerning splitting fields for G.

The following is an immediate consequence of Theorem 1.30(v).

Lemma 4.10. *Let K be a subfield of $\mathbf{C}$. For any K-character ζ and $\chi \in \mathrm{Irr}(G)$, we have*

$$m_K(\chi) | (\zeta, \chi)_G.$$

Theorem 4.11 (Splitting field theorem of Brauer). *Let n be the exponent of G. Then $\mathbf{Q}(\zeta_n)$ is a splitting field for G. Consequently, $\mathbf{Q}(\zeta_{|G|})$ is a splitting field for G, where ζ_m denotes a primitive mth root of unity in $\mathbf{C}$.*

Proof. Let $K = \mathbf{Q}(\zeta_n)$ and let $\chi \in \mathrm{Irr}(G)$. Then

$$\chi = \sum_i a_i \lambda_i^G,$$

where $a_i \in \mathbf{Z}$ and λ_i is a linear character of $E_i \in \mathscr{E}(G)$. Since each λ_i^G is a K-character, it follows from the above lemma that

$$1 = (\chi, \chi) = \sum_i a_i(\lambda_i^G, \chi) \equiv 0 \qquad (\mathrm{mod}\ m_K(\chi)),$$

whence we have $m_K(\chi) = 1$. But since $K = K(\chi)$, this implies that χ is realizable in K. ■

5. Projective Representations

5.1. Projective Representations and Generalized Group Rings

Let K be a field. By a *projective* (K-)*representation* of a group G of *degree* n, we mean a map

$$\mathbf{X}: G \to \mathrm{GL}_n(K) \qquad (x \mapsto X(x))$$

such that

$$X(x)X(y) = \alpha(x, y)X(xy) \tag{5.1}$$

for all $x, y \in G$, where $\alpha(x, y) \in K^\times$.

Exercise 5.1. Prove that the above $\alpha: G \times G \to K^\times((x, y) \mapsto \alpha(x, y))$ is a factor set, where we assume that G acts on $K^\times$ trivially. [Hint: Use the associative law in G and in $\mathrm{GL}_n(K)$.]

The projective representation $\mathbf{X}$ satisfying (5.1) is said to have the factor set α, and the factor set α is said to be *associated with* $\mathbf{X}$. We write $(\mathbf{X}, \alpha)$ if it is necessary to specify the factor set α. If $\alpha = 1$, then $\mathbf{X}$ is just a K-representation of G we have considered so far, which we shall call a *linear representation* of G if it is necessary to distinguish it from projective representations.

Remark. It follows from (5.1) that $X(1)X(1) = \alpha(1, 1)X(1)$, and thus $X(1) = \alpha(1, 1)I$, which is a scalar matrix.

Let $(\mathbf{X}, \alpha)$ be as above. For $P \in \mathrm{GL}_n(K)$ and a map $\gamma: G \to K^\times$, the map

(5.2) $$\mathbf{Y}: G \to \mathrm{GL}_n(K) \qquad (x \mapsto Y(x) = \gamma(x)P^{-1}X(x)P)$$

is again a projective representation of G.

Exercise 5.2. The factor set β of G associated with the above $\mathbf{Y}$ is given by $\beta(x, y) = \alpha(x, y)\gamma(y)\gamma(xy)^{-1}\gamma(x)$.

The projective representation $(\mathbf{Y}, \beta)$ given in (5.2) is said to be *equivalent* to $(\mathbf{X}, \alpha)$, and we write $(\mathbf{X}, \alpha) \sim (\mathbf{Y}, \beta)$. We remark that α and β are cohomologous. Projective representation of a group G is also interpreted as follows: The center of the general linear group $\mathrm{GL}_n(K)$ consists of precisely the scalar matrices. The factor group of $\mathrm{GL}_n(K)$ by its center, denoted by $\mathrm{PGL}_n(K)$, is called the *projective general linear* group over K of degree n. For $X \in \mathrm{GL}_n(K)$, we let here $\bar{X}$ denote the residue class in $\mathrm{PGL}_n(K)$ containing X.

Given a projective representation $\mathbf{X}: G \to \mathrm{GL}_n(K)$, we obtain a group homomorphism

$$\bar{\mathbf{X}}: G \to \mathrm{PGL}_n(K)$$

by assigning $\overline{X(x)}$ to each $x \in G$. If, conversely, we are given a group homomorphism $\bar{\mathbf{X}}$ as above, then we obtain a projective representation $\mathbf{X}: G \to \mathrm{GL}_n(K)(x \mapsto X(x))$, where $X(x)$ is a fixed representative of $\overline{X(x)}$. A different choice of representatives gives rise to an equivalent projective representation.

Also, we note that two projective representations **X**, **Y** of G are equivalent if and only if there exists $P \in \mathrm{GL}_n(K)$ such that

$$\bar{P}^{-1}\overline{X(x)}\bar{P} = \overline{Y(x)} \qquad \text{for all} \quad x \in G.$$

Like the representations of a group G and those of the group ring of G are essentially the same, the projective representations of G correspond to those of the generalized group rings of G in the following sense.

Let

$$A = K^{(\alpha)}G = \bigoplus_{x \in G} Ku_x, \qquad u_x u_y = \alpha(x, y)u_{xy}$$

be the generalized group ring of G with factor set α. Then, given a projective representation $(\mathbf{X}, \alpha)$ of G, there is an associated representation $\mathbf{X}'$ of the K-algebra A defined by $\sum_x \alpha_x u_x \mapsto \sum_x \alpha_x X(x)$. Conversely, any K-representation $\mathbf{X}': A \to M_n(K)$ induces a projective representation X of G with factor set α such that $\mathbf{X}: x \mapsto X'(u_x)$. Thus the representations of A correspond bijectively to the projective representations of G with factor set α. If the basis $\{u_x; x \in G\}$ of A is replaced with another basis $\{v_x = \gamma_x u_x; x \in G, \gamma_x \in K^\times\}$, then the corresponding projective representation $\mathbf{Y}(: x \mapsto \gamma_x X'(u_x))$ is equivalent to $\mathbf{X}$.

With the above notation, we say that a projective representation $\mathbf{X}$ is irreducible if the corresponding representation $\mathbf{X}'$ of A is irreducible. Similar translations will be done for other notions in representation theory.

The following theorem is obvious from the above observation.

Theorem 5.3.

(i) *For any factor set $\alpha: G \times G \to K^\times$, there exist irreducible projective K-representations of G with factor set α, and the number of equivalence classes of them are finite.*

(ii) *If* Char K *does not divide $|G|$, then every projective K-representation is completely reducible* (Chapter 2, 8.17(ii)).

We show the following lemma for later use.

Lemma 5.4. *Assume that K is an algebraically closed field.*

(i) *Let $\alpha: G \times G \to K^\times$ be a normalized factor set such that $\alpha^n \sim 1$. Then there exists a normalized factor set β of G such that $\beta \sim \alpha$ and $\beta(x, y)^n = 1$ for all $x, y \in G$.*

Also, there exists a normalized factor set γ *of* G *such that* $\gamma \sim \alpha$, $\gamma(x, y)^{2n} = 1$ *and* $\gamma(x, x^{-1}) = 1$ *for all* $x, y \in G$.

(ii) $H^2(G, K^\times)$ *is a finite group. If* Char $K = p > 0$, *then it is a* p'*-group.*

(iii) *If* G *is a cyclic group, then* $H^2(G, K^\times) = 1$.

Proof.

(i) By assumption, there exists $\rho: G \to K^\times$ such that $\alpha(x, y)^n = \rho(y)\rho(xy)^{-1}\rho(x)$. Here we remark that $\rho(1) = 1$, since $\alpha(1, y) = 1$. For each $x \in G$, take $\mu(x) \in K^\times$ so that $\mu(x)^n = \rho(x)^{-1}$ with $\mu(1) = 1$, and define β by $\beta(x, y) = \alpha(x, y)\mu(y)\mu(xy)^{-1}\mu(x)$. Then we have $\beta(x, y)^n = 1$, and β is normalized, as required.

Let $\nu: G \to K^\times$ be such that $\nu(1) = 1$, $\nu(x)^2 = \beta(x, x^{-1})^{-1}$ with $\nu(x) = \nu(x^{-1})$ if $x \neq 1$ (notice that $\beta(x, x^{-1}) = \beta(x^{-1}, x)$ as β is normalized). If we set $\gamma(x, y) = \beta(x, y)\nu(y)\nu(xy)^{-1}\nu(x)$, then the factor set γ satisfies the above conditions.

(ii) We have by Chapter 2, 7.3(i) that $\alpha^{|G|} \sim 1$ for all factor sets α. Consequently, each element of $H^2(G, K^\times) = Z^2(G, K^\times)/B^2(G, K^\times)$ is represented by a factor set β satisfying $\beta(x, y)^{|G|} = 1$ (for all $x, y \in G$). Since there are only a finite number of maps from $G \times G$ into the set $\{\varepsilon \in K^\times; \varepsilon^{|G|} = 1\}$, it follows that $H^2(G, K^\times)$ is a finite group.

If Char $K = p > 0$, then we have $\varepsilon^m = 1$ if $\varepsilon^{p^r m} = 1$ in K. Thus each element of $H^2(G, K^\times)$ has an order relatively prime to p, and hence, $H^2(G, K^\times)$ is a p'-group.

(iii) Let $G = \langle x \rangle$ be a cyclic group of order n and α be any factor set of G. In order to simplify the argument, we work with the generalized group ring $K^{(\alpha)}G = \bigoplus_{i=0}^{n-1} Ku_{x^i}(u_{x^i}u_{x^j} = \alpha(x^i, x^j)u_{x^{i+j}}$ with $u_1 = 1)$. We have $u_x^n = \gamma u_1 = \gamma$ for some $\gamma \in K^\times$. Choose $\rho \in K^\times$ so that $\rho^n = \gamma^{-1}$ and set $v_x = \rho u_x$. Then $v_x^n = 1$. If we let $w_1 = 1$, $w_{x^i} = v_x^i (1 \leq i \leq n-1)$, then $w_{x^i}w_{x^j} = w_{x^{i+j}}$. Write $w_{x^i} = \mu(x^i)u_{x^i}$ and define $\mu: G \to K^\times$ by $x^i \mapsto \mu(x^i)$. Then we find that $\alpha(x^i, x^j)\mu(x^i)\mu(x^{i+j})^{-1}\mu(x^j) = 1$, hence $\alpha \sim 1$. ■

Now, let us study irreducible projective **C**-representations of G with factor set $\alpha: G \times G \to \mathbf{C}^\times$. They are considered as irreducible representations of the semisimple algebra $\mathbf{C}^{(\alpha)}G$.

We may assume that α is normalized and that $\alpha(x, y)^{2|G|} = 1$, $\alpha(x, x^{-1}) = 1$ for all $x, y \in G$. If k denotes the field generated over $\mathbf{Q}$ by a primitive $2|G|$th root of unity, then α is considered as the map $\alpha: G \times G \to k^\times$. Thus the generalized group ring $k^{(\alpha)}G$ can be defined, and $\mathbf{C}^{(\alpha)}G$ is obtained from it by extending the coefficient field to $\mathbf{C}$. Take a splitting field K for $k^{(\alpha)}G$ which is a finite extension of k. Thus K is an algebraic number field. Let

$$A = K^{(\alpha)}G = \bigoplus_{i=1}^{l} A_i, \qquad A_i \simeq M_{n_i}(K) \tag{5.3}$$

be the direct sum of simple components A_i, and let

$$\mathbf{F}_i\colon a \mapsto F_i(a) = (\beta_{\mu\nu}^{(i)}(a))_{\mu,\nu} \qquad (a \in A) \tag{5.4}$$

be an absolutely irreducible representation belonging to the simple component A_i, and ζ_i be the character defined by $\mathbf{F}_i$. We write

$$A = \bigoplus_{x \in G} Ku_x, \qquad u_x u_y = \alpha(x, y)u_{xy}$$

and let $\mathbf{\Gamma}$ be the regular representation of A relative to the K-basis $\{u_x\}$ of A. Then the character $\chi_{\mathbf{\Gamma}}$ is expressed as

$$\chi_{\mathbf{\Gamma}} = \sum_{i=1}^{l} \zeta_i(u_1)\zeta_i. \tag{5.5}$$

Observe that $\chi_{\mathbf{\Gamma}}(u_1) = |G|$, and that $\chi_{\mathbf{\Gamma}}(u_x) = 0$ if $x \neq 1$, since then all the diagonal entries of $\mathbf{\Gamma}(u_x)$ are zero. Consequently, it follows from (5.5) that

$$\begin{cases} \displaystyle\sum_{i=1}^{l} \zeta_i(u_1)^2 = |G|, \\ \displaystyle\sum_{i=1}^{l} \zeta_i(u_1)\zeta_i(u_x) = 0, \qquad \text{if } x \neq 1. \end{cases} \tag{5.6}$$

Exercise 5.5. With the notation above, the following holds.

$$\sum_{x \in G} \beta_{\mu\nu}^{(i)}(u_x)\beta_{\rho\sigma}^{(j)}(u_{x^{-1}}) = \delta_{ij}\delta_{\mu\sigma}\delta_{\nu\rho}\frac{|G|}{\zeta_i(u_1)}. \tag{5.7}$$

[*Hint*: *Set*

$$A = (\beta_{\mu\nu}^{(i)}(u_x))_{(i,\mu,\nu),x}, \quad B = \left(\frac{\zeta_i(u_1)}{|G|}\beta_{\nu\mu}^{(i)}(u_{x^{-1}})\right)_{(i,\mu,\nu),x}$$

as in the proof of Theorem 2.2. Then we have ${}^tAB = I$ from (5.6), and hence $B\,{}^tA = I$.]

Let R be the ring of algebraic integers in K, $\mathfrak{p}$ be a prime ideal of R, and $R_{\mathfrak{p}}$ be the set of $\mathfrak{p}$-integers in K. Note that $\alpha(x, y) \in R_{\mathfrak{p}}$. Hence, $A_{\mathfrak{p}} = \bigoplus_x R_{\mathfrak{p}} u_x$ is an $R_{\mathfrak{p}}$-algebra such that $A = K \otimes_{R_{\mathfrak{p}}} A_{\mathfrak{p}}$. We prove the following result by a similar method to the proof of Theorem 2.4.

Theorem 5.6. *The degree of an irreducible projective* **C**-*representation of* G *is a divisor of* $|G|$.

Proof. By Chapter 2, 1.6, we may assume that each $\mathbf{F}_i$ in (5.4) is a representation over $R_{\mathfrak{p}}$, i.e., $\beta_{\mu\nu}^{(i)}(u_x) \in R_{\mathfrak{p}}$. Then we see from (5.7) that $|G|/\zeta_i(u_1) \in R_{\mathfrak{p}}$. Since $\mathfrak{p}$ is an arbitrary prime ideal, it follows that $|G|/\zeta_i(u_1) \in \bigcap_{\mathfrak{p}} R_{\mathfrak{p}} = R$. Thus $|G|/\zeta_i(u_1)$ is an algebraic integer in **Q**, i.e., a rational integer. ■

5.2. Projective Representations and Clifford Theory

Throughout the remainder of this section, we assume that K is an algebraically closed field.

Let $\mathrm{IRR}_K(G)$ be a complete set of representatives for equivalence classes of irreducible K-representations of G.

The projective representations of G are closely related to Clifford theory, as will be described in the following.

Let $G \trianglerighteq H$ and let $\mathbf{Y} \in \mathrm{IRR}_K(H)$. For $x \in G$, we have a representation $\mathbf{Y}^x\colon h \mapsto Y(xhx^{-1})$, which is called a *conjugate* representation of $\mathbf{Y}$. As in the case of modules, $T(\mathbf{Y}) = \{t \in G; \mathbf{Y}^t \sim \mathbf{Y}\}$ is called the *inertial group* of $\mathbf{Y}$. If $\mathbf{X} \in \mathrm{IRR}_K(G)$, then $\mathbf{X}_H$ is completely reducible (cf. Section 3). We denote by $\mathrm{IRR}_K(G|\mathbf{Y})$ the set of all $\mathbf{X} \in \mathrm{IRR}_K(G)$ such that $\mathbf{X}_H$ has $\mathbf{Y}$ as one of its irreducible constituents.

Given $\mathbf{X} \in \mathrm{IRR}_K(G|\mathbf{Y})$, there exists $\mathbf{Z} \in \mathrm{IRR}_K(T(\mathbf{Y})|\mathbf{Y})$ such that $\mathbf{X} = \mathbf{Z}^G$. This of course makes no particular sense if $T(\mathbf{Y}) = G$. But in that case, we shall see that $\mathbf{X}$ is expressed as a tensor product of two irreducible projective representations of G, one of which can be viewed as a projective representation of G/H. This will sometimes allow us to reduce problems under consideration to a smaller degree case (see Theorem 5.8 below). We first prove the following theorem.

Theorem 5.7. *Let $G \rhd H$, and let* $\mathbf{Y}$ *be an (absolutely) irreducible K-representation of H of degree n. If* $T(\mathbf{Y}) = G$, *then the following holds.*

(i) $\mathbf{Y}$ *extends to a projective representation* $\mathbf{X}$ *of G, which satisfies the following two conditions: For* $h \in H$ *and* $x \in G$,

$$\text{(a)} \quad X(hx) = X(h)X(x), \qquad \text{(b)} \quad X(xh) = X(x)X(h).$$

(ii) *If* $\mathbf{X}'$ *is any extension of* $\mathbf{Y}$ *satisfying the above two conditions, then there exists a map* $\bar{\gamma}: \bar{G} = G/H \to K^{\times}$ *such that* $X'(x) = \bar{\gamma}(\bar{x})X(x)$ *for all* $x \in G$.

(iii) *If* α *is a factor set associated with the projective representation* $\mathbf{X}$ *in (i), then* $\alpha(hx, ky) = \alpha(x, y)$ *holds for all* $h, k \in H$. *Consequently,* α *is regarded as a factor set* $\bar{\alpha}$ of $\bar{G}(: (\bar{x}, \bar{y}) \mapsto \alpha(x, y))$, *and the following are true*:

$$\bar{\alpha}^{n|H|} \sim 1, \text{ and if Char } K = 0, \text{ then } \bar{\alpha}^{|H|^2} \sim 1.$$

Furthermore, we have that

$$\mathbf{Y} \text{ extends to a linear representation of } G \Leftrightarrow \bar{\alpha} \sim 1.$$

In particular, if $H^2(\bar{G}, K^{\times}) = 1$, *then* $\mathbf{Y}$ *always extends to a linear representation of G.*

Proof.

(i) Let T be a complete set of coset representatives of H in G containing 1. For each $t \in T$, we have $\mathbf{Y}^t \sim \mathbf{Y}$ by assumption. Thus there exists $X(t) \in \mathrm{GL}_n(K)$ such that $Y(tht^{-1}) = X(t)Y(h)X(t)^{-1}$ for all $h \in H$, where we let $X(1) = I$ (the identity matrix). Since every element of G can be expressed uniquely as ht with $h \in H$ and $t \in T$, it is possible to define $\mathbf{X}: G \to \mathrm{GL}_n(K)$ by $ht \mapsto Y(h)X(t)$. We want to show that $\mathbf{X}$ is a projective representation of G which satisfies (a) and (b). Clearly, $\mathbf{X}$ is an extension of $\mathbf{Y}$. Also, if $h \in H$ and $x = kt$ $(k \in H, t \in T)$, we have

$$X(hx) = X(hkt) = Y(hk)X(t) = Y(h)Y(k)X(t)$$

and

$$X(h)X(x) = Y(h)Y(k)X(t).$$

Thus (a) holds. Furthermore, since $xh = kth = k(tht^{-1})t$, it follows that

$$\begin{aligned} X(xh) &= Y(k)Y(tht^{-1})X(t) = Y(k)X(t)Y(h)X(t)^{-1}X(t) \\ &= Y(k)X(t)Y(h) = X(x)X(h), \end{aligned}$$

so (b) holds. To see that **X** is a projective representation of G, we first observe that $Y(h)X(x) = X(hx) = X(xh^x) = X(x)Y(h^x)$, whence we have

$$X(x)^{-1}Y(h)X(x) = Y(h^x).$$

Consequently, for any $x, y \in G$, we have

$$\begin{aligned} &X(xy)X(y)^{-1}X(x)^{-1}Y(h)X(x)X(y)X(xy)^{-1} \\ &\quad = X(xy)Y(h^{xy})X(xy)^{-1} = Y(h) \qquad \text{(for all } h \in H\text{)}. \end{aligned}$$

But since **Y** is absolutely irreducible, it follows that $X(x)X(y)X(xy)^{-1}$ must be a scalar matrix, say $\alpha(x, y)I$. Thus $X(x)X(y) = \alpha(x, y)X(xy)$.

(ii) Since $X'(x)X(x)^{-1}Y(h)X(x)X'(x)^{-1} = X'(x)Y(h^x)X'(x)^{-1} = Y(h)$ for all $h \in H$, it follows that $X'(x) = \gamma(x)X(x)$ for some $\gamma(x) \in K^\times$. If $h \in H$, then $X'(hx) = Y(h)X'(x) = \gamma(x)Y(h)X(x) = \gamma(x)X(hx)$, while $X'(hx) = \gamma(hx)X(hx)$ by the definition of γ. Thus, $\gamma(hx) = \gamma(x)$. Hence, the map $\bar{\gamma}: \bar{G} \to K^\times (\bar{x} \mapsto \gamma(x))$ is well defined and (ii) holds.

(iii) We have $X(hx)X(ky) = \alpha(hx, ky)X(hxky)$. On the other hand,

$$\begin{aligned} X(hx)X(ky) &= X(h)X(x)X(k)X(y) = X(h)X(k^{x^{-1}})X(x)X(y) \\ &= \alpha(x, y)X(hk^{x^{-1}})X(xy) = \alpha(x, y)X(hk^{x^{-1}}xy) \\ &= \alpha(x, y)X(hxky). \end{aligned}$$

Hence $\alpha(hx, ky) = \alpha(x, y)$, and $\bar{\alpha}: \bar{G} \times \bar{G} \to K^\times$ is defined, which is clearly a factor set of $\bar{G}$. Set $\delta(x) = \det X(x)$. Then it follows from $X(x)X(y) = \alpha(x, y)X(xy)$ that $\alpha(x, y)^n = \delta(x)\delta(y)\delta(xy)^{-1}$. Also, since $X(hx) = X(h)X(x)$, we get $\delta(hx)^{|H|} = \delta(h)^{|H|}\delta(x)^{|H|} = \delta(x)^{|H|}$. We now define $\bar{\gamma}: \bar{G} \to K^\times$ by $\bar{x} \mapsto \delta(x)^{|H|}$ to obtain $\bar{\alpha}(\bar{x}, \bar{y})^{n|H|} = \bar{\gamma}(\bar{x})\bar{\gamma}(\bar{y})(\bar{x}\bar{y})^{-1}$. Thus $\bar{\alpha}^{n|H|} \sim 1$, and if Char $K = 0$, then $\bar{\alpha}^{|H|^2} \sim 1$ because $n \mid |H|$.

If **Y** extends to a linear representation **X**$'$ of G, then we see easily from (ii) that $\bar{\alpha} \sim 1$. Conversely, if there exists some $\bar{\gamma}: \bar{G} \to K^\times$ such that $\bar{\alpha}(\bar{x}, \bar{y}) = \bar{\gamma}(\bar{y})\bar{\gamma}(\bar{x}\bar{y})^{-1}\bar{\gamma}(\bar{x})$, then **X**$': x \mapsto \bar{\gamma}(\bar{x})^{-1}X(x)$ is a linear representation of G. ■

Remark. With the notation of (iii) in the above theorem, **Y** does not necessarily extend to a linear representation of G under the weaker condition that $\alpha \sim 1$ (see Issacs[2], p. 179).

Let $G \rhd H$ and $\bar{G} = G/H$ as above. If $(\bar{\mathbf{X}}, \bar{\alpha})$ is a projective representation of $\bar{G}$, then the induced map $\mathbf{X}: x \to \bar{X}(\bar{x})$ is a projective representation of G with factor set α such that $\alpha: (x, y) \mapsto \bar{\alpha}(\bar{x}, \bar{y})$.

Theorem 5.8. *Let $G \rhd H$ and $\bar{G} = G/H$. Given $\mathbf{Y} \in \mathrm{IRR}_K(H)$ with $T(\mathbf{Y}) = G$, we extend it to a projective representation of G as in Theorem 5.7, which is denoted by the same letter $\mathbf{Y}$ for convenience, and let α be a factor set of G associated with it. Then the following holds.*

(i) *If $\mathbf{X} \in \mathrm{IRR}_K(G|\mathbf{Y})$ and $\mathbf{X}_H \sim e\mathbf{Y}$, then there exists an irreducible projective representation $(\bar{\mathbf{Z}}, \bar{\alpha}^{-1})$ of $\bar{G}$ of degree e such that $\mathbf{X}$ is equivalent to the following representation of G:*

$$(5.8) \qquad \mathbf{Z} \otimes \mathbf{Y}: x \mapsto Z(x) \otimes Y(x), \qquad \text{for } x \in G,$$

where $\bar{\alpha}$ is the factor set of $\bar{G}$ obtained from α, and $\mathbf{Z}$ is the representation of G obtained naturally from $\bar{\mathbf{Z}}$.

(ii) *If, conversely, $(\bar{\mathbf{Z}}, \bar{\alpha}^{-1})$ is any irreducible projective representation of $\bar{G}$, then the above $\mathbf{Z} \otimes \mathbf{Y}$ is an irreducible representation of G, which is equivalent to some representation in $\mathrm{IRR}_K(G|\mathbf{Y})$. Moreover, if e is the degree of $\mathbf{Z}$, then $(\mathbf{Z} \otimes \mathbf{Y})_H \sim e\mathbf{Y}$.*

Proof. Let n be the degree of $\mathbf{Y}$. We understand that $M_e(K) \otimes_K M_n(K) = M_{en}(K)$ by identifying each $A \otimes B \in M_e(K) \otimes_K M_n(K)$ with the tensor product of A and B (cf. Exercise 1.13(i)). We denote by I_f the identity matrix of degree f and abbreviate $\otimes_K$ to $\otimes$.

(i) We may assume that $\mathbf{X}_H = e\mathbf{Y}$, and hence $X(h) = I_e \otimes Y(h)$ for $h \in H$. Thus, if $x \in G$, then

$$X(h^x) = X(x)^{-1}X(h)X(x) = X(x)^{-1}(I_e \otimes Y(h))X(x).$$

On the other hand, using that $Y(h^x) = Y(x)^{-1}Y(h)Y(x)$, we have

$$X(h^x) = I_e \otimes Y(h^x) = (I_e \otimes Y(x))^{-1}(I_e \otimes Y(h))(I_e \otimes Y(x)).$$

Thus $X(x)(I_e \otimes Y(x))^{-1}$ commutes with $I_e \otimes Y(h)$ for all $h \in H$. Note that since Y is absolutely irreducible, $M_n(K)$ is generated by $\{Y(h); h \in H\}$ over K. Therefore $X(x)(I_e \otimes Y(x))^{-1} \in C_{M_{en}(K)}(KI_e \otimes M_n(K)) = M_e(K) \otimes KI_n$ (Chapter 2, 4.1), and there exists $Z(x) \in M_e(K)$ such that

$$X(x) = Z(x) \otimes Y(x).$$

It then follows from $X(x)X(y) = X(xy)$ and $Y(x)Y(y) = \alpha(x, y)Y(xy)$ that $Z(x)Z(y) = \alpha(x, y)^{-1}Z(xy)$. Also, it is easy to see that $Z(hx) = Z(x)$ for all $h \in H$. Hence, the induced map $\bar{\mathbf{Z}}: \bar{x} \to Z(x)$ can be defined, and $(\bar{\mathbf{Z}}, \bar{\alpha}^{-1})$ is a projective representation of $\bar{G}$. If $\mathbf{Z}$ is reducible, then so is $\mathbf{X}$, which contradicts the assumption. Therefore $\mathbf{Z}$ is irreducible.

(ii) Let $B = \bigoplus_{\bar{x} \in \bar{G}} Ku_{\bar{x}}\ (u_{\bar{x}}u_{\bar{y}} = \bar{\alpha}(\bar{x}, \bar{y})^{-1}u_{\overline{xy}})$ be the generalized group ring with factor set $\bar{\alpha}$, and let $A = B \otimes M_n(K)$. Then there is an algebra homomorphism $f: KG \to A$ such that $x \mapsto u_{\bar{x}} \otimes Y(x) (x \in G)$. We claim that f is an epimorphism. To prove this, we may assume $u_{\bar{1}} = 1_B$. Since $f(h) = u_{\bar{1}} \otimes Y(h)$ for all $h \in H$ and the set $\{Y(h); h \in H\}$ generates $M_n(K)$, it follows that Im $f \supset u_{\bar{1}} \otimes M_n(K)$. Therefore $u_{\bar{x}} \otimes I_n = (u_{\bar{x}} \otimes Y(x))(u_{\bar{1}} \otimes Y(x)^{-1}) \in \operatorname{Im} f$ for all $x \in G$, which yields that $\operatorname{Im} f = A$ as claimed.

Now, $\bar{\mathbf{Z}}$ gives an (absolutely) irreducible representation of B, and the identity map $\iota: M_n(K) \to M_n(K)$ is certainly an (absolutely) irreducible representation. Therefore $\bar{\mathbf{Z}} \otimes \iota$ is an irreducible representation of A (Chapter 2, 3.15). Since $(\bar{\mathbf{Z}} \otimes \iota) \circ f: KG \to M_{en}(K)$ gives the representation $\mathbf{Z} \otimes \mathbf{Y}$ of (5.8) and f is epimorphic, $\mathbf{Z} \otimes \mathbf{Y}$ is an irreducible representation of G, which is clearly equivalent to some representation in $\mathrm{IRR}_K(G|\mathbf{Y})$. The last assertion is clear. ■

Corollary 5.9. *In addition to the assumption of Theorem* 5.8, *we assume that* $\mathbf{Y}$ *extends to a linear representation* $\mathbf{X}$ *of* G. *Then for each* $\mathbf{M} \in \mathrm{IRR}_K(G|\mathbf{Y})$, *there exists a unique* $\bar{\mathbf{Z}} \in \mathrm{IRR}_K(\bar{G})$ *such that* $\mathbf{M} \sim \mathbf{Z} \otimes \mathbf{X}$, *where* $\mathbf{Z}$ *is the representation of* G *obtained naturally from* $\bar{\mathbf{Z}}$. *Hence we have*

$$\mathrm{IRR}_K(G|\mathbf{Y}) = \{\mathbf{Z} \otimes \mathbf{X}; \bar{\mathbf{Z}} \in \mathrm{IRR}_K(\bar{G})\}.$$

Proof. It suffices to show that $\mathbf{Z} \otimes \mathbf{X} \sim \mathbf{Z}' \otimes \mathbf{X}$ implies $\mathbf{Z} \sim \mathbf{Z}'$. Now suppose that $P^{-1}(Z(x) \otimes X(x))P = Z'(x) \otimes X(x)$ for all $x \in G$. Thus, if $h \in H$, then $P^{-1}(I_e \otimes Y(h))P = I_e \otimes Y(h)$, and it follows that $P = Q \otimes I_n$. Therefore, $Q^{-1}Z(x)Q = Z'(x)$ and $\mathbf{Z} \sim \mathbf{Z}'$. ■

Let $G = H_1 \times H_2$ be the direct product of two groups H_1 and H_2, and let $\mathbf{X}_i$ be a representation of $H_i (i = 1, 2)$. We obtain a representation $\mathbf{X}_1 \times \mathbf{X}_2$ of G that is defined by $(h_1, h_2) \mapsto X_1(h_1) \otimes X_2(h_2)$ for $h_i \in H_i$. If χ_i denotes the character defined by $\mathbf{X}_i$, then $\mathbf{X}_1 \times \mathbf{X}_2$ defines the character of G, denoted by $\chi_1 \times \chi_2$, such that

$$(\chi_1 \times \chi_2)(h_1, h_2) = \chi_1(h_1)\chi_2(h_2).$$

Corollary 5.10. *If* $G = H_1 \times H_2$, *then we have*

$$\mathrm{IRR}_K(G) = \{\mathbf{X}_1 \times \mathbf{X}_2; \mathbf{X}_i \in \mathrm{IRR}_K(H_i)\}.$$

Proof. If $\mathbf{X}_2 \in \mathrm{IRR}_K(H_2)$, then $T(\mathbf{X}_2) = G$ and it extends naturally to a representation $\mathbf{X}_2$ of G, where $\mathbf{X}_2$: $(h_1, h_2) \mapsto X_2(h_2)$. On the other hand, since $\bar{G} = G/H_2 \simeq H_1$, any irreducible representation of $\bar{G}$ may be considered as that of H_1. Therefore we have $\mathrm{IRR}_K(G|\mathbf{X}_2) = \{\mathbf{X}_1 \times \mathbf{X}_2;\ \mathbf{X}_i \in \mathrm{IRR}_K(H_i)\}$ by Corollary 5.9. This yields the result, since $\mathrm{IRR}_K(G) = \bigcup_{\mathbf{X}_2} \mathrm{IRR}_K(G|\mathbf{X}_2)$, where $\mathbf{X}_2$ runs through $\mathrm{IRR}_K(H_2)$. ■

The next theorem gives some sufficient conditions for a given $\mathbf{Y} \in \mathrm{IRR}_K(H)$ with $T(\mathbf{Y}) = G$ to extend to a representation of G.

Theorem 5.11. *Let $G \rhd H$, and let $\mathbf{Y} \in \mathrm{IRR}_K(H)$ be such that $T(\mathbf{Y}) = G$. Then each one of the following conditions implies that $\mathbf{Y}$ is extendible to a linear representation of G:*

(i) *G/H is cyclic.*

(ii) *Char $K = p > 0$ and G/H is a p-group.*

(iii) (Gallagher) *Char $K = 0$ and $(|H|, |G/H|) = 1$.*

Proof. Let $\bar{G} = G/H$.

(i) Since $H^2(\bar{G}, K^\times) = 1$ by Lemma 5.4(iii), the assertion follows from Theorem 5.7(iii).

(ii) We know that $H^2(\bar{G}, K^\times)$ is a p'-group (Lemma 5.4(ii)). On the other hand, since $\bar{G}$ is a p-group, $H^2(\bar{G}, K^\times)$ is a p-group, yielding $H^2(\bar{G}, K^\times) = 1$. Therefore the assertion holds.

(iii) As was shown in Theorem 5.7(i), $\mathbf{Y}$ extends to a projective representation $(\mathbf{X}, \alpha)$ of G, where α is actually a factor set $\bar{\alpha}$ of $\bar{G}$ such that $\bar{\alpha}^{|H|^2} \sim 1$. But since $\bar{\alpha}^{|\bar{G}|} \sim 1$ in general, we conclude that $\bar{\alpha} \sim 1$, whence the assertion follows. ■

Let us concentrate on complex characters.

Theorem 5.12. *Let $G \rhd H$, $\zeta \in \mathrm{Irr}(H)$ and $T = T(\zeta)$.*

(i) *For any $\chi \in \mathrm{Irr}(G|\zeta)$, the ramification index $e = (\chi_H, \zeta)_H$ is a divisor of $|T{:}H|$.*

(ii) *If ζ extends to an irreducible character η of T, then the following hold.*

(a) $\operatorname{Irr}(T|\zeta) = \{\theta\eta;\, \theta \in \operatorname{Irr}(T/H)\}$,

$\operatorname{Irr}(G|\zeta) = \{(\theta\eta)^G;\, \theta \in \operatorname{Irr}(T/H)\}$.

(b) $\zeta^G = \sum_{\theta \in \operatorname{Irr}(T/H)} \theta(1)(\theta\eta)^G$.

Proof.

(i) There exists $\eta \in \operatorname{Irr}(T|\zeta)$ such that $e = (\eta_H, \zeta)_H$ by Theorem 3.8(ii). But according to Theorem 5.8, e is the degree of some irreducible projective representation of T/H. Hence $e \,|\, |T:H|$ by Theorem 5.6.

(ii) Part(a) is clear by Corollary 5.9 and Theorem 3.8. Since $\theta(1) = ((\theta\eta)_H, \zeta)_H = (\theta\eta, \zeta^T)_T$, we have $\zeta^T = \sum_\theta \theta(1)(\theta\eta)$, whence the assertion is immediate. ■

Part (i) of the above theorem yields the following.

Theorem 5.13. *Let $G \trianglerighteq H$. Then the following hold.*

(i) *If $\zeta \in \operatorname{Irr}(H)$ and $\chi \in \operatorname{Irr}(G|\zeta)$, then $\chi(1) \,|\, |G:H|\zeta(1)$.*

(ii) (Ito) *If H is abelian, then $\chi(1) \,|\, |G:H|$ for all $\chi \in \operatorname{Irr}(G)$.*

Proof.

(i) Let $T = T(\zeta)$. There exists $\eta \in \operatorname{Irr}(T|\zeta)$ such that $\eta^G = \chi$. If $\eta_H = e\zeta$, then, since $e \,|\, |T:H|$, it follows that

$$\chi(1) = |G:T|\eta(1) \,|\, |G:T||T:H|\zeta(1) = |G:H|\zeta(1).$$

(ii) Since H is abelian, we have $\zeta(1) = 1$ for any $\zeta \in \operatorname{Irr}(H)$, whence the result is immediate by (i). ■

5.3. Projective Representations and Central Extensions

If we are given a group epimorphism $\pi\colon \tilde{G} \to G$ such that $\operatorname{Ker} \pi$ is contained in the center $Z(\tilde{G})$ of $\tilde{G}$, we say that $\tilde{G} \xrightarrow{\pi} G$ is a *central extension* of G (by $\operatorname{Ker} \pi$).

Thus, if $\tilde{G} \overset{\pi}{\to} G$ is a central extension with $H = \text{Ker}\ \pi$, then

$$H \leq Z(\tilde{G}), \qquad \tilde{G}/H \simeq G.$$

Let $\tilde{G} = \sum_{a \in G} Ht_a$, where $\pi(t_a) = a$, and write

$$t_a t_b = z(a, b)t_{ab} \qquad \text{with} \quad z(a, b) \in H.$$

Then the following holds.

Exercise 5.14.
(i) It holds that $z(a, b)z(ab, c) = z(b, c)z(a, bc)$. Consequently, $z: G \times G \to H((a, b) \mapsto z(a, b))$ is a factor set with coefficients in H, where H is considered a trivial G-module.
(ii) Let $\{t'_a\}$ be any coset representative of H in G with $\pi(t'_a) = a$ and write $t'_a t'_b = z'(a, b)t'_{ab}$. Then $z \sim z'$.

We shall construct central extensions of G from factor sets.

Lemma 5.15. *Let H be an abelian group. Supposing that a group G acts on H trivially, let $z: G \times G \to H$ be a factor set. Then there exists a central extension $\tilde{G}$ of G by H such that the following holds:*

$$\tilde{G} = \sum_{a \in G} Ht_a, \qquad t_a t_b = z(a, b)t_{ab}. \tag{5.9}$$

Proof. Let $\tilde{G} = H \times G = \{(h, a); h \in H, a \in G\}$ be the direct product of sets, and define the multiplication in it as follows:

$$(h, a)(k, b) = (hkz(a, b), ab).$$

Then $\tilde{G}$ is a group with identity $(z(1, 1)^{-1}, 1)$. Indeed, by using that $z(a, 1) = z(1, a) = z(1, 1)$ (cf. Chapter 2, 7.5), we have

$$(h, a)(z(1, 1)^{-1}, 1) = (z(1, 1)^{-1}, 1)(h, a) = (h, a).$$

We also note that $H \times 1 = \{(h, 1); h \in H\} \leq Z(\tilde{G})$.

The map $\zeta: H \to H \times 1 (h \mapsto (hz(1, 1)^{-1}, 1))$ is an isomorphism, and so we understand that $H \leq Z(\tilde{G})$ by identifying h with $\zeta(h)$. Set $t_a = (1, a)$. Then $ht_a = (hz(1, 1)^{-1}, 1)(1, a) = (hz(1, 1)^{-1}z(1, a), a) = (h, a)$ and $t_a t_b = (z(a, b), ab) = z(a, b)t_{ab}$. Therefore (5.9) holds. ■

Lemma 5.16. *Let $\tilde{G} \xrightarrow{\pi} G$ be a central extension and $\tilde{G} = \sum_{a \in G} Ht_a$, where $H = \operatorname{Ker} \pi$ and $\pi(t_a) = a$. Suppose that we are given a linear representation $\tilde{\mathbf{X}}: \tilde{G} \to \mathrm{GL}_n(K)$ such that $\tilde{X}(h)$ is a scalar matrix $\lambda(h)I$ for all $h \in H$. Then $\mathbf{X}: G \to \mathrm{GL}_n(K)(a \mapsto \tilde{X}(t_a))$ is a projective representation of G such that $\tilde{X}(x) = \lambda(h)X(\pi(x))$ for $x = ht_a$.*

Proof. If $t_a t_b = z(a, b)t_{ab}$, then we have $\tilde{X}(t_a)\tilde{X}(t_b) = \lambda(z(a, b))\tilde{X}(t_{ab})$. Thus $\mathbf{X}$ is a projective representation of G with factor set $(a, b) \mapsto \lambda(z(a, b))$. The latter half is trivial. ■

Remark. The assumption in the above lemma is automatically satisfied provided $\tilde{\mathbf{X}}$ is absolutely irreducible.

Exercise 5.17. If in the above lemma, $\{t'_a\}$ is another set of coset representatives of H in $\tilde{G}$ such that $\pi(t'_a) = a$, then $\mathbf{X}': G \to \mathrm{GL}_n(K)$ $(a \mapsto \tilde{X}(t'_a))$ is a projective representation equivalent to $\mathbf{X}$.

Let $\tilde{G} \xrightarrow{\pi} G$ be a central extension as above, and let $\mathbf{X}: G \to \mathrm{GL}_n(K)$ be a projective representation. If there exist a linear representation $\tilde{\mathbf{X}}$ of $\tilde{G}$ and a map $\lambda: \tilde{G} \to K^\times$ such that

$$\tilde{X}(x) = \lambda(x)X(\pi(x)) \qquad \text{for all } x \in \tilde{G}, \tag{5.10}$$

then $\mathbf{X}$ is said to be *lifted* to the linear representation $\tilde{\mathbf{X}}$ of $\tilde{G}$.

Remark. For $h \in \operatorname{Ker} \pi$, we have $\tilde{X}(h) = \lambda(h)X(1)$ in (5.10), which is a scalar matrix (cf. Remark following Exercise 5.1.)

The above situation is also interpreted as follows. Namely, if $\varphi: \mathrm{GL}_n(K) \to \mathrm{PGL}_n(K)$ is the natural map and if $\bar{\mathbf{X}}: G \to \mathrm{PGL}_n(K)$ is the composite map $\varphi \circ \mathbf{X}$, then $\tilde{\mathbf{X}}$ is a homomorphism that completes the diagram below to a commutative one.

$$\begin{array}{ccc} \tilde{G} & \xrightarrow{\tilde{\mathbf{X}}} & \mathrm{GL}_n(K) \\ {\scriptstyle\pi}\downarrow & \circlearrowright & \downarrow{\scriptstyle\varphi} \\ G & \xrightarrow[\bar{\mathbf{X}}]{} & \mathrm{PGL}_n(K) \end{array}$$

A central extension $\tilde{G} \to G$ is said to have the *projective lifting property* (relative to K) if every projective K-representation of G is lifted to a linear K-representation of $\tilde{G}$.

For every G, as we shall see later, there exists a central extension $\tilde{G} \to G$ which has the projective lifting property. To show this, we need some preparations.

For an abelian group H, let us denote by $L(H)$ the set of all linear K-characters of H. $L(H)$ is an abelian group with the usual multiplication, namely, $(\lambda\mu)(h) = \lambda(h)\mu(h)$ for $\lambda, \mu \in L(H)$; this is called the *character group* of H.

Lemma 5.18. *Let H be an abelian group and suppose that* Char $K \nmid |H|$. *Then the following statements hold.*

(i) *$L(H) \simeq H$.*

(ii) *Let $A \leq H$ and $A^{\perp} = \{\lambda \in L(H);\ A \leq \operatorname{Ker} \lambda\}$. Then every $\lambda_0 \in L(A)$ extends to a linear character λ of H and there hold the following*:

$$\text{(a)} \quad A^{\perp} \simeq L(H/A), \qquad \text{(b)} \quad L(A) \simeq L(H)/A^{\perp}.$$

(iii) *For $h \in H$, define $\chi_h \in L(L(H))$ by $\chi_h\colon \lambda \mapsto \lambda(h)$. Then the map $\chi\colon h \mapsto \chi_h$ induces an isomorphism $H \simeq L(L(H))$.*

Proof.

(i) Let $H = \langle h_i \rangle \times \cdots \times \langle h_r \rangle$ be a decomposition into a direct product of cyclic subgroups $\langle h_i \rangle$ of order n_i. Since each n_i is relatively prime to Char K, K has a primitive n_ith root of unity, say ε_i. Define $\lambda_i \in L(H)$ by $\lambda_i(h_j) = \varepsilon_i^{\delta_{ij}}$. Then we find that $L(H) = \langle \lambda_1 \rangle \times \cdots \times \langle \lambda_r \rangle$, and (i) follows, since $|\langle \lambda_i \rangle| = n_i$ for all i.

(ii) There exists $\lambda \in L(H)$, which contains λ_0 as an irreducible constituent, and then we have $\lambda|_A = \lambda_0$ because λ is a linear character. Consequently the restriction map $\operatorname{Res}_A^H\colon L(H) \to L(A)$ is an epimorphism with $A^{\perp}$ as its kernel by definition, and (b) follows. Clearly, it may be considered that $L(H/A) \supset A^{\perp}$, whence we have (a) by comparing the order of both sides.

(iii) Clearly, χ gives a homomorphism. If $h = h_1^{t_1} \dots h_r^{t_r} \neq 1$, then $h_i^{t_i} \neq 1$ for some i, and we have $\chi_h(\lambda_i) = \lambda_i(h) = \varepsilon_i^{t_i} \neq 1$. Thus χ is a monomorphism and hence an isomorphism because $|H| = |L(L(H))|$. ∎

In view of the above lemma, we may identify $L(L(H))$ with H via the map $h \mapsto \chi_h$ whenever Char $K \nmid |H|$.

The second cohomology group $H^2(G, K^\times)$ is often called the *Schur multiplier* of G (over K) and is denoted by $M_K(G)$. If $\alpha \in Z^2(G, K^\times)$, we shall denote the residue class $\alpha B^2(G, K^\times)$ by α^*.

Lemma 5.19. *Let $\tilde{G} \xrightarrow{\pi} G$ be a central extension and let $H = \text{Ker } \pi$, $\tilde{G} = \sum_{a \in G} Ht_a$ with $\pi(t_a) = a$, $t_a t_b = z(a, b)t_{ab}$.*

(i) *If $\lambda \in L(H)$, then τ_λ: $G \times G \to K^\times((a, b) \mapsto \lambda(z(a, b)))$ is a factor set and we have a homomorphism τ^*: $L(H) \to M_K(G)(\lambda \mapsto \tau_\lambda^*)$, which is free from the choice of coset representatives $\{t_a\}$ of H in G.*

(ii) Ker $\tau^* = (H \cap \tilde{G}')^\perp = \{\lambda \in L(H);\ H \cap \tilde{G}' \leq \text{Ker } \lambda\}$, *where $\tilde{G}'$ is the commutator subgroup of $\tilde{G}$. Consequently, we have the following.*

$$\tau^* \text{ is a monomorphism} \Leftrightarrow H \leq \tilde{G}'.$$

(iii) *A projective representation $(\mathbf{X}, \alpha)$ of G lifts to a linear representation of $\tilde{G}$ if and only if $\alpha^* \in \text{Im } \tau^*$.*

Therefore, the central extension $\tilde{G} \to G$ has the projective lifting property if and only if τ^* is an epimorphism.

Proof.

(i) The first half is easy. For the second half, let $\{t'_a\}$ be any complete set of coset representatives of H in G and let $t'_a = h_a t_a$ with $h_a \in H$. If we write $t'_a t'_b = z'(a, b)t'_{ab}$, then $z'(a, b) = z(a, b)h_a h_{ab}^{-1} h_b$. Thus the associated factor set τ'_λ of G relative to $\{t'_a\}$ satisfies

$$\tau'_\lambda(a, b) = \lambda(z'(a, b)) = \tau_\lambda(a, b)\gamma(b)\gamma(ab)^{-1}\gamma(a),$$

where $\gamma(a) = \lambda(h_a)$. Thus $\tau'_\lambda \sim \tau_\lambda$.

(ii) Suppose that $\lambda \in (H \cap \tilde{G}')^\perp$. Then λ is considered as a linear character of $H/H \cap \tilde{G}'$ and hence it induces a homomorphism $\tilde{\lambda}$: $H\tilde{G}'/\tilde{G}' \to K^\times(h\tilde{G}' \mapsto \lambda(h))$. Extend it to a homomorphism $\tilde{G}/\tilde{G}' \to K^\times$, and we obtain a linear character $\tilde{\lambda}$: $\tilde{G} \to K^\times$ such that $\tilde{\lambda}(h) = \lambda(h)$ for all $h \in H$. Then we have

$$\tilde{\lambda}(t_a)\tilde{\lambda}(t_b) = \lambda(z(a, b))\tilde{\lambda}(t_{ab}),$$

whence it follows that $\tau_\lambda \sim 1$ and $\lambda \in \text{Ker } \tau^*$. Suppose conversely that $\lambda \in \text{Ker } \tau^*$, i.e. $\tau_\lambda \sim 1$, then there exists a map γ: $G \to K^\times$ such that

$$\lambda(z(a, b)) = \gamma(b)\gamma(ab)^{-1}\gamma(a).$$

Let us define $\tilde{\lambda}: \tilde{G} \to K^\times$ by $\tilde{\lambda}(x) = \lambda(h)\gamma(a)$ for $x = ht_a \in G$ with $h \in H$. It is easy to see that $\tilde{\lambda}$ is a homomorphism. Since $t_1 t_1 = z(1, 1)t_1$, we have $t_1 = z(1, 1)$ and hence $\lambda(t_1) = \lambda(z(1, 1)) = \gamma(1)$. Thus $\tilde{\lambda}(h) = \tilde{\lambda}((ht_1^{-1})t_1) = \lambda(ht_1^{-1})\gamma(1) = \lambda(h)\lambda(t_1)^{-1}\gamma(1) = \lambda(h)$, namely, $\tilde{\lambda}$ is an extension of λ. Since $\tilde{G}' \leq \operatorname{Ker} \tilde{\lambda}$, it follows that $H \cap \tilde{G}' \leq \operatorname{Ker} \lambda$ and thus $\lambda \in (H \cap \tilde{G}')^\perp$. The latter half follows easily from the fact that $(H \cap \tilde{G}')^\perp = 1 \Leftrightarrow H \cap \tilde{G}' = H$.

(iii) To show the "only if" part, we may assume that $X(1)$ is the identity matrix by replacing $\mathbf{X}$ with a suitable equivalent representation if necessary. Now suppose that there exist a representation $\tilde{\mathbf{X}}$ of $\tilde{G}$ and a map $\tilde{\lambda}: \tilde{G} \to K^\times$ such that $\tilde{X}(x) = \tilde{\lambda}(x)X(\pi(x))$ for all $x \in \tilde{G}$. Then we see that $\tilde{\lambda}(hx) = \tilde{\lambda}(h)\tilde{\lambda}(x)$ for $h \in H$ and $x \in \tilde{G}$ because $\tilde{X}(hx) = \tilde{X}(h)\tilde{X}(x)$. In particular, if λ denotes the restriction of $\tilde{\lambda}$ to H, then $\lambda \in L(H)$. We observe that

$$\begin{aligned}\tilde{X}(t_a t_b) &= \tilde{X}(z(a, b)t_{ab}) = \lambda(z(a, b))\tilde{\lambda}(t_{ab})X(ab)\\ &= \tau_\lambda(a, b)\tilde{\lambda}(t_{ab})X(ab).\end{aligned}$$

On the other hand,

$$\begin{aligned}\tilde{X}(t_a t_b) &= \tilde{X}(t_a)\tilde{X}(t_b) = \tilde{\lambda}(t_a)\tilde{\lambda}(t_b)X(a)X(b)\\ &= \tilde{\lambda}(t_a)\tilde{\lambda}(t_b)\alpha(a, b)X(ab).\end{aligned}$$

Thus we conclude that $\tau_\lambda(a, b) = \alpha(a, b)\tilde{\lambda}(t_b)\tilde{\lambda}(t_{ab})^{-1}\tilde{\lambda}(t_a)$ and $\tau_\lambda \sim \alpha$. This implies that $\alpha^* \in \operatorname{Im} \tau^*$.

Suppose conversely that $\tau_\lambda \sim \alpha$ for some $\lambda \in L(H)$, then there exists $\gamma: G \to K^\times$ such that $\lambda(z(a, b)) = \alpha(a, b)\gamma(b)\gamma(ab)^{-1}\gamma(a)$. Define $\tilde{\mathbf{X}}: \tilde{G} \to \mathrm{GL}_n(K)$ by $\tilde{X}(ht_a) = \lambda(h)\gamma(a)X(a)$ for $ht_a \in \tilde{G}$ with $h \in H$. We see easily that $\tilde{X}((ht_a)(kt_b)) = \tilde{X}(ht_a)\tilde{X}(kt_b)$, namely, $\tilde{\mathbf{X}}$ is a representation of $\tilde{G}$. And if we set $\mu(x) = \lambda(h)\gamma(a)$ for $x = ht_a$, then $\tilde{X}(x) = \mu(x)X(\pi(x))$. ∎

The homomorphism $\tau^*: L(H) \to M_K(G)$ defined in (i) of the above lemma is called the *transgression map*.

As the final preliminary lemma, we show the following.

Lemma 5.20. *$B^2(G, K^\times)$ is a divisible (abelian) group. Consequently, there exists a subgroup M of $Z^2(G, K^\times)$ such that*

$$Z^2(G, K^\times) = B^2(G, K^\times) \times M, \qquad M \simeq M_K(G).$$

Proof. Let $\delta\gamma: (a, b) \mapsto \gamma(b)\gamma(ab)^{-1}\gamma(a)$ be any element of $B^2(G, K^\times)$, where $\gamma: G \to K^\times$ is a 1-cochain, and let $n(\neq 0)$ be any rational integer. For each

$a \in G$, choose $\rho(a) \in K^\times$ so that $\rho(a)^n = \gamma(a)$. This defines a 1-cochain $\rho: G \to K^\times$, and we have $(\delta\rho)^n = \delta\gamma$. Thus $B^2(G, K^\times)$ is divisible (as a multiplicative group). The latter half is immediate from Chapter 1, 12.4. ■

In the above lemma, we have that Char $K \nmid |M|$ by Lemma 5.4. Therefore $L(M) \simeq M$, and we may identify $L(L(M))$ with M. Also, it will be natural to regard M as the Schur multiplier $M_K(G)$ by identifying $\alpha \in M$ with α^*.

Given $(a, b) \in G \times G$, we have a homomorphism $z(a, b): M \to K^\times$ $(\alpha \mapsto \alpha(a, b))$. This defines a factor set z of G with coefficients in $L(M)$, namely, we have $z: G \times G \to L(M)$ such that $z(a, b)(\alpha) = \alpha(a, b)$ for $\alpha \in M$. So, there is a central extension $E \xrightarrow{\pi} G$ such that Ker $\pi = L(M)$ and

$$E = \sum_{a \in G} L(M)t_a, \qquad t_a t_b = z(a, b)t_{ab}.$$

The transgression map τ^* corresponding to this central extension is interpreted as follows. Given $\alpha \in M = L(L(M))$, we have $\tau_\alpha(a, b) = \alpha(z(a, b)) = z(a, b)(\alpha) = \alpha(a, b)$, that is, $\tau^*: M = L(L(M)) \to M_K(G) = M$ is the identity map. Therefore, we get the following theorem from Lemma 5.19(iii).

Theorem 5.21 (Schur). *Given a group G, there exists a central extension $E \xrightarrow{\pi} G$ with kernel $M_K(G)$ having the projective lifting property. Also, E has minimal order among those central extensions of G that have the projective lifting property.*

A *representation group* E of a group G is a central extension $E \xrightarrow{\pi} G$ of minimal order with the projective lifting property. The order of E equals $|G||M_K(G)|$ by the above theorem.

The next theorem gives a group theoretical characterization of a representation group.

Theorem 5.22. *Let $\tilde{G} \xrightarrow{\pi} G$ be a central extension of G with* Ker $\pi = H$. *Then $\tilde{G}$ is a representation group of G if and only if the following two conditions hold:*

$$\text{(a)} \quad H \le \tilde{G}', \qquad \text{(b)} \quad |H| = |M_K(G)|.$$

If this is the case, we have $H \simeq M_K(G)$.

Proof. Let $\tilde{G} \overset{\pi}{\to} G$ be a central extension and $\tau^*\colon L(H) \to M_K(G)$ be the transgression map. If $\tilde{G}$ is a representation group of G, then τ^* is an epimorphism by Lemma 5.19(iii) and hence an isomorphism by the minimality of the order of $\tilde{G}$. Therefore $H \simeq L(H) \simeq M_K(G)$. Part(a) is immediate from (ii) of the same lemma.

Conversely, if (a) is true, then τ^* is a monomorphism by Lemma 5.19(ii) and hence an isomorphism if (b) is true. Therefore $\tilde{G} \to G$ is a representation group of G. ■

6. Introduction to Modular Representation Theory

By a (p-)*modular representation* we mean a representation of a group over a field of characteristic $p > 0$, and a character defined by such a representation is called a *modular character*.

In this section, we shall establish some elementary facts on p-modular representations for a fixed prime number p.

Let R be a complete discrete valuation ring with quotient field K of characteristic 0. We assume that the residue field $F = R/(\pi)$ has characteristic p, where (π) denotes the unique maximal ideal of R. With this assumption we refer to the triple (K, R, F) as a *p-modular system*. We fix a valuation v of K such that $v(p) = 1$, and denote by α^* the image of $\alpha \in R$ under the natural map $R \to F = R/(\pi)$.

Let K_0 be the algebraic closure of $\mathbf{Q}$ in K. We may assume without loss of generality that $K_0 \subset \mathbf{C}$. For $\alpha \in K_0$, $\bar{\alpha}$ denotes the complex conjugate of α.

In this section, we work with a fixed group G and with a p-modular system (K, R, F) satisfying the following:

K contains all the $|G|$th root of unity.

In particular both K and F are splitting fields of every subgroup of G (Theorem 1.32(ii), Theorem 4.11). By a modular representation of G, we always mean a representation over F.

6.1. *Brauer Characters*

We let $|G|_p = p^a$ throughout this section. For $x \in G$, let $x = x_p x_{p'}$ be the decomposition into the product of the p-part x_p and the p'-part $x_{p'}$ of x, respectively, where $x_p, x_{p'} \in \langle x \rangle$. If $\mathbf{T}: G \to \mathrm{GL}_n(F)$ is a modular representation of G, then, by restricting $\mathbf{T}$ to $\langle x \rangle$, we find $P \in \mathrm{GL}_n(F)$ such that

$$P^{-1}T(x)P = \begin{pmatrix} \lambda_1 & & 0 \\ & \ddots & \\ * & & \lambda_n \end{pmatrix},$$

where each λ_i is an $o(x)$th root of unity. Since 1 is the only p^ath root of unity, we have from the above that

$$P^{-1}T(x_p)P = \begin{pmatrix} 1 & & 0 \\ & \ddots & \\ * & & 1 \end{pmatrix}, \qquad P^{-1}T(x_{p'})P = \begin{pmatrix} \lambda_1 & & 0 \\ & \ddots & \\ * & & \lambda_n \end{pmatrix},$$

and the following equality holds for the character $\chi_{\mathbf{T}}$ defined by $\mathbf{T}$:

$$\chi_{\mathbf{T}}(x) = \chi_{\mathbf{T}}(x_{p'}). \tag{6.1}$$

Let us denote by $G_{p'}$ the set of p'-elements of the group G. As is seen in (6.1), the modular character is determined by its value on $G_{p'}$.

Now, the set of $|G|_{p'}$th roots of unity in K or in F forms a cyclic group of order $|G|_{p'}$ in either case. Consequently, they are isomorphic to each other by the natural map $R \to F = R/(\pi)$ (note that if $(h, p) = 1$, then $x^h - 1$ has no multiple root in F). If $x \in G_{p'}$, then the characteristic roots, say $\{\lambda_1, \ldots, \lambda_n\}$, of $T(x)$ are all $|G|_{p'}$th roots of unity in F. Thus there exists, for each λ_i, a unique $|G|_{p'}$th root of unity μ_i in K such that $\mu_i^* = \lambda_i$. Define

$$\varphi_{\mathbf{T}}(x) = \mu_1 + \cdots + \mu_n.$$

This is called the *Brauer character* defined by the modular representation $\mathbf{T}$. If $x, y \in G_{p'}$ are G-conjugate, then the sets of characteristic roots of $T(x)$ and $T(y)$ are identical, and hence we have $\varphi_{\mathbf{T}}(x) = \varphi_{\mathbf{T}}(y)$. *Namely, the Brauer character is a complex-valued class function defined on $G_{p'}$. It is clear that* $\varphi_{\mathbf{T}}(x^{-1}) = \overline{\varphi_{\mathbf{T}}(x)}$.

The following result is immediate from the definition.

Lemma 6.1. *Let* **S**, **T** *etc., be modular representations of* G.

(i) *If* $\mathbf{S} \sim \mathbf{T}$, *then* $\varphi_{\mathbf{S}} = \varphi_{\mathbf{T}}$.

(ii) *If we define* $\varphi_{\mathbf{T}}^*$ *by* $\varphi_{\mathbf{T}}^*(x) = \varphi_{\mathbf{T}}(x_{p'})^*$, *then* $\varphi_{\mathbf{T}}^*$ *coincides with the modular character defined by* **T**.

(iii) *If*

$$\mathbf{T} \sim \begin{pmatrix} \mathbf{T}' & 0 \\ * & \mathbf{T}'' \end{pmatrix}, \qquad \text{then} \quad \varphi_{\mathbf{T}} = \varphi_{\mathbf{T}'} + \varphi_{\mathbf{T}''}.$$

Let $\mathrm{IRR}_F(G) = \{\mathbf{T}_1, \ldots, \mathbf{T}_l\}$ and let φ_i be the Brauer character defined by $\mathbf{T}_i$. Each φ_i is called an irreducible Brauer character, and $\mathrm{IBr}(G)$ denotes the set of irreducible Brauer characters of G; $\mathrm{IBr}(G) = \{\varphi_1, \ldots, \varphi_l\}$. By (iii) of the above lemma, every Brauer character is expressed as a linear combination of irreducible Brauer characters with coefficients in nonnegative rational integers. Also the set $\{\varphi_1^*, \ldots, \varphi_l^*\}$ gives all the irreducible modular characters of G.

Theorem 6.2. $\mathrm{IBr}(G)$ *is linearly independent over* K.

Proof. Suppose that there exists a nontrivial linear relation

$$\sum_{i=1}^{l} \alpha_i \varphi_i = 0$$

with $\alpha_i \in K$. Let $v(\alpha_j) = \min\{v(\alpha_i); \alpha_i \neq 0\}$. Then, by multiplying both sides of the above equality by α_j^{-1}, we get the following equality with $\beta_i = \alpha_j^{-1}\alpha_i$:

$$\sum_i \beta_i \varphi_i = 0; \qquad \beta_i \in R, \qquad \beta_j = 1.$$

Consequently, we have the following nontrivial linear relation over F:

$$\sum_i \beta_i^* \varphi_i^* = 0.$$

But this is impossible, since $\{\varphi_i^*\}$ is linearly independent (Chapter 2, 3.13). ∎

The following is an immediate consequence of the above theorem.

Corollary 6.3. *For two F-representations* $\mathbf{S}$ *and* $\mathbf{T}$ *of G, we have the following*:

$$\mathbf{S} \leftrightarrow \mathbf{T} \Leftrightarrow \varphi_{\mathbf{S}} = \varphi_{\mathbf{T}}.$$

6.2. Decomposition Numbers

If $\mathbf{X}$ is a K-representation of G, then $\mathbf{X}$ is K-equivalent to some R-representation $\mathbf{X}_1$, and thus we obtain a modular representation $\mathbf{X}_1^*$ of G. If

$x \in G_{p'}$, then the characteristic roots, say $\mu_1, \ldots, \mu_n$, of $X_1(x)$ are all $|G|_{p'}$th roots of unity, which are by definition the roots of the polynomial equation $\det(tI - X_1(x)) = 0$. By reducing the coefficients of this polynomial mod(π), we obtain $\det(tI - X_1^*(x)) = 0$, whose roots $\mu_1^*, \ldots, \mu_n^*$ are precisely the characteristic roots of $X_1^*(x)$. Thus $\chi_{\mathbf{X}_1}(x) = \sum_{i=1}^n \mu_i$ coincides with the value $\varphi_{\mathbf{X}_1^*}(x)$ of the Brauer character $\varphi_{\mathbf{X}_1^*}$. Summarizing the above, we obtain the following.

Lemma 6.4. *Let* **X**, **Y** *be R-representations of G. Then*

(i) $\chi_{\mathbf{X}}|_{G_{p'}} = \varphi_{\mathbf{X}_1^*}$.

(ii) *If* $\mathbf{X} \underset{K}{\sim} \mathbf{Y}$, *then* $\varphi_{\mathbf{X}^*} = \varphi_{\mathbf{Y}^*}$.

Remark. With the notation above, the modular representation $\mathbf{X}_1^*$ is not uniquely determined by **X**, not even within F-equivalence, as remarked in Chapter 2. But the Brauer character defined by $\mathbf{X}_1^*$ is determined only by **X**, and this provides, in our case, an alternative proof to Chapter 2, 1.9.

Let $\mathrm{Irr}(G) = \{\chi_1, \ldots, \chi_k\}$, $\mathrm{IBr}(G) = \{\varphi_1, \ldots, \varphi_l\}$. Then $\chi_i|_{G_{p'}}$ is a Brauer character, which is uniquely expressed as

$$\chi_i|_{G_{p'}} = \sum_{j=1}^{l} d_{ij}\varphi_j, \tag{6.2}$$

where d_{ij} is a nonnegative integer. Each d_{ij} is called the *decomposition number*, and the $k \times l$ matrix $D = (d_{ij})$ is the *decomposition matrix*. We sometimes write $d_{\chi_i \varphi_j}$ for d_{ij}.

Let $\mathrm{Cl}(G_{p'}) = \{C_1, \ldots, C_m\}$ be the set of conjugate classes represented by p'-elements of G and $x_i \in C_i (1 \le i \le m)$. We define

$$X' = (\chi_i(x_j))_{i,j} \qquad (\text{a} \quad k \times l \quad \text{matrix}),$$

$$\mathbf{\Phi} = (\varphi_i(x_j))_{i,j} \qquad (\text{an} \quad l \times m \quad \text{matrix}).$$

Then it follows from (6.2) that

$$X' = D\mathbf{\Phi}. \tag{6.3}$$

Theorem 6.5. $|\mathrm{IBr}(G)| = |\mathrm{Cl}(G_{p'})|$.

Proof. Let $\mathbf{CF}_K(G_{p'})$ be the set of K-valued class functions defined on $G_{p'}$. If ζ_i denotes the characteristic function of C_i, then $\mathbf{CF}_K(G_{p'})$ is a vector space over K with basis $\{\zeta_i; 1 \leq i \leq m\}$ and hence has dimension m. Since $\mathbf{CF}_K(G_{p'})$ contains all $\varphi_i (1 \leq i \leq l)$ and they are linearly independent, we have $l \leq m$.

On the other hand, Theorem 2.9 implies that $\zeta_\rho = \sum_{i=1}^{k} (\overline{\chi_i(x_\rho)}/|C_G(x_\rho)|)\chi_i$, thus $\{\chi_i|_{G_{p'}}; 1 \leq i \leq k\}$ generates $\mathbf{CF}_K(G_{p'})$ over K. It then follows from (6.2) that $\mathrm{IBr}(G)$ generates $\mathbf{CF}_K(G_{p'})$, whence we have $m \leq l$. ■

We henceforth denote $|\mathrm{Cl}(G)|$ and $|\mathrm{Cl}(G_{p'})|$ by k (or $k(G)$) and l (or $l(G)$), respectively. Then the matrix $\mathbf{\Phi}$ of (6.3) is a square matrix of degree l, and since $\{\varphi_i; 1 \leq i \leq l\}$ is linearly independent, we have

(6.4) $$\det \mathbf{\Phi} \not\equiv 0 \pmod{(\pi)}.$$

Thus $\mathbf{\Phi} \in \mathrm{GL}_l(R)$.

6.3. Principal Indecomposable Characters

For $a \in RG$, a^* denotes the image of a under the natural map $RG \xrightarrow{*} RG/\pi(RG) = FG\ (\sum_{x \in G} \alpha_x x \mapsto \sum_{x \in G} \alpha_x^* x)$. The theorem on lifting idempotents (cf. Chapter 1, 14.2) states that $\widetilde{\mathrm{pi}}(RG)$ corresponds bijectively to $\widetilde{\mathrm{pi}}(FG)$, and hence we have $|\widetilde{\mathrm{pi}}(RG)| = l$. If $\{e_1, \ldots, e_l\}$ is a complete set of representatives for $\widetilde{\mathrm{pi}}(RG)$, then so is $\{e_1^*, \ldots, e_l^*\}$ for $\widetilde{\mathrm{pi}}(FG)$. Write $P_i = e_i RG$, and so $P_i^* = e_i^* FG = P_i/\pi P_i$. Then $\{P_i; 1 \leq i \leq l\}$ and $\{P_i^*; 1 \leq i \leq l\}$ give complete sets of representatives for nonisomorphic principal indecomposable RG- and FG-modules, respectively.

Let M be an FG-module and $\mathbf{T}$ be an F-representation defined by M. If there exists an RG-representation module V such that $V^* = V/\pi V \simeq M$, we say that M is *lifted* to V. This is equivalent to that there exists a suitable R-representation $\mathbf{X}$ of G such that $\mathbf{X}^* = \mathbf{T}$. If this is the case, then the Brauer character $\varphi_{\mathbf{T}}$ defined by $\mathbf{T}$ coincides with $\chi_{\mathbf{X}}|_{G_{p'}}$, and we also say that the F-representation $\mathbf{T}$ and the Brauer character $\varphi_{\mathbf{T}}$ are lifted to the R-representation $\mathbf{X}$ and to the ordinary character $\chi_{\mathbf{X}}$, respectively.

As is observed above, every principal indecomposable FG-module is lifted to a principal indecomposable RG-module, and hence every projective FG-module is lifted to a projective RG-module.

We assume henceforth that the irreducible FG-module $P_i^*/P_i^* J(FG)$ defines the Brauer character φ_i for each i.

Let η_i be the R-character defined by $P_i = e_i RG$, which is called a *principal indecomposable character* of G. Thus the Brauer character defined by P_i^* coincides with $\eta_i|_{G_{p'}}$.

We sometimes denote η_i by η_{φ_i} in order to emphasize that it is the principal indecomposable character corresponding to $\varphi_i \in \mathrm{IBr}(G)$.

Exercise 6.6. Let γ_G denote the ordinary character induced by the regular representation of G. Then the following holds:

$$\gamma_G = \sum_{i=1}^{l} \varphi_i(1)\eta_i.$$

The next result is an immediate consequence of Theorem 1.26.

Theorem 6.7. *$p^a | \eta_i(1)$ for all i $(1 \leq i \leq l)$.*

If $\mathbf{P}_i^* \leftrightarrow \sum_j c_{ij}\mathbf{T}_j$, i.e., c_{ij} is the Cartan invariant of FG, then the following holds, where $\mathbf{P}_i$ denotes an R-representation of G defined by P_i.

$$\eta_i|_{G_{p'}} = \sum_{j=1}^{l} c_{ij}\varphi_j \qquad (1 \leq i \leq l). \tag{6.5}$$

Here the square matrix $C = (c_{ij})$ is called the Cartan matrix of G, which is a symmetric matrix (cf. Chapter 2, 8.21).

For an RG-module V, we let $V^* = V/\pi V$.

Theorem 6.8. *Let $D = (d_{ij})$ and $C = (c_{ij})$ be the decomposition and Cartan matrices of G, respectively. Then*

(i) $$\eta_i = \sum_{j=1}^{k} d_{ji}\chi_j \qquad (1 \leq i \leq l).$$

(ii) $$C = {}^tDD.$$

Proof.

(i) Let V_j be an RG-representation module that defines the ordinary irreducible character χ_j. If we write $e_iKG \simeq \bigotimes_j a_{ij}V_j^K$, then $\eta_i = \sum_j a_{ij}\chi_j$, where $a_{ij} = \dim_K \mathrm{Hom}_{KG}(e_iKG, V_j^K) = \dim_K V_j^K e_i$. Since an R-basis of V_je_i is, under the convention that $V_j \subset V_j^K$, also a K-basis of $V_j^Ke_i$, we have

$$a_{ij} = \mathrm{rank}_R\, V_je_i = \dim_F V_j^*e_i^*,$$

which is equal to the multiplicity of $e_i^* FG/e_i^* J(FG)$ in V_j^*. Therefore $a_{ij} = d_{ji}$, and (i) holds.

(ii) Since $\eta_i|_{G_{p'}} = \sum_\nu d_{\nu i}\chi_\nu|_{G_{p'}} = \sum_{j,\nu} d_{\nu i} d_{\nu_j}\varphi_j$ and $\{\varphi_j\}$ is linearly independent over K, we conclude from (6.5) that $c_{ij} = \sum_\nu d_{\nu_i} d_{\nu_j}$. ■

For $x \in G$, we denote by x^G the conjugate class of G containing x. Then the second orthogonality relation of characters (Theorem 2.9) is expressed in the following form:

(6.6)
$$\sum_{i=1}^{k} \overline{\chi_i(x)}\chi_i(y) = \delta_{x^G, y^G}|C_G(x)|.$$

This yields the following theorem.

Theorem 6.9.

(i) *For* $y \in G_{p'}$, *we have*

(6.7)
$$\sum_{i=1}^{l} \overline{\eta_i(x)}\varphi_i(y) = \delta_{x^G, y^G}|C_G(x)|.$$

(ii) *If* $x \in G - G_{p'}$, *then* $\eta_i(x) = 0 \qquad (1 \le i \le l)$.

Proof.

(i) By using $\chi_i(y) = \sum_j d_{ij}\varphi_j(y)$, we deduce (6.7) easily from the equation (6.6).

(ii) If $x \in G - G_{p'}$, then the right-hand side of the equation (6.7) is zero for all $y \in G_{p'}$. So, we have $\sum_i \overline{\eta_i(x)}\varphi_i = 0$, and the linear independence of $\{\varphi_i\}$ implies that $\eta_i(x) = 0$. ■

A *p-regular element* is a p'-element of G, whereas if $x \in G - G_{p'}$, namely, $p|o(x)$, then we call x a *p-singular element.*

Let $\mathrm{Cl}(G_{p'}) = \{C_1, \dots, C_l\}$ and $x_i \in C_i$ $(1 \le i \le l)$ as before. Also, set

$$\mathbf{\Phi} = (\varphi_i(x_j))_{i,j}, \qquad Y = (\eta_i(x_j))_{i,j},$$

and denote by S the diagonal matrix

$$\mathrm{diag}(|C_G(x_1)|, \dots, |C_G(x_l)|),$$

with diagonal entries $|C_G(x_1)|, \ldots, |C_G(x_l)|$. Then we obtain from (6.7) that

$$ {}^t\bar{Y}\mathbf{\Phi} = S, \tag{6.8} $$

thus ${}^t\bar{Y}(\mathbf{\Phi}S^{-1}) = I$ (the identity matrix), and we have $(\mathbf{\Phi}S^{-1})^t\bar{Y} = I$. By evaluating the (i,j)-entries on both sides, we get

$$ \sum_{v=1}^{l} \varphi_i(x_v)\frac{1}{|C_G(x_v)|}\overline{\eta_j(x_v)} = \delta_{ij}. \tag{6.9} $$

Given K-valued class functions f, g defined on G (or on $G_{p'}$), we set

$$ (f, g)' = \frac{1}{|G|}\sum_{x\in G_{p'}} f(x)\overline{g(x)}. $$

Then the above observation yields the following result.

Theorem 6.10.
(i) $(\varphi_i, \eta_j)' = \delta_{ij}$.
(ii) *If* $x \in G_{p'}$, *then* $v(|C_G(x)|) = \min\{v(\eta_i(x));\ 1 \le i \le l\}$.

Proof.
(i) It follows from (6.9) that

$$ \frac{1}{|G|}\sum_{v=1}^{l}\frac{|G|}{|C_G(x_v)|}\varphi_i(x_v)\overline{\eta_j(x_v)} = \delta_{ij}. $$

But the left-hand side of this equation is $(\varphi_i, \eta_j)'$.
(ii) We get $\sum_j \eta_j(x_i^{-1})\varphi_j(x_i) = |C_G(x_i)|$ from (6.8). Noting that $C_G(x) = C_G(x^{-1})$, we have

$$ \begin{aligned} v(|C_G(x_i^{-1})|) &\ge \min\{v(\eta_j(x_i^{-1})\varphi_j(x_i));\ 1 \le j \le l\} \\ &\ge \min\{v(\eta_j(x_i^{-1}));\ 1 \le j \le l\}. \end{aligned} $$

On the other hand, we have ${}^t\bar{Y} = S\mathbf{\Phi}^{-1}$ from (6.8) and $\mathbf{\Phi}^{-1} = (\varphi'_{ij}) \in \mathrm{GL}_l(R)$ from (6.4). Thus $\eta_j(x_i^{-1}) = |C_G(x_i)|\varphi'_{ij}$, and

$$ v(\eta_j(x_i^{-1})) \ge v(|C_G(x_i^{-1})|), $$

which proves (ii). ■

Exercise 6.11. Prove the following assertions.

(i) $c_{ij} = (\eta_i, \eta_j)_G$.

(ii) Set $Z = ((\varphi_i, \varphi_j)')_{i,j}$. Then $CZ = I$ (the identity matrix). Hence $\eta_1, \eta_2, \ldots, \eta_l$ are linearly independent over K.

(iii) $\det C > 0$. (Hint: Use $C = {}^tDD$.)

In the special case that $p \nmid |G|$, we have $G_{p'} = G$ and $l = k$. Since FG is a semisimple algebra, $e_i^* FG = (e_i RG)^*$ is irreducible, and thus the Cartan matrix of G is the identity matrix. Then we see that each $e_i KG$ is irreducible and hence $\mathrm{Irr}(G) = \{\chi_1, \ldots, \chi_k\}$, where χ_i is the character defined by $e_i KG$. Therefore, the decomposition matrix is also the identity matrix. Summarizing the above, we get the following theorem.

Theorem 6.12. *If $p \nmid |G|$, then we have $l = k$ and*

(i) $\chi_i = \eta_i = \varphi_i \qquad (1 \le i \le k)$.

(ii) *Every irreducible modular representation* $\mathbf{T}$ *of* G *is lifted to an* R-*representation* $\mathbf{X}$ *of* G, *i.e.*, $\mathbf{X}^* = \mathbf{T}$.

Using the above theorem, we prove the following useful lemma.

Lemma 6.13. *Let φ be a Brauer character of G, and define $\tilde{\varphi}$ by*

$$\tilde{\varphi}(x) = \varphi(x_{p'}) \qquad \text{for } x \in G. \tag{6.10}$$

Then $\tilde{\varphi}$ is a generalized character of G, and it follows that

$$\tilde{\varphi} = \sum_{i=1}^{k} m_i \chi_i, \qquad \varphi = \sum_{i=1}^{k} m_i (\chi_i|_{G_{p'}}), \quad \text{with} \quad m_i \in \mathbf{Z}.$$

Proof. Let E be any elementary subgroup of G, then $E = P \times Q$, where P is a p-group and Q is a p'-group. From the definition of $\tilde{\varphi}$, we see that $\tilde{\varphi}_E = \mathbf{1}_P \times \varphi_Q$. But by Theorem 6.12, φ_Q is an ordinary character. Now the assertion follows from the fundamental theorem of Brauer (cf. Theorem 4.2(i)). ■

As the first application of the above lemma, we get the following.

Theorem 6.14. *The decomposition matrix D has rank l; the same is true for D^*.*

Proof. Write $\tilde{\varphi}_i = \sum_j m_{ij}\chi_j$ with $m_{ij} \in \mathbf{Z}$ and let $M = (m_{ij})$, an $l \times k$ matrix. Since $\{\varphi_i\}$ is linearly independent, we find that $MD = I$ (the identity matrix), and hence $M^*D^* = I$ as well. Therefore, both D and D^* have rank l. ■

If D is a matrix with entries in $\mathbf{Z}$, the rank of D^* is called the *p-rank* of D. Let $C_i \in \mathrm{Cl}(G_{p'})$. Then the set

$$S_i = \{x \in G;\ x_{p'} \in C_i\}$$

is called the *p′-section* of G containing C_i. This is a union of some conjugate classes of G, and $G = \bigcup_{i=1}^{l} S_i$ (a disjoint union).

As the second application of Lemma 6.13, we show the following.

Theorem 6.15.

(i) *Let $X(G|G_{p'})$ denote the set of generalized characters of G that vanish on $G - G_{p'}$. Then $\{\eta_1, \dots, \eta_l\}$ forms a $\mathbf{Z}$-basis of $X(G|G_{p'})$.*

(ii) *Let $\tilde{\varphi}_i$ be as in (6.10). Then $\{\tilde{\varphi}_1, \dots, \tilde{\varphi}_l\}$ is an R-basis of the R-module of all R-valued class functions that are constant on each S_i.*

(iii) *The characteristic function ζ_i of S_i is an R-linear combination of ordinary irreducible characters.*

Proof. Let $\tilde{\varphi}_i = \sum_j m_{ij}\chi_j$ $(m_{ij} \in \mathbf{Z})$.

(i) Since $\eta_1, \dots, \eta_l \in X(G|G_{p'})$ are linearly independent over K and $\dim_K K \otimes_{\mathbf{Z}} X(G|G_{p'}) \le l$, $\{\eta_1, \dots, \eta_l\}$ forms a K-basis of $K \otimes_{\mathbf{Z}} X(G|G_{p'})$. Thus, each $\theta \in X(G|G_{p'})$ is expressed as $\theta = \sum_i \alpha_i \eta_i$ $(\alpha_i \in K)$, and we have

$$\alpha_i = (\theta, \varphi_i)' = (\theta, \tilde{\varphi}_i)_G = \sum_j m_{ij}(\theta, \chi_j)_G.$$

Hence $\alpha_i \in \mathbf{Z}$, since $(\theta, \chi_j)_G \in \mathbf{Z}$. Therefore $\{\eta_i\}$ is a $\mathbf{Z}$-basis of $X(G|G_{p'})$.

(ii) Let $\mathbf{F}$ be the R-module of all R-valued class functions that are constant on each S_i. Then $K \otimes_R \mathbf{F}$ is of K-dimension l with basis $\{\tilde{\varphi}_i\}$, and therefore each $\theta \in \mathbf{F}$ is expressed as $\theta = \sum_i \alpha_i \tilde{\varphi}_i$ $(\alpha_i \in K)$. Since η_i vanishes on $G - G_{p'}$, we find $(\tilde{\varphi}_j, \eta_i)_G = (\varphi_j, \eta_i)' = \delta_{ij}$. Hence

$$\alpha_i = (\theta, \eta_i)_G = \sum_{\rho=1}^{l} \frac{1}{|C_G(x_\rho)|} \theta(x_\rho)\eta_i(x_\rho^{-1}).$$

But since $\eta_i(x_\rho^{-1})/|C_G(x_\rho)| \in R$ by Theorem 6.10(ii), it follows that $\alpha_i \in R$, proving (ii).

(iii) This follows immediately from (ii) and the fact that $\tilde{\varphi}_i \in X(G)$. ■

6.4. Blocks

Recall that the centers of RG and FG are spanned by class sums:

$$Z(RG) = \bigoplus_{C \in \mathrm{Cl}(G)} R\hat{C}, \qquad Z(FG) = \bigoplus_{C \in \mathrm{Cl}(G)} F\hat{C}.$$

Thus

$$Z(FG) = Z(RG)/\pi Z(RG). \tag{6.11}$$

In other words, the natural map $RG \xrightarrow{*} FG$ induces an epimorphism $Z(RG) \xrightarrow{*} Z(FG)$. This is one of the characteristic properties of group rings.

In particular, the theorem on lifting idempotents is applied to the situation (6.11), and we get the following central primitive idempotent decomposition of 1:

$$1 = \varepsilon_1 + \cdots + \varepsilon_t \qquad (\varepsilon_i \in \mathrm{pi}(Z(RG))),$$

which under the reduction mod(π) gives the central primitive idempotent decomposition of $1 \in FG$:

$$1 = \varepsilon_1^* + \cdots + \varepsilon_t^*.$$

Let $B_i = \varepsilon_i(RG)$. Then we have the indecomposable decomposition of RG as an (RG, RG)-bimodule:

$$RG = B_1 \oplus B_2 \oplus \cdots \oplus B_t.$$

By taking the reduction module (π) of the above, we obtain the following indecomposable decomposition of FG as an (FG, FG)-bimodule:

$$FG = B_1^* \oplus B_2^* \oplus \cdots \oplus B_t^*.$$

Each B_i is called a *(p-)block* of G, and we denote by $\mathrm{Bl}(G)$ the set of (p-)blocks of G. Each ε_i is called the *block idempotent* of B_i, which is also denoted by e_{B_i}. The set of block idempotents is denoted by $\mathrm{bli}(G)$: $\mathrm{bli}(G) = \{e_B;\ B \in \mathrm{Bl}(G)\}$.

Let $B \in \mathrm{Bl}(G)$. An RG-module V is said to *belong* to the block B if $Ve_B = V$. We know that any indecomposable RG-module belongs to some block (Chapter 1, 8.13). If V is a representation module for some R-representation $\mathbf{X}$ of G and V belongs to a block B, we say that $\mathbf{X}$ and the character $\chi_{\mathbf{X}}$ belong to B and write, by abusing the notation, $V \in B$, $\mathbf{X} \in B$, and $\chi_{\mathbf{X}} \in B$. The next lemma is obvious.

Lemma 6.16. *The following three conditions on an R-representation* $\mathbf{X}$ *are equivalent.*

(1) $\mathbf{X} \in B$.
(2) $X(e_B) = I$ *(the identity matrix).*
(3) *If B' is a block of G different from B, then* $X(e_{B'}) = 0$.

For KG-modules, K-representations, and K-characters, the notion of belonging to blocks will be defined in a similar manner.

An FG-module W is said to belong to a block B if $We_B^* = W$, in which case we also write $W \in B$. The same terminology and notation will be used for the Brauer and modular characters defined by W.

$\mathrm{Irr}(B)$ denotes the set of ordinary irreducible characters belonging to the block B, and $\mathrm{IBr}(B)$ denotes the set of irreducible Brauer characters belonging to B. Then $\mathrm{Irr}(G)$ and $\mathrm{IBr}(G)$ are expressed as disjoint unions of $\mathrm{Irr}(B)$ and $\mathrm{IBr}(B)$, respectively:

$$\mathrm{Irr}(G) = \bigcup_{B \in \mathrm{Bl}(G)} \mathrm{Irr}(B), \qquad \mathrm{IBr}(G) = \bigcup_{B \in \mathrm{Bl}(G)} \mathrm{IBr}(B).$$

Theorem 6.17. *Let χ_i, $\eta_j \in B$. If φ_ν does not belong to B, then the following holds:*

$$d_{i\nu} = 0, \qquad c_{j\nu} = 0.$$

Proof. Let V be a representation module that defines χ_i or η_j. Then it holds that $Ve_B = V$ and hence $V^*e_B^* = V^*$. Therefore, all irreducible constituents of V^* belong to B. ■

Now, we arrange the rows and columns of the decomposition D and Cartan matrices C of G so that $\{\chi_i\}$, $\{\eta_j\}$, and $\{\varphi_\nu\}$ are distributed into blocks. Then D and C split into the following forms from the above theorem:

$$D = \begin{pmatrix} D_{B_1} & & 0 \\ & \ddots & \\ 0 & & D_{B_t} \end{pmatrix}, \qquad C = \begin{pmatrix} C_{B_1} & & 0 \\ & \ddots & \\ 0 & & C_{B_t} \end{pmatrix}.$$

For $B \in \mathrm{Bl}(G)$, we set $k(B) = |\mathrm{Irr}(B)|$, $l(B) = |\mathrm{IBr}(B)|$. Then D_B and C_B are $k(B) \times l(B)$ and $l(B) \times l(B)$ matrices, respectively, and

$$(6.12) \qquad {}^tD_B D_B = C_B$$

holds. The next result is clear from Theorem 6.14.

Theorem 6.18. *For $B \in \mathrm{Bl}(G)$, the rank and p-rank of D_B are both equal to $l(B)$. In particular, it follows that $l(B) \leq k(B)$. Also, we have that* $\det C_B > 0$.

As observed in Chapter 1, 8.15, it is impossible to arrange the rows and columns of C_B so that it splits into the form:

$$\left(\begin{array}{c|c} * & 0 \\ \hline 0 & * \end{array}\right)$$

Hence, the same is true for D_B. We shall record this fact in terms of characters in the next theorem. We write, for $\chi, \chi' \in \mathrm{Irr}(G)$, $\chi - \chi'$, if there exists $\varphi \in \mathrm{IBr}(G)$ such that $d_{\chi\varphi} \neq 0 \neq d_{\chi'\varphi}$. And we define $\chi \sim \chi'$ if there exists a finite chain $\chi = \chi_{i1} - \cdots - \chi_{ir} = \chi'$, linking χ to χ'. Similarly, for φ, $\varphi' \in \mathrm{IBr}(G)$, we write $\varphi - \varphi'$ if there exists $\chi \in \mathrm{Irr}(G)$ such that $d_{\chi\varphi} \neq 0 \neq d_{\chi\varphi'}$, and define $\varphi \sim \varphi'$ in a similar way as above. With this notation, we have the following result.

Theorem 6.19. *Let $\chi, \chi' \in \mathrm{Irr}(G)$. Then they belong to the same block of G if and only if $\chi \sim \chi'$. The analogous statement holds for irreducible Brauer characters.*

Exercise 6.20.

(i) If $\varphi \in \mathrm{IBr}(B)$, then

$$\varphi = \sum_{\chi \in \mathrm{Irr}(B)} m_{\varphi\chi} \chi|_{G_{p'}} \qquad (m_{\varphi\chi} \in \mathbf{Z}).$$

[Hint: Use Lemma 6.13, Theorem 6.17 and the linear independence of $\mathrm{IBr}(G)$.]

(ii) If two irreducible Brauer characters φ and φ' belong to different blocks, then $(\varphi, \varphi')' = 0$. [Hint: Use the nonsingularity of C_B and Theorem 6.10(i).]

If $\varphi \in \mathrm{IBr}(B)$, then we have from Theorem 6.17 that

$$\eta_\varphi(x) = \sum_{\chi \in \mathrm{Irr}(G)} d_{\chi\varphi}\chi(x) = \sum_{\chi \in \mathrm{Irr}(B)} d_{\chi\varphi}\chi(x),$$

and this vanishes if x is p-singular. This gives the following lemma, which refines part of the second orthogonality relation of characters.

Lemma 6.21. *Let $B \in \mathrm{Bl}(G)$. If x is p-regular and y is p-singular, then*

$$\sum_{\chi \in \mathrm{Irr}(B)} \chi(x)\overline{\chi(y)} = 0.$$

Proof.

$$\begin{aligned}\sum_{\chi \in \mathrm{Irr}(B)} \chi(x)\overline{\chi(y)} &= \sum_{\chi \in \mathrm{Irr}(B)} \sum_{\varphi \in \mathrm{Irr}(B)} d_{\chi\varphi}\varphi(x)\overline{\chi(y)} \\ &= \sum_{\varphi \in B} \varphi(x) \sum_{\chi \in B} d_{\chi\varphi}\overline{\chi(y)} \\ &= \sum_{\varphi \in B} \varphi(x)\overline{\eta_\varphi(y)} = 0. \qquad \blacksquare\end{aligned}$$

We show the following facts on block idempotents.

Theorem 6.22. *Let $B \in \mathrm{Bl}(G)$. Then*

(i)
$$e_B = \sum_{\chi \in \mathrm{Irr}(B)} e_\chi,$$

where e_χ is the central primitive idempotent of KG corresponding to χ, namely,

$$e_\chi = \frac{\chi(1)}{|G|} \sum_{x \in G} \overline{\chi(x)}x.$$

(ii) (Osima) *Each e_B is an R-linear combination of $\{\hat{C};\, C \in \mathrm{Cl}(G_{p'})\}$:*

$$e_B = \sum_{C \in \mathrm{Cl}(G_{p'})} \beta_B(C)\hat{C} \qquad (\beta_B(C) \in R).$$

Proof.

(i) Write $e_B = \sum_{\chi \in T} e_\chi$. Then

$$\chi \in T \Leftrightarrow \chi(e_B) = \chi(1) \Leftrightarrow \chi \in B.$$

Thus, $T = \mathrm{Irr}(B)$.

(ii) Write $e_B = \sum_C \beta_B(C)\hat{C}$. If $x \in C \subset G - G_{p'}$, then by (i) and Lemma 6.21, we get

$$\beta_B(C) = \frac{1}{|G|} \sum_{\chi \in \mathrm{Irr}(B)} \chi(1)\overline{\chi(x)} = 0. \qquad \blacksquare$$

Exercise 6.23.

(i) If $C \in \mathrm{Cl}(G_{p'})$ and $x \in C$, then

$$\beta_B(C) = \frac{1}{|G|} \sum_{\varphi \in B} \varphi(1)\overline{\eta_\varphi(x)} = \frac{1}{|G|} \sum_{\varphi \in B} \eta_\varphi(1)\overline{\varphi(x)}. \tag{6.13}$$

(ii) Every block has a principal indecomposable character η_φ such that $\nu(\eta_\varphi(1)) = \nu(|G|)$. [Hint: Use (6.13) and the fact that $e_B^* \neq 0$. Remember that $\nu(\eta_\varphi(1)) \geq \nu(|G|)$.]

Let $B \in \mathrm{Bl}(G)$. Then e_B^* is the unique idempotent of $Z(B^*)$, $Z(B^*)$ is a local ring, and hence $Z(B^*)/J(Z(B^*))$ is a field. Hence

$$Z(FG)/J(Z(FG)) = \bigoplus_{B \in \mathrm{Bl}(G)} Z(B^*)/J(Z(B^*))$$

is the decomposition into the direct sum of simple components. In particular, $Z(FG)$ has exactly $|\mathrm{Bl}(G)|$ irreducible F-representations.

On the other hand, if $\chi \in \mathrm{Irr}(G)$ and $x \in C \in \mathrm{Cl}(G)$, then

$$\omega_\chi(\hat{C}) = \frac{|G|\chi(x)}{|C_G(x)|\chi(1)}$$

is an algebraic integer (Theorem 2.25). Thus ω_χ gives rise to a linear representation $\omega_\chi\colon Z(RG) \to R$ and hence a linear F-representation $\omega_\chi^*\colon Z(FG) \to F(\hat{C} \mapsto \omega_\chi(\hat{C})^*)$. Furthermore, we see from Theorem 6.22(i) that if $\chi \in \mathrm{Irr}(B)$, then $\omega_\chi(e_B) = 1$, and $\omega_\chi(e_{B'}) = 0$ if $B' \neq B$.

From the above, we get the following theorem.

Theorem 6.24.

(i) *Let $\chi, \chi' \in \mathrm{Irr}(G)$. Then χ and χ' belong to the same block if and only if the following holds for all $C \in \mathrm{Cl}(G_{p'})$.*

$$\omega_\chi(\hat{C}) \equiv \omega_{\chi'}(\hat{C}) \qquad \mathrm{mod}(\pi). \tag{6.14}$$

(ii) *Each block B determines ω_χ^* uniquely, i.e., ω_χ^* is independent of the choice of $\chi \in \mathrm{Irr}(B)$. If we denote ω_χ^* by ω_B^*, then the set $\{\omega_B^*; B \in \mathrm{Bl}(G)\}$ gives all the irreducible F-representations of $Z(FG)$.*

(iii) *$Z(B^*)/J(Z(B^*)) = F$ for all $B \in \mathrm{Bl}(G)$.*

Proof.

(i) If χ belongs to $B \in \mathrm{Bl}(G)$, then, as remarked above, ω_χ^* is an irreducible representation of $Z(FG)$ belonging to the simple component $Z(B^*)/J(Z(B^*))$ of $Z(FG)/J(Z(FG))$. Thus $\omega_\chi^* = \omega_{\chi'}^*$ if $\chi' \in B$, which proves (6.14). On the other hand, if $\chi' \notin B$, then $\omega_{\chi'}^*(e_B^*) = 0$. But since $\omega_\chi^*(e_B^*) = 1$, there is $C \in \mathrm{Cl}(G_{p'})$ such that $\omega_\chi^*(\hat{C}) \neq \omega_{\chi'}^*(\hat{C})$, which establishes (i).

(ii) The first statement is shown above, whereas the rest is clear, since the number of irreducible representations of $Z(FG)$ equals $|\mathrm{Bl}(G)|$.

(iii) Since ω_B^* induces a linear F-representation of $Z(B^*)$, the assertion is obvious. ■

6.5. The Defect of Blocks

For $B \in \mathrm{Bl}(G)$, we define

$$d(B) = \max\{\nu(|G|/\chi(1)); \chi \in \mathrm{Irr}(B)\},$$

which is called the *defect* of the block B. So $d(B) \geq a - \nu(\chi(1))$ for all $\chi \in \mathrm{Irr}(B)$ (where $a = \nu(|G|)$), i.e., $\nu(\chi(1)) \geq a - d(B)$. Thus we may write

$$\nu(\chi(1)) = a - d(B) + \mathrm{ht}(\chi) \qquad (\mathrm{ht}(\chi) \geq 0).$$

The above integer $\mathrm{ht}(\chi)$ is called the *height* of χ. By definition, B has an irreducible character of height 0.

According to Theorem 6.17 and Exercise 6.20, for every $\chi \in \mathrm{Irr}(B)$, $\chi|_{G_{p'}}$ is a $\mathbf{Z}$-linear combination of $\mathrm{IBr}(B)$, and conversely every $\varphi \in \mathrm{IBr}(B)$ is a $\mathbf{Z}$-linear combination of $\{\chi|_{G_{p'}}; \chi \in \mathrm{Irr}(B)\}$. From this we get the following theorem.

Theorem 6.25. *If $\varphi \in \mathrm{IBr}(B)$, then $\nu(\varphi(1)) \geq a - d(B)$. Moreover, there exists $\varphi \in \mathrm{IBr}(B)$ such that $\nu(\varphi(1)) = a - d(B)$.*

Exercise 6.26. Let $\chi \in \mathrm{Irr}(B)$ and $\varphi \in \mathrm{IBr}(B)$. Then, for any $x \in G$ and $y \in G_{p'}$, the following statements hold:

$$\nu(\chi(x)) \geq \nu(|C_G(x)|) - d(B).$$

$$\nu(\varphi(y)) \geq \nu(|C_G(y)|) - d(B).$$

[Hint: Use the fact that $\omega_\chi(\hat{C}) \in R$ and Exercise 6.20.]

Let $C \in \mathrm{Cl}(G)$. By a *defect group* of C, denoted by $\delta(C)$, we mean a Sylow p-subgroup of $C_G(x)$ for some $x \in C$. This is unique up to G-conjugacy. If $|\delta(C)| = p^c$, i.e., $c = \nu(|C_G(x)|)$, then c is called the *defect* of the conjugate class C, which is denoted by $d(C)$.

Lemma 6.27.

(i) *If* $e_B = \sum_{C \in \mathrm{Cl}(G_{p'})} \beta_B(C)\hat{C}$, *then*

$$\nu(\beta_B(C)) \geq d(C) - d(B).$$

In particular $\beta_B(C)^* = 0$ *if* $d(B) < d(C)$.

(ii) *If* $d(C) < d(B)$, *then* $\omega_B^*(\hat{C}) = 0$.

Consequently, the following holds:

$$\omega_B^*(e_B^*) = \sum_{\substack{C \in \mathrm{Cl}(G_{p'}) \\ d(C) = d(B)}} \beta_B(C)^* \omega_B^*(\hat{C}). \tag{6.15}$$

In particular, there exists $C \in \mathrm{Cl}(G_{p'})$ *such that* $d(B) = d(C)$ *and* $\beta_B^*(C) \neq 0$, $\omega_B^*(\hat{C}) \neq 0$.

Proof. Let $x \in C$.

(i) It follows from Theorem 6.10(ii) that

$$\nu(\beta_B(C)) \geq \nu(\eta_\varphi(x^{-1})) - d(B) \geq \nu(|C_G(x)|) - d(B) = d(C) - d(B).$$

(ii) Let $\chi \in \mathrm{Irr}(B)$ be such that $\mathrm{ht}(\chi) = 0$. If $d(C) < d(B)$, then

$$\begin{aligned} \nu(\omega_\chi(\hat{C})) &= a + \nu(\chi(x)) - d(C) - (a - d(B)) \\ &= d(B) - d(C) + \nu(\chi(x)) > 0. \end{aligned}$$

Therefore, $\omega_B^*(\hat{C}) = 0$ and by combining this with (i), we obtain (6.15). The latter half follows from $\omega_B^*(e_B^*) \neq 0$. ■

From the above lemma, we obtain the following, which refines Theorem 6.24(i).

Theorem 6.28. *Assume that* $\chi, \chi' \in \mathrm{Irr}(G)$ *belong to blocks of the same defect* d. *Then* χ *and* χ' *belong to the same block if and only if it holds that* $\omega_\chi(\hat{C}) \equiv \omega_{\chi'}(\hat{C}) \pmod{(\pi)}$ *for any* $C \in \mathrm{Cl}(G_{p'})$ *with* $d(C) = d$.

Proof. We only need to prove the "if" part. Let $\chi \in B$ and $\chi' \in B'$. Then, since $d(B) = d(B') = d$, we have that $\omega_{\chi'}^*(e_B^*) = \sum_{d(c)=d} \beta_B(C)^* \omega_{\chi'}^*(\hat{C}) = \omega_\chi^*(e_B^*) = 1$, and therefore $B' = B$. ∎

As for blocks of defect zero, we have the following result.

Theorem 6.29. *Let $B \in \mathrm{Bl}(G)$ and $\chi \in \mathrm{Irr}(B)$. Then the following six conditions are equivalent to each other.*

(1) $d(B) = 0$.
(2) $\nu(\chi(1)) = a(= \nu(|G|))$.
(3) $k(B) = 1$.
(4) $C_B = D_B = (1)$ *(the identity matrix of degree* 1).
(5) $\chi(x) = 0$ *for all p-singular elements x of G.*
(6) $\chi(x) = 0$ *for all p-elements $x \neq 1$ of G.*

Proof.
(1) ⇒ (2). This is clear since $\nu(\chi(1)) \leq a$ in general.
(2) ⇒ (3). It follows from the assumption that $e_\chi \in Z(RG)$, thus it must be equal to the block idempotent e_B. Hence $\mathrm{Irr}(B) = \{\chi\}$.
(3) ⇒ (4). Since $k(B) = l(B) = 1$, both C_B and D_B are of degree 1. Thus, if $\mathrm{Irr}(B) = \{\chi\}$ and $\mathrm{IBr}(B) = \{\varphi\}$, then $\eta_\varphi = d\chi (0 < d \in \mathbf{Z})$. Consequently, χ vanishes on $G - G_{p'}$ and hence is written as $\chi = \sum_i m_i \eta_{\varphi_i} (m_i \in \mathbf{Z})$ by Theorem 6.15(i). If $\varphi_1 = \varphi$ in this expression, then $m_1 d = 1$ and hence $d = 1$. Namely, $D_B = (1)$, and hence $C_B = (1)$.
The implications "(4) ⇒ (5)" and "(5) ⇒ (6)" are trivial.
(6) ⇒ (1). If $P \in \mathrm{Syl}_p(G)$, then

$$(\chi_P, \mathbf{1}_P)_P = \frac{1}{|P|} \sum_{x \in P} \chi(x) = \frac{\chi(1)}{|P|} \in \mathbf{Z}.$$

thus, $\nu(\chi(1)) = a$ and $e_B = e_\chi$. Therefore, $\mathrm{Irr}(B) = \{\chi\}$ and $d(B) = 0$. ∎

We say that $\chi \in \mathrm{Irr}(G)$ is an *irreducible character of defect zero* if $\nu(\chi(1)) = \nu(|G|)$.

***Exercise 6.30* (Gallagher).** Let π be a set of primes and $\chi \in \mathrm{Irr}(G)$. If $|G|_\pi \mid \chi(1)$, then χ vanishes on all π-elements other than the identity.

6.6. Cartan Matrices

We set

$$X_{\mathrm{Br}}(G) = \bigoplus_{\varphi \in \mathrm{IBr}(G)} \mathbf{Z}\varphi, \qquad Y(G) = \bigoplus_{\varphi \in \mathrm{IBr}(G)} \mathbf{Z}(\eta_\varphi|_{G_{p'}}).$$

Then $X_{\mathrm{Br}}(G) \supset Y(G)$, and if $\{a_1, \ldots, a_l\}$ are the elementary divisors of the Cartan matrix C of G, we have, noting that $\det C > 0$,

$$X_{\mathrm{Br}}(G)/Y(G) \simeq \mathbf{Z}/(a_1) \oplus \cdots \oplus \mathbf{Z}/(a_l),$$
$$|X_{\mathrm{Br}}(G)/Y(G)| = \det C.$$

(cf. Hungerford [1], Chapter IV, Section 6.)

For a function θ defined on G or $G_{p'}$, we set

$$\hat{\theta}(x) = \begin{cases} p^a\theta(x) & \text{if } x \in G_{p'}, \\ 0 & \text{if } x \in G - G_{p'}, \end{cases} \tag{6.16}$$

where $a = \nu(|G|)$. Then we have the following.

Lemma 6.31.

(i) *If $\theta \in X(G)$ or $\theta \in X_{\mathrm{Br}}(G)$, then $\hat{\theta} \in X(G)$.*

(ii) *If $\theta \in X_{\mathrm{Br}}(G)$, then $p^a\theta \in Y(G)$.*

(iii) *The elementary divisors of the Cartan matrix C are powers of p.*

Proof.

(i) Let E be an arbitrary elementary subgroup of G and let $E = P \times Q$, where P is a p-group of order p^b and Q is a p'-group. If γ_P is the character defined by the regular representation of P, then $\hat{\theta}_E = p^{a-b}\gamma_P \times \theta_Q$. But since $\theta_Q \in X(Q)$, we find that $\hat{\theta}_E \in X(E)$. Thus we conclude $\hat{\theta} \in X(G)$ from the fundamental theorem of Brauer.

(ii) It follows from the definition that $\hat{\theta}|_{G_{p'}} = p^a\theta$. But since $\hat{\theta}$ is a generalized character that vanishes on $G - G_{p'}$, it is expressed as $\hat{\theta} = \sum_\varphi m_\varphi \eta_\varphi$ with $m_\varphi \in \mathbf{Z}$. Thus $p^a\theta = \sum_\varphi m_\varphi(\eta_\varphi|_{G_{p'}}) \in Y(G)$.

(iii) From (ii) it follows that $X_{\mathrm{Br}}(G)/Y(G)$ is a p-group. Therefore the assertion is obvious. ∎

Theorem 6.32. *Let* $\mathrm{Cl}(G_{p'}) = \{C_1, \ldots, C_l\}$ *and* $x_i \in C_i$ $(1 \le i \le l)$. *Then the elementary divisors of the Cartan matrix* C *of* G *are given by*

$$\{|C_G(x_1)|_p, \ldots, |C_G(x_l)|_p\}.$$

Proof. Since the elementary divisors of C are powers of p, they are also the elementary divisors of C considered as the matrix over R. From (6.8), we have

$${}^t\bar{\mathbf{\Phi}}C\mathbf{\Phi} = \mathrm{diag}\{|C_G(x_1)|, \ldots, |C_G(x_l)|\},$$

whence the result follows, since $\mathbf{\Phi} \in \mathrm{GL}_l(R)$. ■

For a block B, we set

$$X_{\mathrm{Br}}(B) = \bigoplus_{\varphi \in \mathrm{IBr}(B)} \mathbf{Z}\varphi, \qquad Y(B) = \bigoplus_{\varphi \in \mathrm{IBr}(B)} \mathbf{Z}(\eta_\varphi|_{G_{p'}}).$$

If the Cartan matrix C_B corresponding to B has the elementary divisors $\{a_1, \ldots, a_{l(B)}\}$, then

$$X_{\mathrm{Br}}(B)/Y(B) \simeq \mathbf{Z}/(a_1) \oplus \cdots \oplus \mathbf{Z}/(a_{l(B)}),$$
$$|X_{\mathrm{Br}}(B)/Y(B)| = \det C_B.$$

In order to study the elementary divisors of C_B, we set

$$X(B) = \bigoplus_{\chi \in \mathrm{Irr}(B)} \mathbf{Z}\chi,$$

and first prove the following lemma.

Lemma 6.33. *Let* $d = d(B)$. *Given* $\theta \in X(B)$ *or* $\theta \in X_{\mathrm{Br}}(B)$, *we have the following.*

(i) $\hat{\theta} \in X(B)$. *Furthermore, if* $\hat{\theta} = \sum_{\chi \in \mathrm{Irr}(B)} n_{\theta\chi}\chi$ *with* $n_{\theta\chi} \in \mathbf{Z}$, *then*

$$\nu(n_{\theta\chi}) \ge \nu(\chi(1)) \ge a - d.$$

In particular $p^{a-d} | n_{\theta\chi}$.

(ii) *If* $\theta \in X_{\mathrm{Br}}(B)$, *then* $p^d\theta \in Y(B)$.

(iii) *If* $p^e\theta \in Y(B)$, *then* $p^{-(a-e)}\hat{\theta} \in X(B)$.

Proof.

(i) Since $\hat{\theta}$ is a generalized character vanishing on $G - G_{p'}$, it is expressed as $\hat{\theta} = \sum_{\varphi} m_{\varphi}\eta_{\varphi}$ $(m_{\varphi} \in \mathbf{Z})$. If $\varphi \notin \mathrm{IBr}(B)$, then $(\hat{\theta}, \varphi)' = (p^a\theta, \varphi)' = 0$ by Exercise 6.20(ii), which implies that $m_{\varphi} = 0$ and $\hat{\theta} \in X(B)$. Also, for $x_i \in C_i$, where $\mathrm{Cl}(G_{p'}) = \{C_i; 1 \leq i \leq l\}$ as above, we have

$$
\begin{aligned}
n_{\theta\chi} &= (\chi, \hat{\theta})_G = (\chi, p^a\theta)' \\
&= \frac{1}{|G|_{p'}} \sum_{x \in G_{p'}} \chi(x)\overline{\theta(x)} = \frac{1}{|G|_{p'}} \sum_{i=1}^{l} |C_i|\chi(x_i)\overline{\theta(x_i)} \\
&= \frac{\chi(1)}{|G|_{p'}} \sum_{i=1}^{l} \omega_{\chi}(\hat{C}_i)\overline{\theta(x_i)},
\end{aligned}
$$

whence $\nu(n_{\theta\chi}) \geq \nu(\chi(1))$. The remaining assertion is clear from this.

(ii) As shown above, $p^{-(a-d)}\hat{\theta}$ is a generalized character vanishing on $G - G_{p'}$, and hence $p^{-(a-d)}\hat{\theta} = \sum_{\varphi} l_{\varphi}\eta_{\varphi}$ $(l_{\varphi} \in \mathbf{Z})$, where $l_{\varphi} = 0$ for $\varphi \notin B$. Thus we have that $p^{-(a-e)}\hat{\theta}|_{G_{p'}} = p^d\theta \in Y(B)$.

(iii) $p^{-(a-e)}\hat{\theta}$ vanishes on $G - G_{p'}$, and by assumption, we have

$$p^{-(a-e)}\hat{\theta}|_{G_{p'}} = p^e\theta = \sum\nolimits_{\varphi \in B} a_{\varphi}\eta_{\varphi}|_{G_{p'}}$$

with $a_{\varphi} \in \mathbf{Z}$. Therefore, $p^{-(a-e)}\hat{\theta} = \sum_{\varphi \in B} a_{\varphi}\eta_{\varphi} \in X(B)$. ■

For $\theta \in X(B)$ or $\theta \in X_{\mathrm{Br}}(B)$, let $\dot{\theta} = p^{-(a-d(B))}\hat{\theta}$. Thus,

$$
\dot{\theta}(x) = \begin{cases} p^{d(B)}\theta(x), & \text{if } x \in G_{p'}, \\ 0, & \text{otherwise.} \end{cases}
$$

Furthermore, with the notation of the above lemma, let $a_{\theta\chi} = n_{\theta\chi}/p^{a-d}$. Then $a_{\theta\chi} \in \mathbf{Z}$, if $\chi \in \mathrm{Irr}(B)$, and it holds that

$$\dot{\theta} = \sum_{\chi \in \mathrm{Irr}(B)} a_{\theta\chi}\chi \in X(B).$$

Concerning the integer $a_{\theta\chi}$, we have the following.

Lemma 6.34. *Let $\theta \in X(B)$ or $\theta \in X_{\mathrm{Br}}(B)$.*

(i) *If $\chi, \chi' \in \mathrm{Irr}(B)$, then*

$$a_{\theta\chi} \equiv a_{\theta\chi'}\frac{\chi(1)}{\chi'(1)} \pmod{p^{\mathrm{ht}(\chi)+1}}. \tag{6.17}$$

(ii) *Suppose that* $\nu(\theta(1)) = a - d$, *where* $d = d(B)$. *Then* $a_{\theta\chi} \neq 0$ *for all* $\chi \in \mathrm{Irr}(B)$. *Moreover, if* $\chi_0 \in \mathrm{Irr}(B)$ *is of height zero, then*

$$a_{\theta\chi_0} \not\equiv 0 \qquad (\mathrm{mod}\ p).$$

Proof.
(i) As is seen in the proof of Lemma 6.33(ii), we have

$$\begin{aligned}\frac{n_{\theta\chi}}{\chi(1)} &= \frac{1}{|G|_{p'}} \sum_i \omega_\chi(\hat{C}_i)\overline{\theta(x_i)} \\ &\equiv \frac{1}{|G|_{p'}} \sum_i \omega_{\chi'}(\hat{C}_i)\overline{\theta(x_i)} = \frac{n_{\theta\chi'}}{\chi'(1)} \qquad (\mathrm{mod}(\pi)).\end{aligned}$$

Therefore, $\nu(n_{\theta\chi}/\chi(1) - n_{\theta\chi'}/\chi'(1)) > 0$, and by adding $\nu(\chi(1)/p^{a-d})$ to both sides, we obtain

$$\nu(a_{\theta\chi} - a_{\theta\chi'}\chi(1)/\chi'(1)) > \mathrm{ht}(\chi).$$

(ii) If $a_{\theta\chi'} = 0$ for some $\chi' \in \mathrm{Irr}(B)$, then it follows from (6.17) that $p \mid a_{\theta\chi}$ for all $\chi \in \mathrm{Irr}(B)$. Also, if $p \mid a_{\theta\chi_0}$ for some $\chi_0 \in \mathrm{Irr}(B)$ with $\mathrm{ht}(\chi_0) = 0$, then by letting $\chi' = \chi_0$ in (6.17), we find that $p \mid a_{\theta\chi}$ for all $\chi \in \mathrm{Irr}(B)$. Therefore it holds that $p^{-1}\dot{\theta} \in X(B)$ in either case, and thus

$$p^{-1}\dot{\theta} = \sum_\varphi m_\varphi \eta_\varphi \qquad (m_\varphi \in \mathbf{Z}).$$

Since $p^a \mid \eta_\varphi(1)$, we get

$$\nu(p^{-1}\dot{\theta}(1)) = d - 1 + \nu(\theta(1)) \geq a,$$

and hence $\nu(\theta(1)) \geq a - d + 1$, which contradicts the assumption. ■

Theorem 6.35. *The greatest elementary divisor of* C_B *is* $p^{d(B)}$ *and it appears only once.*

Proof. We shall use the same notation as in Lemmas 6.33 and 6.34. First of all, by Lemma 6.33(ii), the elementary divisors of C_B are not larger than p^d. Next, we take $\varphi \in \mathrm{IBr}(B)$ so that $\nu(\varphi(1)) = a - d$. Then $\nu(a_{\varphi\chi_0}) = 0$ for any $\chi_0 \in \mathrm{Irr}(B)$ of height zero (Lemma 6.34(ii)). Hence, by Lemma 6.33(iii), $p^{-1}\dot{\varphi} = p^{-(a-d+1)}\hat{\varphi}$ does not lie in $X(B)$ and $p^{d-1}\varphi \notin Y(B)$. This implies that $\varphi + Y(B)$ has order p^d in $X_{\mathrm{Br}}(B)/Y(B)$. Therefore p^d is an elementary divisor of C_B. In order to complete the proof, it suffices to show that the p-rank of the matrix $p^dC_B^{-1}$ is at most 1.

Let $\mathrm{IBr}(B) = \{\varphi_i; 1 \leq i \leq l(B)\}$ and $\mathrm{Irr}(B) = \{\chi_j; 1 \leq j \leq k(B)\}$. Since $p^d\varphi_i \in Y(B)$, there exist $l_{i\nu} \in \mathbf{Z}$ such that $p^d\varphi_i = \sum_\nu l_{i\nu}(\eta_\nu|_{G_{p'}}) = \sum_{\nu,j} l_{i\nu}c_{\nu j}\varphi_j$. Let $L = (l_{ij})$, the square matrix of degree $l(B)$; thus $LC_B = p^dI$ and hence $L = p^dC_B^{-1}$.

We write a_{ij} for $a_{\varphi_i\chi_j}$, i.e., $\dot{\varphi}_i = \sum_\nu a_{i\nu}\chi_\nu$, and $\tilde{\varphi}_j = \sum_i m_{ji}\chi_i$ with $m_{ji} \in \mathbf{Z}$ (cf. Lemma 6.13). Thus we have $l_{ij} = (p^d\varphi_i, \varphi_j)' = (\dot{\varphi}_i, \tilde{\varphi}_j)_G = \sum_\nu a_{i\nu}m_{j\nu}$. Then, if we set $A = (a_{ij})$ and $M = (m_{ij})$,

$$p^dC_B^{-1} = L = A{}^tM.$$

Since the p-rank of A is at most 1 by (6.17), this implies that the p-rank of $p^dC_B^{-1}$ is at most 1. ■

As an immediate consequence of the above theorem, we get:

Corollary 6.36. *For a natural number d, we have*

$$\#\{B \in \mathrm{Bl}(G); d(B) = d\} \leq \#\{C \in \mathrm{Cl}(G_{p'}); d(C) = d\}.$$

Also, we have the following characterization of blocks of defect zero.

Theorem 6.37. *The following three conditions on $B \in \mathrm{Bl}(G)$ are equivalent.*

(1) $d(B) = 0$.
(2) $k(B) = l(B)$.
(3) *B^* is a simple ring.*

Proof.
(1) ⇒ (2). This is immediate from $1 = k(B) \geq l(B) \geq 1$.

(2) ⇒ (3). Since D_B^* is nonsingular, $C_B^* = {}^tD_B^*D_B^*$ is also nonsingular, and hence the elementary divisors of C_B are all equal to 1. Consequently, $C_B = (1)$ by Theorem 6.35. In other words, B^* has exactly one nonisomorphic principal indecomposable module, which is irreducible. Thus B^* is simple.

(3) ⇒ (1). Since $C_B = (1)$, it follows from the above theorem that $d(B) = 0$. ■

For a block B of positive defect, the following holds.

Corollary 6.38. *If $d(B) > 0$, then $k(B) > l(B)$ and $c_{ii} \geq 2$ for every diagonal Cartan invariant c_{ii} of C_B.*

Proof. We only need to show the latter half. We shall show that no principal indecomposable FG-module belonging to B is irreducible. Once this is shown, then the result is clear from Chapter 2, 8.21.

On the contrary, suppose that there exists a principal indecomposable character η belonging to B such that $\eta|_{G_{p'}} = \varphi \in \mathrm{IBr}(B)$. Then since the Cartan matrix C_B is symmetric, it splits into the form:

$$C_B = \left(\begin{array}{c|c} 1 & 0 \\ \hline 0 & * \end{array}\right),$$

which is impossible. ■

Finally we establish the following.

Theorem 6.39 **(Brauer and Feit[1]).** *Let $B \in \mathrm{Bl}(G)$ and $d = d(B)$. Then*

$$k(B) \leq \frac{1}{4} p^{2d(B)} + 1.$$

Proof. Let $\chi_0 \in \mathrm{Irr}(B)$ be such that $\mathrm{ht}(\chi_0) = 0$ and denote by a_χ the integer $a_{\chi_0 \chi}$ in Lemma 6.34. Thus,

$$\dot{\chi}_0 = \sum_{\chi \in \mathrm{Irr}(B)} a_\chi \chi,$$

and $a_\chi \neq 0$ for all $\chi \in \mathrm{Irr}(B)$ by Lemma 6.34(ii). Thus,

$$\sum_{\chi \in \mathrm{Irr}(B)} a_\chi^2 = (\dot{\chi}_0, \dot{\chi}_0)_G = p^d(\dot{\chi}_0, \chi_0)' = p^d(\dot{\chi}_0, \chi_0)_G = p^d a_{\chi_0}, \tag{6.18}$$

and since $a_\chi \neq 0$, we obtain that $k(B) - 1 + a_{\chi_0}^2 \leq p^d a_{\chi_0}$, namely,

$$k(B) \leq -a_{\chi_0}^2 + p^d a_{\chi_0} + 1.$$

But since the values of the polynomial function $-t^2 + p^d t + 1$ are bounded by $\frac{1}{4}p^{2d} + 1$, the result is clear. ∎

Exercise 6.40. Let $\mathrm{Irr}(B) = \{\chi_i; 1 \le i \le k(B)\}$, and write a_{ij} for $a_{\chi_i \chi_j}$ in Lemma 6.34. Then the following holds for the matrix $A_B = (a_{ij})_{i,j}$.

(i) $A_B = p^{d(B)} D_B C_B^{-1}\, {}^t D_B$.

(ii) $A_B^2 = p^{d(B)} A_B$.

(iii) $\mathrm{tr}\, A_B = p^{d(B)} l(B)$.

[Hint: (i) $a_{ij} = (p^d \chi_i, \chi_j)'$. (iii) Since A_B satisfies the polynomial equation $x^2 - p^{d(B)} x = 0$, the characteristic roots of A_B are either $p^{d(B)}$ or 0. Use also rank $A_B = l(B)$.]

Problems

R is a commutative ring and F is a field of characteristic $p > 0$.

1. $R[G \times H] \simeq RG \otimes_R RH$ $((x, y) \mapsto x \otimes y,\ x \in G,\ y \in H)$.

2. (Tensor induction) Let $H \le G$. For W_{RH}, define $W^{\otimes G} = \bigotimes_{t \in H\backslash G} (W \otimes_{RH} t)$ to be the tensor product of the $(W \otimes_{RH} t)$'s over R. Show that $W^{\otimes G}$ becomes a right RG-module via the natural action of G.

3. Let P be a p-subgroup of G, and let $V = F \otimes_{FP} FG$. Then any irreducible FG-module appears as a direct summand of both $\mathrm{soc}(V)$ and $\mathrm{hd}(V)$.

4. Assume that F is a splitting field for G. Let $\mathrm{IRR}(FG) = \{W_1, \ldots, W_l\}$ and U_i be the projective cover of W_i, where we assume $W_1 = F_G$. Then

(i) $U_i | U_1 \otimes_F W_i$ $(1 \le i \le l)$.

(ii) $\sum_{i=1}^{l} (\dim_F W_i)^2 \ge |G|/\dim_F U_1$.

5. Define $\delta\colon FG \to FG$ by $\delta(\sum_{x \in G} a_x x) = \sum_x a_x x^{-1}$ $(a_x \in F)$. Then there holds an FG-isomorphism: $\delta(e)FG \simeq (eFG)^\wedge$ for any $e \in \mathrm{pi}(FG)$.

6. Let χ be a $\mathbf{Q}$-character of G defined by a transitive permutation representation of G. Then it holds that

$$|G : C_G(x)|\chi(x)/\chi(1) \in \mathbf{Z}, \qquad \text{for all } x \in G.$$

7. Let $G \rhd H$ and $\bar{G} = G/H$. Then $|C_{\bar{G}}(\bar{x})| \le |C_G(x)|$ for all $x \in G$. [Hint: Use the second orthogonality relation of characters.]

8. Let K be a subfield of $\mathbf{C}$, ζ a primitive $|G|$th root of unity, and $\mathfrak{G} = \mathrm{Gal}(K(\zeta)/K)$. Given $\sigma \in \mathfrak{G}$, we choose $m_\sigma \in \mathbf{Z}$ so that $\zeta^\sigma = \zeta^{m_\sigma}$. (Note that m_σ is unique within mod $|G|$.)

(i) $\chi(x)^\sigma = \chi(x^{m_\sigma})$ for all $x \in G$.

(ii) $\mathfrak{G}$ acts on $\mathrm{Cl}(G)$ as a permutation group via the map $x \mapsto x^\sigma = x^{m_\sigma}$ $(x \in G)$.

(iii) The number of the irreducible K-characters of G equals the number of $\mathfrak{G}$-orbits on $\mathrm{Cl}(G)$.

(iv) The number of the irreducible $\mathbf{Q}$-characters of G equals the number of non-G-conjugate cyclic subgroups of G.

9. (**Burnside**) If χ is an irreducible $\mathbf{C}$-character of G with $\chi(1) > 1$, then there exists $x \in G$ such that $\chi(x) = 0$. [Hint: The Cauchy–Schwartz inequality:

$$\frac{1}{|G| - 1} \sum_{x \neq 1} |\chi(x)|^2 \geq \sqrt[|G|-1]{\prod_{x \neq 1} |\chi(x)|^2},$$

in which we have $\prod_{x \neq 1} |\chi(x)|^2 \in \mathbf{Z}$.]

10. No simple group has an irreducible $\mathbf{C}$-representation of degree 2.

11. Let $G \rhd H$, $\chi \in \mathrm{Irr}(G)$, $\zeta \in \mathrm{Irr}(H)$ and $(\chi_H, \zeta) = e \geq 1$.

(i) If G/H is cyclic, then $e = 1$.

(ii) If χ vanishes on $G - H$, then $\zeta^G = e\chi$.

(iii) If there exists a linear character λ of G such that $\lambda\chi = \chi$, then there exists $\theta \in \mathrm{Irr}(\mathrm{Ker}\ \lambda)$ such that $\chi = \theta^G$.

12. With the assumption of the preceding problem, assume that $T(\zeta) = G$, and let α be a factor set of G/H associated with ζ. Then

(i) $e^2 \leq |G{:}H|$.

(ii) The following four conditions are equivalent.

(1) $e^2 = |G{:}H|$.

(2) $\zeta^G = e\chi$.

(3) $\mathrm{Irr}(G|\zeta) = \{\chi\}$.

(4) $\mathbf{C}^{(\alpha^{-1})}[G/H]$ is a simple ring.

13. Let $G = SH \rhd H$ and suppose that $(|S|, |H|) = 1$. Let $\mathbf{Y}$ be an irreducible $\mathbf{C}$-representation of H such that $T(\mathbf{Y}) = G$. Then there exists $\mathbf{X} \in \mathrm{IRR}_{\mathbf{C}}(G)$ satisfying the following two conditions, and $\mathbf{X}$ is unique up to equivalence.

(1) $\mathbf{X}_H = \mathbf{Y}$.

(2) $\det X(s) = 1$ for all $s \in S$.

14. Let $G \rhd H$ and suppose that K is an algebraically closed field. Let $\mathbf{Y}$ be an irreducible K-representation of H such that $T(\mathbf{Y}) = G$. Then $\mathbf{Y}$ extends to a (linear) representation of G if and only if for every prime p dividing $|G|$ and $S_p \in \mathrm{Syl}_p(G)$, $\mathbf{Y}$ extends to a representation of S_pH.

15. If F is algebraically closed, then $M_F(G) \simeq M_{\mathbf{C}}(G)/M_{\mathbf{C}}(G)_p$.

16. Let R be a discrete valuation ring with quotient field K of characteristic zero. If U, V are finitely generated projective RG-modules such that $K \otimes_R U \simeq K \otimes_R V$, then $U \simeq V$.

17. Let R, K be the same as in the preceding problem and let F be the residue field of R. If K is a splitting field for G, then so is F.

18. (Swan) Let R be a Dedekind domain with quotient field K of characteristic zero and assume that no prime divisor of $|G|$ is a unit in R. If V is a finitely generated projective RG-module, then for any prime ideal $\mathfrak{p}$ of R, $V_{\mathfrak{p}} = R_{\mathfrak{p}} \otimes_R V$ is a free $R_{\mathfrak{p}}G$-module. Prove this in the following way.

(i) By Problem 16, it suffices to show that $K \otimes_R V$ is KG-free (the case of $\mathfrak{p} = 0$). Thus, we have to show that $\chi(x) = 0$ for any $x \in G - \{1\}$, where χ denotes the R-character defined by V.

(ii) With the notation above, let p be a prime divisor of $o(x)$ and let $\mathfrak{p}$ be a prime ideal of R containing p. Since V is $R_{\mathfrak{p}}G$-projective, χ vanishes on the p-singular element x of G.

19. (Heller) Let R be a discrete valuation ring, $\tilde{R}$ the completion of R, and let K, $\tilde{K}$ be the quotient fields of R, $\tilde{R}$, respectively.

(i) Given $S \in \mathrm{GL}_n(\tilde{K})$, there exists $T \in \mathrm{GL}_n(K)$ such that $TS \in \mathrm{GL}_n(\tilde{R})$. [Hint: Apply the π-adic expansions of the elements of $\tilde{K}$ (cf., Chapter 1, 13.20), where (π) denotes the valuation ideal of R.]

(ii) Let A be an R-algebra and let $\mathbf{X}$ be an $\tilde{R}$-representation of A. If $\mathbf{X}$, viewed as a $\tilde{K}$-representation, is realized in K, then so is $\mathbf{X}$ viewed as an $\tilde{R}$-representation in R.

20.

(i) Theorem 6.12(ii) remains valid without assuming that R is complete.

(ii) Let $G = \langle x \rangle$ be a cyclic group of order $p - 1$, where $p \geq 5$. Then G has only two irreducible $\mathbf{Z}/(p)$-representations which are lifted to $\mathbf{Z}_p$, namely, the trivial one and $\mathbf{T}: G \to \mathbf{Z}/(p)$ with $T(x) = -1$.

21. Let L be an algebraically closed field. If Char L does not divide $|G|$, then the degree of every irreducible projective L-representation of G is a divisor of $|G|$.

22. Let $a = \nu(|G|)$. Then the following holds:

$$\#\{B \in \mathrm{Bl}(G);\ d(B) = a\} = \#\{C \in \mathrm{Cl}(G_{p'});\ d(C) = a\}.$$

[Hint: Immediate by Theorems 6.32 and 6.35.]

23. (Brauer and Tuan) Suppose that $Z(G) = \{1\}$ and let $P \in \mathrm{Syl}_p(G)$. If G has a faithful irreducible ordinary character χ of degree $p^r (r \geq 1)$, then χ belongs to a block B of G with $d(B) \leq \nu(|P{:}Z(P)|)$.
[Hint: Show that $\chi(x) = 0$ for all $x(\neq 1) \in Z(P)$.]

24. Let $B \in \mathrm{Bl}(G)$ with $d(B) = d \geq 1$ and $\chi \in \mathrm{Irr}(B)$. With the notation of Lemmas 6.33 and 6.34, the following statements hold.

(i) $a_{\theta\chi} = a_{\chi\theta}$, $0 < a_{\chi\chi} < p^d$.
(ii) If $\nu(\theta(1)) = a - d$, then $\nu(a_{\theta\chi}) = \mathrm{ht}(\chi)$.
(iii) If $\theta \in \mathrm{Irr}(B)$ and $\mathrm{ht}(\theta) > 0$, then $\nu(a_{\theta\chi}) > \mathrm{ht}(\chi)$.
(iv) If $\mathrm{ht}(\chi) > 0$, then $d - 1 \geq \nu(a_{\chi\chi}) \geq \mathrm{ht}(\chi) + 1$.
(v) If there exists $\chi \in \mathrm{Irr}(B)$ such that $\nu(\chi(1)) = a - 1$, then $d = 1$.

4 Indecomposable Modules

This chapter presents Green's theory on indecomposable modules of group rings over complete discrete valuation rings. Our first goal is the vertex theory of indecomposable modules in Section 3, which is fundamental throughout the theory. The Green correspondence in Section 4 is one of the main themes of this chapter. In this connection, we show a result due to Burry and Carlson (Theorem 4.6) and establish some functorial properties of it (Theorems 4.5 and 5.4). In Section 6 we study the endomorphism rings of induced modules and show a Clifford-type theorem on indecomposable modules (Corollary 6.8). The Green indecomposability theorem in Section 7 is one of the main results of this chapter, and in the proof of it, Theorem 13.27 of Chapter 1 concerning the extension of valuation will play an important role. In Section 8 we discuss Scott modules.

1. Trace Maps

Let G be a group, R a commutative ring, and V an RG-module. For $H \leq G$, $\mathrm{Inv}_H(V)$ denotes, as in the preceding chapter, the set of H-invariant elements of V, namely,

$$\mathrm{Inv}_H(V) = \{v \in V;\, vx = v \text{ for all } x \in H\}.$$

It is clear that if $H \leq K \leq G$, then $\mathrm{Inv}_K(V) \subset \mathrm{Inv}_H(V)$, and, in particular, $\mathrm{Inv}_G(V) \subset \mathrm{Inv}_H(V)$.

Exercise 1.1. $\mathrm{Inv}_{H^x}(V) = \mathrm{Inv}_H(V)x$ for $x \in G$.

For $H \leq G$, define $\mathrm{Tr}_H^G\colon \mathrm{Inv}_H(V) \to \mathrm{Inv}_G(V)$ by

$$\mathrm{Tr}_H^G(v) = \sum_{t \in H\backslash G} vt \qquad (v \in \mathrm{Inv}_H(V)). \tag{1.1}$$

Exercise 1.2. Prove that $\mathrm{Tr}_H^G(v)$ is independent of the choice of coset representatives of H in G and indeed belongs to $\mathrm{Inv}_G(V)$.

Tr_H^G is clearly an R-linear map, and it is called a *trace map*.

For the sake of simplicity, we write $\mathrm{Tr}_H^G(W) = \mathrm{Tr}_H^G(\mathrm{Inv}_H(W))$ for an RH-submodule W of V.

If H, K are subgroups of G, then we write $H =_G K$ if H and K are conjugate in G, and $H \leq_G K$ if $H^x \leq K$ for some $x \in G$. We also write $x \in_G K$ if x is conjugate in G to some element of K.

Theorem 1.3. *Let $H, K \leq G$ and V be an RG-module.*

(i) *If $H \leq K \leq G$, then $\mathrm{Tr}_H^G = \mathrm{Tr}_K^G \circ \mathrm{Tr}_H^K$.*

(ii) $\mathrm{Tr}_H^G(v) = \mathrm{Tr}_{H^x}^G(vx)$ *for all $v \in \mathrm{Inv}_H(V)$ and $x \in G$. Hence $\mathrm{Tr}_H^G(V) = \mathrm{Tr}_{H^x}^G(V)$.*

(iii) *If $H \leq_G K$, then $\mathrm{Tr}_H^G(V) \subset \mathrm{Tr}_K^G(V)$.*

Proof.

(i) Let $H\backslash K = \{s_i\}$, $K\backslash G = \{t_j\}$. Then $H\backslash G = \{s_i t_j\}$, and the assertion is clear from the definition.

(ii) We see that $G = \sum_{t \in H\backslash G} H^x x^{-1} t$ and $vx \in \mathrm{Inv}_{H^x}(V)$ for $v \in \mathrm{Inv}_H(V)$. Thus we have $\mathrm{Tr}_H^G(v) = \sum_t vt = \sum_t (vx)x^{-1}t = \mathrm{Tr}_{H^x}^G(vx)$.

(iii) This is clear from (i) and (ii). ■

Also, we have the following theorem.

Theorem 1.4. *Let $H, K \le G$. Then the following statements hold for V_{RG}.*

(i) **(Mackey decomposition)** $\mathrm{Tr}_H^G(v) = \sum_{t \in H\backslash G/K} \mathrm{Tr}_{H^t \cap K}^K(vt)$ *for* $v \in \mathrm{Inv}_H(V)$.

(ii) $\mathrm{Tr}_H^G(V) \subset \sum_{t \in H\backslash G/K} \mathrm{Tr}_{H^t \cap K}^K(V)$.

Proof.

(i) If $G = \sum_t HtK$, $K = \sum_j (H^t \cap K)s_{tj}$, then it follows that $G = \sum_{t,j} Hts_{tj}$ and $vt \in \mathrm{Inv}_{H^t}(V) \subset \mathrm{Inv}_{H^t \cap K}(V)$. Therefore we have

$$\mathrm{Tr}_H^G(v) = \sum_t \left(\sum_j (vt)s_{tj} \right) = \sum_t \mathrm{Tr}_{H^t \cap K}^K(vt).$$

(ii) This is clear from (i). ■

We assume in the following that

$$H\backslash G \ni 1, \qquad H\backslash G/K \ni 1.$$

Let W be an RH-module. The RH-submodule $W \otimes 1$ of W^G is isomorphic to W via the map $w \otimes 1 \mapsto w$. Thus we understand that $W \subset W^G$ by identifying w with $w \otimes 1$.

With the above convention we have $wx = (w \otimes 1)x = w \otimes x$ for $w \in W$, $x \in G$, and if $w \in \mathrm{Inv}_H(W)$, then

$$\mathrm{Tr}_H^G(w) = \sum_{t \in H\backslash G} wt = \sum_{t \in H\backslash G} w \otimes t.$$

Since this coincides with $\mathrm{Tr}(w)$ defined in Chapter 3, 1.20, we have

Lemma 1.5. *Let W be an RH-module. The trace map Tr_H^G gives rise to an isomorphism:*

$$\mathrm{Tr}_H^G\colon \mathrm{Inv}_H(W) \xrightarrow{\sim} \mathrm{Inv}_G(W^G) \left(w \mapsto \sum_{t \in H\backslash G} w \otimes t \right).$$

Let A be an R-algebra and let $\mathrm{Aut}(A)$ be the group of all R-algebra automorphisms of A. We say that A is a G-algebra (over R) if there is given a homomorphism $\varphi\colon G \to \mathrm{Aut}(A)$. For $a \in A$ and $x \in G$, a^x denotes the image of a under the map $\varphi(x) \in \mathrm{Aut}(A)$. Also, we write A^H to denote $\mathrm{Inv}_H(A)$:

$$A^H = \{a \in A;\ a^h = a \quad \textit{for all} \quad h \in H\}.$$

Lemma 1.6. *Let A be a G-algebra and $H, K \leq G$.*

(i) *For $a \in A^G$, $b \in A^H$ we have*

$$\operatorname{Tr}_H^G(ab) = a \operatorname{Tr}_H^G(b), \qquad \operatorname{Tr}_H^G(ba) = \operatorname{Tr}_H^G(b)a.$$

In particular $\operatorname{Tr}_H^G(A)$ is an ideal of A^G.

(ii) *For $a \in A^H$, $b \in A^K$ we have*

$$\operatorname{Tr}_H^G(a)\operatorname{Tr}_K^G(b) = \sum_{t \in H\backslash G/K} \operatorname{Tr}_{H^t \cap K}^G(a^t b),$$

and, in particular

$$\operatorname{Tr}_H^G(A)\operatorname{Tr}_K^G(A) \subset \sum_{t \in H\backslash G/K} \operatorname{Tr}_{H^t \cap K}^G(A).$$

Proof.

(i) $\operatorname{Tr}_H^G(ab) = \sum_{t \in H\backslash G} a^t b^t = a(\sum_t b^t) = a \operatorname{Tr}_H^G(b)$. The second equality can be proved similarly.

(ii) $$\operatorname{Tr}_H^G(a)\operatorname{Tr}_K^G(b) = \operatorname{Tr}_K^G(\operatorname{Tr}_H^G(a)b) = \operatorname{Tr}_K^G\Bigg(\sum_{t \in H\backslash G/K} \operatorname{Tr}_{H^t \cap K}^K(a^t)b\Bigg)$$

$$= \sum_t \operatorname{Tr}_K^G(\operatorname{Tr}_{H^t \cap K}^K(a^t b)) = \sum_t \operatorname{Tr}_{H^t \cap K}^G(a^t b). \qquad \blacksquare$$

For the sake of simplicity, we use the following abbreviated notation for RG-modules U and V:

$$_R(U, V) = \operatorname{Hom}_R(U, V), \qquad _{RG}(U, V) = \operatorname{Hom}_{RG}(U, V),$$
$$E_{RG}(V) = \operatorname{End}_{RG}(V).$$

As was mentioned in Chapter 3, Section 1.3, $_R(U, V)$ becomes an RG-module if we define f^x, for $f \in {}_R(U, V)$ and $x \in G$, by

$$f^x(u) = f(ux^{-1})x \qquad \text{for} \quad u \in U.$$

Observe that with this action it holds that

$$\operatorname{Inv}_G(_R(U, V)) = {}_{RG}(U, V).$$

In particular $E_R(V)$ is a G-algebra and $\operatorname{Inv}_G(E_R(V)) = E_{RG}(V)$.

Exercise 1.7. Let $H, K \leq G$ and U, V, W be RG-modules. For $f \in {}_{RK}(U, V)$ and $g \in {}_{RH}(V, W)$, the following statements hold.

(i) If f is an RG-homomorphism, then $\operatorname{Tr}_H^G(g \circ f) = \operatorname{Tr}_H^G(g) \circ f$.

(ii) If g is an RG-homomorphism, then $\mathrm{Tr}_K^G(g \circ f) = g \circ \mathrm{Tr}_K^G(f)$.
(iii) $\mathrm{Tr}_H^G(g) \circ \mathrm{Tr}_K^G(f) = \sum_{t \in H\backslash G/K} \mathrm{Tr}_{H^t \cap K}^G(g^t \circ f)$.
[Hint: See the proof of Lemma 1.6.]

Let W be an RH-module. Since $W \subset W^G$, we may assume that

(1.2) $$_R(V, W) \subset {}_R(V, W^G).$$

Let Tr_H^G: $_{RH}(V, W^G) \to {}_{RG}(V, W^G)$ be the trace map. If $f \in {}_{RH}(V, W)$, then

(1.3) $$\mathrm{Tr}_H^G(f)(v) = \sum_{t \in H\backslash G} f^t(v) = \sum_{t \in H\backslash G} f(vt^{-1}) \otimes t.$$

Now, let π_W: $W^G \to W$ be the projection w.r.t. the RH-decomposition

$$W^G = W \oplus \left(\sum_{\substack{t \in H\backslash G \\ t \neq 1}} W \otimes t \right).$$

Then

$$\pi_W^*: {}_R(W, V) \to {}_R(W^G, V) \quad (g \mapsto g \circ \pi_W)$$

is an RH-monomorphism, and by identifying g with $g \circ \pi_W$, we obtain that

(1.4) $$_R(W, V) \subset {}_R(W^G, V).$$

Namely, each $g \in {}_R(W, V)$ is regarded as an element of $_R(W^G, V)$ such that

(1.5) $$g\left(\sum_{t \in H\backslash G} w_t \otimes t \right) = g(w_1).$$

Thus, if Tr_H^G: $_{RH}(W^G, V) \to {}_{RG}(W^G, V)$ is the trace map, then, for $g \in {}_{RH}(W, V)$ and $u = \sum_{t \in H\backslash G} w_t \otimes t \in W^G$, we have

(1.6) $$\mathrm{Tr}_H^G(g)(u) = \sum_{t \in H\backslash G} g(ut^{-1})t = \sum_{t \in H\backslash G} g(w_t)t.$$

Exercise 1.8. Check that, with the above identifications, both of the isomorphisms given in Chapter 3, 1.19 coincide with Tr_H^G.

Moreover, we get

Theorem 1.9. *For V_{RG} and W_{RH}, let $_R(V_H, W)^G = {}_R(V_H, W) \otimes_{RH} RG$ and $_R(W, V_H)^G = (W, V_H) \otimes_{RH} RG$. Then we have the following RG-isomorphisms:*

(i) $$\varphi: {}_R(V_H, W)^G \cong (V, W^G)$$

such that for $f = \sum_{t \in H\backslash G} f_t \otimes t \quad (f_t \in {}_R(V, W))$,

(1.7) $$\varphi(f)(v) = \sum_t f_t(vt^{-1}) \otimes t \qquad (v \in V),$$

and

(ii) $$\psi\colon {}_R(W, V_H)^G \simeq {}_R(W^G, V)$$

such that for $g = \sum_{t \in H\backslash G} g_t \otimes t \quad (g_t \in {}_R(W, V))$,

(1.8) $$\psi(g)\left(\sum_t w_t \otimes t\right) = \sum_t g_t(w_t)t \qquad (w \in W).$$

Proof.

(i) It is clear by (1.7) that φ is an R-monomorphism. Let $x \in G$ and $tx = h_t t'$ $(h_t \in H,\ t' \in H\backslash G)$. We have $fx = \sum_t f_t^{h_t} \otimes t'$ and $\varphi(f)^x(v) = \varphi(f)(vx^{-1})x$. Then

$$\varphi(fx)(v) = \sum_t f_t^{h_t}(vt'^{-1}) \otimes t' = \sum_t f_t(vt'^{-1}h_t^{-1})h_t \otimes t'$$
$$= \sum_t f_t(vx^{-1}t^{-1}) \otimes tx = \varphi(f)^x(v).$$

Thus φ is an RG-homomorphism. To see that φ is an epimorphism, write $f'(v) = \sum_t f_t'(v) \otimes t$ for $f' \in {}_R(V, W^G)$ and $v \in V$. We define $f_t \in {}_R(V, W)$ by $f_t(v) = f'(vt)$ and set $f = \sum_t f_t \otimes t$. Then we have $\varphi(f) = f'$. ∎

Exercise 1.10. Prove (ii) of the above theorem.

2. *H*-Projective Modules

Throughout the remainder of this chapter, unless otherwise stated explicitly, we let

(K, R, F) *denote a p-modular system,* $\mathfrak{o}$ *denote* R *or* F, *and* (π) *be the unique maximal ideal of* R.

Furthermore, we assume that

all $\mathfrak{o}$*-modules are finitely generated.*

Hence the Krull–Schmidt–Azumaya theorem holds for $\mathfrak{o}G$-modules.

Let $G \geq H$. An $\mathfrak{o}G$-module V is said to be *H-projective* if there is an $\mathfrak{o}H$-module W such that $V \mid W^G$.

The following lemma is immediate from the definition.

Lemma 2.1. *Suppose that V is H-projective.*

(i) *If $H \le T \le G$, then V is T-projective.*

(ii) *V is H^x-projective for all $x \in G$.*

(iii) *If U is an $\mathfrak{o}G$-module, then $U \otimes_{\mathfrak{o}} V$, ${}_{\mathfrak{o}}(U, V)$ and ${}_{\mathfrak{o}}(V, U)$ are all H-projective.*

Proof. There is $W_{\mathfrak{o}H}$ such that $V \mid W^G$.

(i) This is clear because $V \mid W^G = (W^T)^G$.

(ii) This is clear because W^x is an $\mathfrak{o}H^x$-module and $(W^x)^G \simeq W^G$.

(iii) Since $U \otimes_{\mathfrak{o}} V \mid U \otimes_{\mathfrak{o}} W^G \simeq (U_H \otimes_{\mathfrak{o}} W)^G$ (Chap. 3, 1.15(ii)), it follows that $U \otimes_{\mathfrak{o}} V$ is H-projective. The remaining assertions follow similarly from Theorem 1.9. ∎

Theorem 2.2. *Let $G \ge H$. The following conditions on $V_{\mathfrak{o}G}$ are equivalent.*

(1) *V is H-projective.*

(2) *There exists $\theta \in E_{\mathfrak{o}H}(V)$ such that $\mathrm{Tr}_H^G(\theta) = \mathrm{id}_V$.*

(3) *$E_{\mathfrak{o}G}(V) = \mathrm{Tr}_H^G(E_{\mathfrak{o}}(V))$.*

(4) *For the following diagram of $\mathfrak{o}G$-modules and $\mathfrak{o}G$-homomorphisms f, μ with an exact row*

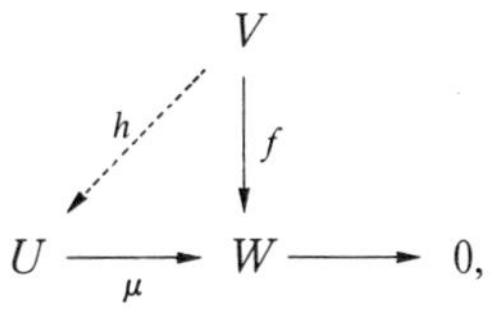

there exists an $\mathfrak{o}G$-homomorphism $h\colon V \to U$ such that $f = \mu \circ h$, provided there exists an $\mathfrak{o}H$-homomorphism $g\colon V \to U$ such that $f = \mu \circ g$.

(5) *Every epimorphism $\mu\colon U \to V$ of $\mathfrak{o}G$-modules splits if it splits as an $\mathfrak{o}H$-homomorphism.*

(6) *The $\mathfrak{o}G$-epimorphism $\lambda\colon (V_H)^G \to V$ ($\sum_t v_t \otimes t \mapsto \sum_t v_t t$) splits.*

(7) *$V \mid (V_H)^G$.*

Proof.

(1) ⇒ (2). We may asume that $W^G = V \oplus U$ for some $W_{\mathfrak{o}H}$ and $U_{\mathfrak{o}G}$. Let $\varepsilon \in E_{\mathfrak{o}G}(W^G)$ be the projection on V and $\tau \in E_{\mathfrak{o}H}(W^G)$ be the projection on $W \otimes 1$ w.r.t. the $\mathfrak{o}H$-decomposition

$$W = W \otimes 1 \oplus \left(\bigoplus_{\substack{t \in H\backslash G \\ t \neq 1}} W \otimes t \right).$$

Then we see that $\mathrm{Tr}_H^G(\tau) = \mathrm{id}_{W^G}$ and $\varepsilon\, \mathrm{Tr}_H^G(\tau)\varepsilon = \mathrm{Tr}_H^G(\varepsilon\tau\varepsilon) = \varepsilon\, \mathrm{id}_{W^G}\, \varepsilon$. Therefore we get $\mathrm{Tr}_H^G(\theta) = \mathrm{id}_V$, where $\theta = (\varepsilon\tau\varepsilon)|_V \in E_{\mathfrak{o}H}(V)$.

(2) ⇔ (3). This is clear, since $\mathrm{Tr}_H^G(E_{\mathfrak{o}}(V))$ is an ideal of $E_{\mathfrak{o}G}(V)$.

(2) ⇒ (4). Let θ be as in (2). Then $\mu \circ \mathrm{Tr}_H^G(g \circ \theta) = \mathrm{Tr}_H^G(\mu \circ g \circ \theta) = \mathrm{Tr}_H^G(f \circ \theta) = f \circ \mathrm{Tr}_H^G(\theta) = f$. Thus, it suffices to set $h = \mathrm{Tr}_H^G(g \circ \theta)$

(4) ⇒ (5). In (4) let $W = V$ and $f = \mathrm{id}_V$.

(5) ⇒ (6). Observe that $\nu: V \to (V_H)^G (v \mapsto v \otimes 1)$ is an $\mathfrak{o}H$-homomorphism such that $\lambda \circ \nu = \mathrm{id}_V$. Hence, from the assumption, λ splits as an $\mathfrak{o}G$-homomorphism.

(6) ⇒ (7). This is trivial.

(7) ⇒ (1). This is clear from the definition of the H-projectivity. ■

Remark. Condition (2) in Theorem 2.2 is extremely useful; this is sometimes referred to as *Higman's criterion.*

Exercise 2.3. Let V be H-projective. Then every $\mathfrak{o}G$-monomorphism $\nu: V \to U$ splits if it splits as an $\mathfrak{o}H$-homomorphism. [Hint: Let $\mu: U \to V$ be an $\mathfrak{o}H$-homomorphism such that $\mu \circ \nu = \mathrm{id}_V$ and choose $\theta \in E_{\mathfrak{o}H}(V)$ so that $\mathrm{Tr}_H^G(\theta) = \mathrm{id}_V$. Then $\mathrm{Tr}_H^G(\theta \circ \mu) \circ \nu = \mathrm{id}_V$.]

Exercise 2.4. Assume that V is $\mathfrak{o}$-free. Then V is $\mathfrak{o}G$-projective if and only if it is $\{1\}$-projective. [Hint: $(V_{\{1\}})^G$ is $\mathfrak{o}G$-free.]

Theorem 2.5. *If $|G:H|$ is prime to p, then every $\mathfrak{o}G$-module V is H-projective. In particular, V is P-projective if $P \in Syl_p(G)$.*

Proof. Since $|G:H|$ is a unit of $\mathfrak{o}$, we have $\mathrm{id}_V = \mathrm{Tr}_H^G(|G:H|^{-1}\, \mathrm{id}_V)$. ■

Remark. If $(|G|, p) = 1$, then every $\mathfrak{o}$-free $\mathfrak{o}G$-module is $\mathfrak{o}G$-projective by the above theorem. If $\mathfrak{o}$ is F, this is just Maschke's theorem (cf. Chapter 3, 1.22).

Exercise 2.6. Let V be H-projective. Then
(i) V is Q-projective if $Q \in \mathrm{Syl}_p(H)$.
(ii) Every direct summand of V is H-projective.

Now let us give a brief generalization of H-projectivity. Let $\mathfrak{X}$ be a set of subgroups of G, and let V be an $\mathfrak{o}G$-module.

V is called *$\mathfrak{X}$-projective* provided there is a decomposition $V = \bigoplus_{i=1}^{n} V_i$ such that each V_i is X_i-projective for some $X_i \in \mathfrak{X}$.

Exercise 2.7. Let V be $\mathfrak{X}$-projective. Then the following statements hold.
(i) Every indecomposable component of V is X-projective for some $X \in \mathfrak{X}$.
(ii) Every direct summand of V is $\mathfrak{X}$-projective.

Let $f: U \to V$ be an $\mathfrak{o}G$-homomorphism. We say *f factors through W* if there exist an $\mathfrak{o}G$-module W and $\mathfrak{o}G$-homomorphisms $\lambda: U \to W$, $\mu: W \to V$ such that $f = \mu \circ \lambda$.

Also, f is said to be $\mathfrak{X}$-projective, provided it factors through some $\mathfrak{X}$-projective module.

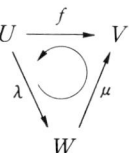

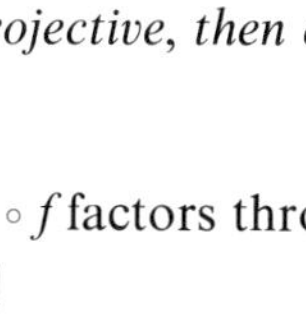

Lemma 2.8. *If U or V is $\mathfrak{X}$-projective, then every $f: U \to V$ is $\mathfrak{X}$-projective.*

Proof. Since $f = f \circ \mathrm{id}_U = \mathrm{id}_V \circ f$ factors through an $\mathfrak{X}$-projective module U or V we have the result. ■

For $\mathfrak{o}G$-module V we define

$$\mathrm{Tr}_{\mathfrak{X}}^{G}(V) = \sum_{X \in \mathfrak{X}} \mathrm{Tr}_{X}^{G}(V).$$

Theorem 2.9. *Let $f: U \to V$ be an $\mathfrak{o}G$-homomorphism. Then the following two conditions are equivalent.*

(1) *f is $\mathfrak{X}$-projective.*
(2) *$f \in \mathrm{Tr}_{\mathfrak{X}}^{G}({}_{\mathfrak{o}}(U, V))$.*

Proof.
(1) $\Rightarrow$ (2). Suppose that there are an $\mathfrak{X}$-projective module W and a sequence $U \xrightarrow{\lambda} W \xrightarrow{\mu} V$ of $\mathfrak{o}G$-homomorphisms such that $f = \mu \circ \lambda$. Thus, we have $W = \bigoplus_{i=1}^{n} W_i$, where each W_i is X_i-projective for some $X_i \in \mathfrak{X}$. Hence there exists $\theta_i \in E_{\mathfrak{o}X_i}(W_i)$ such that $\mathrm{id}_{W_i} = \mathrm{Tr}_{X_i}^{G}(\theta_i)$. Let $\pi_i: W \to W_i$ and $\iota_i: W_i \to W$ be the projection and the inclusion map w.r.t. the above decomposition respectively. Then

$$\mathrm{id}_W = \sum_i \iota_i \circ \pi_i = \sum_i \iota_i \circ \mathrm{Tr}_{X_i}^{G}(\theta_i) \circ \pi_i = \sum_i \mathrm{Tr}_{X_i}^{G}(\iota_i \circ \theta_i \circ \pi_i),$$

and thus,

$$f = \mu \circ \left(\sum_i \mathrm{Tr}_{X_i}^{G}(\iota_i \circ \theta_i \circ \pi_i) \right) \circ \lambda = \sum_i \mathrm{Tr}_{X_i}^{G}(\mu \circ \iota_i \circ \theta_i \circ \pi_i \circ \lambda).$$

Therefore, f belongs to $\mathrm{Tr}_{\mathfrak{X}}^{G}({}_{\mathfrak{o}}(U, V))$.

(2) $\Rightarrow$ (1). Let $f = \sum_{X \in \mathfrak{X}} \mathrm{Tr}_X^G(g_X)$, $g_X \in {}_{\mathfrak{o}X}(U, V)$ and $W = \bigoplus_{X \in \mathfrak{X}} (V_X)^G$. Since W is $\mathfrak{X}$-projective, it suffices to show that f factors through W. To do it, we set, for each $X \in \mathfrak{X}$,

$$\lambda_X: U \to (V_X)^G \quad \left(u \mapsto \sum_{t \in X \backslash G} g_X(ut^{-1}) \otimes t \right),$$

and

$$\mu_X: (V_X)^G \to V \quad \left(\sum_{t \in X \backslash G} v_t \otimes t \mapsto \sum_i v_t t \right).$$

Both λ_X and μ_X are $\mathfrak{o}G$-homomorphisms and $\mathrm{Tr}_X^G(g_X) = \mu_X \circ \lambda_X$. Define

$$\lambda = \sum_X \lambda_X: U \to W \left(u \mapsto \sum_X \lambda_X(u) \right),$$

and

$$\mu = \bigoplus_X \mu_X: W \to V \left(\sum_X w_X \mapsto \sum_X \mu_X(w_X), \quad w_X \in (V_X)^G \right).$$

Then we get $\mu \circ \lambda = \sum_X \mu_X \circ \lambda_X = \sum_X \mathrm{Tr}_X^G(g_X) = f$. ∎

Theorem 2.10. *The following two conditions on $V_{\mathfrak{o}G}$ are equivalent.*

(1) *V is $\mathfrak{X}$-projective.*

(2) $E_{\mathfrak{o}G}(V) = \mathrm{Tr}_{\mathfrak{X}}^{G}(E_{\mathfrak{o}}(V))$.

Proof.

(1) $\Rightarrow$ (2). This follows from Lemma 2.8 and Theorem 2.9.

(2) $\Rightarrow$ (1). By Theorem 2.9, id_V factors through an $\mathfrak{X}$-projective module W. Therefore, $V \mid W$ and it follows that V is $\mathfrak{X}$-projective. ∎

3. Vertices and Sources

Throughout this section, we identify all modules with their isomorphic ones and denote by $\mathrm{IND}(\mathfrak{o}G)$ the set of all indecomposable $\mathfrak{o}G$-modules.

Lemma 3.1. *Let $G \geq H$.*

(i) *If $V \in \mathrm{IND}(\mathfrak{o}G)$ and $V \mid M^G$ for some $\mathfrak{o}H$-module M, then $V \mid W^G$ for some indecomposable component W of $M_{\mathfrak{o}H}$.*

(ii) *If $W \in \mathrm{IND}(\mathfrak{o}H)$ and $W \mid L_H$ for some $\mathfrak{o}G$-module L, then $W \mid V_H$ for some indecomposable component V of $L_{\mathfrak{o}G}$.*

Proof.

(i) If $M = \bigoplus_i W_i$ is an indecomposable decomposition, then $V \mid W_i^G$ for some i, since V is an indecomposable component of $M^G = \bigoplus_i W_i^G$.

(ii) If $L = \bigoplus_i V_i$ is an indecomposable decomposition, then we have $W \mid (V_i)_H$ for some i, since W is an indecomposable component of $L_H = \bigoplus_i (V_i)_H$. ∎

Lemma 3.2. *Let $G \geq H$, P and let S be an $\mathfrak{o}P$-module.*

(i) *If W is an indecomposable component of $(S^G)_H$, then there is $t \in G$ such that $W \mid ((S^t)_{P^t \cap H})^H$. In particular, W is $P^t \cap H$-projective.*

(ii) *Let V be an indecomposable component of S^G, which is H-projective. Then $V \mid ((S^t)_{P^t \cap H})^G$ for some $t \in G$. In particular, V is $P^t \cap H$-projective.*

Proof.
(i) This is clear, since $(S^G)_H = \bigoplus_{t \in P\backslash G/H} ((S^t)_{P^t \cap H})^H$.
(ii) Observe that $V | (V_H)^G | ((S^G)_H)^G = \bigoplus_t ((S^t)_{P^t \cap H})^G$. ■

The above lemma yields the following theorem.

Theorem 3.3. *Given $V \in \mathrm{IND}(\mathfrak{o}G)$, there exists a p-subgroup P of G uniquely determined up to G-conjugacy such that the following statements are true.*

(a) *V is P-projective.*
(b) *If V is H-projective, then $P \leq_G H$.*

Proof. Choose $P \leq G$ with minimal order, subject to V being P-projective. Thus, there exists $S_{\mathfrak{o}P}$ such that $V | S^G$. If V is H-projective, V is $P^t \cap H$-projective for some $t \in G$ by the above lemma, and the minimality of $|P|$ implies that $P^t = P^t \cap H \leq H$, hence $P \leq_G H$.

This applies in particular to $H = Q \in \mathrm{Syl}_p(G)$, yielding that $P \leq_G Q$. Therefore P is a p-group. Finally, if $P' \leq G$ satisfies both (a) and (b), then it follows that $P \leq_G P'$, $P' \leq_G P$, and thus $P =_G P'$. ■

Any p-subgroup P of G satisfying the two conditions in the above theorem is called a *vertex* of V and is denoted by $P = \mathrm{vx}(V)$. Clearly, we have the following:

$$V \in \mathrm{IND}(\mathfrak{o}G) \text{ is } H\text{-projective} \Leftrightarrow \mathrm{vx}(V) \leq_G H.$$

Lemma 3.4. *Let $H \leq G$ and $V \in \mathrm{IND}(\mathfrak{o}G)$, $W \in \mathrm{IND}(\mathfrak{o}H)$.*
(i) *If $V | W^G$, then $\mathrm{vx}(V) \leq_G \mathrm{vx}(W)$.*
(ii) *If $W | V_H$, then $\mathrm{vx}(W) \leq_G \mathrm{vx}(V)$.*

Proof. Set $P = \mathrm{vx}(V)$, $Q = \mathrm{vx}(W)$ and let $S_{\mathfrak{o}P}$, $T_{\mathfrak{o}Q}$ be such that $V | S^G$, $W | T^H$.
(i) Since $V | W^G | (T^H)^G = T^G$, V is Q-projective. Hence $P \leq_G Q$.
(ii) Since $W | V_H | (S^G)_H$, W is $P^t \cap H$-projective for some $t \in G$ by Lemma 3.2(i), and therefore $Q \leq_H P^t \cap H \leq_G P$. ■

Lemma 3.5. *Let $H \leq G$. Then the following statements hold.*

(i) *For $W \in \mathrm{IND}(\mathfrak{o}H)$, there exists $V \in \mathrm{IND}(\mathfrak{o}G)$, which satisfies the following three conditions*:

$$\text{(a)}\quad V \mid W^G, \qquad \text{(b)}\quad W \mid V_H, \qquad \text{(c)}\quad \mathrm{vx}(W) =_G \mathrm{vx}(V).$$

(ii) *If $V \in \mathrm{IND}(\mathfrak{o}G)$ is H-projective, then there exists $W \in \mathrm{IND}(\mathfrak{o}H)$, which satisfies the three conditions above.*

Proof. Note that (c) readily follows once (a) and (b) are shown.

(i) Since $W \mid (W^G)_H$, there exists $V \in \mathrm{IND}(\mathfrak{o}G)$ satisfying (a) and (b) by Lemma 3.1.

(ii) Since $V \mid (V_H)^G$, there exists $W \in \mathrm{IND}(\mathfrak{o}H)$ satisfying (a) and (b) by Lemma 3.1. ■

Theorem 3.6. *Let $V \in \mathrm{IND}(\mathfrak{o}G)$, and set $\mathrm{vx}(V) = P$.*

(i) *There exists $S \in \mathrm{IND}(\mathfrak{o}P)$ that satisfies the following three conditions*:

$$\text{(a)}\quad V \mid S^G, \qquad \text{(b)}\quad S \mid V_P, \qquad \text{(c)}\quad \mathrm{vx}(S) = P.$$

(ii) *Let $T \in \mathrm{IND}(\mathfrak{o}P)$ be such that $V \mid T^G$. Then there exists $t \in N_G(P)$ such that $T^t = S$. Also, we have $T \mid V_P$ and $\mathrm{vx}(T) = P$.*

Proof.

(i) This is clear from Lemma 3.5(ii).

(ii) We have $S \mid V_P \mid (T^G)_P$ and thus $S \mid ((T^t)_{P^t \cap P})^P$ for some $t \in G$ by Lemma 3.2(i). Since $\mathrm{vx}(S) = P$, this implies that $P^t = P$, namely $t \in N_G(P)$. Also, it follows that $S \mid T^t$, and thus $S = T^t$, since $(T^t)_{\mathfrak{o}P}$ is indecomposable. Therefore $T = S^{t^{-1}} \mid (V^{t^{-1}})_P = V_P$. It is clear that $\mathrm{vx}(T) = P$. ■

Let V be an indecomposable $\mathfrak{o}G$-module with vertex P. An indecomposable $\mathfrak{o}P$-module S is called a *source* of V if it satisfies the three conditions in (i) of Theorem 3.6. Once P is fixed, S is uniquely determined up to $N_G(P)$-conjugacy by (ii) of the above theorem. Since the vertices of V are mutually G-conjugate, a source of V is determined uniquely up to G-conjugacy. A source of V is denoted by $\mathrm{s}(V)$.

Exercise 3.7. Let $G \geq H$. Let $V \in \mathrm{IND}(\mathfrak{o}G)$ with $\mathrm{vx}(V) = P$, $\mathrm{s}(V) = S$, and let $W \in \mathrm{IND}(\mathfrak{o}H)$ with $\mathrm{vx}(W) = P'$, $\mathrm{s}(W) = S' \in \mathrm{IND}(\mathfrak{o}P')$.

(i) If $W \mid V_H$, then there exists $a \in G$ such that $P' \leq P^a$ and $S' \mid (S^a)_{P'}$.

(ii) If $V \mid W^G$, then there exists $b \in G$ such that $P \leq P'^b$ and $S \mid (S'^b)_P$. [Hint: (i) $S' \mid (S^G)_{P'}$. (ii) $S \mid (S'^G)_P$.]

The following lemma will be needed in the next section.

Lemma 3.8. *Let V be an indecomposable $\mathfrak{o}G$-module with vertex P and $P \leq H \leq G$. Then, for any two of the following three conditions, there exists $W \in \mathrm{IND}(\mathfrak{o}H)$, which satisfies them.*

$$\text{(a)} \quad V \mid W^G, \qquad \text{(b)} \quad W \mid V_H, \qquad \text{(c)} \quad \mathrm{vx}(W) =_H P.$$

Proof. As was shown in Lemma 3.5(ii), there exists W, which satisfies both (a) and (b). Let $\mathrm{s}(V) = S$. Then $V \mid S^G$, $S \mid V_P$.

There exists W satisfying (a) *and* (c): Since $V \mid S^G = (S^H)^G$, there exists $W \in \mathrm{IND}(\mathfrak{o}H)$ such that $V \mid W^G$, $W \mid S^H$. Thus

$$\mathrm{vx}(V) = P \leq_G \mathrm{vx}(W) \leq_H \mathrm{vx}(S) = P,$$

whence we have $\mathrm{vx}(W) =_H P$ by comparing the order of both sides.

There exists W satisfying (b) *and* (c): Since $S \mid V_P = (V_H)_P$, there exists $W \in \mathrm{IND}(\mathfrak{o}H)$ such that $S \mid W_P$, $W \mid V_H$. Thus

$$\mathrm{vx}(S) = P \leq_H \mathrm{vx}(W) \leq_G \mathrm{vx}(V) = P,$$

whence we have $P =_H \mathrm{vx}(W)$. ■

Remark. It should be emphasized that the condition (c) in the above lemma asserts that $\mathrm{vx}(W)$ and P *are conjugate in* H (cf. Lemma 3.5 (ii)). Actually, as will be shown in the next section, by making use of the Green correspondence, there is W, which satisfies all of the three conditions in the above lemma (see Corollary 4.9 in the next section).

Exercise 3.9. With the assumption of the above lemma, consider the set $\{P^x;\ P^x \leq H,\ x \in G\}$ and let Ω be a set of representatives of H-conjugacy classes of this set. Then V_H has at least $|\Omega|$ nonisomorphic indecomposable components. [Hint: Apply Lemma 3.8 with $P^x = \mathrm{vx}(V)$.]

4. The Green Correspondence

Let $\mathfrak{X}$ be a set of subgroups of G. For $G \geq H$, we write $H \in_G \mathfrak{X}$ if $H^x \in \mathfrak{X}$ for some $x \in G$.

Also, for an $\mathfrak{o}G$-module V, we write

$$V \equiv U \oplus O(\mathfrak{X})$$

if $V = U \oplus L$ is a direct sum of $\mathfrak{o}G$-modules U, L, with L being $\mathfrak{X}$-projective.

In particular, $V \equiv O(\mathfrak{X})$ means that V is $\mathfrak{X}$-projective. Furthermore, we let

$$\begin{aligned} \mathrm{IND}(\mathfrak{o}G|\mathfrak{X}) &= \{V \in \mathrm{IND}(\mathfrak{o}G); \quad \mathrm{vx}(V) \in_G \mathfrak{X}\}, \\ \mathrm{IND}(\mathfrak{o}G|Q) &= \{V \in \mathrm{IND}(\mathfrak{o}G); \quad \mathrm{vx}(V) =_G Q\}, \end{aligned}$$

where Q is a p-subgroup of G.

Let $G \geq H \geq P$, and define the following sets of subgroups of H:

$$(4.1) \qquad \begin{cases} \mathfrak{X} = \mathfrak{X}(G, P, H) = \{Q; Q \leq P^x \cap P \quad \text{for some} \quad x \in G - H\}, \\ \mathfrak{Y} = \mathfrak{Y}(G, P, H) = \{Q; Q \leq P^x \cap H \quad \text{for some} \quad x \in G - H\}, \\ \mathfrak{A} = \mathfrak{A}(G, P, H) = \{Q; Q \leq P, \quad Q \notin_G \mathfrak{X}\}. \end{cases}$$

It is clear that $\mathfrak{X} \subset \mathfrak{Y}$.

In this section, we mainly consider those sets under the following condition:

(4.2) *P is a p-subgroup and $N_G(P) \leq H$.*

If this condition is satisfied, then $\mathfrak{X}$ consists of proper subgroups of P and we have $P \in \mathfrak{A}$.

Lemma 4.1. *Under the assumption* (4.2), *the following statements holds.*

(i) *If $Q, Q' \in \mathfrak{A}$, then*

$$Q =_G Q' \Leftrightarrow Q =_H Q'.$$

(ii) *If $Q \leq P$, then the following four conditions are equivalent:*

$$(1) \quad Q \in_G \mathfrak{X}, \qquad (2) \quad Q \in \mathfrak{X}, \qquad (3) \quad Q \in \mathfrak{Y}, \qquad (4) \quad Q \in_H \mathfrak{Y}.$$

In particular, if $Q \in \mathfrak{A}$, then $Q \notin_H \mathfrak{Y}$.

Proof.

(i) ($\Leftarrow$) is trivial. ($\Rightarrow$): Suppose that $Q' = Q^x$ for some $x \in G$. If $x \notin H$, then $Q' \leq P \cap P^x$, and hence $Q' \in \mathfrak{X}$, which is a contradiction. Thus $x \in H$.

(ii) (1) $\Rightarrow$ (2). If $Q \in_G \mathfrak{X}$, then $Q^t \leq P^x \cap P$ for some $t \in G$ and $x \in G - H$. Therefore $Q \leq P^{xt^{-1}} \cap P^{t^{-1}} \cap P$. Since either $t^{-1} \notin H$ or $xt^{-1} \notin H$, it follows that $Q \in \mathfrak{X}$.

(2) $\Rightarrow$ (3) and (3) $\Rightarrow$ (4) are trivial.

(4) $\Rightarrow$ (1). If $Q \in_H \mathfrak{Y}$, then $Q^h \in P^x \cap H$ for some $h \in H$ and $x \in G - H$. Thus, $Q \leq P^{xh^{-1}} \cap H^{h^{-1}} \cap P$. This implies that $Q \in \mathfrak{X}$, since $xh^{-1} \notin H$. ■

Before proving the theorem on the Green correspondence, we show the following lemma.

Lemma 4.2.

(i) *Let $G \geq H \geq P$ and W be a P-projective $\mathfrak{o}H$-module. Then*

$$(W^G)_H \equiv W \oplus O(\mathfrak{Y}).$$

(ii) *With the assumption* (4.2), *the following statements hold.*

(a) *If $W \in \mathrm{IND}(\mathfrak{o}H \mid \mathfrak{A})$, then W is a unique indecomposable component of $(W^G)_H$ with vertex in $\mathfrak{A}$.*

(b) *If an $\mathfrak{o}G$-module V is P-projective, then*

$$V \equiv O(\mathfrak{X}) \Leftrightarrow V_H \equiv O(\mathfrak{Y}).$$

Proof.

(i) There exists an $\mathfrak{o}P$-module M such that $M^H = W \oplus W_1$. If we let $(W^G)_H = W \oplus W'$, $(W_1^G)_H = W_1 \oplus W'_1$, then

$$(M^G)_H = (W^G)_H \oplus (W_1^G)_H = M^H \oplus W' \oplus W'_1. \tag{4.3}$$

On the other hand, it follows from the Mackey decomposition theorem that

$$(M^G)_H = M^H \oplus \left(\bigoplus_{\substack{t \in P\backslash G/H \\ t \notin H}} (M^t_{P^t \cap H})^H \right). \tag{4.4}$$

By comparing (4.3) and (4.4), we see that if $W' = \bigoplus_i X_i$ is an indecomposable decomposition, there exists $t \notin H$ for each X_i such that $X_i \mid (M^t_{P^t \cap H})^H$. Therefore $\mathrm{vx}(X_i) \in_H \mathfrak{Y}$ and $W' \equiv O(\mathfrak{Y})$.

(ii) (a) We have $(W^G)_H = W \oplus (\bigoplus_i X_i)$ with $\mathrm{vx}(X_i) \in_H \mathfrak{Y}$, as is shown above. Suppose that $\mathrm{vx}(X_i) \in_H \mathfrak{A}$ for some i. We may assume that $\mathrm{vx}(X_i) \in \mathfrak{A}$, hence $\mathrm{vx}(X_i) \leq P$. Then since $\mathrm{vx}(X_i) \notin \mathfrak{X}$, it holds that $\mathrm{vx}(X_i) \notin_H \mathfrak{Y}$ by Lemma 4.1 (ii), which is a contradiction.

(b) We may assume that V is indecomposable and that $\mathrm{vx}(V) = Q \leq P$.

($\Rightarrow$). We have $Q \in_G \mathfrak{X}$ by assumption and hence, by Lemma 3.8, there is $W \in \mathrm{IND}(\mathfrak{o}H)$ such that $V | W^G$, $\mathrm{vx}(W) =_H Q$. Thus $V_H | (W^G)_H \equiv W \oplus O(\mathfrak{Y})$, and it follows that $V_H \equiv O(\mathfrak{Y})$, since $\mathrm{vx}(W) \in_H \mathfrak{Y}$.

($\Leftarrow$). There exists $W \in \mathrm{IND}(\mathfrak{o}H)$ such that $W | V_H$, $\mathrm{vx}(W) =_H Q$ by Lemma 3.8. Since $Q \in_H \mathfrak{Y}$ by assumption, we get $Q \in_G \mathfrak{X}$, i.e., $V \equiv O(\mathfrak{X})$. ■

We now prove the main theorem of this section.

Theorem 4.3 **(Green).** *Suppose that P is a p-subgroup of G, and let $G \geq H \geq N_G(P)$.*

(i) *There is a bijection $f\colon \mathrm{IND}(\mathfrak{o}G|\mathfrak{A}) \to \mathrm{IND}(\mathfrak{o}H|\mathfrak{A})$ satisfying the following.*

(a) *If $V \in \mathrm{IND}(\mathfrak{o}G|\mathfrak{A})$, then $V_H \equiv fV \oplus O(\mathfrak{Y})$.*
(b) *If $W \in \mathrm{IND}(\mathfrak{o}H|\mathfrak{A})$, then $W^G \equiv f^{-1}W \oplus O(\mathfrak{X})$.*
(c) *If $V \in \mathrm{IND}(\mathfrak{o}G|\mathfrak{A})$ and $\mathrm{vx}(V) \in \mathfrak{A}$, then $\mathrm{vx}(fV) =_H \mathrm{vx}(V)$.*

(ii) *For $V \in \mathrm{IND}(\mathfrak{o}G|\mathfrak{A})$ and $W \in \mathrm{IND}(\mathfrak{o}H|\mathfrak{A})$, the following holds.*

$$W | V_H \Leftrightarrow W = fV \Leftrightarrow V | W^G.$$

Proof.
(i) Let $V \in \mathrm{IND}(\mathfrak{o}G|\mathfrak{A})$ with $\mathrm{vx}(V) = Q \in \mathfrak{A}$. Then there exists $W \in \mathrm{IND}(\mathfrak{o}H)$ such that $V | W^G$, $\mathrm{vx}(W) =_H Q$. Hence $W \in \mathrm{IND}(\mathfrak{o}H|\mathfrak{A})$. Since $V_H | (W^G)_H \equiv W \oplus O(\mathfrak{Y})$, we see that $V_H \equiv W \oplus O(\mathfrak{Y})$. In fact, if W is not a direct summand of V_H, then $V_H \equiv O(\mathfrak{Y})$ and hence $V \equiv O(\mathfrak{X})$, contradicting the assumption. Define $f\colon \mathrm{IND}(\mathfrak{o}G|\mathfrak{A}) \to \mathrm{IND}(\mathfrak{o}H|\mathfrak{A})$ by $fV = W$. Then (a) and (c) are clear. To see (b), let $W^G = V \oplus U$. Then $(W^G)_H = V_H \oplus U_H \equiv W \oplus U_H \oplus O(\mathfrak{Y})$. On the other hand, since W is P-projective, we have $(W^G)_H \equiv W \oplus O(\mathfrak{Y})$ by Lemma 4.2. This, together with the above, yields that $U_H \equiv O(\mathfrak{Y})$ and hence $U \equiv O(\mathfrak{X})$ because U is P-projective. Therefore, f is injective and (b) holds as well.

We next show that f is surjective. Let $W \in \mathrm{IND}(\mathfrak{o}H|\mathfrak{A})$ be arbitrary. There exists $V \in \mathrm{IND}(\mathfrak{o}G)$ such that $V|W^G$, $W|V_H$ and $\mathrm{vx}(V) =_G \mathrm{vx}(W)$. Then $V \in \mathrm{IND}(\mathfrak{o}G|\mathfrak{A})$, and hence $V_H \equiv fV \oplus O(\mathfrak{Y})$. Therefore $fV = W$.

(ii) The first equivalence is immediate from part (a) of (i), and the second one follows from part (b) of (i). ■

The bijection defined above is called the *Green correspondence* w.r.t. (G, P, H) and denoted by $f(G, P, H)$ or simply by f (under the assumption of (4.2) of course).

From the property (c) of the Green correspondence, we obtain the following theorem.

Theorem 4.4. *$f(G, P, H)$ gives rise to a bijection between $\mathrm{IND}(\mathfrak{o}G|P)$ and $\mathrm{IND}(\mathfrak{o}H|P)$.*

Exercise 4.5. Let $f = f(G, P, H)$ and $V \in \mathrm{IND}(\mathfrak{o}G|P)$. Prove the following.

(i) $\mathrm{s}(V)$ is also a source of fV. [Hint: Exercise 3.7].

(ii) If V is $\mathfrak{o}$-free and $V^\wedge$ is the contragredient module of V, then

$$\text{(a)}\quad \mathrm{vx}(V) =_G \mathrm{vx}(V^\wedge), \qquad \text{(b)}\ (fV)^\wedge \simeq fV^\wedge.$$

[Hint: Use Chapter 3, 1.15(i) and the fact that $(fV)^\wedge | V_H^\wedge$.]

Now let us show the following remarkable result concerning the Green correspondence.

Theorem 4.6 (Burry and Carlson [1]). *Let f be the Green correspondence w.r.t. (G, P, H) satisfying (4.2). Let $V \in \mathrm{IND}(\mathfrak{o}G)$, $W \in \mathrm{IND}(\mathfrak{o}H)$. Then the following statements hold.*

(i) *If $W|V_H$ and $\mathrm{vx}(W) =_H P$, then $\mathrm{vx}(V) =_G P$ and $fV = W$.*

(ii) *$\mathrm{vx}(V) =_G P \Leftrightarrow V_H$ has an indecomposable component with vertex P.*

Proof.

(i) It follows from the assumption that $P \leq_G \mathrm{vx}(V)$. Hence it suffices to show that V is P-projective, i.e., $\mathrm{End}_{\mathfrak{o}G}(V) = \mathrm{Tr}_P^G(\mathrm{End}_{\mathfrak{o}}(V))$. By assumption, we have $V_H = W \oplus W'$. Let $\pi'_W\colon V \to W$ and $\iota_W\colon W \to V$ be the projection on W and the inclusion map, respectively, and let $\pi_W = \iota_W \circ \pi'_W \in \mathrm{End}_{\mathfrak{o}H}(V)$. Since W is P-projective, $\pi'_W = \mathrm{Tr}_P^H(\sigma')$ for some $\sigma' \in {}_{\mathfrak{o}P}(V, W)$ and hence $\pi_W = \mathrm{Tr}_P^H(\sigma)$, where $\sigma = \iota_W \circ \sigma'$. Let $\rho = \mathrm{Tr}_P^G(\sigma)$. Then by Theorem 1.4 (i), we have

$$\rho = \sum_{t \in P\backslash G/H} \mathrm{Tr}_{P^t \cap H}^H(\sigma^t) = \mathrm{Tr}_P^H(\sigma) + \theta = \pi_W + \theta, \tag{4.5}$$

where $\theta \in \mathrm{Tr}_{\mathfrak{Y}}^H(E_{\mathfrak{o}}(V))$ and $\mathfrak{Y} = \mathfrak{Y}(G, P, H)$. Note that $\pi_W \notin \mathrm{Tr}_{\mathfrak{Y}}^H(E_{\mathfrak{o}}(V))$. In fact, if $\pi_W \in \mathrm{Tr}_{\mathfrak{Y}}^H(E_{\mathfrak{o}}(V))$, then

$$\mathrm{id}_W = \pi'_W \circ \pi_W \circ \iota_W \in \mathrm{Tr}_{\mathfrak{Y}}^H(\pi'_W \circ E_{\mathfrak{o}}(V) \circ \iota_W) \subset \mathrm{Tr}_{\mathfrak{Y}}^H(E_{\mathfrak{o}}(W)),$$

and hence W is $\mathfrak{Y}$-projective, which is a contradiction.

Now let $I = E_{\mathfrak{o}G}(V) \cap \mathrm{Tr}_{\mathfrak{Y}}^H(E_{\mathfrak{o}}(V))$. Then I is an ideal of $E_{\mathfrak{o}G}(V)$, and there is a natural monomorphism

$$\varphi\colon \bar{E}_G = E_{\mathfrak{o}G}(V)/I \to \bar{E}_H = E_{\mathfrak{o}H}(V)/\mathrm{Tr}_{\mathfrak{Y}}^H(E_{\mathfrak{o}}(V)).$$

As was observed above, $\varphi(\bar{\rho}) = \bar{\pi}_W(\neq 0)$ is an idempotent of $\bar{E}_H$. Therefore $\bar{\rho}$ is an idempotent of $\bar{E}_G$. But this implies that $\rho = \mathrm{Tr}_P^G(\sigma)$ is a unit of $E_{\mathfrak{o}G}(V)$, because $E_{\mathfrak{o}G}(V)$ is a local ring. Therefore $E_{\mathfrak{o}G}(V) = \mathrm{Tr}_P^G(E_{\mathfrak{o}}(V))$.

(ii) ($\Leftarrow$) is obvious by (i), whereas ($\Rightarrow$) is trivial, since fV satisfies the condition. ∎

For an $\mathfrak{o}G$-module V, let

$$\mathrm{Comp}(V) = \{V_1, \ldots, V_r\}$$

be a complete set of nonisomorphic indecomposable components of V. Thus $V = \oplus_{i=1}^r m_i V_i$ with positive integer m_i, which is called the *multiplicity* of V_i in V. Let P be a p-subgroup of G. In $\mathrm{Comp}(V)$ we define

$$\mathrm{Comp}(V|P) = \{V_i;\ \mathrm{vx}(V_i) =_G P\}.$$

The behavior of $\mathrm{Comp}(V|P)$ under the restriction or the induction of modules will be well described via the Green correspondence. Namely, we have the following.

Theorem 4.7 **(Burry [3]).** *Let $G \geq H \geq P$ and P be a p-subgroup.*

(i) *For an $\mathfrak{o}G$-module V, the Green correspondence $f_H = f(H, P, N_H(P))$ induces a multiplicity-preserving bijection*

$$f_H\colon \mathrm{Comp}(V_H|P) \to \mathrm{Comp}(V_{N_H(P)}|P).$$

(ii) For an $\mathfrak{o}H$-module W, the Green correspondence $f_G = f(G, P, N_G(P))$ induces a multiplicity-preserving bijection

$$(4.6) \quad f_G\colon \mathrm{Comp}(W^G|P) \to \mathrm{Comp}\left(\bigoplus_{t\in H\backslash G/N_G(P),\, P\le H^t} (W^t_{N_{H^t}(P)})^{N_G(P)}|P\right).$$

(iii) *If* $W \in \mathrm{IND}(\mathfrak{o}H|P)$, *then the right-hand side of* (4.6) *may be replaced with* $\mathrm{Comp}((W_{N_H(P)})^{N_G(P)}|P)$.

Proof.

(i) Let $V_H = \bigoplus_{i=1}^r m_i W_i$ be an indecomposable decomposition with $\mathrm{Comp}(V_H|P) = \{W_i; 1 \le i \le s\}$. Then, for $1 \le i \le s$, $(W_i)_{N_H(P)}$ has a unique indecomposable component with vertex P, which is of course $f_H W_i$. On the other hand, if $s < j \le r$, then $(W_j)_{N_H(P)}$ has no indecomposable component with vertex P by Theorem 4.6. This proves (i).

(ii) From the above, we get a multiplicity-preserving bijection

$$f_G\colon \mathrm{Comp}(W^G|P) \to \mathrm{Comp}((W^G)_{N_G(P)}|P).$$

Since

$$(W^G)_{N_G(P)} = \bigoplus_{t\in H\backslash G/N_G(P)} (W^t_{H^t\cap N_G(P)})^{N_G(P)},$$

it suffices to show that if $U \in \mathrm{IND}(\mathfrak{o}N_G(P)|P)$ and $U|(W^t_{H^t\cap N_G(P)})^{N_G(P)}$, then $P \le H^t$. However this is obvious as $\mathrm{vx}(U) = P \le_{N_G(P)} H^t \cap N_G(P)$.

(iii) We need only to show that if $(W^t_{N_{H^t}(P)})^{N_G(P)}$ has an indecomposable component, say U, with vertex P, then $t \in HN_G(P)$. Take an indecomposable component M of $W^t_{N_{H^t}(P)}$ such that $U|M^{N_G(P)}$. Then

$$\mathrm{vx}(U) = P \le_{N_G(P)} \mathrm{vx}(M) \le_{H^t} \mathrm{vx}(W^t) = P^t.$$

Therefore $P^{ht} = P^t$ for some $h \in H$, and thus $t^{-1}h \in N_G(P)$. This proves (iii). ∎

We refer to the normalizers of the nonidentity p-subgroups as the *p-local subgroups* in general. In the theory of finite groups, it often turns out to be an important method to reduce problems to the case of p-local subgroups. The next theorem, as well as Theorem 4.7, can be considered as reducing certain problems concerning indecomposable components of restricted modules or induced modules to p-local subgroups via the Green correspondence.

Theorem 4.8 **(Burry [1]).** *Let $G \geq H \geq P$ and P be a p-subgroup. Let f_G and f_H denote $f(G, P, N_G(P))$ and $f(H, P, N_H(P))$, respectively. Then the following statements hold for $V \in \mathrm{IND}(\mathfrak{o}G|P)$ and $W \in \mathrm{IND}(\mathfrak{o}H|P)$.*

(i) $W|V_H \Leftrightarrow f_H W|(f_G V)_{N_H(P)}$.

(ii) $V|W^G \Leftrightarrow f_G V|(f_H W)^{N_G(P)}$.

Proof. Set $N_G = N_G(P)$, $N_H = N_H(P)$.

(i) ($\Rightarrow$). Observe that $f_H W|W_{N_H}|V_{N_H} = (V_{N_G})_{N_H}$ and $V_{N_G} = f_G V \oplus (\bigoplus_i Y_i)$, where $\mathrm{vx}(Y_i) \leq_{N_G} P^{x_i} \cap N_G$ $(x_i \notin N_G)$. If $f_H W|(Y_i)_{N_H}$, then $\mathrm{vx}(f_H W) = P \leq_{N_G} \mathrm{vx}(Y_i) \leq_{N_G} P^{x_i}$, and hence $P = P^{x_i}$, i.e., $x_i \in N_G$, which is a contradiction. Therefore, $f_H W|(f_G V)_{N_H}$.

($\Leftarrow$). $f_H W|(f_G V)_{N_H}|(V_{N_G})_{N_H} = V_{N_H} = (V_H)_{N_H}$. Therefore there exists

$$U \in \mathrm{IND}(\mathfrak{o}H)$$

such that $U|V_H$, $f_H W|U_{N_H}$. It then follows from Theorem 4.6 that $U = f_H^{-1}(f_H W) = W$ and $W|V_H$.

(ii) ($\Rightarrow$). $V|W^G|(f_H W)^G = ((f_H W)^{N_G})^G$. Hence there exists

$$M \in \mathrm{IND}(\mathfrak{o}N_G)$$

such that $M|(f_H W)^{N_G}$, $V|M^G$. Then $\mathrm{vx}(V) =_G P \leq_G \mathrm{vx}(M) \leq_{N_G} \mathrm{vx}(f_H W) = P$, which implies that $\mathrm{vx}(M) =_{N_G} P$. Therefore, by Theorem 4.3(ii), $M = f_G V$.

($\Leftarrow$). $V|(f_G V)^G|(f_H W)^G = ((f_H W)^H)^G$ and $(f_H W)^H = W \oplus (\bigoplus_i U_i)$ with $\mathrm{vx}(U_i) \leq_H P^{h_i} \cap P$ $(h_i \in H - N_H)$. In particular we have $|\mathrm{vx}(U_i)| < |P|$. If $V|U_i^G$ for some i, then it follows that $\mathrm{vx}(V) =_G P \leq_G \mathrm{vx}(U_i)$, and this contradicts the above. Therefore $V|W^G$. ∎

As one of the applications of Theorem 4.8, we show the following.

Corollary 4.9. *In Lemma* 3.8, *there exists $W \in \mathrm{IND}(\mathfrak{o}H)$, which enjoys all of the three conditions.*

Proof. We use the same notation as in Theorem 4.8. By Lemma 3.5 (ii), there exists $U \in \mathrm{IND}(\mathfrak{o}N_H)$ such that $f_G V|U^{N_G}$, $U|(f_G V)_{N_H}$ and $\mathrm{vx}(U) =_{N_G} \mathrm{vx}(f_G V) =_{N_G} P$. Thus $\mathrm{vx}(U) = P$. Letting $W = f_H^{-1} U (\in \mathrm{IND}(\mathfrak{o}H|P))$, we have $\mathrm{vx}\ (W) =_H P$. Also, by Theorem 4.8, we have $V|W^G$ and $W|V_H$. ∎

Finally, we show the following theorem concerning the relationship between the Green correspondence and the Heller operator in the case $\mathfrak{o} = F$.

Theorem 4.10. *Let* $f\colon \mathrm{IND}(FG|\mathfrak{A}) \to \mathrm{IND}(FH|\mathfrak{A})$ *be the Green correspondence with respect to* (G, P, H) *satisfying* (4.2). *Then the following statements hold for* $V \in \mathrm{IND}(FG)$.

(i) $\mathrm{vx}(V) =_G \mathrm{vx}(\Omega(V))$.

(ii) *If* $\mathrm{vx}(V) \in \mathfrak{A}$, *then* $\Omega(fV) \simeq f\Omega(V)$.

Proof. We may assume that V is nonprojective. Recall that $\Omega(V)$ is indecomposable (Chapter 2, 8.26(iii)). Let $\mathrm{vx}\,(V) = Q$.

(i) There exists an FQ-module W such that $V | W^G$. Hence $\Omega(V)|\Omega(W^G)$. Since FG is free as a left FH-module, we get an exact sequence

$$0 \to \Omega(W)^G \to P(W)^G \to W^G \to 0$$

from the exact sequence $0 \to \Omega(W) \to P(W) \to W \to 0$. Note that $P(W)^G$ is projective because $P(W)$ is projective. Thus $\Omega(W^G)|\Omega(W)^G$ by Chapter 1, 10.18, and hence we have $\Omega(V)|\Omega(W)^G$. Therefore, $\Omega(V)$ is Q-projective, and we get $\mathrm{vx}(\Omega(V)) \leq_G \mathrm{vx}(V) = Q$.

On the other hand, it follows from Exercise 4.5 and Chapter 2, 8.25 that

$$\mathrm{vx}(\Omega^{-1}(V)) =_G \mathrm{vx}(\Omega(V^\wedge)^\wedge) =_G \mathrm{vx}(\Omega(V^\wedge)) \leq_G \mathrm{vx}(V^\wedge) =_G \mathrm{vx}(V).$$

Replacing V with $\Omega(V)$ in the above, we have

$$\mathrm{vx}(V) = \mathrm{vx}(\Omega^{-1}(\Omega(V))) \leq_G \mathrm{vx}(\Omega(V)).$$

(ii) Let $Q = \mathrm{vx}(V) \in \mathfrak{A}$. Since $fV | V_H$, it follows that $\Omega(fV)|\Omega(V_H)$. On the other hand, from the exact sequence $0 \to \Omega(V)_H \to P(V)_H \to V_H \to 0$, it follows that $\Omega(V_H)|\Omega(V)_H$ by Chapter 1, 10.18, because $P(V)_H$ is FH-projective. Therefore,

$$\Omega(fV)|\Omega(V)_H \equiv f\Omega(V) \oplus O(\mathfrak{Y}),$$

and we have $\Omega(fV) \simeq f\Omega(V)$, because $\mathrm{vx}(\Omega(V)) =_G Q$ and

$$\mathrm{vx}(\Omega(fV)) =_H \mathrm{vx}(fV) =_H Q. \qquad \blacksquare$$

5. Green Correspondences and Endomorphism Rings

There is a remarkable relationship between the endomorphism ring of an indecomposable module and that of its Green correspondent. To show it, we need some preparation.

Let $\mathfrak{X}$ be a set of subgroups of G. For $G \geq H$, we define

$$\mathfrak{X} \wedge H = \{X^x \cap H; \qquad X \in \mathfrak{X}, \quad x \in G\}.$$

For an $\mathfrak{o}H$-module W, we identify W with $W \otimes 1 \subset W^G$ as in Section 1.

Lemma 5.1 (Dade). *Let $G \geq H$ and W be an $\mathfrak{o}H$-module. Then the following isomorphism holds:*

$$\mathrm{Tr}_H^G\colon \mathrm{Tr}_{\mathfrak{X}\wedge H}^H(W) \cong \mathrm{Tr}_{\mathfrak{X}}^G(W^G).$$

Proof. First, we show that $\mathrm{Tr}_H^G(u) \in \mathrm{Tr}_{\mathfrak{X}}^G(W^G)$ if $u \in \mathrm{Tr}_{\mathfrak{X}\wedge H}^H(W)$. In fact, if $u = \mathrm{Tr}_{X^x \cap H}^H(w)$ ($X \in \mathfrak{X}$, $x \in G$, and $w \in \mathrm{Inv}_{X^x \cap H}(W)$), then

$$\mathrm{Tr}_H^G(u) = \mathrm{Tr}_{X^x \cap H}^G(w) \in \mathrm{Tr}_{X^x}^G(W^G) = \mathrm{Tr}_X^G(W^G).$$

Thus Tr_H^G induces a map $\mathrm{Tr}_{\mathfrak{X}\wedge H}^H(W) \to \mathrm{Tr}_{\mathfrak{X}}^G(W^G)$, which certainly is a monomorphism by Lemma 1.5. To see that this is an epimorphism, let $X \in \mathfrak{X}$. Then

$$(W^G)_X = \bigoplus_{t \in H\backslash G/X} ((W \otimes t)_{H^t \cap X})^X,$$

and by Lemma 1.5,

$$\mathrm{Inv}_X((W \otimes t)_{H^t \cap X})^X = \mathrm{Tr}_{H^t \cap X}^X(W \otimes t).$$

Thus

$$\mathrm{Tr}_X^G(W^G) = \bigoplus_t \mathrm{Tr}_X^G(\mathrm{Tr}_{H^t \cap X}^X(W \otimes t)) = \bigoplus_t \mathrm{Tr}_{H^t \cap X}^G(W \otimes t) = \bigoplus_t \mathrm{Tr}_{H \cap X^{t^{-1}}}^G(W)$$

$$= \mathrm{Tr}_H^G\left(\bigoplus_t \mathrm{Tr}_{X^{t^{-1}} \cap H}^H(W)\right) \subset \mathrm{Tr}_H^G(\mathrm{Tr}_{\mathfrak{X}\wedge H}^H(W)),$$

which proves the assertion. ■

Remark. When $\mathfrak{X} = \{G\}$, the above lemma implies Lemma 1.5.

Let V be an $\mathfrak{o}G$-module and W be an $\mathfrak{o}H$-module. Recall that with the identification $W = W \otimes 1$, we have

(5.1) $\qquad {}_{\mathfrak{o}}(V, W) \subset {}_{\mathfrak{o}}(V, W^G) \qquad$ and $\qquad {}_{\mathfrak{o}}(W, V) \subset {}_{\mathfrak{o}}(W^G, V).$

Then we get the following result from the above lemma, Theorem 1.9, and Lemma 1.5.

Corollary 5.2. *Let V and W be as above. Then the trace map Tr_H^G gives rise to the following isomorphisms.*

(i) $$\mathrm{Tr}_{\mathfrak{X}\wedge H}^H({}_{\mathfrak{o}}(V, W)) \simeq \mathrm{Tr}_{\mathfrak{X}}^G({}_{\mathfrak{o}}(V, W^G)),$$
$$ {}_{\mathfrak{o}H}(V, W)/\mathrm{Tr}_{\mathfrak{X}\wedge H}^H({}_{\mathfrak{o}}(V, W)) \simeq {}_{\mathfrak{o}G}(V, W^G)/\mathrm{Tr}_{\mathfrak{X}}^G({}_{\mathfrak{o}}(V, W^G)).$$

(ii) $$\mathrm{Tr}_{\mathfrak{X}\wedge H}^H({}_{\mathfrak{o}}(W, V)) \simeq \mathrm{Tr}_{\mathfrak{X}}^G({}_{\mathfrak{o}}(W^G, V)),$$
$$ {}_{\mathfrak{o}H}(W, V)/\mathrm{Tr}_{\mathfrak{X}\wedge H}^H({}_{\mathfrak{o}}(W, V)) \simeq {}_{\mathfrak{o}G}(W^G, V)/\mathrm{Tr}_{\mathfrak{X}}^G({}_{\mathfrak{o}}(W^G, V)).$$

Remark.

(i) When $\mathfrak{X} = \{G\}$, the above result yields Theorem 1.19 of Chapter 3.

(ii) Lemma 5.1, Corollary 5.2, and Lemma 5.3(i) below hold for a general commutative ring $\mathfrak{o}$.

Let U, W be $\mathfrak{o}H$-modules. We get the following inclusion from (5.1):

$$ {}_{\mathfrak{o}}(U, W) \subset {}_{\mathfrak{o}}(U, W^G) \subset {}_{\mathfrak{o}}(U^G, W^G).$$

Consequently, by restricting Tr_H^G to ${}_{\mathfrak{o}H}(U, W)$, we obtain a homomorphism

(5.2) $$\mathrm{Tr}_H^G\colon {}_{\mathfrak{o}H}(U, W) \to {}_{\mathfrak{o}G}(U^G, W^G),$$

such that the following holds: for $f \in {}_{\mathfrak{o}H}(U, W)$ and $\sum_{t\in H\backslash G} u_t \otimes t \in U^G$,

(5.3) $$\mathrm{Tr}_H^G(f)\left(\sum_t u_t \otimes t\right) = \sum_t f(u_t) \otimes t.$$

Lemma 5.3.

(i) *The above homomorphism* (5.2) *induces the following monomorphism*:

(5.4) $$\overline{\mathrm{Tr}_H^G}\colon {}_{\mathfrak{o}H}(U, W)/\mathrm{Tr}_{\mathfrak{X}\wedge H}^H({}_{\mathfrak{o}}(U, W)) \to {}_{\mathfrak{o}G}(U^G, W^G)/\mathrm{Tr}_{\mathfrak{X}}^G({}_{\mathfrak{o}}(U^G, W^G)).$$

If $U = W$, then this is a ring homomorphism.

(ii) *Let $H \geq P$ and $\mathfrak{X} = \mathfrak{X}(G, P, H)$ be as in* (4.1). *If U is P-projective, then $\mathrm{Tr}_{\mathfrak{X}\wedge H}^H({}_{\mathfrak{o}}(U, W)) = \mathrm{Tr}_{\mathfrak{X}}^H({}_{\mathfrak{o}}(U, W))$. If, furthermore, W is P-projective, then* (5.4) *is an isomorphism*:

$$\overline{\mathrm{Tr}_H^G}\colon {}_{\mathfrak{o}H}(U, W)/\mathrm{Tr}_{\mathfrak{X}}^H({}_{\mathfrak{o}}(U, W)) \simeq {}_{\mathfrak{o}G}(U^G, W^G)/\mathrm{Tr}_{\mathfrak{X}}^G({}_{\sigma}(U^G, W^G)).$$

Proof.

(i) By Corollary 5.2(ii), Tr_H^G induces the following isomorphism:

$$\overline{\mathrm{Tr}_H^G}\colon {}_{\mathfrak{o}H}(U, W^G)/\mathrm{Tr}_{\mathfrak{X}\wedge H}^H({}_{\mathfrak{o}}(U, W^G)) \simeq {}_{\mathfrak{o}G}(U^G, W^G)/\mathrm{Tr}_{\mathfrak{X}}^G({}_{\mathfrak{o}}(U^G, W^G)).$$

Therefore, it suffices to show that

$$(5.5) \qquad {}_{oH}(U, W) \cap \mathrm{Tr}^H_{\mathfrak{X} \wedge H}({}_o(U, W^G)) = \mathrm{Tr}^H_{\mathfrak{X} \wedge H}({}_o(U, W)).$$

The inclusion "$\supset$" is clear. To show the reverse inclusion, let f be an element of the left-hand side of (5.5), thus

$$f = \sum_{Q \in \mathfrak{X} \wedge H} \mathrm{Tr}^H_Q(g_Q) \qquad (g_Q \in {}_{oQ}(U, W^G)).$$

Let $\pi_W : W^G = W \oplus (\bigoplus_{t \neq 1} W \otimes t) \to W$ be the projection. Then

$$f = \pi_W \circ f = \sum_Q \mathrm{Tr}^H_Q(\pi_W \circ g_Q) \qquad (\pi_W \circ g_Q \in {}_{oQ}(U, W)),$$

and, therefore, f belongs to the right-hand side of (5.5), as required.

If $U = W$, then it follows from (5.3) that $\mathrm{Tr}^G_H(f) \circ \mathrm{Tr}^G_H(g) = \mathrm{Tr}^G_H(f \circ g)$ for f, $g \in E_{oH}(W)$, and hence $\overline{\mathrm{Tr}^G_H}$ is a ring homomorphism.

(ii) Clearly, we have $\mathfrak{X} \subset \mathfrak{X} \wedge H$, and thus

$$\mathrm{Tr}^H_{\mathfrak{X}}({}_o(U, W)) \subset \mathrm{Tr}^H_{\mathfrak{X} \wedge H}({}_o(U, W)).$$

To show the reverse inclusion, let $f = \mathrm{Tr}^H_D(\mu) \in \mathrm{Tr}^H_{\mathfrak{X} \wedge H}({}_o(U, W))$, where $D \in \mathfrak{X} \wedge H$ and $\mu \in {}_{oD}(U, W)$, and write $D = Q^x \cap H$ with $x \in G$ and $Q \leq P^t \cap P$ for some $t \in G - H$. Since U is P-projective by assumption, it follows that $\mathrm{id}_U = \mathrm{Tr}^H_P(\lambda)$ with $\lambda \in {}_{oP}(U, U)$. Therefore by Exercise 1.7(iii), we have

$$f = f \circ \mathrm{id}_U = \mathrm{Tr}^H_D(\mu) \circ \mathrm{Tr}^H_P(\lambda) = \sum_{h \in D \backslash H / P} \mathrm{Tr}^H_{D^h \cap P}(\mu^h \circ \lambda).$$

Since $D \leq Q^x \leq P^{tx} \cap P^x$, it follows that $D^h \cap P \leq P^{txh} \cap P^{xh} \cap P$, where we have that either $xh \notin H$ or $txh \notin H$ because $t \notin H$. Therefore, $D^h \cap P \in \mathfrak{X}$, and we get $f \in \mathrm{Tr}^H_{\mathfrak{X}}({}_o(U, W))$ as required.

Next we show that $\overline{\mathrm{Tr}^G_H}$ is an isomorphism if W is also P-projective. From the proof of (i), it suffices to show that

$${}_{oH}(U, W^G) = {}_{oH}(U, W) + \mathrm{Tr}^H_{\mathfrak{X} \wedge H}({}_o(U, W^G)).$$

Let $(W^G)_H = W \oplus W'$, where $W' = \bigoplus_{\substack{s \in H \backslash G / H \\ s \neq 1}} W \otimes s$. Then we have

$${}_{oH}(U, W^G) = {}_{oH}(U, W) \oplus {}_{oH}(U, W').$$

Thus, it suffices to show that ${}_{oH}(U, W') \subset \mathrm{Tr}^H_{\mathfrak{X} \wedge H}({}_o(U, W^G))$.

If $\mathfrak{Y} = \mathfrak{Y}(G, P, H)$ is the same as in (4.1), then W' is $\mathfrak{Y}$-projective by Lemma 4.2(i). In particular, every $f \in {}_{oH}(U, W')$ can be expressed as

$f = \sum_{Y \in \mathfrak{Y}} \mathrm{Tr}_Y^H(v_Y)$ with $v_Y \in {}_{\mathfrak{o}Y}(U, W')$ (cf. Lemma 2.8 and Theorem 2.9). Hence, if $\iota_{W'}: W' \to (W^G)_H$ is the inclusion map, then

$$f = \iota_{W'} \circ f \circ \mathrm{id}_U = \sum_{Y \in \mathfrak{Y}} \mathrm{Tr}_Y^H(\iota_{W'} \circ v_Y) \circ \mathrm{Tr}_P^H(\lambda)$$

$$= \sum_{Y \in \mathfrak{Y}} \sum_{h \in Y \backslash H / P} \mathrm{Tr}_{Y^h \cap P}^H(\iota_{W'} \circ (v_Y)^h \circ \lambda).$$

If $Y \leq P^a \cap H (a \notin H)$, then $Y^h \cap P \leq P^{ah} \cap P$. Thus $Y^h \cap P \in \mathfrak{X}$ as $ah \notin H$, and we get $f \in \mathrm{Tr}_{\mathfrak{X} \wedge H}^H({}_{\mathfrak{o}}(U, W^G))$. ■

We conclude this section with the following result on the Green correspondence.

Theorem 5.4. *Let f be the Green correspondence* w.r.t. (G, P, H), *satisfying* (4.2), *and let $\mathfrak{X}$, $\mathfrak{Y}$, $\mathfrak{A}$ be the same as in* (4.1). *Then the following isomorphism holds for $U, V \in \mathrm{IND}(\mathfrak{o}G|\mathfrak{A})$:*

$${}_{\mathfrak{o}H}(fU, fV)/\mathrm{Tr}_{\mathfrak{X}}^H({}_{\mathfrak{o}}(fU, fV)) \simeq {}_{\mathfrak{o}G}(U, V)/\mathrm{Tr}_{\mathfrak{X}}^G({}_{\mathfrak{o}}(U,V)).$$

If $U = V$, then the following ring isomorphism holds:

$$E_{\mathfrak{o}H}(fV)/\mathrm{Tr}_{\mathfrak{X}}^H(E_{\mathfrak{o}}(fV)) \simeq E_{\mathfrak{o}G}(V)/\mathrm{Tr}_{\mathfrak{X}}^G(E_{\mathfrak{o}}(V)).$$

Proof. Since both fU and fV are P-projective, Tr_H^G gives rise to the following isomorphism by Lemma 5.3(ii).

(5.6) $${}_{\mathfrak{o}H}(fU, fV)/\mathrm{Tr}_{\mathfrak{X}}^H({}_{\mathfrak{o}}(fU, fV))) \simeq {}_{\mathfrak{o}G}((fU)^G, (fV)^G)/\mathrm{Tr}_{\mathfrak{X}}^G({}_{\mathfrak{o}}((fU)^G, (fV)^G)).$$

We write $(fU)^G = U \oplus U'$, $(fV)^G = V \oplus V'$, and let $\iota_U: U \to (fU)^G$ and $\pi_V: (fV)^G \to V$ be the inclusion map and the projection, respectively. Since U', V' are both $\mathfrak{X}$-projective, it holds that

$${}_{\mathfrak{o}G}(f(U)^G, (fV)^G) = {}_{\mathfrak{o}G}(U, V) + \mathrm{Tr}_{\mathfrak{X}}^G({}_{\mathfrak{o}}((fU)^G, (fV)^G)).$$

Thus, in view of (5.6), it suffices to show that

$${}_{\mathfrak{o}G}(U, V) \cap \mathrm{Tr}_{\mathfrak{X}}^G({}_{\mathfrak{o}}((fU)^G, (fV)^G)) = \mathrm{Tr}_{\mathfrak{X}}^G({}_{\mathfrak{o}}(U, V)).$$

Clearly, the left-hand side contains the right side. Take φ from the left-hand side. Then φ is expressed as $\varphi = \sum_{Q \in \mathfrak{X}} \mathrm{Tr}_Q^G(\mu_Q)$ $(\mu_Q \in {}_{0Q}((fU)^G, (fV)^G)$, and it follows that

$$\varphi = \pi_V \circ \varphi \circ \iota_U = \sum_{Q \in \mathfrak{X}} \mathrm{Tr}_Q^G(\pi_V \circ \mu_Q \circ \iota_U) \in \mathrm{Tr}_{\mathfrak{X}}^G({}_0(U, V)),$$

whence we have the reverse inclusion. ■

6. Endomorphism Rings of Induced Modules

6.1. G-graded Algebras

Assume, for the time being, that R is an arbitrary commutative ring. An R-algebra A is called a *G-graded algebra* (over R), provided there exists a set of R-submodules $\{A_x : x \in G\}$ of A, satisfying the following two conditions:

(6.1) $$A = \bigoplus_{x \in G} A_x,$$

(6.2) $$A_x A_y \subset A_{xy}.$$

We first observe that

(6.3) $A_1 \ni 1(=1_A)$, *and* A_1 *is a subalgebra of* A.

Indeed if $1_A = \sum_x a_x$ $(a_x \in A_x)$, then $a_x = a_x 1_A = \sum_y a_x a_y$, and hence $a_x = a_x a_1$. On the other hand since $a_1 = 1_A a_1 = \sum_x a_x a_1$, we have that $a_x a_1 = 0$ and $a_x = 0$, unless $x = 1$. Hence, $1_A = a_1 \in A_1$. Clearly, A_1 is a subalgebra of A by (6.2).

A G-graded algebra A is said to be *strongly graded*, provided the following holds:

(6.4) A_x contains a unit of A for all $x \in G$.

Example 6.1.

(i) Let $H \lhd G$ and let $\bar{G} = G/H$. Then $A = RG = \bigoplus_{\bar{x} \in \bar{G}} A_{\bar{x}}$ is a strongly $\bar{G}$-graded algebra, where $A_{\bar{x}} = x(RG)$.

(ii) A twisted group ring of G over a commutative ring S:

$$A = (S, G, \alpha) = \bigoplus_{x \in G} u_x S \quad (u_x u_y = u_{xy}\alpha(x, y), \quad \gamma u_x = u_x \gamma^x \ (\gamma \in S)),$$

defined in Section 7 of Chapter 2 is a strongly G-graded algebra over the invariant ring $R = S^G$ with $A_x = u_x S$.

Lemma 6.2. *Let $A = \bigoplus_{x \in G} A_x$ be a strongly G-graded algebra and let $u_x \in A_x$ be a unit of A with $u_1 = 1$. Then the following statements hold.*

(i) $u_x^{-1} \in A_{x^{-1}}$.

(ii) $A_x = u_x A_1 = A_1 u_x$.

(iii) $u_x u_y = u_{xy}\alpha(x, y), \quad \alpha(x, y) \in A_1^{\times}$.

(iv) $A_x A_y = A_{xy}$.

(v) $u_x^{-1} A_1 u_x = A_1$.

Proof.

(i) If $u_x^{-1} = \sum_y a_y \quad (a_y \in A_y)$, then

$$A_1 \ni 1 = u_x u_x^{-1} = \sum_y u_x a_y, \qquad u_x a_y \in A_{xy},$$

whence we have $u_x a_y = 0$, and hence $a_y = 0$, unless $y = x^{-1}$. Thus $u_x^{-1} = a_{x^{-1}} \in A_{x^{-1}}$.

(ii) $u_x A_1 \subset A_x A_1 \subset A_x$ and $u_x^{-1} A_x \subset A_{x^{-1}} A_x \subset A_1$. Hence $u_x A_1 = A_x$. The second equality follows similarly.

(iii) $u_{xy}^{-1} u_x u_y$ is a unit of A, hence of A_1 by (i).

(iv) This is clear by (ii) and (iii).

(v) This is clear by (ii). ■

6.2. Endomorphism Rings of Induced Modules

Let $G \rhd H$ and $\bar{G} = G/H$. For an RH-module W, let, as in Section 3 of Chapter 3,

$$T(W) = \{x \in G;\ W^x \simeq W \quad (\text{as } RH\text{-modules})\}$$

be the inertial group of W in G. Also, we let $E = E_{RG}(W^G)$, and for $x \in G$,

$$E_x = \{\sigma \in E;\ \sigma W \subset Wx\},$$

where $Wx = W \otimes x \subset W^G$.

Lemma 6.3.

(i) $\bar{x} = \bar{y} \Rightarrow E_x = E_y$.

(ii) $E_x \ni \sigma \Rightarrow \sigma(Wy) \subset Wxy$.

Proof.

(i) If $y = hx$ $(h \in H)$, then $Wy = Wx$ and hence $E_x = E_y$.

(ii) $\sigma(Wy) = (\sigma W)y \subset Wxy$. ■

In view of (i) above, we may write $E_{\bar{x}}$ for E_x. We shall also use such an expression in the following.

Lemma 6.4.

(i) $E_x \simeq {}_{RH}(W, Wx)$ *(as R-modules). Also* $E_1 \simeq E_{RH}(W)$ *as R-algebras.*

(ii) $E = \bigoplus_{\bar{x} \in \bar{G}} E_{\bar{x}}, \quad E_{\bar{x}} E_{\bar{y}} \subset E_{\overline{xy}}$.

Namely, E is a $\bar{G}$*-graded algebra.*

Proof.

(i) For $\varphi \in {}_{RH}(W, Wx)$, define $\hat{\varphi}: W^G \to W^G$ by

$$\hat{\varphi}\left(\sum_{t \in H\backslash G} w_t t\right) = \sum_{t \in H\backslash G} \varphi(w_t)t. \tag{6.5}$$

We claim that $\hat{\varphi} \in E_x$. In fact, given $y \in G$, we write $ty = h_t t'$ with $h_t \in H$ and $t' \in H\backslash G$. Then

$$\hat{\varphi}\left(\left(\sum_t w_t t\right)y\right) = \hat{\varphi}\left(\sum_t (w_t h_t)t'\right) = \sum_t \varphi(w_t h_t)t' = \sum_t \varphi(w_t)h_t t'$$
$$= \left(\sum_t \varphi(w_t)t\right)y = \hat{\varphi}\left(\sum_t w_t t\right)y,$$

which shows that $\hat{\varphi}$ is an RG-homomorphism. It is clear that $\hat{\varphi}(W) = \varphi(W) \subset Wx$, and therefore $\hat{\varphi} \in E_x$, as claimed. Thus we obtain an R-homomorphism ${}_{RH}(W, Wx) \hookrightarrow E_x$ $(\varphi \mapsto \hat{\varphi})$. This is monomorphic because $\hat{\varphi}|_W = \varphi$. On the other hand, if $\sigma \in E_x$, then $\sigma|_W \in {}_{RH}(W, Wx)$, and $\widehat{\sigma|_W} = \sigma$ as σ is an RG-homomorphism. Therefore the map $\wedge$ is an isomorphism.

If $x = 1$, then the map $\wedge$ is an algebra isomorphism as is easily seen.

(ii) Clearly, $\sum_{\bar{x} \in \bar{G}} E_{\bar{x}}$ is a direct sum. If $E_{\bar{x}} \ni \sigma$, $E_{\bar{y}} \ni \rho$, then $\sigma\rho W \subset \sigma(Wy) \subset Wxy$ and hence $\sigma\rho \in E_{\overline{xy}}$. To see that $E = \sum_{\bar{x} \in \bar{G}} E_{\bar{x}}$, let $\pi_x: W \to Wx$ be the projection on Wx w.r.t. the decomposition $W^G = \bigoplus_{x \in H\backslash G} Wx$. Hence, $\pi_x \in E_{RH}(W^G)$. For $\sigma \in E$, define $\sigma_x \in {}_{RH}(W, Wx)$ by $\sigma_x = \pi_x(\sigma|_W)$. Then $\hat{\sigma}_x \in E_x$ and $\sum_{x \in H\backslash G} \hat{\sigma}_x$ coincides with σ on W, whence it follows that $\sigma = \sum_x \hat{\sigma}_x$, because W generates W^G as an RG-module. This proves that $E = \sum_{\bar{x}} E_{\bar{x}}$. ■

Lemma 6.5.

(i) *$x \in T(W) \Leftrightarrow E_x$ contains a unit of E.*

(ii) *Set $T = T(W)$, $\bar{T} = T/H$, and let*

$$E_T = \bigoplus_{\bar{t} \in \bar{T}} E_{\bar{t}}.$$

Then E_T is a strongly $\bar{T}$-graded algebra, which is isomorphic to $E_{RT}(W^T)$ as R-algebras.

Proof.

(i) ($\Rightarrow$). If $x \in T(W)$, then there exists an RH-isomorphism $\varphi: W \to Wx$, and $\hat{\varphi} \in E_x$ defined in (6.5) is a unit of E.

($\Leftarrow$). If $\sigma \in E_x$ is a unit of E, then $\sigma|_W: W \to Wx$ is an RH-isomorphism, and hence $x \in T(W)$.

(ii) The first assertion is clear by (i). Let $E' = E_{RT}(W^T)$ and $E'_t = \{\sigma \in E'; \sigma W \subset Wt\}$. Then $E' = \bigoplus_{\bar{t} \in \bar{T}} E'_{\bar{t}}$. Since every element of E_T stabilizes W^T, we have a homomorphism $f: E_T \to E'(\sigma \mapsto \sigma|_{W^T})$. As is shown in the proof of Lemma 6.4(i), f induces an isomorphism $E_t \xrightarrow{\sim} E'_t$. Consequently, $f: E_T \xrightarrow{\sim} E'$ is an isomorphism. ∎

If, in particular, $T(W) = G$, then E is a strongly G-graded algebra:

$$E = \bigoplus_{\bar{x} \in \bar{G}} \mu_{\bar{x}} E_1,$$

where $\mu_{\bar{x}} \in E_{\bar{x}}$ is a unit of E. Moreover, $E \otimes_{E_1} W$ becomes an RG-module with the action of $x \in G$ defined by

$$(6.6) \qquad (\sigma \otimes w)x = \sigma\mu_{\bar{x}} \otimes \mu_{\bar{x}}^{-1}(wx) \qquad (\sigma \in E, w \in W).$$

Now, the following theorem holds.

Theorem 6.6 **(Cline [1]).** *If $T(W) = G$, then there holds an (E, RG)-isomorphism,*

$$f: E \otimes_{E_1} W \xrightarrow{\sim} W^G \qquad (\sigma \otimes w \mapsto \sigma w).$$

Proof. It is almost clear from (6.6) that f is in fact an (E, RG)-homomorphism. Also, f induces an isomorphism $\mu_{\bar{x}} \otimes W \xrightarrow{\sim} Wx$, whence the result follows, because $E \otimes_{E_1} W = \bigoplus_{x \in H \backslash G} \mu_{\bar{x}} \otimes W$ and $W^G = \bigoplus_{x \in H \backslash G} Wx$. ∎

Now let us assume again that (K, R, F) is a p-modular system, and let $\mathfrak{o} = R$, or F. Also (π) denotes the unique maximal ideal of R. Then if $W \in \mathrm{IND}(\mathfrak{o}H)$, $E_1(\simeq E_{\mathfrak{o}H}(W))$ is a local ring, that is, $E_1/J(E_1)$ is a division ring.

Theorem 6.7. *Let $G \rhd H$ and $\bar{G} = G/H$. Suppose that $W \in \mathrm{IND}(\mathfrak{o}H)$ and let $T = T(W)$, $E = E_{\mathfrak{o}G}(W^G)$, $E_1 = E_{\mathfrak{o}H}(W)$.*

(i) *Let $L = (\bigoplus_{\bar{t} \in \bar{T}} J(E_1)E_{\bar{t}}) \oplus (\bigoplus_{\bar{x} \notin \bar{T}} E_{\bar{x}})$. Then L is an ideal of E that is contained in $J(E)$.*

(ii) *Set $E_T = \bigoplus_{\bar{t} \in \bar{T}} E_{\bar{t}}$ $(\simeq E_{\mathfrak{o}T}(W^T))$. Then*

$$E/L \simeq E_T/J(E_1)E_T.$$

(iii) $E/J(E) \simeq E_T/J(E_T)$.

Proof. For $t \in T$, let μ_t be a unit of E contained in E_t. Hence $E_t = \mu_t E_1 = E_1 \mu_t$ by Lemmas 6.5 and 6.2(ii).

(i) Since $\mu_t^{-1} E_1 \mu_t = E_1$, $\mu_t^{-1} J(E_1) \mu_t = J(E_1)$, and it follows that $J(E_1)E_t = E_t J(E_1)$. Therefore LE_t, $E_t L \subset L$. Thus, in order to prove that L is an ideal of E, it suffices to show that $E_x E_y \subset L$ for $x, y \notin T$. This is clear if $xy \notin T$. Let us show that $E_x E_y \subset J(E_1)E_{xy}$ when $xy \in T$. Let $\sigma \in E_x$, $\rho \in E_y$. In the following sequence of $\mathfrak{o}H$-homomorphisms,

$$W \xrightarrow{\mu_{y^{-1}x^{-1}}} Wy^{-1}x^{-1} \xrightarrow{\rho} Wx^{-1} \xrightarrow{\sigma} W,$$

$(Wx^{-1})_{\mathfrak{o}H}$ is indecomposable, and σ is not an isomorphism by Lemma 6.5(i). This implies that $\sigma\rho\mu_{y^{-1}x^{-1}}(\in E_1)$ cannot be an isomorphism (cf. Remark below), and hence $\sigma\rho\mu_{y^{-1}x^{-1}} \in J(E_1)$. Thus we have that $\sigma\rho \in J(E_1)\mu_{xy} \subset J(E_1)E_{xy}$.

We next prove that $L \subset J(E)$. It suffices to show that $L^* \subset J(E^*)$, where $E^* = E/\pi E$. If $t \in T$, then every element of $J(E_1)^* E_t^*$ is nilpotent. Now take any $\sigma \in E_x$, where $x \notin T$. Then $\sigma^n \in E_{x^n} = E_1$ if $n = o(x)$. If σ^n is a unit, then so is σ, contradicting Lemma 6.5(i). Thus $\sigma^n \in J(E_1)$, and σ^* is nilpotent. From the above, we see that L^* is spanned by nilpotent elements over F and hence nilpotent by Chapter 2, 3.7.

(ii) This is clear by (i).

(iii) This is clear by (ii), since $L \subset J(E)$ and $J(E_1)E_T \subset J(E_T)$. ∎

Remark. Let A be a ring and let U, V, W be A-modules. Suppose that there are A-homomorphisms $U \xrightarrow{f} V \xrightarrow{g} W$ such that the composite map $g \circ f$ is an isomorphism. Then, if V is indecomposable, g must be an isomorphism. This can be easily shown by using Chapter 1, 9.2.

Corollary 6.8. *With the assumption of Theorem* 6.7, *let*

$$W^T = \bigoplus_{i=1}^{r} U_i \tag{6.7}$$

be an indecomposable decomposition. Then

$$W^G = \bigoplus_{i=1}^{r} U_i^G \tag{6.8}$$

is also an indecomposable decomposition, and it holds that

$$U_i \simeq U_j \Leftrightarrow U_i^G \simeq U_j^G. \tag{6.9}$$

In particular, W^G is indecomposable whenever $T(W) = H$.

Proof. Since $E_T \simeq E_{\mathfrak{o}T}(W^T)$, there exists a primitive idempotent decomposition $1 = \sum_{i=1}^{r} \varepsilon_i$ in E_T corresponding to (6.7), i.e., $\varepsilon_i(W^T) = U_i$. Observe that each ε_i is also primitive in E by Theorem 6.7(iii) and the theorem on lifting idempotents. Therefore, each $U_i^G = \varepsilon_i(W^G)$ must be indecomposable. Also, by Theorem 6.7 and Chapter 1, 14.2(iii), ε_i and ε_j are equivalent in E_T if and only if they are equivalent in E. Thus (6.9) holds. ■

7. The Green Indecomposability Theorem and Its Applications

Let (K, R, F) be a p-modular system as before. If (K', R', F') is another p-modular system, we write $(K', R', F') \geq (K, R, F)$, provided $K' \supset K$, and the valuation v' of K' is an extension of the valuation v of K. We call $\mathfrak{o}' = R'$(respectively, F') an extension of $\mathfrak{o} = R$(respectively, F).

An indecomposable $\mathfrak{o}G$-module V is called *absolutely indecomposable* if $V^{\mathfrak{o}'} = \mathfrak{o}' \otimes V$ is indecomposable for any extension $\mathfrak{o}'$ of $\mathfrak{o}$.

For $V \in \mathrm{IND}(\mathfrak{o}G)$, we let $D(V) = E_{\mathfrak{o}G}(V)/J(E_{\mathfrak{o}G}(V))$. Hence $D(V)$ is a division algebra over F.

Lemma 7.1. *Let $V \in \mathrm{IND}(\mathfrak{o}G)$ as above.*

(i) *If $D(V) = F$, then V is absolutely indecomposable.*

(ii) *If F is algebraically closed, then V is absolutely indecomposable.*

Proof.

(i) Let $E = E_{\mathfrak{o}G}(V)$ and $J(\mathfrak{o}) = (\pi)$. Then $E^{\mathfrak{o}'} \simeq E_{\mathfrak{o}'G}(V^{\mathfrak{o}'})$ by Chapter 1, 11.12, and we see that $J(E)^{\mathfrak{o}'} \subset J(E^{\mathfrak{o}'})$, because $J(E)^{\mathfrak{o}'}/\pi E^{\mathfrak{o}'}$ is nilpotent and $\pi E^{\mathfrak{o}'} \subset J(E^{\mathfrak{o}'})$. Moreover, it holds that

$$E^{\mathfrak{o}'}/J(E)^{\mathfrak{o}'} \simeq \mathfrak{o}' \otimes_{\mathfrak{o}} D(V) = \mathfrak{o}' \otimes_{\mathfrak{o}} F \simeq \mathfrak{o}'/\pi\mathfrak{o}'.$$

Therefore, $E^{\mathfrak{o}'}/J(E^{\mathfrak{o}'}) \simeq \mathfrak{o}'/J(\mathfrak{o}') = F'$. Hence, $E^{\mathfrak{o}'}$ is a local ring and $V^{\mathfrak{o}'}$ is indecomposable.

(ii) This is clear, since $D(V) = F$ from the assumption. ■

The following theorem is important. On its proof, we shall use Theorem 13.27 in Chapter 1.

Theorem 7.2 (Green). *Let $G \rhd H$, and suppose that $\bar{G} = G/H$ is a p-group. Let W be an absolutely indecomposable $\mathfrak{o}H$-module. Then W^G is absolutely indecomposable.*

Proof. We may assume $|\bar{G}| = p$ by way of induction on the order of G. Take an arbitrary p-modular system (K', R', F') such that $(K', R', F') \geq (K, R, F)$ and let $\mathfrak{o}' = R'$ or F'. If $\tilde{F}'$ is the algebraic closure of F', there exists a p-modular system $(\tilde{K}', \tilde{R}', \tilde{F}')$ such that $(\tilde{K}', \tilde{R}', \tilde{F}') \geq (K', R', F')$. Let $\tilde{\mathfrak{o}}' = \tilde{R}'$ or $\tilde{F}'$, and $\tilde{W} = W^{\tilde{\mathfrak{o}}'}$. Then $\tilde{W}_{\tilde{\mathfrak{o}}'H}$ is indecomposable by assumption. Now it suffices to show that $\tilde{W}^G (\simeq (W^G)^{\tilde{\mathfrak{o}}'})$ is indecomposable. Since $|\bar{G}| = p$, $T(\tilde{W})$ is equal to either H or G. In the first case, $\tilde{W}^G$ is indecomposable by Corollary 6.8. Suppose that $T(\tilde{W}) = G$ and let $E = E_{\tilde{\mathfrak{o}}'G}(\tilde{W}^G)$. Then

$$E = \bigoplus_{\bar{x} \in \bar{G}} E_{\bar{x}} \qquad (E_{\bar{x}} = \mu_{\bar{x}} E_1)$$

is a strongly $\bar{G}$-graded algebra, where $\mu_{\bar{x}}$ is a unit of E. Furthermore, $E_1/J(E_1) = \tilde{F}'$, since $E_1 \simeq E_{\tilde{\mathfrak{o}}'H}(\tilde{W})$ is a local ring. Consequently,

$$E/J(E_1)E \simeq \bigoplus_{\bar{x} \in \bar{G}} \mu_{\bar{x}} \tilde{F}'$$

is a generalized group ring of $\bar{G}$ over $\tilde{F}'$ and hence is isomorphic, by Chapter 3, 5.4(iii), to the group ring $\tilde{F}'\bar{G}$, which is local by Chapter 3, 1.24. This implies that E is local as $J(E_1)E \subset J(E)$. Therefore $\tilde{W}^G$ must be indecomposable. ■

Corollary 7.3. *Suppose that G is a p-group, and let H be a subgroup of G. If W is an absolutely indecomposable $\mathfrak{o}H$-module, then W^G is absolutely indecomposable.*

Proof. Since G is nilpotent, there exists a chain of subgroups

$$H = H_0 < H_1 < \cdots < H_n = G$$

such that $H_{i-1} \lhd H_i$ $(i = 1, 2, \ldots, n)$. Therefore the assertion follows by successive use of Theorem 7.2. ■

As applications of Green's theorem, we show the following two theorems.

Theorem 7.4. *Let P be a p-subgroup of G. Let V be a P-projective $\mathfrak{o}G$-module that is $\mathfrak{o}$-free, and χ be the $\mathfrak{o}$-character defined by V. Then the following holds for $x \in G$:*

$$x_p \notin_G P \Rightarrow \chi(x) = 0.$$

Proof. Let $\tilde{F}$ be the algebraic closure of F. There is a p-modular system $(\tilde{K}, \tilde{R}, \tilde{F})$ such that $(K, R, F) \le (\tilde{K}, \tilde{R}, \tilde{F})$. Let $\tilde{\mathfrak{o}} = \tilde{R}$ or $\tilde{F}$. Then $V^{\tilde{\mathfrak{o}}}$ is also P-projective. Therefore, we may assume without loss of generality that F is algebraically closed. In particular, for $H \le G$, every indecomposable $\mathfrak{o}H$-module is absolutely indecomposable by Lemma 7.1.

Let $H = \langle x \rangle$, $M = \langle x^p \rangle$. Since V is P-projective, we have $V \mid (V_P)^G$ and

$$((V_P)^G)_H = \bigoplus_{t \in P\backslash G/H} ((V \otimes t)_{P^t \cap H})^H$$

by the Mackey decomposition theorem. Since $x_p \notin_G P$, we have $x_p \notin P^t \cap H$, and hence $P^t \cap H \le M$. If $\bigoplus_{t \in P\backslash G/H} ((V \otimes t)_{P^t \cap H})^M = \bigoplus_{i=1}^n Y_i$ is an indecomposable decomposition, it follows that

$$((V_P)^G)_H = \bigoplus_{i=1}^{n} Y_i^H$$

is an indecomposable decomposition by Green's theorem, and we may write

$$V_H \simeq \bigoplus_{i=1}^{r} Y_i^H.$$

Since the character of H defined by Y_i^H vanishes at x as $x \notin M$, we get $\chi(x) = 0$ from the above, as asserted. ■

Theorem 7.5. *Let P be a p-subgroup of G and let V be a P-projective $\mathfrak{o}G$-module. Then the following holds.*

$$|G\colon P|_p \,|\, \dim_F V/\pi V,$$

In particular, if V is $\mathfrak{o}$-free, then

$$|G\colon P|_p \,|\, \mathrm{rank}_{\mathfrak{o}} V.$$

Proof. We only need to show the first statement. Note that $V/\pi V$ is also P-projective. Therefore we may assume that $\mathfrak{o} = F$ and that F is algebraically closed as in the proof of the preceding theorem.

Let S be a Sylow p-subgroup of G containing P, then $|G\colon P|_p = |S\colon P|$. Since V is P-projective, we have $V \,|\, (V_P)^G$ and

$$((V_P)^G)_S = \sum_{t \in P\backslash G/S} ((V \otimes t)_{P^t \cap S})^S.$$

Let $(V \otimes t)_{P^t \cap S} = \bigoplus_i Y_i^{(t)}$ be an indecomposable decomposition. Then by Corollary 7.3, each $(Y_i^{(t)})^S$ is indecomposable, and hence

$$((V_P)^G)_S = \bigoplus_{t,i} (Y_i^{(t)})^S$$

is an indecomposable decomposition. Consequently, V_S is isomorphic to a direct sum of some $(Y_i^{(t)})^S$'s as $V_S \,|\, ((V_P)^G)_S$. This yields our assertion, since $|S\colon P^t \cap S| \,|\, \dim_F (Y_i^{(t)})^S$ and $|S\colon P| \,|\, |S\colon P^t \cap S|$. ■

For each $n \geq 1$, let us consider $\mathfrak{o}/(\pi^n)$ as a G-module with trivial action. For convenience, we let $(\pi^\infty) = 0$.

Corollary 7.6. $\mathrm{vx}(\mathfrak{o}/(\pi^n))$ *is a Sylow p-subgroup of G.*

Proof. This is obvious by the above theorem, since $\dim_F \mathfrak{o}/(\pi) = 1$. ■

The above corollary yields the following two results.

Theorem 7.7. *Every p-subgroup P of G is a vertex of some indecomposable* $\mathfrak{o}G$*-module.*

Proof. Since $\mathrm{vx}(\mathfrak{o}_P) = P$, the result follows easily from Lemma 3.5(i). ■

As introduced in Chapter 3, we let, for an $\mathfrak{o}G$-module V,

$$\mathrm{Ker}_G V = \{x \in G;\ vx = v \text{ for all } v \in V\} \lhd G.$$

Theorem 7.8. *Let P be a p-subgroup of G and* $V \in \mathrm{IND}(\mathfrak{o}G)$. *Then the following statements hold.*

(i) $P \leq \mathrm{Ker}_G V \Rightarrow P \leq_G \mathrm{vx}(V)$.

In particular $\mathrm{vx}(V) \cap \mathrm{Ker}_G V \in \mathrm{Syl}_p(\mathrm{Ker}_G V)$.

(ii) *Suppose that* $P \lhd G$ *and that V is irreducible. Then* $P \leq \mathrm{Ker}_G(V)$ *and hence* $P \leq \mathrm{vx}(V)$.

Proof.

(i) Since P acts trivially on V and $\mathfrak{o}$ is a principal ideal domain, $V_P \simeq \bigoplus_i \mathfrak{o}/(\pi^{n_i})$. Therefore, $P = \mathrm{vx}((\mathfrak{o}/(\pi^{n_i}))_{\mathfrak{o}P}) \leq_G \mathrm{vx}(V)$. To show the second, let $Q \in \mathrm{Syl}_p(\mathrm{Ker}_G V)$. Then $Q^x \leq \mathrm{vx}(V)$ for some $x \in G$ from the above, and since $Q^x \leq \mathrm{Ker}_G V$, we have that $Q^x = \mathrm{vx}(V) \cap \mathrm{Ker}_G V \in \mathrm{Syl}_p(\mathrm{Ker}_G V)$.

(ii) Since V is irreducible, we have $\pi V = 0$. Thus, V may be regarded as an irreducible FG-module. Then by Chapter 3, 3.1 (i) and 1.24 (ii), we have $V_P = F \oplus \cdots \oplus F$, whence the result is clear. ■

8. Scott Modules

We use the same notation as in the preceding section. Let $G \geq H$. Let us denote $\mathfrak{o}$ by $\mathfrak{o}_H$ when it is considered as a trivial $\mathfrak{o}H$-module. Then the induced module

$$(\mathfrak{o}_H)^G = \bigoplus_{t \in H\backslash G} \mathfrak{o} \otimes t$$

defines a transitive permutation representation of G. This is called a *transitive permutation module* (over $\mathfrak{o}$). A *permutation module* is a finite direct sum of transitive permutation modules, and, more generally, a direct summand of a permutation module is called a *quasi-permutation module*. The main purpose of this section is to show that every transitive permutation module has a special component called a Scott module.

As in the preceding sections, we assume that a set of coset representatives of a subgroup in G contains the identity.

The following fact holds for indecomposable components of permutation modules in general.

Theorem 8.1. *Let $V \in \mathrm{IND}(\mathfrak{o}G)$ and $\mathrm{vx}(V) = P$. Then V is a component of some transitive permutation module if and only if a source $\mathrm{s}(V)_{\mathfrak{o}P}$ of V is isomorphic to $\mathfrak{o}_P$.*

Proof. If $\mathrm{s}(V) = \mathfrak{o}_P$, then $V|(\mathfrak{o}_P)^G$. To show the converse, suppose that $V|(\mathfrak{o}_H)^G$. Then

$$V_P|((\mathfrak{o}_H)^G)_P = \bigoplus_{t \in H\backslash G/P} (\mathfrak{o}_{H^t \cap P})^P,$$

and hence $\mathrm{s}(V)|(\mathfrak{o}_{H^t \cap P})^P$ for some t as $\mathrm{s}(V)|V_P$. But since $\mathrm{vx}(\mathrm{s}(V)) = P$, we have $H^t \cap P = P$ and thus $\mathrm{s}(V) = \mathfrak{o}_P$. ■

An indecomposable module V with vertex P is said to be a *trivial source module* if $\mathrm{s}(V)_{\mathfrak{o}P} = \mathfrak{o}_P$.

Exercise 8.2. Let V be an $\mathfrak{o}G$-module.

(i) V is a quasi-permutation module if and only if every indecomposable component of V is a trivial source module.

(ii) If V is a quasi-permutation module, then so is V_H for $H \leq G$. [Hint: Consider the Mackey decomposition.]

The next lemma will be needed later.

Lemma 8.3. *Let $H, L \leq G$. Then the following equality holds:*

$$\mathrm{rank}_{\mathfrak{o}}({}_{\mathfrak{o}G}((\mathfrak{o}_H)^G, (\mathfrak{o}_L)^G)) = |L\backslash G/H|.$$

Proof. By Chapter 3, 1.19, we have

$$_{\mathfrak{o}G}((\mathfrak{o}_H)^G, (\mathfrak{o}_L)^G) \simeq {}_{\mathfrak{o}H}(\mathfrak{o}_H, ((\mathfrak{o}_L)^G)_H) \simeq \mathrm{Inv}_H(((\mathfrak{o}_L)^G)_H),$$

and by the Mackey decomposition theorem,

$$((\mathfrak{o}_L)^G)_H = \bigoplus_{t \in L\backslash G/H} (\mathfrak{o}_{L^t \cap H})^H,$$

whence the result follows, since $\mathrm{Inv}_H((\mathfrak{o}_{L^t \cap H})^H) \simeq \mathrm{Inv}_{L^t \cap H}(\mathfrak{o}_{L^t \cap H}) = \mathfrak{o}$ by Lemma 1.5. ■

First let us show the existence of the Scott module in the case of $\mathfrak{o} = F$ following Burry [2].

Theorem 8.4 (Scott–Alperin). *Let $G \geq H$ and $Q \in \mathrm{Syl}_p(H)$.*

(i) *There exists an indecomposable component S of $(F_H)^G$ that satisfies the following three conditions:*

(a) $F_G | \mathrm{soc}(S)$.

(b) $F_G | \mathrm{hd}(S)$.

(c) $\mathrm{vx}(S) =_G Q$. *And if $f = f(G, Q, N = N_G(Q))$ denotes the Green correspondence, then fS can be considered as an $F[N/Q]$-module, which is a projective cover of $F_{N/Q}$.*

(ii) *For any indecomposable decomposition of $(F_H)^G$, there exists a unique indecomposable component S of $(F_H)^G$ that satisfies any one of the three conditions above.*

Proof. Let

$$(8.1) \qquad (F_H)^G = V_1 \oplus \cdots \oplus V_r$$

be an indecomposable decomposition. Consider the element $u = \sum_t 1 \otimes t$ of $(F_H)^G = \bigoplus_{t \in H\backslash G} F \otimes t$. Then $u \in \mathrm{Inv}_G((F_H)^G)$, and we have $\mathrm{Inv}_G((F_H)^G) = Fu$, since $\mathrm{Inv}_G((F_H)^G) \simeq \mathrm{Inv}_H(F_H) = F$. On the other hand, it follows from (8.1) that $\mathrm{Inv}_G((F_H)^G) = \bigoplus_i \mathrm{Inv}_G(V_i)$, and hence there exists a unique V_i such that $V_i \supset Fu$. Let $S = V_i$. Then, as is shown above, S is a unique indecomposable component in (8.1) that satisfies (a).

Second, we show that the above S satisfies (c). Since $Q \in \mathrm{Syl}_p(H)$, F_H is Q-projective and hence $F_H|(F_Q)^H$. Therefore $S|(F_H)^G|(F_Q)^G$, and it follows that $\mathrm{vx}(S) \leq_G Q$. Let $U = \bigoplus_{\substack{t \in H\backslash G \\ t \neq 1}} F \otimes t$. Then

$$((F_H)^G)_Q = Fu \oplus U$$

as an FQ-module. Since $S \supset Fu$, we have $S = Fu \oplus (S \cap U)$ and thus $F_Q|S_Q$. Then $\mathrm{vx}(F_Q) = Q \leq_G \mathrm{vx}(S)$, and from the above, we conclude that $\mathrm{vx}(S) =_G Q$. Also, we have $\mathrm{s}(S) = F_Q$ by Theorem 8.1.

Now, since the Green correspondence preserves vertices and sources, we see that $\mathrm{vx}(fS) = Q$ and $\mathrm{s}(fS) = F_Q$. Hence $fS|(F_Q)^N$. However, since $Q \lhd N$, it follows that $(F_Q)^N \simeq F[N/Q]$, and hence fS can be regarded as an $F[N/Q]$-module, which is a principal indecomposable module. On the other hand, since

$$0 \neq {}_{FG}(F_G, S) \subset {}_{FG}(F_G, (fS)^G) \simeq {}_{FN}(F_N, fS),$$

it follows that $\mathrm{soc}(fS) \simeq F_N$. Therefore fS is a projective cover of the trivial module $F_{N/Q}$. Moreover, S is a unique component of $(F_H)^G$ that satisfies (c) because f is bijective.

To show that S satisfies (b), let $(fS)^\wedge$ be the contragredient module of fS. Then $fS \simeq (fS)^\wedge$ by Chapter 3, 1.12, and $(fS)^\wedge \simeq fS^\wedge$ by Exercise 4.5, whence it follows that $S \simeq S^\wedge$. Therefore, $F_G|\mathrm{hd}(S^\wedge) \simeq \mathrm{hd}(S)$, because $F_G|\mathrm{soc}(S)$. The uniqueness assertion subject to (b) in (ii) follows from ${}_{FG}((F_H)^G, F_G) \simeq {}_{FH}(F_H, F_H) \simeq F$. ∎

The above FG-module S is called the *Scott module* and is denoted by $S(G, H)$ or simply by $S(H)$ for brevity. $S(G, H)$ is uniquely determined by G and H up to isomorphism.

From the uniqueness assertion subject to (c) in the above theorem, we obtain the following.

Corollary 8.5. *Let $G \geq H, H'$ and $Q \in \mathrm{Syl}_p(H)$, $Q' \in \mathrm{Syl}_p(H')$. Then*

$$S(H) \simeq S(H') \Leftrightarrow Q =_G Q'.$$

In particular, it holds that $S(H) \simeq S(Q)$.

If Q is a p-subgroup of G, then $S(Q)$ is called a *Scott module with vertex Q*. Scott modules over G and those over subgroups of G are related as follows.

Theorem 8.6. *Let $G \geq H \geq Q$ and Q be a p-subgroup. Let $S = S(G, Q)$ and $S_1 = S(H, Q)$. Then the following statements hold.*

(i) $S | S_1^G$.

(ii) $S_1 | S_H$.

Proof. Let $f_G = f(G, Q, N_G)$ and $f_H = f(H, Q, N_H)$ be Green correspondences, where $N_G = N_G(Q)$, $N_H = N_H(Q)$. By Theorem 8.4(c), there is an epimorphism $f_H S_1 \to F$, which yields an epimorphism $(f_H S_1)^{N_G} = (f_H S_1)^{N_G/Q} \xrightarrow{\lambda} (F_{N_H/Q})^{N_G/Q}$. Combining this with $(F_{N_H/Q})^{N_G/Q} \to F_{N_G/Q}$ $(\sum_{\bar{t}} \alpha_{\bar{t}} \otimes \bar{t} \mapsto \sum_{\bar{t}} \alpha_{\bar{t}})$, we get an epimorphism $(f_H S_1)^{N_G} \to F_{N_G/Q}$. Then we find that $f_G S | (f_H S_1)^{N_G}$ from Chapter 1, 10.18, since $(f_H S_1)^{N_G}$ is projective as an $F[N_G/Q]$-module. Therefore, we get (i) from Theorem 4.8.

Also, there exists an epimorphism $f_G S \to F_{N_G/Q}$, and hence an epimorphism $(f_G S)_{N_H} \to F_{N_H/Q}$. Since $(f_G S)_{N_H}$ is projective as an $F[N_H/Q]$-module, we get $f_H S_1 | (f_G S)_{N_H}$ by the same reason as above. Therefore, (ii) holds by Theorem 4.8. ■

We have also the following theorem.

Theorem 8.7. *Let $G \geq H \geq Q$ with Q a p-subgroup and $S = S(G, Q)$. Let W_{FH} be a trivial source module. If $S | W^G$, then $W = S(H, Q^x)$ for some $x \in G$.*

Proof. Let $Q' = \mathrm{vx}(W)$. Then $W | (F_{Q'})^H$ and hence $S | W^G | (F_{Q'})^G$. This means that $S = S(G, Q')$, and then $Q' = Q^x$ for some $x \in G$ by Corollary 8.5. Moreover, since

$$0 \neq {}_{FG}(F_G, S) \subset {}_{FG}(F_G, W^G) \simeq {}_{FH}(F_H, W),$$

we get $W = S(H, Q') = S(H, Q^x)$. ■

Now we consider the case $\mathfrak{o} = R$. If U, V are RG-modules, any RG-homomorphism $\lambda: U \to V$ induces an FG-homomorphism $\lambda_*: U^* \to V^* (u^* \mapsto \lambda(u)^*)$, where $U^* = U/\pi U$, $V^* = V/\pi V$. Hence there exists a homomorphism

$$\Phi: {}_{RG}(U, V) \to {}_{FG}(U^*, V^*) \quad (\lambda \mapsto \lambda_*).$$

Lemma 8.8. *With the above notation, let V be R-torsion-free. Then*

(8.2) $$\operatorname{Ker} \Phi = \pi \cdot {}_{RG}(U, V).$$

Consequently, Φ induces the following monomorphism:

(8.3) $$\Phi^*: {}_{RG}(U, V)^* = {}_{RG}(U, V)/\pi \cdot {}_{RG}(U, V) \to {}_{FG}(U^*, V^*).$$

Proof. Clearly, the left-hand side contains the right side in (8.2). If $\lambda \in \operatorname{Ker} \Phi$, then $\lambda(U) \subset \pi V$, so that, for each $u \in U$, there exists a unique element $\lambda_1(u)$ of V such that $\lambda(u) = \pi\lambda_1(u)$. Clearly, $\lambda_1 \in {}_{RG}(U, V)$, and $\lambda = \pi\lambda_1$ holds. This proves the reverse inclusion. ■

For quasi-permutation modules, we have the following.

Theorem 8.9.

(i) *If U, V are both quasi-permutation modules of G over R, then the above Φ^* is an isomorphism:*

(8.4) $$\Phi^*: {}_{RG}(U, V)^* \xrightarrow{\sim} {}_{FG}(U^*, V^*).$$

In particular, it holds that $E_{RG}(V)^* \simeq E_{FG}(V^*)$.

(ii) *Let V be as above. Then*

$$V \text{ is indecomposable} \Leftrightarrow V^* \text{ is indecomposable.}$$

If this is the case, then $\operatorname{vx}(V) = \operatorname{vx}(V^*)$.

(iii) *Let T be a quasi-permutation module of G over F. Then T lifts to a quasi-permutation module V of G over R, i.e., we have $V^* = T$. Moreover, such a V is unique up to isomorphisms.*

Proof.

(i) In the transitive case where $U = (R_H)^G$, $V = (R_L)^G$, the F-dimensions of both sides in (8.4) are identical, namely, $|L\backslash G/H|$ by Lemma 8.3. Thus the monomorphism Φ^* is an isomorphism. The general case follows easily from this.

(ii) Let $E = E_{RG}(V)$. Then $E^* = E_{FG}(V^*)$, and E is local if and only if E^* is local. Therefore, the first assertion is obvious.

To show the second assertion, let $\operatorname{vx}(V) = P$, $\operatorname{vx}(V^*) = Q$. Then, since $V \mid (V_P)^G$, we have $V^* \mid (V_P^*)^G$ and hence $Q \leq_G P$. On the other hand, there is

$\gamma \in E_{FQ}(V^*)$ such that $\mathrm{id}_{V^*} = \mathrm{Tr}_Q^G(\gamma)$, and we have $\lambda^* = \gamma$ for some $\lambda \in E_{RQ}(V)$ by (i). Consequently,

$$\mathrm{id}_V \equiv \mathrm{Tr}_Q^G(\lambda) \mod \pi E_{RG}(V),$$

whence we have $E_{RG}(V) = \mathrm{Tr}_Q^G(E_{RQ}(V)) + \pi E_{RG}(V) = \mathrm{Tr}_Q^G(E_{RQ}(V))$. Therefore V is Q-projective and $P \leq_G Q$.

(iii) To show the first assertion, we may assume that $T|(F_H)^G$. Let $(F_H)^G = T \oplus T'$, let τ be the projection on T, and let $E = E_{RG}((R_H)^G)$. Since $((R_H)^G)^* = (F_H)^G$, it follows that $E^* = \mathrm{End}_{FG}((F_H)^G)$. Thus τ lifts to an idempotent, say ε, of E, i.e., $\varepsilon^* = \tau$. Then $V = \varepsilon(R_H)^G$ satisfies $V^* = T$.

To show the uniqueness, let U, V be quasi-permutation modules over R, and suppose that there exists an FG-isomorphism $\gamma: U^* \xrightarrow{\sim} V^*$. Then there exists $\lambda: U \to V$ such that $\lambda^* = \gamma$. This means that $V = \lambda(U) + \pi V$, whence $V = \lambda(U)$, and λ is an epimorphism. However, since U and V are free over the principal ideal domain R and have the same R-rank by assumption, we find that $\mathrm{Ker}\,\lambda = 0$ (c.f. Chapter 1, 11.8(i)). Thus λ is an isomorphism. ∎

Now let $S(Q)$ be the Scott module having a p-subgroup Q as a vertex. Then $(F_Q)^G$ has an indecomposable decomposition

$$(F_Q)^G = S(Q) \oplus \left(\bigoplus_i T_i\right), \tag{8.5}$$

where $S(Q)$ is a unique indecomposable component such that ${}_{FG}(S(Q), F_G) \neq 0$. By Theorem 8.9, the above decomposition lifts to $(R_Q)^G$, namely,

$$(R_Q)^G = S'(Q) \oplus \left(\bigoplus_i V_i\right) \qquad (S'(Q)^* = S(Q),\ V_i^* = T_i).$$

It turns out that $S'(Q)$ is the unique indecomposable component in the above decomposition such that ${}_{RG}(S'(Q), R_G) \neq 0$. Also $\mathrm{vx}(S'(Q)) = Q$. Such an $S'(Q)$ is also referred to as the Scott module (over R) with vertex Q.

Problems

Let (K, R, F) be a p-modular system, and $\mathfrak{o}$ denotes R or F as before.

1. Let $P = \langle x \rangle$ be a cyclic p-group. Define $\mathbf{X}_n: P \to GL_n(F)$ by

$$X_n(x^i) = \begin{pmatrix} 1 & 1 & & 0 \\ & \ddots & \ddots & \\ & & 1 & 1 \\ 0 & & & 1 \end{pmatrix}^i.$$

Then $\{\mathbf{X}_n; 1 \le n \le |P|\}$ gives a complete set of nonequivalent indecomposable F-representations of P. (Consider the Jordan canonical forms.)

2. Let $P = \langle x \rangle \times \langle y \rangle$ $(x^p = y^p = 1)$. For $\alpha \in F$, define $\mathbf{X}_\alpha: P \to \mathrm{GL}_2(F)$ by

$$X_\alpha(x) = \begin{pmatrix} 1 & 1 \\ 0 & 1 \end{pmatrix}, \qquad X_\alpha(y) = \begin{pmatrix} 1 & \alpha \\ 0 & 1 \end{pmatrix}.$$

If $\alpha \ne \beta$, then $\mathbf{X}_\alpha$ and $\mathbf{X}_\beta$ are not equivalent.

3. Let $P \in \mathrm{Syl}_p(G)$.

(i) If P is cyclic, then $|\mathrm{IND}(FG)| < \infty$.

(ii) Prove that $|\mathrm{IND}(FG)| = \infty$, unless P is cyclic in the following way.

(a) Use Problem 2 when $|F| = \infty$.

(b) Use Problem 8(ii) in Chapter 2 when F is finite.

4. (Thompson) Let V_0 be an R-free RG-module, and suppose that there exists a projective FG-module U such that $V_0^* = U \oplus W$, where $V_0^* = V_0/\pi V_0$ and W is an FG-module. Then there exist a projective RG-module U_0 and an RG-module W_0 such that

$$V_0 \simeq U_0 \oplus W_0, \qquad U_0^* \simeq U, \qquad W_0^* \simeq W.$$

Prove this in the following way.

(i) Since R is complete, we have $U_0^* \simeq U$ for some projective RG-module U_0 with the following commutative diagram:

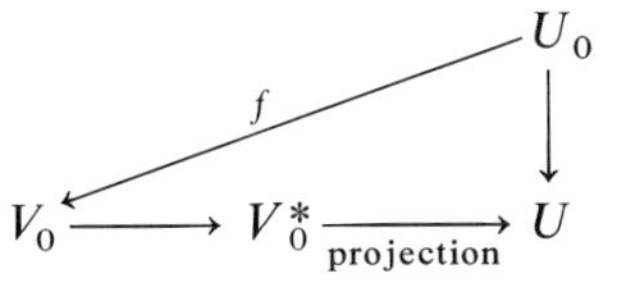

(ii) $\mathrm{Im}\, f \cap \pi V_0 = \pi\, \mathrm{Im}\, f$.

(iii) f is injective and splits as an R-homomorphism.

(iv) Use Exercise 2.3 to get $U_0 | V_0$.

5. Let $U = eFG$ be a projective cover of F_G, where $e \in \mathrm{pi}(FG)$.

(i) $\mathrm{Inv}_G(FG) = F\sigma = \mathrm{Tr}_1^G(U) = \mathrm{soc}(U)$, where $\sigma = \sum_{x \in G} x$.

(ii) Let V be an FG-module. Then $\dim_F \mathrm{Tr}_1^G(V)$ is equal to the multiplicity of U in an indecomposable decomposition of V.

(iii) Let H be a p'-subgroup of G and let $f = (1/|H|) \sum_{x \in H} x$. Then f is an idempotent of FG, and it holds that $U | fFG$.

6. Let $G \rhd H$ and $\bar{G} = G/H$. Let $V \in \mathrm{IND}(\mathfrak{o}G)$.

(i) If V is H-projective, then there exists $W \in \mathrm{IND}(\mathfrak{o}H)$ such that

$$V_H \simeq a\Big(\bigoplus_{t \in T(W)\backslash G} W^t\Big) \qquad (a \ge 1).$$

(ii) If $\mathrm{Ker}_G V \supset H$, then $\mathrm{vx}(V_{\mathfrak{o}\bar{G}}) = \mathrm{vx}(V)H/H$.

(iii) If V_H is absolutely indecomposable, then $\mathrm{vx}(V)H/H \in \mathrm{Syl}_p(\bar{G})$.

(iv) Let Y be an FH-module. Then $\Omega^m(Y^G) \simeq \Omega^m(Y)^G$ for all $m \in \mathbf{Z}$.

(v) If $(p, |G{:}H|) = 1$, then $J(FG) = J(FH)FG$.

(vi) Suppose that $\bar{G}$ is a p-group and that F is algebraically closed. Let $W \in \mathrm{IRR}(FH)$. Then $W^G \leftrightarrow |T(W){:}H|U$ for some $U \in \mathrm{IRR}(FG)$.

7. Let $G \geq H$. If $|G{:}H|$ is a power of p, then $(\mathfrak{o}_H)^G$ is indecomposable.

8. Let $(K', R', F') \geq (K, R, F)$. If $V \in \mathrm{IND}(RG)$, then $\mathrm{vx}(U) =_G \mathrm{vx}(V)$ for all $U \in \mathrm{Comp}(R' \otimes_R V)$.

9. Suppose that F is algebraically closed. Let $G = G_1 \times G_2$ and V_i be an indecomposable $\mathfrak{o}G_i$-module with vertex $P_i (i = 1, 2)$. Then

(i) $V_1 \otimes_{\mathfrak{o}} V_2 \in \mathrm{IND}(\mathfrak{o}G)$.

(ii) $\mathrm{vx}(V_1 \otimes_{\mathfrak{o}} V_2) = P_1 \times P_2$.

10. Let P be a p-subgroup of G and $N = N_G(P)$. Let $F[N/P] = \bigoplus_{i=1}^m a_i W_i$ be a decomposition into the direct sum of principal indecomposable modules W_i $(a_i \geq 1)$. Then

(i) $\mathrm{vx}((W_i)_{FN}) = P \quad (1 \leq i \leq m)$.

(ii) Let $f = f(G, P, N)$ and $V_i = f^{-1}(W_i)$. Then $(F_P)^G \simeq \bigoplus_{i=1}^m a_i V_i \oplus M$, where we have $\mathrm{vx}(U) <_G P$ for any $U \in \mathrm{Comp}(M)$.

11. Let $G \rhd H$, and suppose that F is algebraically closed. Then the following holds for any $W \in \mathrm{IND}(\mathfrak{o}H)$.

$$W^G \in \mathrm{IND}(\mathfrak{o}G) \Leftrightarrow T(W)/H \text{ is a } p\text{-group.}$$

12. Let G be a p-group. Then a generalized group ring $F^{(\alpha)}G$ is a local ring.

13. Let $A = \bigoplus_{x \in G} u_x A_1$ be a strongly G-graded algebra over a field $F_0 (u_1 = 1)$. We assume that A_1 is a division algebra. Let $E = Z(A_1)$, and let $H = \{x \in G;$ the map $A_1 \to A_1 (\gamma \mapsto u_x^{-1}\gamma u_x)$ is an inner automorphism$\}$.

Then the following statements hold.

(i) H is a normal subgroup of G.

(ii) Let $A_H = \bigoplus_{x \in H} u_x A_1$ and $C = C_A(A_1)$. Then C is a generalized group ring of H over E and $A_H = C \otimes_E A_1$.

(For the proof, we shall use an argument in the proof of Chapter 5, 7.2.)

(iii) For an ideal I of A and an ideal L of C, we have $I = (C \cap I)A$ and $LA \cap C = L$. Consequently, the ideals of A and the G-invariant ideals of C are in natural bijective correspondence, where an ideal L of C is said to be G-invariant if $u_x^{-1} L u_x = L$ for all $x \in G$.

(iv) If $\mathrm{Char}\, F_0 = p$ and G is a p-group, then $A/J(A)$ is a simple ring.

14. Let $G \rhd H$ and let $W \in \mathrm{IND}(\mathfrak{o}H)$. If $T(W)/H$ is a p-group, then there exist $V \in \mathrm{IND}(\mathfrak{o}G)$ and a positive integer e such that $W^G \simeq eV$.

15. Let $G \geq H, T$ and $|G:H| = |G:T| = n$.

(i) If $(\mathfrak{o}_H)^G \simeq (\mathfrak{o}_T)^G$, then the Sylow p-subgroups of H and T are conjugate in G.

(ii) The converse of the above is true if n is a power of p.

16. Let G be a p'-group. Then, for every $\chi \in \mathrm{Irr}(G)$, its Schur index over $\tilde{\mathbf{Q}}_p$ is equal to 1.

17. (Alperin) We say that an $\mathfrak{o}G$-module is a *local module* if it is the direct sum of modules induced from p-local subgroups. If V is an $\mathfrak{o}G$-module, then there exist local modules U, U' and projective $\mathfrak{o}G$-modules P, P' such that

$$V \oplus U \oplus P \simeq U' \oplus P'.$$

[Hint: V may be assumed to be indecomposable. Use induction on the order of a vertex of V.]

18. If V, U are FG-modules and $n \in \mathbf{Z}$, then $\Omega^n(V) \otimes_F U \simeq \Omega^n(V \otimes_F U) \oplus P$, with P being projective.

19. An FG-module M is said to be *periodic* if there is a natural number n such that $\Omega^n(M) \simeq M$. Let $G \geq H$.

(i) If an FH-module W is periodic, then W^G is a direct sum of periodic modules and projective modules.

(ii) If an FG-module V is periodic, then V_H is a direct sum of periodic modules and projective modules.

20. Let P be a p-group and let $\Phi(P)$ be the Frattini subgroup of P.

(i) $J(F[\Phi(P)]) \subset J(FP)^2$.

(ii) Considering $P/\Phi(P)$ as a module over the prime field $\mathbf{Z}/(p) = F_0$, we have the following isomorphism:

$$F \otimes_{F_0} P/\Phi(P) \simeq J(FP)/J(FP)^2 \qquad (\alpha \otimes \bar{x} \mapsto \alpha(x-1) + J(FP)^2).$$

(iii) $\Omega^2(F_P) \simeq F_P \Leftrightarrow P$ is cyclic.

21. Let $(K, R, F) \leq (K', R', F')$. Let A be an R-algebra and let V, W be A-modules that are finitely generated over R.

(i) If $V^{R'} | W^{R'}$, then $V | W$.

(ii) If $V^{R'} \simeq W^{R'}$, then $V \simeq W$.

[Hint: (i) Assume first that V is indecomposable. Use 11.12 in Chapter 1.]

22. (Maranda) Let R be a discrete valuation ring and assume that the characteristic of the quotient field of R does not divide $|G|$. Let V, W be RG-modules that are R-free of finite rank. Let $\tilde{R}$ be the completion of R. Then we have that

$$\tilde{R} \otimes_R V \simeq \tilde{R} \otimes_R W \Rightarrow V \simeq W.$$

We prove this in the following way. Let (π) be the unique maximal ideal of R. Then we have a natural isomorphism $R/\pi^n R \simeq \tilde{R}/\pi^n \tilde{R}$ for all $n \geq 0$, by applying Chapter 1, 13.20 with $\pi_i = \pi^i$, for instance, and hence

$$(*) \qquad V/\pi^n V \simeq W/\pi^n W \qquad \text{for all} \quad n \geq 0.$$

Thus it suffices to show that $(*)$ implies $V \simeq W$.

(i) Let n be a fixed positive integer. Since V, W are R-free, the isomorphism $V/\pi^n V \simeq W/\pi^n W$ lifts to an R-isomorphism $f\colon V \simeq W$. Hence the following holds for all $x \in G$:

$$f^x - f \equiv 0 \qquad \text{mod } \pi^n \operatorname{Hom}_R(V, W).$$

(ii) Let $g(x) = \pi^{-n}(f^x - f)$ for $x \in G$. Then g is a 1-cocycle of G, namely, $g \in Z^1(G, \operatorname{Hom}_R(V, W))$.

(iii) Let $|G|R = (\pi^k)$. Then, by Chapter 2, 7.3(i), there exists

$$h \in \operatorname{Hom}_R(V, W)$$

such that the following is true for all $x \in G$:

$$\pi^k g(x) = h^x - h.$$

(iv) Let $n > k$. Then it holds that $(f - \pi^{n-k}h)^x = f - \pi^{n-k}h$ for all $x \in G$. In other words, $f - \pi^{n-k}h$ is an RG-isomorphism from V onto W.

5 Theory of Blocks

In this chapter, we shall establish Brauer's first, second, and third main theorems on blocks, Fong's theory of block covering, and related results. We shall give new proofs to the second and third main theorems, which are due to Green, Tsushima, Watanabe, and Juhász. For the Brauer correspondence, we shall start with the standard definition and then in Section 10 explain the module-theoretic version of it given by Alperin and Burry. In the final section, we shall generalize the argument of Alperin on the Glauberman correspondence (following the method of Issacs).

Throughout this chapter G will always denote a finite group. We fix a p-modular system (K, R, F), in which K will be assumed, unless otherwise specified, to have a primitive $|G|$th root of unity. The maximal ideal of R will be denoted by (π), and if V is an R-module, then the image of $v \in V$ by the natural map $V \to V/\pi V$ will be denoted by v^*. Also we write $V^* = V/\pi V$. As in the preceding chapter, we let $\mathfrak{o}$ denote either R or F. So, if $\mathfrak{o} = F$, then $(\pi) = 0$. All modules and algebras will be assumed to be finitely generated over the coefficient rings under consideration.

1. Defect Groups of a Block

For the time being, we do not assume that K contains a primitive $|G|$th root of unity.

The following fact, which is referred to as *Rosenberg's lemma* is easy but useful.

Lemma 1.1. *Let A be an $\mathfrak{o}$-algebra and let $I_1, \ldots, I_r$ be ideals of it. If $e \in \mathrm{pi}(A)$ belongs to $I_1 + \cdots + I_r$, then $e \in I_i$ for some i.*

Proof. Since eAe is a local ring with e as identity, we find that $eI_ie \in J(eAe)$, unless $eI_ie = eAe$. Our assumption implies that $e \in eI_1e + \cdots + eI_re \in eAe$. So, if $eI_ie \neq eAe$ for all i, then $e \in J(eAe)$, which is impossible. Therefore, $e \in eAe = eI_ie \in I$ for some i. ■

Now let A be a G-algebra over $\mathfrak{o}$. For $G \geq H$, let us denote $\mathrm{Inv}_H A$ by A^H. Also, if $\mathrm{Tr}_H^G \colon A^H \to A^G$ is the trace map, then we abbreviate the image $\mathrm{Tr}_H^G(A^H)$ to $\mathrm{Tr}_H^G(A)$ and set

$$A_H^G = \mathrm{Tr}_H^G(A) + \pi A^G.$$

By Chapter 4, 1.6, A_H^G is an ideal of A^G, and if $G \geq H$, L, then

$$A_H^G A_L^G \subset \sum_{t \in H\backslash G/L} A_{H^t \cap L}^G. \tag{1.1}$$

Theorem 1.2. *Given $e \in \mathrm{pi}(A^G)$, there exists a p-subgroup D of G satisfying the following two conditions* (a) *and* (b). *Moreover, D is unique up to G-conjugacy.*

(a) $e \in A_D^G$.
(b) $e \in A_H^G \Rightarrow D \leq_G H$.

Proof. There always exists a subgroup D of G satisfying (a) ($D = G$, for instance). Choose D with $|D|$ minimal such that $e \in A_D^G$. If $e \in A_H^G$ for some $H \leq G$, then by (1.1),

$$e \in A_D^G A_H^G \subset \sum_t A_{D^t \cap H}^G,$$

and hence $e \in A^G_{D^t \cap H}$ for some t by Rosenberg's lemma. Then by the minimality of $|D|$, we have that $D^t \cap H = D^t \leq H$ and (b) follows. It is clear by (b) that D is unique up to G-conjugacy. Moreover, if $P \in \mathrm{Syl}_p(G)$, then $e = 1/|G:P| \operatorname{Tr}^G_P(e) \in A^G_P$. Therefore, $D \leq_G P$, and hence D is a p-group. ■

With the notation of the above theorem, we call D a *defect group* of e and write $D = \delta(e)$.

Exercise 1.3. For $e \in \mathrm{pi}(A^G)$, the following holds:

$$e \in A^G_H \Leftrightarrow e \in \operatorname{Tr}^G_H(A).$$

Let us first consider the case where the G-algebra A is the $\mathfrak{o}$-endomorphism ring of an $\mathfrak{o}G$-module V.

Theorem 1.4. *Let V be an $\mathfrak{o}G$-module and let $E = E_{\mathfrak{o}}(V)$. If ε is a primitive idempotent of $E^G = E_{\mathfrak{o}G}(V)$, then the following holds:*

$$\delta(\varepsilon) =_G \mathrm{vx}(\varepsilon V).$$

Proof. ε is the identity of $\varepsilon E \varepsilon = E_{\mathfrak{o}}(\varepsilon V)$, and $\varepsilon \in \operatorname{Tr}^G_H(E)$ if and only if $\varepsilon \in \operatorname{Tr}^G_H(\varepsilon E \varepsilon)$. Hence, by Chapter 4, 2.2, $\delta(\varepsilon)$ is a vertex of εV. ■

Let next $A = \mathfrak{o}G$. This is a G-algebra by conjugation and $A^G = Z(\mathfrak{o}G)$, the center of $\mathfrak{o}G$.

Lemma 1.5. *Let $G \geq H$ and $x \in G$. If C and C' respectively, denote the G-conjugate and H-conjugate classes containing x, then the following holds:*

$$\operatorname{Tr}^G_H(\hat{C}') = |C_G(x):C_H(x)|\hat{C}.$$

Proof. Set $G_x = C_G(x)$, $H_x = C_H(x)$. Then

$$\hat{C} = \operatorname{Tr}^G_{G_x}(x), \qquad \hat{C}' = \operatorname{Tr}^H_{H_x}(x),$$

and hence

$$\begin{aligned}\mathrm{Tr}_H^G(\hat{C}') &= \mathrm{Tr}_{H_x}^G(x) = \mathrm{Tr}_{G_x}^G(\mathrm{Tr}_{H_x}^{G_x}(x)) \\ &= |G_x : H_x|\,\mathrm{Tr}_{G_x}^G(x) \\ &= |G_x : H_x|\hat{C}. \qquad \blacksquare\end{aligned}$$

Recall that a defect group of $C \in \mathrm{Cl}(G)$, denoted by $\delta(C)$, is a Sylow p-subgroup of $C_G(x)$ for some $x \in C$ (cf. Chapter 3, Section 6.5).

Exercise 1.6. With the notation of Lemma 1.5, the following holds:

$$|C_G(x) : C_H(x)| \not\equiv 0 \pmod{p} \text{ for some } x \in C \Leftrightarrow \delta(C) \leq_G H.$$

Let

$$RG = \bigoplus_{B \in \mathrm{Bl}(G)} B$$

be the block decomposition of RG, and let e_B be the block idempotent of B (cf. Chapter 3, Section 6.4). We write

$$e_B = \sum_{C \in \mathrm{Cl}(G)} \beta_B(C)\hat{C} \qquad (\beta_B(C) \in R). \tag{1.2}$$

For a p-subgroup Q of G, let

$$Z_Q(\mathfrak{o}G) = \sum_{\delta(C) \leq_G Q} \mathfrak{o}\hat{C} + \pi Z(\mathfrak{o}G) \qquad (C \in \mathrm{Cl}(G)), \tag{1.3}$$

then the following holds.

Lemma 1.7. *Let Q be a p-subgroup of G. Then*

(i) $(\mathfrak{o}G)_Q^G = Z_Q(\mathfrak{o}G)$.

(ii) $(FG)_Q^G = (RG)_Q^G/\pi Z(RG)$.

(iii) *For a block idempotent e_B in RG, it holds that*

$$e_B \in (RG)_Q^G \Leftrightarrow e_B^* \in (FG)_Q^G,$$

and we have $\delta(e_B) = \delta(e_B^*)$.

(iv) *With the notation of (1.2), if $\beta_B(C)^* \neq 0$, then $\delta(C) \leq_G \delta(e_B)$. Moreover, there exists $C \in \mathrm{Cl}(G)$ such that $\beta_B(C)^* \neq 0$ and $\delta(C) =_G \delta(e_B)$.*

Proof.

(i) Since $(\mathfrak{o}G)_Q^G = \sum_{C'} \mathfrak{o}\,\mathrm{Tr}_Q^G(\hat{C}') + \pi Z(\mathfrak{o}G)$, with C' running through the Q-conjugate classes in G, it follows from Lemma 1.5 and Exercise 1.6 that $(\mathfrak{o}G)_Q^G \subset Z_Q(\mathfrak{o}G)$. On the other hand, if $C \in \mathrm{Cl}(G)$ satisfies $\delta(C) \leq_G Q$, then there exists $x \in C$ such that $|C_G(x):C_Q(x)|$ is prime to p. Let C' be the Q-conjugate class in G containing x. Then it follows that $\hat{C} = |C_G(x):C_Q(x)|^{-1}\,\mathrm{Tr}_Q^G(\hat{C}')$, and hence $Z_Q(\mathfrak{o}G) \subset (\mathfrak{o}G)_Q^G$.

Parts (ii) and (iii) are immediate from (i) and (ii), respectively.

(iv) Set $D = \delta(e_B)$. Then $e_B^* \in (FG)_D^G = Z_D(FG)$, which implies the first assertion. If the second assertion is false, then $e_B^* \in \sum_{Q<D} Z_Q(FG)$ and $e_B^* \in Z_Q(FG)$ for some $Q < D$ by Rosenberg's lemma, which is a contradiction. ■

If B is a block of G, then a defect group of e_B is called a *defect group of* B and is denoted by $\delta(B)$.

For $G \geq H$, $\mathrm{Cl}_H(G)$ denotes the set of H-conjugate classes in G. With this notation, we have $(\mathfrak{o}G)^H = \bigoplus_{C' \in \mathrm{Cl}_H(G)} \mathfrak{o}\hat{C}'$, and if $x \in C' \in \mathrm{Cl}_H(G)$, then $\hat{C}' = \mathrm{Tr}_{C_H(x)}^H(x)$.

The following lemma will be crucial in our proof of Brauer's second main theorem.

Lemma 1.8 (Green–Tsushima). *Let* $G \geq H$, $\mathrm{Cl}_H(G) = \{C_1', \ldots, C_r'\}$ *and* $x_i \in C_i'$. *Let V be an $\mathfrak{o}$-free $\mathfrak{o}G$-module and W an indecomposable $\mathfrak{o}H$-module that is a direct summand of* V_H. *If* $a = \sum_{i=1}^r \alpha_i \hat{C}_i' \in (\mathfrak{o}G)^H$ *satisfies* $Wa = W$, *then there exists* x_i *such that*

$$\alpha_i \not\equiv 0 \pmod{\pi} \text{ and } W \text{ is } C_H(x_i)\text{-projective.}$$

Proof. Every $c \in \mathfrak{o}G$ induces an $\mathfrak{o}$-homomorphism $\gamma_c\colon V \to V (v \mapsto vc)$, and if $x \in G$, we have $(\gamma_c)^x = \gamma_{x^{-1}cx}$. Hence $\mathrm{Tr}_{C_H(x_i)}^H(\gamma_{x_i}) = \sum_{x \in C_i'} \gamma_x$, and in particular, we have $\gamma_a = \sum_i \alpha_i\,\mathrm{Tr}_{C_H(x_i)}^H(\gamma_{x_i})$. Now if γ_a' denotes the restriction of γ_a to W, then $\gamma_a'(W) = W$ by assumption. But since W is $\mathfrak{o}$-free, the kernel of the epimorphism $\gamma_a'\colon W \to W$ must be zero, and hence γ_a' is an $\mathfrak{o}H$-automorphism of W.

Let $V_H = W \oplus W'$ and π_W be the projection on W. Then we have

$$\mathrm{id}_W = \gamma_a'^{-1}\pi_W\gamma_a' = \sum_{i=1}^r \alpha_i\,\mathrm{Tr}_{C_H(x_i)}^H(\gamma_a'^{-1}\pi_W\gamma_{x_i}'),$$

and since $\mathrm{End}_{\mathfrak{o}H}(W)$ is a local ring, some $\alpha_i \, \mathrm{Tr}^H_{C_H(x_i)}(\gamma_a'^{-1}\pi_W\gamma_{x_i}')$ must be a unit of it, by Chapter 1, 5.8. This implies that $\alpha_i \not\equiv 0(\mathrm{mod}(\pi))$, and we have $E_{\mathfrak{o}H}(W) = \mathrm{Tr}^H_{C_H(x_i)}(E_{\mathfrak{o}}(W))$. Thus W is $C_H(x_i)$-projective by Chapter 4, 2.2. ■

Now we show the following result on the vertex of an indecomposable module belonging to a block B of G.

Theorem 1.9 (Green). *Let V be an indecomposable $\mathfrak{o}G$-module that is $\mathfrak{o}$-free and let B be a block of G to which V belongs. Then*

(i) $\mathrm{vx}(V) \leq_G \delta(B)$.

(ii) $\nu(\mathrm{rank}_{\mathfrak{o}} V) \geq \nu(|G|) - \nu(|\delta(B)|)$, *and the equality sign holds only if* $\mathrm{vx}(V) =_G \delta(B)$.

Proof.

(i) With the notation of (1.2), we have $Ve_B = V$. Hence, by applying the above lemma with $H = G$, we find $C \in \mathrm{Cl}(G)$ such that $\beta_B(C)^* \neq 0$ and V is $\delta(C)$-projective. Then $\mathrm{vx}(V) \leq_G \delta(C) \leq_G \delta(B)$ (cf. Lemma 1.7) and (i) follows. Part (ii) is immediate by Chapter 4, 7.5. ■

Corollary 1.10. *Let V be an R-free RG-module such that V^K is an irreducible KG-module belonging to a block B of G. If $\chi \in \mathrm{Irr}(B)$ is the character defined by V^K, then the following statement holds for $x \in G$.*

$$x_p \notin_G \delta(B) \Rightarrow \chi(x) = 0.$$

Proof. Since V is an indecomposable RG-module belonging to B, $\mathrm{vx}(V) \leq_G \delta(B)$ by the above theorem. Then the assertion follows readily from Chapter 4, 7.4. ■

For the remainder of this section we assume that

*K contains a primitive $|G|$*th *root of unity.*

As in Chapter 3, Section 6.4, ω_B^* denotes the linear function of $Z(FG)$ associated with B, $G_{p'}$, the set of p-regular elements of G, and $\mathrm{Cl}(G_{p'})$, the set of p-regular classes of G. According to Chapter 3, 6.22(ii), we have

$$\beta_B(C) \neq 0 \Rightarrow C \in \mathrm{Cl}(G_{p'}).$$

This, together with Lemma 1.7(iv), yields the assertion (i) of the next theorem.

Theorem 1.11. *Let* $B \in \mathrm{Bl}(G)$.

(i) $\beta_B(C)^* \neq 0 \Rightarrow \delta(C) \leq_G \delta(B)$.

(ii) $\omega_B(\hat{C})^* \neq 0 \Rightarrow \delta(B) \leq_G \delta(C)$.

Moreover there exists $C \in \mathrm{Cl}(G_{p'})$ *such that* $\omega_B^*(\hat{C}) \neq 0$ *and* $\delta(C) =_G \delta(B)$.

(iii) $|\delta(B)| = p^{d(B)}$.

(iv) *Let* $\chi \in \mathrm{Irr}(B)$ *be defined by an* RG-*module* V. *Then the following holds*:

$$\mathrm{ht}(\chi) = 0 \Rightarrow \mathrm{vx}(V) =_G \delta(B).$$

Proof.

(ii) If $\omega_B^*(\hat{C}) \neq 0$, then $\omega_B^*(\hat{C}) = \omega_B^*(e_B^*\hat{C}) \neq 0$, and hence $e_B^*\hat{C}$ is a unit of $Z(B^*)$ since $Z(B^*)$ is a local ring. Hence $e_B^* \in Z(B^*) = Z(B^*)\hat{C}$. Thus, if we put $Q = \delta(C)$, then $\hat{C} \in Z_Q(FG)$, and it follows that $e_B^* \in Z_Q(FG)$, since $Z_Q(FG)$ is an ideal of $Z(FG)$. Therefore $\delta(B) \leq_G Q$.

Moreover, since $1 = \omega_B^*(e_B^*) = \sum_C \beta_B(C)^* \omega_B^*(\hat{C})$, there is $C \in \mathrm{Cl}(G_{p'})$ such that $\beta_B(C)^* \neq 0$ and $\omega_B^*(\hat{C}) \neq 0$. It is then clear that $\delta(C) =_G \delta(B)$.

(iii) Choose C so that $\beta_B(C)^* \neq 0 \neq \omega_B^*(\hat{C})$. Then $\delta(C) =_G \delta(B)$, and hence $|\delta(B)| = |\delta(C)| = p^{d(C)}$. But we know that $d(C) = d(B)$ (Chapter 3, 6.27). Thus, $\delta(B) = p^{d(B)}$.

Part (iv) is immediate from (iii) above and Theorem 1.9(ii). ■

As is shown in the above proof, given $B \in \mathrm{Bl}(G)$, there exists $C \in \mathrm{Cl}(G_{p'})$ satisfying

$$\beta_B(C)^* \neq 0 \neq \omega_B^*(\hat{C}).$$

The above C is said to be a *defect class* for the block B (though C is not unique).

Corollary 1.12. *Let* B, $B' \in \mathrm{Bl}(G)$ *and suppose that* $\delta(B) =_G \delta(B')$. *Then* $B = B'$ *if and only if* $\omega_B^*(\hat{C}) = \omega_{B'}^*(\hat{C})$ *for all* $C \in \mathrm{Cl}(G_{p'})$ *such that* $\delta(C) =_G \delta(B)$.

Proof. It suffices to show the "if" part. Using $\delta(B') =_G \delta(B)$ and Theorem 1.11, we observe

$$\omega_{B'}^*(e_B^*) = \sum_{\substack{C \in \mathrm{Cl}(G_{p'}) \\ \delta(C) =_G \delta(B)}} \beta_B(C)^* \omega_{B'}^*(\hat{C}) = \omega_B^*(e_B^*) = 1,$$

whence we have $B' = B$. ■

2. The Brauer Homomorphism and the First Main Theorem

In this section, we do not assume that K contains a primitive $|G|$th root of unity.

2.1. The Brauer Homomorphism

Set $\mathrm{Cl}(G) = \{C_\alpha;\ 1 \le \alpha \le k\}$ and let $x_\alpha \in C_\alpha$. If we write

$$\hat{C}_\alpha \hat{C}_\beta = \sum_\gamma t_{\alpha\beta\gamma} \hat{C}_\gamma,$$

with $t_{\alpha\beta\gamma} \in \mathbf{Z}$, then

$$t_{\alpha\beta\gamma} = \#\{(a, b) \in C_\alpha \times C_\beta;\ ab = x_\gamma\}.$$

Lemma 2.1. *If* $t_{\alpha\beta\gamma} \not\equiv 0 \pmod{p}$, *then* $\delta(C_\gamma) \le_G \delta(C_\alpha)$ *and* $\delta(C_\gamma) \le_G \delta(C_\beta)$.

Proof. Let $Q = \delta(C_\gamma) \in \mathrm{Syl}_p(C_G(x_\gamma))$. Clearly, by conjugation, Q acts on the set $\{(a, b) \in C_\alpha \times C_\beta;\ ab = x_\gamma\}$, and the assumption implies that there is a Q-orbit of length 1, say $\{(a, b)\}$, on it. Then $Q \le C_G(a) \cap C_G(b)$. ■

In the following, let $G \ge Q$ be a p-subgroup and let, for $C \in \mathrm{Cl}(G)$, $C^0 = C \cap C_G(Q)$. Note that C^0 is $N_G(Q)$-stable under conjugation, and hence is a union of $N_G(Q)$-conjugate classes in G.

Exercise 2.2. $C^0 \ne \varnothing \Leftrightarrow Q \le_G \delta(C)$.

Theorem 2.3. *If* $C_G(Q) \le H \le N_G(Q)$, *the following holds in* $Z(FH)$:

$$\hat{C}_\alpha^0 \hat{C}_\beta^0 = \sum_\gamma t_{\alpha\beta\gamma}^* \hat{C}_\gamma^0.$$

Consequently, we obtain the following F-algebra homomorphism

$$\mathrm{Br}_Q\colon Z(FG) \to Z(FH) \qquad (\hat{C} \mapsto \hat{C}^0),$$

where we set $\hat{C}^0 = 0$ *if* $C^0 = \varnothing$.

Proof. Let $x_\gamma \in C_\gamma^0 = C_\gamma \cap C_G(Q)$. It suffices to show that

$$\#\{(a,b) \in C_\alpha^0 \times C_\beta^0;\ ab = x_\gamma\} \equiv t_{\alpha\beta\gamma} \pmod p.$$

Set $C'_\alpha = C_\alpha - C_\alpha^0$. Then

$$C_\alpha \times C_\beta = C_\alpha^0 \times C_\beta^0 \cup C_\alpha^0 \times C'_\beta \cup C'_\alpha \times C_\beta^0 \cup C'_\alpha \times C'_\beta,$$

and we see that neither $C_\alpha^0 C'_\beta$ nor $C'_\alpha C_\beta^0$ includes x_γ. Thus, if we set $X = \{(a,b) \in C_\alpha^0 \times C_\beta^0;\ ab = x_\gamma\}$ and $Y = \{(a',b') \in C'_\alpha \times C'_\beta;\ a'b' = x_\gamma\}$, then $t_{\alpha\beta\gamma} = |X| + |Y|$. But Q acts on Y (by conjugation) and thus, if $Y \neq \varnothing$, every Q-orbit has a length divisible by p. Consequently, $|Y| \equiv 0 (\mathrm{mod}\ p)$ and hence $|X| \equiv t_{\alpha\beta\gamma} (\mathrm{mod}\ p)$. ■

Exercise 2.4. Prove that $\mathrm{Ker}\ \mathrm{Br}_Q = \sum_{Q \not\leq \delta(C)} F\hat{C}$.

The homomorphism Br_Q in the above theorem is called the *Brauer homomorphism* (with respect to (G, Q, H)). The defect group of a block can be characterized in terms of the Brauer homomorphism as follows.

Theorem 2.5. *Let B be a block with block idempotent* e_B. *Then*

$$\mathrm{Br}_Q(e_B^*) \neq 0 \Leftrightarrow Q \leq_G \delta(B).$$

Namely, the defect group $\delta(B)$ *is the maximal one (up to G-conjugacy) of the p-subgroups Q of G such that* $\mathrm{Br}_Q(e_B^*) \neq 0$.

Proof. By Lemma 1.7(iv), we may write

$$e_B^* = \sum_{\delta(C) \leq_G \delta(B)} \beta_B(C)^* \hat{C}.$$

($\Rightarrow$). If $Q \not\leq_G \delta(B)$, then $Q \not\leq_G \delta(C)$ for every $C \in \mathrm{Cl}(G)$ such that $\delta(C) \leq_G \delta(B)$, whence it follows that $e_B^* \in \mathrm{Ker}\ \mathrm{Br}_Q$.

($\Leftarrow$) There exists $C \in \mathrm{Cl}(G)$ such that $\beta_B(C)^* \neq 0$ and $\delta(C) =_G \delta(B)$. Thus, if $Q \leq_G \delta(B)$, then $e_B^* \notin \mathrm{Ker}\ \mathrm{Br}_Q$ by Exercise 2.4. ■

Suppose that $C_G(Q) \leq H \leq N_G(Q)$. If $\mathrm{Br}_Q(e_B^*) \neq 0$, then this is an idempotent of $Z(FH)$, and hence we have a primitive idempotent decomposition

$$\mathrm{Br}_Q(e_B^*) = e_{b_1}^* + \cdots + e_{b_r}^*$$

in $Z(FH)$. We say that each $b_i \in \mathrm{Bl}(H)(1 \leq i \leq r)$ is *associated* with B (relative to (Q, H)).

2.2. *The First Main Theorem*

Let us denote by $\mathrm{Bl}(G|Q)$ the set of blocks of G with defect group Q: $\mathrm{Bl}(G|Q) = \{B \in \mathrm{Bl}(G);\ \delta(B) =_G Q\}$. The first main theorem of Brauer (cf. Theorem 2.15 below) asserts that $\mathrm{Bl}(G|Q)$ and $\mathrm{Bl}(N_G(Q)|Q)$ correspond bijectively via Br_Q. In order to show this, we need some preparation.

Let $G \trianglerighteq H$ and $\bar{G} = G/H$. The natural map

$$\mu_H\colon FG \to F\bar{G} \quad \left(\sum_{x \in G} \alpha_x x \mapsto \sum_{x \in G} \alpha_x \bar{x}\right)$$

is an F-algebra homomorphism. If in particular $G = H$, we have

$$\mu_G\colon FG \to F \quad \left(\sum_{x \in G} \alpha_x x \mapsto \sum_{x \in G} \alpha_x\right).$$

This is called the *augmentation map* of FG, and the Ker μ_G is called the *augmentation ideal* of FG, which is denoted by $I(G)$. It is easy to see that

$$I(G) = \mathrm{Ker}\,\mu_G = \bigoplus_{x \in G} F(x - 1).$$

Lemma 2.6. *Let $G \trianglerighteq H$ and $\bar{G} = G/H$. Then the following statements hold.*

(i) Ker $\mu_H = I(H) \cdot FG = FG \cdot I(H)$.

(ii) *For $\mathrm{Cl}(G) \ni C \ni x$, let $\bar{C}$ denote the conjugate class in $\bar{G}$ that contains $\bar{x}$. Then*

$$\mu_H(\hat{C}) = |C_{\bar{G}}(\bar{x}) : \overline{C_G(x)}|\,|H : C_H(x)|\,\hat{\bar{C}}.$$

(iii) *With the notation of the above* (ii), $\mu_H(\hat{C}) \neq 0$ *if and only if the following two conditions* (a) *and* (b) *hold*:

(a) *$\delta(C)$ contains a Sylow p-subgroup of H.*

(b) *$\overline{C_G(x)} = HC_G(x)/H$ contains a Sylow p-subgroup of $C_{\bar{G}}(\bar{x})$.*

Proof.

(i) This is obvious since, for $a = \sum_{t \in H\backslash G} \sum_{h \in H} \alpha_{ht} ht \in FG$, we have

$$\mu_H(a) = \sum_{\bar{t} \in \bar{G}} \left(\sum_{h \in H} \alpha_{ht} \right) \bar{t}.$$

(ii) Note that $\mu_H(\hat{C}) = (|C|/|\bar{C}|)\hat{\bar{C}}$, because the number of $y \in C$, such that $\bar{y} = \bar{x}$, is independent of the choice of $\bar{x}$, which is equal to $|C|/|\bar{C}|$. Set $C_G(\bar{x}) = \{a \in G; \bar{a}^{-1}\bar{x}\bar{a} = \bar{x}\}$. Then $C_{\bar{G}}(\bar{x}) = C_G(\bar{x})/H$, and it follows that

$$\begin{aligned} |C|/|\bar{C}| &= |G:C_G(x)|/|G:C_G(\bar{x})| = |C_G(\bar{x}):C_G(x)| \\ &= |C_G(\bar{x}):HC_G(x)||HC_G(x):C_G(x)| \\ &= |C_G(\bar{x})/H:HC_G(x)/H||H:H \cap C_G(x)| \\ &= |C_{\bar{G}}(\bar{x}):\overline{C_G(x)}||H:C_H(x)|. \end{aligned}$$

(iii) This follows immediately from the fact that $\mu_H(\hat{C}) \neq 0$ if and only if $|C|/|\bar{C}| \not\equiv 0 \pmod{p}$. ■

Lemma 2.7. *Let Q be a normal p-subgroup of G. Then*

(i) Ker $\mu_Q \subset J(FG)$.

(ii) *For $C \in \mathrm{Cl}(G)$ the following hold*:

$$Q \leq \delta(C) \Leftrightarrow C \subset C_G(Q), \tag{2.3}$$

$$Q \not\leq \delta(C) \Rightarrow \hat{C} \in J(Z(FG)). \tag{2.4}$$

(iii) *If P is a p-subgroup of G such that $Q \not\leq P$, then $Z_P(FG)$ is a nilpotent ideal of $Z(FG)$.*

Proof.

(i) Since Q is a p-group, $I(Q)$ is a nilpotent ideal of FQ. Hence Ker $\mu_Q = I(Q) \cdot FG = FG \cdot I(Q)$ is also a nilpotent ideal.

(ii) The assertion (2.3) is clear from $C_G(Q) \lhd G$. If, on the other hand, $Q \not\leq \delta(C)$, then the condition (a) of Lemma 2.6(iii) is not satisfied for $H = Q$, and hence $\hat{C} \in$ Ker $\mu_Q \cap Z(FG) \subset J(Z(FG))$.

(iii) We have $Z_P(FG) = \sum_{\delta(C) \leq_G P} F\hat{C}$. If $\delta(C) \leq_G P$, then $Q \not\leq \delta(C)$, and it follows from (2.4) that $Z_P(FG) \subset J(Z(FG))$. ■

Theorem 2.8. *Let Q be a normal p-subgroup of G. Then the following statements hold for each block B of G.*

(i) $Q \leq \delta(B)$.

(ii) $e_B \in (RC_G(Q))^G (= \mathrm{Inv}_G(RC_G(Q)) \subset Z(RC_G(Q)))$.

Proof.

(i) Set $D = \delta(B)$. We have $e_B^* \in Z_D(FG)$. Suppose to the contrary that $Q \nleq D$. Then $Z_D(FG)$ is nilpotent by (iii) of the above lemma, which is a contradiction.

(ii) Let $e_B = e_1 + e_2$, where $e_1 = \sum_{Q \leq \delta(C)} \beta_B(C)\hat{C}$ and

$$e_2 = \sum_{Q \nleq \delta(C)} \beta_B(C)\hat{C}.$$

By the above lemma, e_2^* is nilpotent, and so $e_2^{*p^m} = 0$ for some m. Then $e_B^* = e_B^{*p^m} = e_1^{*p^m} + e_2^{*p^m} = e_1^{*p^m} \in Z(FC_G(Q))$. Since $Z(RC_G(Q))^* = Z(FC_G(Q))$, we find a unique idempotent ε of $Z(RC_G(Q))$ such that $\varepsilon^* = e_B^*$ (cf. Chapter 1, 14.2 and 4.6). Hence $\varepsilon^x \in Z(RC_G(Q))$ for any $x \in G$ and $(\varepsilon^x)^* = (\varepsilon^*)^x = e_B^{*x} = e_B^*$, whence it follows that $\varepsilon^x = \varepsilon$, i.e., $\varepsilon \in Z(RG)$. Therefore we conclude that $e_B = \varepsilon \in (RC_G(Q))^G$ by Chapter 1, 4.6. ■

The block of G to which the principal character $\mathbf{1}_G$ belongs is said to be the *principal block*, which is denoted by $B_0(G)$ or simply by B_0.

Exercise 2.9. Let $B \in \mathrm{Bl}(G)$. Then $\delta(B) \in \mathrm{Syl}_p(G)$ if and only if there exists $\chi \in \mathrm{Irr}(B)$ such that $(\chi(1), p) = 1$. In particular, $\delta(B_0)$ is a Sylow p-subgroup of G.

Exercise 2.10. If there exists a normal p-subgroup Q such that $C_G(Q)$ is a p-group, then B_0 is a unique block of G [Hint: Theorem 2.8].

For a ring A not necessarily having identity, we also consider the set $\mathrm{pi}(A)$ of all its primitive idempotents. It is clear that if I is an ideal of A, then $\mathrm{pi}(I) = \mathrm{pi}(A) \cap I$.

Lemma 2.11. *Let A be a commutative Artinian ring, $I_0 \subset I$ ideals of A, and set $\bar{A} = A/I_0$. Then the set $\{e \in \mathrm{pi}(A);\ e \in I - I_0\}$ corresponds bijectively to $\mathrm{pi}(\bar{I})$ under the natural map $\varphi\colon A \to \bar{A}$.*

Proof. If $e \in \mathrm{pi}(A)$ does not lie in I_0, then $eA(=eAe)$ is a local ring and $eI_0 \subsetneqq eA$. Consequently, $\bar{e}\bar{A} \simeq eA/eI_0$ is also a local ring and hence $\bar{e} \in \mathrm{pi}(\bar{A})$. In particular, if $1 = \sum_{e \notin I_0} e + \sum_{e \in I_0} e$ is the primitive idempotent decomposition of 1, then $\bar{1} = \sum_{e \notin I_0} \bar{e}$ is a primitive idempotent decomposition of $\bar{1}$, and thus φ gives rise to a bijection between the set $\{e \in \mathrm{pi}(A);\ e \notin I_0\}$ and $\mathrm{pi}(\bar{A})$. Thus the assertion follows. ■

Let $G \geq Q$ be a p-subgroup as before and let

$$Z_Q^0(FG) = \sum_{\delta(C) =_G Q} F\hat{C}, \qquad Z_Q^1(FG) = \sum_{\delta(C) <_G Q} F\hat{C},$$

where $\delta(C) <_G Q$ means that $\delta(C)$ is conjugate in G to a proper subgroup of Q. Thus,

$$Z_Q(FG) = Z_Q^0(FG) \oplus Z_Q^1(FG), \tag{2.5}$$

and $Z_Q^1(FG) = \sum_{X < Q} Z_X(FG)$ is an ideal of $Z(FG)$.

Lemma 2.12. *Let Q be a p-subgroup of G. Then*

(i) $Z_Q(FG) \cap \mathrm{Ker}\,\mathrm{Br}_Q = Z_Q^1(FG)$.

(ii) *It holds that* $\{e_B^*;\ B \in \mathrm{Bl}(G|Q)\} = \{e_B^*;\ e_B^* \in Z_Q(FG) - Z_Q^1(FG)\}$, *which corresponds bijectively to* $\mathrm{pi}(Z_Q(FG)/Z_Q^1(FG))$ *via the natural map.*

Proof.

(i) Clear by Exercise 2.4.

(ii) The first half is obvious. For the second half, apply Lemma 2.11 with $A = Z(FG)$, $I = Z_Q(FG)$, and $I_0 = Z_Q^1(FG)$. ■

Lemma 2.13. *Let Q be a normal p-subgroup of G. Then the following statements hold.*

(i) $Z_Q^1(FG)$ *is a nilpotent ideal of* $Z(FG)$.

(ii) $\mathrm{Br}_Q(Z_Q(FG)) = Z_Q^0(FG)$, *and this is an F-algebra not necessarily having the identity element.*

(iii) $\mathrm{pi}(Z_Q(FG)) = \mathrm{pi}(Z_Q^0(FG)) = \{e_B^*;\ B \in \mathrm{Bl}(G|Q)\}$.

Proof.

(i) Clear by Lemma 2.7.

(ii) From (2.5) and Lemma 2.12(i) we get $\mathrm{Br}_Q(Z_Q(FG)) = \mathrm{Br}_Q(Z_Q^0(FG))$. And we have $\mathrm{Br}_Q(Z_Q^0(FG)) = Z_Q^0(FG)$ by the definition of Br_Q.

(iii) By Theorem 2.8 (ii), $\mathrm{Br}_Q(e_B^*) = e_B^*$ for all $B \in \mathrm{Bl}(G)$. Thus, if $e_B^* \in Z_Q(FG)$, then $e_B^* \in Z_Q^0(FG)$ by (ii) above, and so $\{e_B^*;\ B \in \mathrm{Bl}(G|Q)\} = \mathrm{pi}(Z_Q^0(FG))$. The first equality is immediate from (2.5) because $Z_Q^1(FG)$ is a nilpotent ideal. ■

The final preliminary lemma is a group-theoretical fact.

For a p-subgroup Q of G, let

$$\mathrm{Cl}(G|Q) = \{C \in \mathrm{Cl}(G);\ \delta(C) =_G Q\}.$$

Lemma 2.14. *Set $N = N_G(Q)$. If $C \in \mathrm{Cl}(G|Q)$, then $C^0 = C \cap C_G(Q)$ forms a single conjugate class in N. And $\mathrm{Cl}(G|Q)$ corresponds to $\mathrm{Cl}(N|Q)$ bijectively via $C \mapsto C^0$.*

Proof. If $x, y \in C^0$, then there exists $a \in G$ such that $y = x^a$. Since Q is a Sylow p-subgroup of both $C_G(x)$ and $C_G(y)$, Q^a is also a Sylow p-subgroup of $C_G(x)^a = C_G(y)$, and this implies that $Q^{ac} = Q$ for some $c \in C_G(y)$. Then $ac \in N = N_G(Q)$ and $x^{ac} = y$, that is, x and y are conjugate in N. Thus C^0 is a conjugate class of N.

It is clear that $C^0 \in \mathrm{Cl}(N|Q)$ and the map $C \mapsto C^0$ is an injection. To show that it is a surjection, take any $C' \in \mathrm{Cl}(N|Q)$. Let $x \in C'$, and let C be a conjugate class of G containing C'. If $Q \notin \mathrm{Syl}_p(C_G(x))$, there exists $P \in \mathrm{Syl}_p(C_G(x))$ properly containing Q. Then, as is well known, we have $Q < N_P(Q)(\leq N \cap C_G(x))$. But this contradicts $\delta(C') = Q$. Therefore $\delta(C) = {}_G Q$, and we have $C' = C \cap C_G(Q)$ as has been shown in the first half. ■

We are now ready to prove the following.

Theorem 2.15 (The first main theorem). *Let Q be a p-subgroup of G. Then there exists a bijection*

$$\varphi\colon \mathrm{Bl}(G|Q) \to \mathrm{Bl}(N_G(Q)|Q)$$

such that $\mathrm{Br}_Q(e_B^*) = e_{\varphi(B)}^*$ for $B \in \mathrm{Bl}(G|Q)$.

Proof. Set $N = N_G(Q)$. Consider the Brauer homomorphism Br_Q (with respect to (G, Q, N)) and restrict it to $Z_Q(FG)$ so that we obtain

$$\mathrm{Br}'_Q\colon Z_Q(FG) = Z_Q^0(FG) \oplus Z_Q^1(FG) \to Z_Q(FN).$$

Here we know that $\mathrm{Ker}\, \mathrm{Br}'_Q = Z_Q^1(FG)$ and $\mathrm{Im}\, \mathrm{Br}'_Q = Z_Q^0(FN)$ from Lemma 2.14, and hence Br'_Q induces the following isomorphism:

$$Z_Q(FG)/Z_Q^1(FG) \xrightarrow{\sim} Z_Q^0(FN),$$

whence the result follows from Lemmas 2.12 and 2.13. ■

Corollary 2.16. *Let Q be a p-subgroup of G and $N_G(Q) \leq H \leq G$. Then there is a bijection*

$$\psi\colon \mathrm{Bl}(G|Q) \to \mathrm{Bl}(H|Q)$$

such that $\mathrm{Br}_Q(e_B^*) = \mathrm{Br}_Q(e_{\psi(B)}^*)$ *for* $B \in \mathrm{Bl}(G|Q)$.

Proof. $N = N_G(Q) = N_H(Q)$ by assumption, and we have the following bijections:

$$\varphi_G\colon \mathrm{Bl}(G|Q) \to \mathrm{Bl}(N|Q), \qquad \varphi_H\colon \mathrm{Bl}(H|Q) \to \mathrm{Bl}(N|Q).$$

Thus we obtain the result by setting $\psi = \varphi_H^{-1} \circ \varphi_G$. ■

We denote by $O_p(G)$ the maximal normal p-subgroup of G. Also, we denote by $O_{p'}(G)$ the maximal normal p'-subgroup of G.

Exercise 2.17. If Q is a defect group of some block of G, then it holds that $O_p(N_G(Q)) = Q$. [Hint: Use the first main theorem and Theorem 2.8(i).]

3. The Brauer Correspondence

Let $G \geq H$. Define an R-homomorphism

$$s_H\colon Z(RG) \to Z(RH)$$

by $s_H(\hat{C}) = \sum_{x \in C \cap H} x$ and denote by s_H^* the induced map $Z(FG) \to Z(FH)$.

Then, given $b \in \mathrm{Bl}(H)$, we have an F-homomorphism

$$\omega_b^* \circ s_H^*\colon Z(FG) \to F.$$

If this is an F-algebra homomorphism, there exists $B \in \mathrm{Bl}(G)$ such that $\omega_b^* \circ s_H^* = \omega_B^*$. In this case, we say that b^G is *defined* and write $B = b^G$. If b^G is defined, the map $b \mapsto b^G$ is called the *Brauer correspondence*.

Lemma 3.1. *Let $G \geq H$, $b \in \mathrm{Bl}(H)$ and $\theta \in \mathrm{Irr}(b)$.*

(i) *For $C \in \mathrm{Cl}(G)$ the following holds:*

$$\omega_\theta \circ s_H(\hat{C}) = \frac{\theta^G(\hat{C})}{\theta^G(1)} \tag{3.1}$$

(ii) *If θ^G is irreducible, then b^G is defined and $\theta^G \in \mathrm{Irr}(b^G)$.*

Proof.

(i) $\omega_\theta \circ s_H(\hat{C}) = \omega_\theta(s_H(\hat{C})) = \theta(s_H(\hat{C}))/\theta(1)$. On the other hand, if $x \in C$, then

$$\theta^G(x) = \frac{1}{|H|}\sum_{t \in G} \hat{\theta}(txt^{-1}) = \frac{|C_G(x)|}{|H|}\,\theta(s_H(\hat{C})),$$

where $\hat{\theta}$ coincides with θ on H and vanishes outside H. Hence

$$\frac{\theta^G(\hat{C})}{\theta^G(1)} = \frac{|G|}{|C_G(x)|}\,\frac{\theta^G(x)}{|G:H|\theta(1)} = \frac{\theta(s_H(\hat{C}))}{\theta(1)},$$

and (3.1) holds.

(ii) If θ^G is irreducible and belongs to B, then $\omega_b^* \circ s_H^* = \omega_{\theta^G}^*$ by (3.1) and hence $\omega_b^* \circ s_H^* = \omega_B^*$. Therefore b^G is defined and $b^G = B$. ∎

Given $B \in \mathrm{Bl}(G)$ and a class function $\varphi = \sum_{\chi \in \mathrm{Irr}(G)} \alpha_\chi \chi$, define φ_B by

$$\varphi_B = \sum_{\chi \in B} \alpha_\chi \chi, \tag{3.2}$$

which is called the *B-component* of φ. Since $\chi(xe_B)$ is $\chi(x)$ or 0, according to whether $\chi \in B$ or not, we have the following:

$$\varphi_B(x) = \varphi(xe_B), \qquad \text{in particular} \quad \varphi_B(1) = \varphi(e_B). \tag{3.3}$$

Corollary 3.2. *Let $G \geq H$, $b \in \mathrm{Bl}(H)$, and $\theta \in \mathrm{Irr}(b)$. If b^G is defined, then the following statements hold for $B \in \mathrm{Bl}(G)$.*

(i) *If $B \neq b^G$, then $\nu((\theta^G)_B(1)) > \nu(\theta^G(1))$.*

(ii) *If $B = b^G$, then $\nu((\theta^G)_B(1)) = \nu(\theta^G(1))$.*

Proof. It follows from (3.1) that

$$\frac{(\theta^G)_B(1)}{\theta^G(1)} = \frac{\theta^G(e_B)}{\theta^G(1)} = \omega_\theta(s_H(e_B)) \in R.$$

Since b^G is defined, $\omega_\theta(s_H(e_B))^* = \omega_{b^G}^*(e_B^*)$, which equals 0 or 1 according to whether $B \neq b^G$ or $b^G = B$. Thus the assertions hold. ∎

We note the following result on defect groups of b^G.

Lemma 3.3. *Let* $b \in \mathrm{Bl}(H)$ *and suppose that* b^G *is defined. Then*

$$\delta(b) \leq_G \delta(b^G).$$

Proof. Let C be a defect class for the block $B = b^G$. Then $\delta(C) =_G \delta(B)$, and

$$0 \neq \omega_B^*(\hat{C}) = \omega_b^*(s_H^*(\hat{C})).$$

Consequently, there exists a conjugate class C' of H contained in $C \cap H$ with $\omega_b^*(\hat{C}') \neq 0$. Therefore, $\delta(b) \leq_H \delta(C') \leq_G \delta(C) =_G \delta(B)$. ∎

Lemma 3.4. *Let* $L \leq H \leq G$, $b \in \mathrm{Bl}(L)$ *and suppose that* b^H *is defined. If one of* b^G *and* $(b^H)^G$ *is defined, then the other is also defined and it holds that* $(b^H)^G = b^G$.

Proof. Define $t_L^*: Z(FH) \to Z(FL)$ by $\hat{C}' \mapsto \sum_{x \in C' \cap L} x$. Since b^H is defined by assumption, we have the following commutative diagram:

$$\begin{array}{ccccc} Z(FG) & \xrightarrow{s_H^*} & Z(FH) & \xrightarrow{\omega_{b^H}^*} & F \\ & \searrow^{s_L^*} & \downarrow^{t_L^*} & \nearrow_{\omega_b^*} & \\ & & Z(FL) & & \end{array}$$

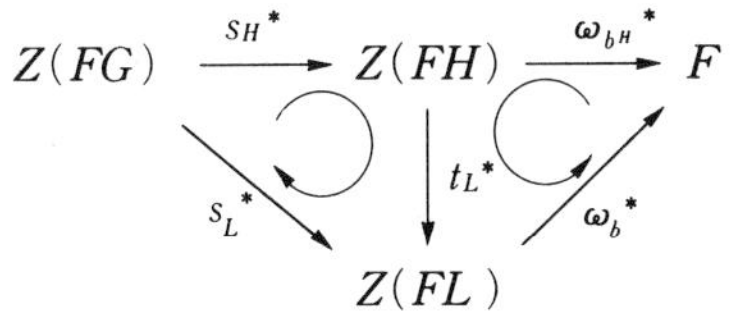

Therefore

$$\omega_b^* \circ s_L^* = \omega_b^* \circ t_L^* \circ s_H^* = \omega_{b^H}^* \circ s_H^*.$$

This is an algebra homomorphism if and only if one of b^G and $(b^H)^G$ is defined, then the other is also defined, and we have $(b^H)^G = b^G$. ∎

The next result gives a typical situation where b^G is always defined. In it the Brauer correspondence is given via the Brauer homomorphism.

Theorem 3.5. *Let Q be a p-subgroup of G and suppose that $QC_G(Q) \leq H \leq N_G(Q)$. Then the following statements hold.*

(i) *For every $b \in \mathrm{Bl}(H)$, we have*

$$\omega_b^* \circ s_H^* = \omega_b^* \circ \mathrm{Br}_Q.$$

Consequently, b^G is defined.

(ii) *If $B \in \mathrm{Bl}(G)$, then $\{b \in \mathrm{Bl}(H);\ b^G = B\}$ is the set of all blocks of H associated with B (relative to (Q, H)).*

Proof.

(i) For $C \in \mathrm{Cl}(G)$, set $C_0 = C \cap C_G(Q)$ and $C_1 = C \cap (H - C_G(Q))$. Thus $s_H^*(\hat{C}) = \hat{C}_0 + \hat{C}_1$. Since $Q \lhd H$, it follows from Lemma 2.7(ii) that $\omega_b^*(\hat{C}_1) = 0$. Hence, using $\hat{C}_0 = \mathrm{Br}_Q(\hat{C})$, we obtain $\omega_b^* \circ s_H^*(\hat{C}) = \omega_b^*(\hat{C}_0) = \omega_b^* \circ \mathrm{Br}_Q(\hat{C})$ as asserted. Moreover, since $\mathrm{Br}_Q\colon Z(FG) \to Z(FH)$ is an algebra homomorphism, so is $\omega_b^* \circ s_H^*$, and hence b^G is defined.

(ii) If $\mathscr{S}$ denotes the set of blocks of H associated with B, then

$$b \in \mathscr{S} \Leftrightarrow 1 = \omega_b^* \circ \mathrm{Br}_Q(e_B^*) = \omega_{b^G}^*(e_B^*) \Leftrightarrow b^G = B,$$

and the assertion (ii) follows. ■

Let $G \geq H$ and $b \in \mathrm{Bl}(H)$. If

$$C_G(\delta(b)) \leq H, \qquad (3.4)$$

then b is said to be $(G\text{-})$*admissible*.

Remark. Recall that $\delta(b)$ is unique up to H-conjugacy. Therefore, if one of the defect groups of b satisfies (3.4), then all others do.

Theorem 3.6. *Let $G \geq H$ and $b \in \mathrm{Bl}(H)$. If b is admissible, then b^G is defined.*

Proof. Set $Q = \delta(b)$. We get $H \geq L = QC_G(Q) = QC_H(Q)$ from the assumption. Since $\mathrm{Br}_Q(e_b^*) \neq 0$, there exists a block, say b_1, of L that is associated with b. Then, by Theorem 3.5, we have $b_1^H = b$, and b_1^G is defined. Therefore, by Lemma 3.4, $(b_1^H)^G = b^G$ is defined. ■

The above theorem will often be applied in the following situation.

Corollary 3.7. *Let Q be a p-subgroup and $QC_G(Q) \leq H \leq G$. If b is a block of H such that $Q \leq_H \delta(b)$, then b is admissible.*

Proof. Since $C_G(\delta(b)) \leq_H C_G(Q) \leq H$ from the assumption, b must be admissible. ■

Now Corollary 2.16, which is a refinement of the first main theorem, may be restated as follows.

Theorem 3.8. *Let Q be a p-subgroup of G and $N_G(Q) \leq H \leq G$. Then the map*

$$\mathrm{Bl}(H|Q) \to \mathrm{Bl}(G|Q) \qquad (b \mapsto b^G)$$

is a bijection.

Proof. Given $b \in \mathrm{Bl}(H|Q)$, we take, with the notation of Corollary 2.16, $B \in \mathrm{Bl}(G|Q)$ so that $\psi(B) = b$. Then there exists a block, say b_1, in $\mathrm{Bl}(N_G(Q)|Q)$ associated with both B and b. Hence $b_1^G = B$ and $b_1^H = b$ by Theorem 3.5, and we have $b^G = (b_1^H)^G = b_1^G = B$ by Lemma 3.4. ■

If $G \geq H$, the R-module $R[G - H] = \sum_{x \in G-H} Rx$ is an H-module by conjugation, and we write $R[G - H]^H$ for $\mathrm{Inv}_H(R[G - H])$.

The next lemma will be, like Lemma 1.8, of fundamental importance in our arguments that follow.

Lemma 3.9 (Juhász–Watanabe). *Let $G \geq H$, $b \in \mathrm{Bl}(H)$, and suppose that $b^G = B$ is defined. Then, for the block idempotents e_B and e_b, we have the following.*

(i) *$e_b s_H(e_B)$ is a unit of $e_b Z(RH)$, and there exists $z \in e_b Z(RH)$ such that*

$$ze_b e_B = e_b + c, \qquad \text{with} \quad c \in R[G - H]^H. \tag{3.5}$$

In particular, it holds that $e_b e_B \neq 0$.

(ii) *$e_b e_B = e_b + (1 - e_B)c$, with $c \in R[G - H]^H$.*

Proof.

(i) We have a decomposition $(RG)^H = Z(RH) \oplus R[G - H]^H$, and hence

$$e_B = s_H(e_B) + u, \qquad u \in R[G - H]^H.$$

Consequently,

$$e_b e_B = e_b s_H(e_B) + e_b u. \tag{3.6}$$

Now using $\omega_B^* = \omega_b^* \circ s_H^*$, we have

$$1 = \omega_B^*(e_B^*) = \omega_b^*(s_H(e_B)^*) = \omega_b^*((e_b s_H(e_B))^*).$$

Thus $e_b s_H(e_B)$ is a unit because $e_b Z(RH)$ is a local ring. Therefore, there exists $z \in e_b Z(RH)$ such that $ze_b s_H(e_B) = e_b$, and it follows from (3.6) that

$$ze_b e_B = e_b + c, \qquad \text{where} \quad c = ze_b u \in R[H - H]^H.$$

(ii) Since $ze_b e_B = (e_b + c)e_B = e_b e_B + ce_B$, we have $e_b + c = e_b e_B + e_B c$, then $e_b e_B = e_b + (1 - e_B)c$. ∎

For $B \in \mathrm{Bl}(G)$, we set

$$\mathrm{IND}(B) = \{V \in \mathrm{IND}(\mathfrak{o}G);\ V \in B\}.$$

Theorem 3.10 (Conlon). *Let $G \geq H$, $b \in \mathrm{Bl}(H)$, and suppose that $b^G = B$ is defined. If $W \in \mathrm{IND}(b)$, then*

$$W \mid (W^G e_B)_H.$$

Consequently, there exists $V \in \mathrm{IND}(B)$ satisfying

$$W \mid V_H \text{ and } V \mid W^G.$$

Proof. With the same notation as in (3.5) set $f = ze_b e_B = e_b + c$. Then

$$\lambda\colon W \to W^G e_B = W \otimes_{RH} (RG)e_B \qquad (w \mapsto w \otimes f)$$

and

$$\mu\colon W^G = \bigoplus_{t \in H\backslash G} W \otimes t \to W \qquad \left(\sum_t w_t \otimes t \mapsto w_1\right)$$

are RH-homomorphisms, where we assume that $H\backslash G \ni 1$. If μ' denotes the restriction of μ to $W^G e_B$, then for any $w \in W$,

$$\mu' \circ \lambda(w) = \mu'(w \otimes f) = \mu'(w \otimes e_b + w \otimes c) = we_b = w,$$

and thus $\mu' \circ \lambda = \mathrm{id}_W$. Therefore $W \mid (W^G e_B)_H$. In particular, there exists $V \in \mathrm{IND}(B)$ such that $V \mid W^G e_B$ and $W \mid V_H$. ■

There is a relationship between the Brauer and Green correspondences as follows.

Corollary 3.11 (Green). *Let P be a p-subgroup of G and $N = N_G(P)$. Let $\mathfrak{A} = \mathfrak{A}(G, P, N)$ be as in Chapter 4, (4.1) and let $f\colon \mathrm{IND}(\mathfrak{o}G \mid \mathfrak{A}) \to \mathrm{IND}(\mathfrak{o}N \mid \mathfrak{A})$ be the Green correspondence. If $V \in \mathrm{IND}(\mathfrak{o}G \mid \mathfrak{A})$ belongs to a block B of G, then fV belongs to the block b of N such that $b^G = B$.*

Proof. Let $fV \in b$ and $B' = b^G$. Apply Theorem 3.10 to $W = fV$ and we find $V' \in \mathrm{IND}(B')$ such that $\mathrm{vx}(V') = \mathrm{vx}(fV) = \mathrm{vx}(V)$ and $fV \mid V'_N$. It then follows that $V' = f^{-1}(fV) = V$, and hence $B' = B$. ■

Using Lemmas 1.8 and 3.9, we obtain the following.

Theorem 3.12 (Nagao–Green). *Let $G \geq H$ and $b \in \mathrm{Bl}(H)$. Let V be an $\mathfrak{o}$-free $\mathfrak{o}G$-module belonging to a block B of G and $W \in \mathrm{IND}(b)$ be a direct summand of V_H. If $C_G(\mathrm{vx}(W)) \leq H$, then $b^G = B$.*

Proof. Let $\mathfrak{o} = R$. Since $\mathrm{vx}(W) \leq_H \delta(b)$, it follows from the assumption that b is admissible and hence b^G is defined. If $b^G = B'$, then we have from Lemma 3.9 that

$$e_b e_{B'} = e_b + (1 - e_{B'})c', \qquad \text{with} \quad c' \in R[G - H]^H.$$

Suppose $B' \neq B$. Then by multiplying the above by e_B, we get

$$0 = e_b e_B + e_B c',$$

and hence $W = Wc'$ because $W = We_b = We_B$ from the assumption. Hence there exists, by Lemma 1.8, $y \in G - H$ such that W is $C_H(y)$-projective. Thus, $\mathrm{vx}(W) \leq_H C_H(y)$ and it follows that $y \in_H C_G(\mathrm{vx}(W)) \leq H$, which is a contradiction. Therefore $B' = B$. ■

4. Generalized Decomposition Numbers and the Second Main Theorem

We begin with the following theorem that implicitly implies the main theorem of Brauer on generalized decomposition numbers (see Theorem 4.2 below).

Theorem 4.1. *Let x be a p-element of G and assume that $C_G(x) \leq H \leq G$. Let V be an R-free RG-module belonging to a block B of G, and set $\mathrm{Bl}(H|B) = \{b \in \mathrm{Bl}(H);\ b^G = B\}$. Then, for $y \in C_G(x)_{p'}$, the following holds:*

$$\chi_V(xy) = \sum_{b \in \mathrm{Bl}(H|B)} \chi_{Ve_b}(xy).$$

Proof. Since $V_H = \bigoplus_{b \in \mathrm{Bl}(H)} Ve_b$, it suffices to show that $\chi_{Ve_b}(xy) = 0$ if $b \notin \mathrm{Bl}(H|B)$. This is true, by Corollary 1.10, provided $x \notin \delta(b)$. Thus assume that $Q = \langle x \rangle \leq \delta(b)$. It then holds that $C_G(Q) \leq H$ and hence b^G is defined. However, if $b^G \neq B$, then, for any indecomposable component W of Ve_b, we have $C_G(\mathrm{vx}(W)) \not\leq H$ by Theorem 3.12, and hence $Q \not\leq \mathrm{vx}(W)$, i.e., $x \notin \mathrm{vx}(W)$. Therefore $\chi_W(xy) = 0$, and thus $\chi_{Ve_b}(xy) = 0$ (cf. Chapter 4, 7.4). ■

Remark. If $\mathrm{Bl}(H|B) = \varnothing$ in the above theorem, then $\chi_V(xy) = 0$ for all $y \in C_G(x)_{p'}$.

Let x be a p-element of G and let $H = C_G(x)$. For $\chi \in \mathrm{Irr}(G)$, we write

(4.1)
$$\chi_H = \sum_{\zeta \in \mathrm{Irr}(H)} r_{\chi\zeta}\zeta.$$

Since $x \in Z(H)$, we see that $\zeta(xy) = \lambda_\zeta \lambda(y)$ for all $y \in H_{p'}$, where λ_ζ is a p^mth root of unity if $p^m = o(x)$. If we denote by $d_{\zeta\mu}^{(x)}$ the decomposition number in H associated with $\zeta \in \mathrm{Irr}(H)$ and $\mu \in \mathrm{IBr}(H)$, then

$$\begin{aligned}\chi(xy) &= \sum_{\mu \in \mathrm{IBr}(H)} \left(\sum_{\zeta} \lambda_\zeta r_{\chi\zeta} d_{\zeta\mu}^{(x)} \right) \mu(y)\\ &= \sum_{\mu \in \mathrm{IBr}(H)} d_{\chi\mu}^{x} \mu(y),\end{aligned}$$

where $d_{\chi\mu}^x = \sum_{\zeta \in \mathrm{Irr}(H)} \lambda_\zeta r_{\chi\zeta} d_{\zeta\mu}^{(x)}$. We call $d_{\chi\mu}^x$ the *generalized decomposition number*. If ρ is a primitive p^mth root of unity, then $d_{\chi\mu}^x \in \mathbf{Z}[\rho]$.

Remember that, for every $b \in \mathrm{Bl}(C_G(x))$, b^G is defined since $\langle x \rangle C_G(x) = C_G(x)$ (cf. Theorem 3.5).

The following result is referred to as the *second main theorem* of Brauer, which has various significant applications.

Theorem 4.2 (The second main theorem). *Let x be a p-element of G, $B \in \mathrm{Bl}(G)$, and $\chi \in \mathrm{Irr}(B)$. If $d_{\chi\mu}^x \neq 0$ and μ belongs to a block b of $C_G(x)$, then $b^G = B$. Hence we have*

$$\chi(xy) = \sum_{b^G = B} \sum_{\mu \in \mathrm{IBr}(b)} d_{\chi\mu}^x \mu(y)$$

for all $y \in C_G(x)_{p'}$.

Proof. Let V be an R-free RG-module which gives χ. From Theorem 4.1, we get

$$\chi(xy) = \sum_{b^G = B} \sum_{\zeta \in \mathrm{Irr}(b)} r_{\chi\zeta} \zeta(xy),$$

whence the assertion is immediate. ∎

In the following, we shall show some applications of the second main theorem.

Given a p-element x of G, we define

$$\mathfrak{S}(x) = \{z \in G;\, z_p =_G x\}$$

and call it the *p-section* (containing x). In particular, $\mathfrak{S}(1) = G_{p'}$, and it is clear that $\mathfrak{S}(x)$ is a union of some conjugate classes of G. We denote by $\tilde{\mathfrak{S}}(x)$ the set of conjugate classes contained in $\mathfrak{S}(X)$.

Exercise 4.3.

(i) Let x be a p-element of G and let $\{y_1, \ldots, y_{l_x}\}$ be a complete set of representatives of $C_G(x)$-conjugate classes contained in $C_G(x)_{p'}$. Then $\{xy_1, \ldots, xy_{l_x}\}$ is a complete set of representatives of G-conjugate classes in $\tilde{\mathfrak{S}}(x)$.

(ii) Let $\{x_1, \ldots, x_r\}$ be a complete set of representatives of G-conjugate classes of the p-elements of G and let $\{y_1^{(i)}, \ldots, y_{l_i}^{(i)}\}$ be a complete set of representatives of $C_G(x_i)$-conjugate classes in $C_G(x_i)_{p'}$. Then the set $\{x_i y_j^{(i)};\, 1 \le i \le r,\, 1 \le j \le l_i\}$ is a complete set of representatives of $\mathrm{Cl}(G)$.

Lemma 4.4. *Let $\varphi = \sum_{\chi \in \mathrm{Irr}(G)} \alpha_\chi \chi$ be a K-valued class function on G, and let x be a given p-element. If φ vanishes on the p-section $\mathfrak{S}(x)$, so does φ_B for any $B \in \mathrm{Bl}(G)$ (see (3.2) for the definition of φ_B).*

Proof. Set $H = C_G(x)$. From the assumption, we have, for any $y \in C_G(x)_{p'}$,

$$0 = \varphi(xy) = \sum_{B \in \mathrm{Bl}(G)} \sum_{\chi \in B} \alpha_\chi \chi(xy) = \sum_B \sum_{\chi \in B} \sum_\mu \alpha_\chi d^x_{\chi\mu} \mu(y),$$

and hence by the second main theorem,

$$\sum_{b \in \mathrm{Bl}(H)} \sum_{\mu \in \mathrm{IBr}(b)} \left(\sum_{\chi \in b^G} \alpha_\chi d^x_{\chi\mu} \right) \mu(y) = 0.$$

Thus, from the linear independence of $\mathrm{IBr}(H)$, it follows that

$$\sum_{\chi \in b^G} \alpha_\chi d^x_{\chi\mu} = 0$$

for any $\mu \in \mathrm{IBr}(b)$. Therefore,

$$\begin{aligned} \varphi_B(xy) &= \sum_{\chi \in B} \sum_{b \in \mathrm{Bl}(H|B)} \sum_{\mu \in \mathrm{IBr}(b)} \alpha_\chi d^x_{\chi\mu} \mu(y) \\ &= \sum_{b \in \mathrm{Bl}(H|B)} \sum_{\mu \in \mathrm{IBr}(b)} \left(\sum_{\chi \in b^G = B} \alpha_\chi d^x_{\chi\mu} \right) \mu(y) = 0. \qquad \blacksquare \end{aligned}$$

From the above lemma, we obtain the following result, which may be considered more or less as a refinement of the second orthogonality relation of characters.

Theorem 4.5. *Let $u, v \in G$ and suppose that $u_p \neq_G v_p$. Then the following holds for any $B \in \mathrm{Bl}(G)$:*

$$\sum_{\chi \in \mathrm{Irr}(B)} \chi(u)\overline{\chi(v)} = 0.$$

Proof. From the assumption and the second orthogonality relation of characters, it follows that $\varphi = \sum_{\chi \in \mathrm{Irr}(G)} \overline{\chi(v)}\chi$ vanishes on $\mathfrak{S}(u_p)$. Hence $\varphi_B = \sum_{\chi \in \mathrm{Irr}(B)} \overline{\chi(v)}\chi$ vanishes on $\mathfrak{S}(u_p)$, proving the theorem. ∎

In order to show the refinement of the first orthogonality relation, we need the following lemma.

Lemma 4.6. *Let $\mathrm{Cl}(G) = \{C_1, \ldots, C_k\}$ and $x_i \in C_i$. Let $B \in \mathrm{Bl}(G)$ be fixed and suppose that*

$$\sum_{i=1}^{k} \alpha_i \chi(x_i) = 0, \qquad \text{for all } \chi \in \mathrm{Irr}(B).$$

Then for any p-section $\mathfrak{S}(x)$, it holds that

$$\sum_{C_i \in \tilde{\mathfrak{S}}(x)} \alpha_i \chi(x_i) = 0, \qquad \text{for all } \chi \in \mathrm{Irr}(B).$$

Proof. By assumption, we have

$$0 = \sum_{\chi \in B} \left| \sum_{i=1}^{k} \alpha_i \chi(x_i) \right|^2 = \sum_{i,j}^{k} \alpha_i \bar{\alpha}_j \left(\sum_{\chi \in B} \chi(x_i)\overline{\chi(x_j)} \right).$$

Note that $\sum_{\chi \in B} \chi(x_i)\overline{\chi(x_j)} = 0$, if x_i and x_j belong to different p-sections of G. Thus, if $\{x_1, \ldots, x_r\}$ are representatives of the conjugate classes of p-elements of G, then

$$\begin{aligned} 0 &= \sum_{\nu=1}^{r} \sum_{C_i, C_j \in \tilde{\mathfrak{S}}(x_\nu)} \alpha_i \bar{\alpha}_j \left(\sum_{\chi \in B} \chi(x_i)\overline{\chi(x_j)} \right) \\ &= \sum_{\nu=1}^{r} \sum_{\chi \in B} \left| \sum_{C_i \in \tilde{\mathfrak{S}}(x_\nu)} \alpha_i \chi(x_i) \right|^2, \end{aligned}$$

and the assertion follows. ∎

We now prove the following result, which may be viewed as a refinement of the first orthogonality relation of characters.

Theorem 4.7 (Brauer–Osima). *Let $\chi, \chi' \in \mathrm{Irr}(G)$ belong to different blocks of G. Then, for any p-section $\mathfrak{S}(x)$, we have*

$$\sum_{z \in \mathfrak{S}(x)} \chi(z)\overline{\chi'(z)} = 0.$$

Proof. If $\chi \in B$, then it follows from the first orthogonality relation that

$$\sum_{i=1}^{k} |C_i|\overline{\chi'(x_i)}\chi_1(x_i) = 0$$

for all $\chi_1 \in B$, because $\chi' \neq \chi_1$ by assumption. Hence by Lemma 4.6, we get

$$0 = \sum_{C_i \in \tilde{\mathfrak{S}}(x)} |C_i|\overline{\chi'(x_i)}\chi_1(x_i) = \sum_{z \in \mathfrak{S}(x)} \overline{\chi'(z)}\chi_1(z). \qquad \blacksquare$$

As a direct consequence of the above theorem, we have

Corollary 4.8. *Let $\chi \in \mathrm{Irr}(G)$ and suppose that χ does not belong to the principal block of G. Then it holds, for any p-section $\mathfrak{S}(x)$, that*

$$\sum_{z \in \mathfrak{S}(x)} \chi(z) = 0.$$

Proof. Let $\chi' = \mathbf{1}_G$ in the above theorem. $\blacksquare$

Set $R[\tilde{\mathfrak{S}}(x)] = \bigoplus_{C \in \tilde{\mathfrak{S}}(x)} R\hat{C}$.

Theorem 4.9. *Let $\{x_1 = 1, \ldots, x_r\}$ be a complete set of representatives of the conjugate classes of p-elements of G. If $B \in \mathrm{Bl}(G)$, then*

$$e_B R[\tilde{\mathfrak{S}}(x_i)] \subset R[\tilde{\mathfrak{S}}(x_i)] \qquad \textit{for all } i (1 \leq i \leq r).$$

Consequently, $Z(RG)$ is expressed in the direct sum as follows:

$$Z(RG) = \bigoplus_{i=1}^{r} \bigoplus_{B \in \mathrm{Bl}(G)} e_B R[\tilde{\mathfrak{S}}(x_i)].$$

Proof. For the first assertion, it suffices to show that if $\tilde{\mathfrak{S}}(x_i) \ni C$, then $e_B \hat{C} \in R[\tilde{\mathfrak{S}}(x_i)]$. Let $x \in C$. Then

$$\begin{aligned} e_B \hat{C} &= \sum_{\chi \in B} e_\chi \omega_\chi(\hat{C}) = \sum_{\chi \in B} \left(\sum_{y \in G} \frac{\chi(1)}{|G|} \overline{\chi(y)} y \right) \frac{|C|}{\chi(1)} \chi(x) \\ &= \frac{|C|}{|G|} \sum_{y \in G} \left(\sum_{\chi \in B} \chi(x) \overline{\chi(y)} \right) y, \end{aligned}$$

where the parenthetic value in the last term is zero unless $y \in \mathfrak{S}(x_i)$. Thus we have $e_B \hat{C} \in R[\tilde{\mathfrak{S}}(x_i)]$ as asserted.

The latter half is obvious because $Z(RG) = \bigoplus_i R[\tilde{\mathfrak{S}}(x_i)]$ and $R[\tilde{\mathfrak{S}}(x_i)] = \bigoplus_B e_B R[\tilde{\mathfrak{S}}(x_i)]$. ■

Given a p-element x of G, we write $d_{\zeta\mu}^{(x)}$ and $c_{\mu\nu}^{(x)}$ to denote the decomposition numbers and Cartan invariants of $C_G(x)$, respectively.

Caution. Be careful not to confuse $d_{\chi\mu}^{(x)}$ with the generalized decomposition number $d_{\chi\mu}^{x}$.

Lemma 4.10. *Let x be a p-element and $H = C_G(x)$. For $\mu \in \mathrm{IBr}(H)$, we let $\eta_\mu^{(x)}$ denote the principal indecomposable character of H corresponding to μ. Define the class function $\hat{\eta}_\mu^{(x)}$ on G by*

$$(4.2) \quad \hat{\eta}_\mu^{(x)}(z) = \begin{cases} \eta_\mu^{(x)}(y) & \text{if } z \in \mathfrak{S}(x) \text{ and } z =_G xy \text{ with } y \in H_{p'}. \\ 0 & \text{if } z \notin \mathfrak{S}(x). \end{cases}$$

Then $\hat{\eta}_\mu^{(x)}$ is expressed as

$$\hat{\eta}_\mu^{(x)} = \sum_{\chi \in \mathrm{Irr}(G)} \overline{d_{\chi\mu}^{x}} \chi.$$

In particular, it holds that

$$z \notin \mathfrak{S}(x) \Rightarrow \sum_{\chi \in \mathrm{Irr}(G)} \overline{d_{\chi\mu}^{x}} \chi(z) = 0.$$

Proof. If $\{y_1, \ldots, y_m\}$ is a complete set of representatives of H-conjugate classes of p'-elements of H, then $\{xy_1, \ldots, xy_m\}$ is a complete set of representatives of G-conjugate classes in $\mathfrak{S}(x)$. Using $C_G(xy_i) = C_H(y_i)$, we have

$$
\begin{aligned}
(\hat{\eta}_\mu^{(x)}, \chi)_G &= \frac{1}{|G|} \sum_{z \in \mathfrak{S}(x)} \eta_\mu^{(x)}(z_{p'})\overline{\chi(z)} \\
&= \frac{1}{|G|} \sum_{i=1}^{m} \frac{|G|}{|C_G(xy_i)|} \eta_\mu^{(x)}(y_i)\overline{\chi(xy_i)} \\
&= \frac{1}{|H|} \sum_{i=1}^{m} \frac{|H|}{|C_H(y_i)|} \eta_\mu^{(x)}(y_i)\Big(\sum_\nu \overline{d_{\chi\nu}^x}\, \overline{\nu(y_i)}\Big) \\
&= \sum_\nu \overline{d_{\chi\nu}^x}(\eta_\mu^{(x)}, \nu)'_H = \overline{d_{\chi\mu}^x},
\end{aligned}
$$

which proves the assertion. ■

Theorem 4.11. *Let x, x' be p-elements of G. Then*

$$
\sum_{\chi \in \mathrm{Irr}(G)} \overline{d_{\chi\mu}^x} d_{\chi\mu'}^{x'} = \begin{cases} c_{\mu\mu'}^{(x)} & \text{if } \; x = x' \\ 0 & \text{if } \; x \neq_G x'. \end{cases}
$$

Proof. It follows from the above lemma that

$$
(\hat{\eta}_\mu^{(x)}, \hat{\eta}_{\mu'}^{(x')})_G = \sum_\chi \overline{d_{\chi\mu}^x} d_{\chi\mu'}^{x'}.
$$

Now, according to (4.2), the left-hand side of this equality is zero if $x \neq_G x'$, whereas $(\eta_\mu^{(x)}, \eta_{\mu'}^{(x)})'_H = c_{\mu\mu'}^{(x)}$ if $x = x'$. ■

With the notation of Exercise 4.3, we arrange $\{x_i y_j^{(i)};\ 1 \le i \le r,\ 1 \le j \le l_i\}$, a complete set of representatives of conjugate classes of G, in lexicographic order relative to (i, j). Consider the $k(G) \times k(G)$ matrix

$$
\hat{D} = (d_{\chi\mu}^{x_i})_{\chi, (x_i, \mu)}, \tag{4.3}
$$

with generalized decomposition numbers as entries, where the row and column indices are taken from $\mathrm{Irr}(G) = \{\chi\}$ and $\{(x_i, \mu);\ \mu \in \mathrm{IBr}(C_G(x_i))\},\ 1 \le i \le r\}$, respectively. Then by Theorem 4.11, we have the following.

Corollary 4.12.

$$ {}^t\bar{\hat{D}}\hat{D} = \begin{pmatrix} C^{(x_1)} & & & 0 \\ & C^{(x_2)} & & \\ & & \ddots & \\ 0 & & & C^{(x_r)} \end{pmatrix}, $$

where $C^{(x_i)}$ denotes the Cartan matrix of $C_G(x_i)$, which is a nonsingular matrix of degree l_i.

In particular, $\hat{D}$ is nonsingular.

Let $\mathrm{Bl}(G) = \{B_1, \ldots, B_t\}$. We arrange rows of $\hat{D}$ so that characters in B_1 come first, then B_2, and so on. We also rearrange columns of $\hat{D}$ so that those (x_i, μ) with $\mu \in b \in \mathrm{Bl}(C_G(x_i))$ and $b^G = B_1$ come first, then $B_2, \ldots, B_t$. Thus we obtain a new matrix $\hat{D}_1$, which splits into the following form by the second main theorem:

$$ \hat{D}_1 = \begin{pmatrix} \hat{D}_{B_1} & & 0 \\ & \ddots & \\ 0 & & \hat{D}_{B_t} \end{pmatrix}, \tag{4.4} $$

where $\hat{D}_{B_j} = (d^{x_i}_{\chi\mu})$, for which χ and μ run through $\mathrm{Irr}(B_j)$ and the set of irreducible Brauer characters belonging to those blocks b of $C_G(x_i)$ with $b^G = B_j$, respectively.

Since $\hat{D}_1$ is nonsingular, we have the following.

Theorem 4.13. *Let $B \in \mathrm{Bl}(G)$. Then*

(i) *$\hat{D}_B$ is a nonsingular matrix of degree $k(B)$.*

(ii) *$k(B) = l(B) + \sum_{i=2}^r \sum_{b \in \mathrm{Bl}(C_G(x_i)|B)} l(b)$.*

Remark. In (ii) of the above theorem, we may only work with those x_i such that $x_i \in_G \delta(B)$. In fact, if $b \in \mathrm{Bl}(C_G(x_i))$ and $b^G = B$, then $x_i \in \delta(b) \leq_G \delta(B)$ (cf. Theorem 2.8 and Lemma 3.3).

In connection with the above remark, we show the following lemma.

Lemma 4.14. *Let x be a p-element of G and $B \in \mathrm{Bl}(G)$.*

(i) *If there exist $\chi \in \mathrm{Irr}(B)$ and $y \in C_G(x)_{p'}$ such that $\chi(xy) \neq 0$, then $x \in_G \delta(B)$.*

(ii) *Let $\chi \in \mathrm{Irr}(B)$ with $\mathrm{ht}(\chi) = 0$. If $x \in_G \delta(B)$, then there exists $y \in C_G(x)_{p'}$ such that*

$$\chi(xy) \not\equiv 0 \pmod{\pi}.$$

Proof.

(i) Since $(xy)_p = x$, the result follows immediately from Corollary 1.10(i).

(ii) Let C be a defect class for B and $y \in C$, then

$$\omega_\chi(\hat{C}) = \frac{|G|}{|C_G(y)|}\frac{\chi(y)}{\chi(1)} \not\equiv 0 \pmod{\pi},$$

whence we have $\chi(y) \not\equiv 0 (\mathrm{mod}\ \pi)$, since $\mathrm{ht}(\chi) = 0$ and $\delta(C) =_G \delta(B)$. After replacing y with a suitable element of C if necessary, we may assume that $x \in \delta(B) \in \mathrm{Syl}_p(C_G(y))$. Thus $y \in C_G(x)_{p'}$ and

$$\chi(xy) \equiv \chi(y) \not\equiv 0 \pmod{\pi}. \qquad \blacksquare$$

Let ε_n be a primitive nth root of unity and let $\mathbf{Q}_n = \mathbf{Q}(\varepsilon_n)$ be the field generated by ε_n over the field $\mathbf{Q}$ of rationals. Then $\mathbf{Q}_n$ is an abelian extension of $\mathbf{Q}$ with Galois group isomorphic to $(\mathbf{Z}/(n))^\times$ (cf. Hungerford [1], Chapter 5, Section 8).

Exercise 4.15. If $(m, n) = 1$, then the following statements hold.

(i) $\mathbf{Q}_{mn} = \mathbf{Q}_m\mathbf{Q}_n$,

(ii) $\mathbf{Q}_m \cap \mathbf{Q}_n = \mathbf{Q}$.

[Hint: (ii) Use (i) and the fact that

$$|\mathrm{Gal}(\mathbf{Q}_{mn}/\mathbf{Q})| = |\mathrm{Gal}(Q_m/\mathbf{Q})||\mathrm{Gal}(\mathbf{Q}_n/\mathbf{Q})|.]$$

Now, let $|G| = g$ and $L = \mathbf{Q}_g$. For $\chi \in \mathrm{Irr}(G)$ and $\sigma \in \mathrm{Gal}(L/\mathbf{Q})$, the function χ^σ defined by $\chi^\sigma(x) = \chi(x)^\sigma$ is also an irreducible character of G, which is called an *algebraic conjugate* of χ. On the other hand, there exists m with $(m, g) = 1$ such that $\varepsilon_g^\sigma = \varepsilon_g^m$. Note that m is unique up to mod g. If $C \in \mathrm{Cl}(G)$, then $\{x^m; x \in C\}$, denoted by C^σ, is a conjugate class of G, which we also call an algebraic conjugate of the conjugate class C. Thus $\mathrm{Gal}(L/\mathbf{Q})$ acts on both $\mathrm{Irr}(G)$ and $\mathrm{Cl}(G)$, and it holds that $\chi^\sigma(x) = \chi(x^\sigma)$ where $x \in C$, and x^σ denotes a representative of C^σ. Consequently we obtain the following result from Brauer's permutation lemma (Chapter 3, 2.18).

Theorem 4.16. *The number of the algebraic conjugate classes in* Irr(G) *is equal to that in* Cl(G).

Proof. See Chapter 3, Problem 8.

Let $|G| = g = p^a h$ with $(h, p) = 1$ and let $\mathfrak{G} = \mathrm{Gal}(L/\mathbf{Q}_h)(\simeq \mathrm{Gal}(\mathbf{Q}_{p^a}/\mathbf{Q}))$. If $\sigma \in \mathfrak{G}$, then χ^σ is said to be *p-conjugate* to χ. It is clear that χ^σ and χ are identical on $G_{p'}$, and hence

$$d_{\chi^\sigma\varphi} = d_{\chi\varphi}, \text{ for all } \sigma \in \mathfrak{G} \text{ and } \varphi \in \mathrm{IBr}(G).$$

In particular, χ and χ^σ belong to the same block of G.

If $\chi^\sigma = \chi$ for all $\sigma \in \mathfrak{G}$, i.e., $\chi(x) \in \mathbf{Q}_h$ for all $x \in G$, then χ is said to be *p-rational.*

Exercise 4.17. Prove that $\chi(\in \mathrm{Irr}(G))$ is p-rational if and only if $d^x_{\chi\mu} \in \mathbf{Z}$ for all p-elements x and $\mu \in \mathrm{IBr}(C_G(x))$.

Let $\sigma \in \mathfrak{G}$ and write $\varepsilon_g^\sigma = \varepsilon_g^m$ with $(m, g) = 1$. Then, as $\varepsilon_g^{p^a}$ lies in $\mathbf{Q}_h$, it is σ-invariant, and it follows that $m \equiv 1 (\mathrm{mod}\ h)$. Therefore $y^m = y$ for all $y \in G_{p'}$. In particular, if $y \in C_G(x)_{p'}$, where x is a p-element, then

$$\chi^\sigma(xy) = \chi(x^m y^m) = \chi(x^m y) = \sum_\mu d^{x^m}_{\chi\mu}\,\mu(y),$$

where μ runs through the set of irreducible Brauer characters of $C_G(x) = C_G(x^m)$. Therefore we have

$$d^x_{\chi^\sigma\mu} = d^{x^m}_{\chi\mu}. \tag{4.5}$$

Using this, we prove the following.

Theorem 4.18. *Let B be a block of G with positive defect.*

(i) Irr(B) *contains at least* ($l(B) + 1$) *characters that are not p-conjugate to each other.*

(ii) *If either $p > 2$ or $p^a = 2^a \leq 4$, then* Irr(B) *contains at least* $l(B)$ *p-rational characters.*

Proof.

(i) With the notation so far, $\mathfrak{G}$ acts on rows and columns of the nonsingular matrix $\hat{D}_B = (d^x_{\chi\mu})_{\chi,(x,\mu)}$, in which $\sigma \in \mathfrak{G}$ induces the row permutation such that $\chi \mapsto \chi^\sigma$, whereas it induces the column permutation such that $(x, \mu) \mapsto (x^m, \mu)$. Hence by virtue of (4.5), Brauer's permutation lemma is applied, yielding that the numbers of $\mathfrak{G}$-orbits on the rows and that on the columns are the same. However, those $l(B)$ columns indexed by $(1, \mu)$ with $\mu \in \mathrm{IBr}(B)$ are $\mathfrak{G}$-invariant, and since $k(B) > l(B)$ as $d(B) > 0$, there are at least $(l(B) + 1)$ $\mathfrak{G}$-orbits on columns. As remarked above, this proves our assertion.

(ii) $\mathfrak{G}$ is cyclic from the assumption (Hungerford [1], Chapter 5, Section 8 Exercise 4). Hence the number of $\mathfrak{G}$-invariant rows is equal to that of $\mathfrak{G}$-invariant columns. But, as was observed above, there are at least $l(B)$ $\mathfrak{G}$-invariant columns, and the result follows. ■

5. Blocks and Normal Subgroups

Let $G \geq H$ and $x \in G$. If b is a block of RH, then $b^x = x^{-1}bx$ is a block of RH^x, which is called a *G-conjugate block* of b. It is clear that $(e_b)^x = e_{b^x}$ and $\delta(b^x) =_{H^x} \delta(b)^x$.

5.1. Covers

We assume until the end of the subsection 5.3 *that* $G \trianglerighteq H$.

Let b be a block of H. Then b^x is a block of H for all $x \in G$, and

$$T(b) = \{x \in G; b^x = b\}$$

is a subgroup of G containing H, which is called the *inertial group* of b. If we set

$$f_b = \sum_{x \in T(b)\backslash G} (e_b)^x,$$

then f_b is a central idempotent of RG, and we have a primitive idempotent decomposition

(5.1) $$f_b = \sum_{i=1}^{r} e_{B_i} \qquad (B_i \in \mathrm{Bl}(G))$$

in $Z(RG)$. Then we say that B_i *covers* the block b of H $(1 \leq i \leq r)$.

Exercise 5.1. Let $B \in \mathrm{Bl}(G)$ and $b \in \mathrm{Bl}(H)$. Then B covers b if and only if $e_B e_b \neq 0$.

Let us consider the above f_b as an idempotent of $RH \cap Z(RG) = (RH)^G$. Then the following holds.

Lemma 5.2. *f_b is a primitive idempotent of $(RH)^G$. Hence, if $\{b_1, \dots, b_m\}$ denotes a complete set of representatives of G-conjugate classes of* $\mathrm{Bl}(H)$, *then*

$$1 = f_{b_1} + \cdots + f_{b_m} \tag{5.2}$$

is the primitive idempotent decomposition of 1 *in* $(RH)^G$.

Proof. Suppose to the contrary that f_b has an idempotent decomposition $f_b = e_1 + e_2$ in $(RH)^G$. We may assume that $e_1 e_b \neq 0$. Then $0 \neq (e_1 e_b)^x = e_1 (e_b)^x$, and hence $(e_b)^x$ appears in the primitive idempotent decomposition of e_1 for all $x \in G$. This implies that $e_1 = f_b$, which is a contradiction. Therefore f_b is primitive in $(RH)^G$. The latter half is immediate from $1 = \sum_{b \in \mathrm{Bl}(H)} e_b$. ■

Comparing the central idempotent decomposition (5.2) with the central primitive idempotent decomposition $1 = \sum_{B \in \mathrm{Bl}(G)} e_B$, we deduce the following result.

Lemma 5.3. *Every block B of G covers some $b \in$* $\mathrm{Bl}(H)$. *Moreover the blocks of H covered by B form a G-conjugate class of* $\mathrm{Bl}(H)$.

Lemma 5.4. *Let $B \in$* $\mathrm{Bl}(G)$, *$b \in$* $\mathrm{Bl}(H)$, *and suppose that B covers b. If V is a nonzero RG-module such that $Ve_B = V$, then*

$$V_H = \bigoplus_{x \in T(b) \backslash G} Ve_{b^x}, \qquad \textit{with} \quad Ve_{b^x} \neq 0 \quad \textit{for all } x \in G.$$

Proof. Since $e_B = e_B f_b = \sum_x e_B e_{b^x}$, it follows that $V = Ve_B = \bigoplus_x Ve_{b^x}$. If $Ve_{b^x} = 0$ for some $x \in G$, then $Ve_{b^x} = Vx^{-1} e_b x = Ve_b x = 0$ and hence $Ve_b = 0$. Therefore $Ve_b x = 0$ for all $x \in G$ and $V = 0$, contradicting the assumption. ■

As before, let us denote by $\mathrm{Cl}_G(H)$ the set of G-conjugate classes of G contained in H.

Theorem 5.5 (Passman). *Let $B \in \mathrm{Bl}(G)$ and $b \in \mathrm{Bl}(H)$. Then B covers b if and only if the following holds for all $C \in \mathrm{Cl}_G(H)$.*

$$\omega_B^*(\hat{C}) = \omega_b^*(\hat{C}).$$

Proof. Since $(FH)^G = \bigoplus_{C \in \mathrm{Cl}_G(H)} F\hat{C} = Z(FG) \cap Z(FH)$, both ω_B^* and ω_b^* induce an algebra homomorphism from $(FH)^G$ onto F. Hence $\omega_B^* = \omega_b^*$ on $(FH)^G$ if and only if it holds, with the notation of Lemma 5.2 that $\omega_B^*(f_{b_i}^*) = \omega_b^*(f_{b_i}^*)$ for all $i (1 \le i \le m)$. Suppose that $b_1 = b$. Then $\omega_b^*(f_{b_i}^*) = \delta_{i1}$. And B covers b if and only if $\omega_B^*(f_{b_i}^*) = \delta_{i1}$. Therefore the assertion holds. ■

Corollary 5.6. *Suppose that G/H is a p-group. Then every $b \in \mathrm{Bl}(H)$ is covered by a unique $B \in \mathrm{Bl}(G)$.*

Proof. Suppose that b is covered by $B, B' \in \mathrm{Bl}(G)$. If $C \in \mathrm{Cl}(G_{p'})$, then $C \subset H$ by assumption. Therefore, $\omega_B^*(\hat{C}) = \omega_b^*(\hat{C}) = \omega_{B'}^*(\hat{C})$ for all $C \in \mathrm{Cl}(G_{p'})$, whence the result follows from Chapter 3, 6.24. ■

Lemma 5.7. *Let $B \in \mathrm{Bl}(G)$ and $b \in \mathrm{Bl}(H)$.*

(i) *Assume B covers b, and let χ be an irreducible (ordinary or modular) character belonging to B. Then every irreducible constituent of χ_H belongs to some $b^x (x \in G)$, and there exists an irreducible constituent of χ_H belonging to b.*

(ii) *If there exists $\chi \in \mathrm{Irr}(B)$ (or $\mathrm{IBr}(B)$, respectively) such that χ_H has an irreducible constituent belonging to $\mathrm{Irr}(b)$ (or $\mathrm{IBr}(b)$, respectively), then B covers b.*

Proof.

(i) This is clear from Lemma 5.4.

(ii) If B does not cover b, then $e_B e_b = 0$, whence we easily deduce a contradiction. ■

Lemma 5.8. *Let $B \in \mathrm{Bl}(G)$, $b \in \mathrm{Bl}(H)$ and suppose that B covers b.*

(i) *For every principal indecomposable RH (respectively, FH)-module W belonging to b, there exists a principal indecomposable RG (respectively, FG)-module V such that $W | V_H$.*

(ii) *For every $\zeta \in \mathrm{Irr}(b)$ (respectively, $\mathrm{IBr}(b)$), there exists $\chi \in \mathrm{Irr}(B)$ (respectively, $\mathrm{IBr}(B)$) such that ζ is an irreducible constituent of χ_H.*

Proof.

(i) Principal indecomposable RH-modules are in one-to-one correspondence to principal indecomposable FH-modules via the reduction mod π (up to isomorphism). Thus it suffices to prove the assertion in the case where W is a principal indecomposable FH-module.

We first note that if Y is a principal indecomposable FH-module and $X | Y^G$ for a principal indecomposable FG-module X, then $Y | X_H$. This is because we have $X_H | (Y^G)_H$, and $(Y^G)_H = \bigoplus_x Y \otimes x$ is an indecomposable decomposition. Hence $Y \otimes x | X_H$ for some $x \in G$ and thus $Y | X_H$.

Now, if U is any principal indecomposable FG-module belonging to B, then by Lemma 5.4, there exists a principal indecomposable FH-module W' belonging to b such that $W' | U_H$. Since $W, W' \in b$, there is a sequence of principal indecomposable FH-modules

$$W' = W_1, W_2, \ldots, W_r = W$$

such that W_i and W_{i+1} have an irreducible constituent in common for each i. Thus it suffices to find V satisfying the conclusion of (i) in case of $r = 2$. In particular, we may assume that there exists a principal indecomposable FH-module $Y \in b$ such that

$$\mathrm{Hom}_{FH}(Y, W') \neq 0 \neq \mathrm{Hom}_{FH}(Y, W).$$

Since

$$\mathrm{Hom}_{FG}(Y^G, U) \simeq \mathrm{Hom}_{FH}(Y, U_H) \supset \mathrm{Hom}_{FH}(Y, W') \neq 0,$$

there exists a principal indecomposable FG-module $X \in B$ such that $X | Y^G$, and it then follows that $Y | X_H$, as remarked above. Consequently,

$$0 \neq \mathrm{Hom}_{FH}(Y, W) \subset \mathrm{Hom}_{FH}(X_H, W) \simeq \mathrm{Hom}_{FG}(X, W^G),$$

and hence there exists a principal indecomposable FG-module $V \in B$ such that $V | W^G$. This implies that $W | V_H$.

(ii) ζ is an irreducible constituent of a character given by some principal indecomposable RH-module. Hence the assertion is immediate by (i). A similar argument works for Brauer characters. ∎

As a consequence of the above lemma, we have the following.

Theorem 5.9. *Let $\chi \in \mathrm{Irr}(B)$ and $\chi' \in \mathrm{Irr}(B')$. Then B and B' cover the same block of H if and only if there exists a sequence of irreducible characters of G,*

$$\chi = \chi_1, \chi_2, \ldots, \chi_r = \chi',$$

such that, for each i, either χ_i and χ_{i+1} belong to the same block, or $(\chi_i)_H$ and $(\chi_{i+1})_H$ have an irreducible constituent in common. An analogous statement holds for Brauer characters.

Proof. If B and B' cover the same block, say b, of H, then given $\zeta \in \mathrm{Irr}(b)$, there exist, by Lemma 5.8, $\chi_1 \in \mathrm{Irr}(B)$ and $\chi'_1 \in \mathrm{Irr}(B')$ such that ζ appears as a common irreducible constituent of $(\chi_1)_H$ and $(\chi'_1)_H$. Then the sequence χ, χ_1, χ'_1, χ' satisfies the desired condition.

Suppose conversely that there exists a sequence of characters with the above property and let $\chi_i \in B_i$. If $(\chi_i)_H$ and $(\chi_{i+1})_H$ have the same irreducible constituent ζ, then B_i and B_{i+1} cover the block of H to which ζ belongs. Our assertion is clear from this. ∎

5.2. Inertial Groups

Let $b \in \mathrm{Bl}(H)$ and let $\zeta \in \mathrm{Irr}(b)$. It is clear that $T(\zeta) \leq T(b)$. We set $T = T(b)$ for simplicity and denote by $\mathrm{Bl}(G|b)$ the set of blocks of G that cover b. Now, the following theorem of Clifford type holds for $\mathrm{Bl}(G|b)$ and $\mathrm{Bl}(T|b)$.

Theorem 5.10 (Fong–Reynolds).

(i) *If $\tilde{B} \in \mathrm{Bl}(T|b)$, then $\tilde{B}^G$ is defined. The map $\tilde{B} \mapsto \tilde{B}^G$ gives rise to a one-to-one correspondence between $\mathrm{Bl}(T|b)$ and $\mathrm{Bl}(G|b)$.*

(ii) *Let $\tilde{B} \in \mathrm{Bl}(T|b)$. If $\tilde{\chi} \in \mathrm{Irr}(\tilde{B})$, then $\tilde{\chi}^G \in \mathrm{Irr}(\tilde{B}^G)$. The map $\tilde{\chi} \mapsto \tilde{\chi}^G$ gives rise to a one-to-one correspondence between $\mathrm{Irr}(\tilde{B})$ and $\mathrm{Irr}(\tilde{B}^G)$. The corresponding statement holds for $\mathrm{IBr}(\tilde{B})$ and $\mathrm{IBr}(\tilde{B}^G)$.*

(iii) *If $\tilde{B} \in \mathrm{Bl}(T|b)$, then $\tilde{B}$ and $\tilde{B}^G$ have the same decomposition matrix and Cartan matrix. Namely if $\tilde{\chi} \in \mathrm{Irr}(\tilde{B})$ and $\tilde{\varphi}$, $\tilde{\varphi}' \in \mathrm{IBr}(\tilde{B})$, then it holds that*

$$d_{\tilde{\chi}\tilde{\varphi}} = d_{\tilde{\chi}^G\tilde{\varphi}^G}, \qquad c_{\tilde{\varphi}\tilde{\varphi}'} = c_{\tilde{\varphi}^G\tilde{\varphi}'^G}.$$

(iv) *If $\tilde{B} \in \mathrm{Bl}(T|b)$, then $\delta(\tilde{B}) =_G \delta(\tilde{B}^G)$.*

(v) *If $B \in \mathrm{Bl}(G|b)$, then $\delta(B) \leq_G T(b)$.*

Proof. Set $\mathrm{Irr}(\mathrm{Bl}(G|b)) = \bigcup_{B\in \mathrm{Bl}(G|b)} \mathrm{Irr}(B)$. By Lemma 5.7, we have

$$\mathrm{Irr}(\mathrm{Bl}(G|b)) = \bigcup_{\zeta\in \mathrm{Irr}(b)} \mathrm{Irr}(G|\zeta),$$

where $\mathrm{Irr}(G|\zeta) = \{\chi \in \mathrm{Irr}(G);\ (\chi_H, \zeta)_H \neq 0\}$. By defining $\mathrm{IBr}(\mathrm{Bl}(G|b))$ and $\mathrm{IBr}(G|\xi)$ analogously for $\xi \in \mathrm{IBr}(H)$, we obtain a similar fact for Brauer characters.

Now, if $\zeta \in \mathrm{Irr}(b)$, then $T(\zeta) \leq T$, and by Clifford's theorem (Chapter 3, 3.4), we have a one-to-one correspondence $\mathrm{Irr}(T|\zeta) \to \mathrm{Irr}(G|\zeta)(\tilde{\chi} \mapsto \tilde{\chi}^G)$. Thus

$$\mathrm{Ind}_T^G\colon \mathrm{Irr}(\mathrm{Bl}(T|b)) \to \mathrm{Irr}(\mathrm{Bl}(G|b)) \qquad (\tilde{\chi} \mapsto \tilde{\chi}^G)$$

is a bijection. Likewise, we have a bijection

$$\mathrm{Ind}_T^G\colon \mathrm{IBr}(\mathrm{Bl}(T|b)) \to \mathrm{IBr}(\mathrm{Bl}(G|b)) \qquad (\tilde{\varphi} \mapsto \tilde{\varphi}^G).$$

With respect to the above bijections, we clearly have

$$d_{\tilde{\chi}\tilde{\varphi}} = d_{\tilde{\chi}^G\tilde{\varphi}^G}. \tag{5.3}$$

In particular, $\tilde{\chi}$, $\tilde{\chi}' \in \mathrm{Irr}(\mathrm{Bl}(T|b))$ belong to the same block if and only if $\tilde{\chi}^G$ and $\tilde{\chi}'^G$ belong to the same block of G.

If $\tilde{B} \in \mathrm{Bl}(T|b)$ and $\tilde{\chi} \in \mathrm{Irr}(\tilde{B})$, then $\tilde{B}^G$ is defined and $\tilde{\chi}^G \in \tilde{B}^G$ by Lemma 3.1 because $\tilde{\chi}^G$ is irreducible. Therefore, from the above remark, Ind_T^G induces a bijection $\mathrm{Irr}(\tilde{B}) \to \mathrm{Irr}(\tilde{B}^G)$, and we have $D_{\tilde{B}} = D_{\tilde{B}^G}$, $C_{\tilde{B}} = C_{\tilde{B}^G}$. It is clear that $d(\tilde{B}) = d(\tilde{B}^G)$. Thus we have shown all the statements from (i) to (iii).

For (iv), remember that, in general, $\delta(\tilde{B}) \leq_G \delta(\tilde{B}^G)$. Hence we have

$$\delta(\tilde{B}) =_G \delta(\tilde{B}^G)$$

because $d(\tilde{B}) = d(\tilde{B}^G)$. Part(v) is immediate from this. ■

Corollary 5.11. *Let $\tilde{B} \in \mathrm{Bl}(T|b)$, $B = \tilde{B}^G$, and $T = T(b)$.*

(i) *If $x, y \in G$ and $x \not\equiv y \pmod{T}$, then $(e_{\tilde{B}})^x(e_{\tilde{B}})^y = 0$.*

(ii) $e_B = \sum_{x\in T\backslash G} (e_{\tilde{B}})^x$.

Proof. Set $\mathrm{Bl}(T|b) = \{\tilde{B}_1, \ldots, \tilde{B}_r\}$ and let $\tilde{e}_i$ denote the block idempotent of $\tilde{B}_i$. Then $e_b = \sum_{i=1}^r \tilde{e}_i$.

(i) Since $\tilde{e}_i = e_b\tilde{e}_i = \tilde{e}_i e_b$, the result is clear from $(e_b)^x(e_b)^y = 0$, unless $x \equiv y \pmod{T}$.

(ii) Let $g_i = \sum_{x\in T\backslash G}(\tilde{e}_i)^x$. Thus, g_i is a central idempotent of RG, and

$$f_b\left(= \sum_{x\in T\backslash G} (e_b)^x\right) = \sum_{i=1}^r g_i \tag{5.4}$$

is a central idempotent decomposition of f_b. On the other hand, if we let $B_i = \tilde{B}_i^G$ and e_i denote the block idempotent of B_i, then $\mathrm{Bl}(G|b) = \{B_1, \ldots, B_r\}$ by Theorem 5.10, and

$$f_b = \sum_{i=1}^{r} e_i \tag{5.5}$$

is a central primitive idempotent decomposition of f_b. Here we know from Lemma 3.9(i) that $\tilde{e}_i e_i \neq 0$ and hence $g_i e_i \neq 0$. Thus by comparing (5.4) with (5.5), we conclude $g_i = e_i$. ■

Let $\mathfrak{M}_B$ stand for a set of nonisomorphic RG-modules belonging to the block B of G.

Theorem 5.12. *Let $\tilde{B} \in \mathrm{Bl}(T|b)$.*

(i) *If $\tilde{V} \in \mathfrak{M}_{\tilde{B}}$, then $\tilde{V}^G \in \tilde{B}^G$. And the map $\tilde{V} \mapsto \tilde{V}^G$ gives rise to a bijection*

$$\Phi \colon \mathfrak{M}_{\tilde{B}} \to \mathfrak{M}_{\tilde{B}^G}$$

whose inverse map is given by $V \mapsto Ve_{\tilde{B}}$.

(ii) *If $\tilde{V}_1, \tilde{V}_2 \in \mathfrak{M}_{\tilde{B}}$, then the following isomorphism holds:*

$$\mathrm{Hom}_{RT}(\tilde{V}_1, \tilde{V}_2) \simeq \mathrm{Hom}_{RG}(\tilde{V}_1^G, \tilde{V}_2^G).$$

(*Similar statements hold for FG-modules.*)

Proof. Set $B = \tilde{B}^G$ and denote by e and $\tilde{e}$ the block idempotent of B and $\tilde{B}$, respectively.

(i) We write $\tilde{V}^G = \bigoplus_{x \in T\backslash G} \tilde{V} \otimes x$. Then

$$\tilde{v} \otimes x = \tilde{v}\tilde{e} \otimes x = \tilde{v} \otimes \tilde{e}x = (\tilde{v} \otimes x)\tilde{e}^x \qquad (\tilde{v} \in \tilde{V}), \tag{5.6}$$

and from the above corollary, we obtain

$$(\tilde{v} \otimes x)e = (\tilde{v} \otimes x)\tilde{e}^x e = (\tilde{v} \otimes x)\tilde{e}^x = \tilde{v} \otimes x,$$

whence it follows that $\tilde{V}^G \in \mathfrak{M}_B$.

To complete the proof, it suffices to show that the map $\Psi \colon \mathfrak{M}_B \to \mathfrak{M}_{\tilde{B}}$ $(V \mapsto V\tilde{e})$ is the inverse of Φ.

In $\tilde{V}^G$, it holds that $\tilde{V} \otimes x = (\tilde{V} \otimes x)\tilde{e}^x$ as observed in (5.6) and hence $(\tilde{V} \otimes x)\tilde{e} = 0$ unless $x \in T$, which means that $(\tilde{V}^G)\tilde{e} = \tilde{V}$. Therefore $\Psi \circ \Phi$ is the identity. Next we define, for $V \in \mathfrak{M}_B$, the following maps λ and μ:

$$(V\tilde{e})^G = \bigoplus_{x \in T\backslash G} V\tilde{e} \otimes x \underset{\mu}{\overset{\lambda}{\rightleftarrows}} V$$

by $\lambda(v\tilde{e} \otimes x) = v\tilde{e}x$, $\mu(v) = \sum_{x \in T\backslash G} vx^{-1}\tilde{e} \otimes x$. It is easy to verify that both λ and μ are RG-homomorphisms. We have $\lambda \circ \mu(v) = \sum_x v\tilde{e}^x = ve = v$. On the other hand, we have $\mu \circ \lambda(v\tilde{e} \otimes x) = \sum_{y \in T\backslash G} v\tilde{e}xy^{-1}\tilde{e} \otimes y$, which reduces to $v\tilde{e} \otimes x$, since if $xy^{-1} \notin T$, then $\tilde{e}xy^{-1}\tilde{e} = xy^{-1}\tilde{e}^{xy^{-1}}\tilde{e} = 0$. Thus $\lambda \circ \mu$ and $\mu \circ \lambda$ are both identity maps. Therefore $(V\tilde{e})^G \simeq V$, and hence $\Phi \circ \Psi$ is the identity map.

(ii) By abbreviating $\otimes_{RT}$ to $\otimes$, we have

$$((\tilde{V}_2)^G)_T = \bigoplus_{y \in T\backslash G/T} \tilde{V}_2 \otimes R[TyT].$$

If $y \notin T$, then we have $\tilde{e}x\tilde{e} = x\tilde{e}^x\tilde{e} = 0$ for any $x \in TyT$, and it follows that

$$(\tilde{V}_2 \otimes R[TyT])\tilde{e} = \tilde{V}_2 \otimes \tilde{e}R[TyT]\tilde{e} = 0.$$

In particular, $\mathrm{Hom}_{RT}(\tilde{V}_1, \tilde{V}_2 \otimes R[TyT]) = 0$. Thus we have

$$\mathrm{Hom}_{RT}(\tilde{V}_1, \tilde{V}_2) \simeq \mathrm{Hom}_{RT}(\tilde{V}_1, (\tilde{V}_2^G)_T) \simeq \mathrm{Hom}_{RG}(\tilde{V}_1^G, \tilde{V}_2^G). \qquad \blacksquare$$

5.3. Regular Blocks

We continue to assume that $G \trianglerighteq H$. Let $B \in \mathrm{Bl}(G)$. If there exists $b \in \mathrm{Bl}(H)$ such that $b^G = B$, then $\omega_B^* = \omega_b^* \circ s_H^*$, and it follows that

(5.7) $\qquad \omega_B^*(\hat{C}) = 0$, for all $C \in \mathrm{Cl}(G)$ with $C \cap H = \varnothing$.

In general, a block B of G is said to be *regular* (relative to H) if the condition (5.7) is satisfied. Using this notion, we can characterize the blocks of the form b^G with $b \in \mathrm{Bl}(H)$. Namely, we have the following.

Theorem 5.13. *Let $B \in \mathrm{Bl}(G)$ and $b \in \mathrm{Bl}(H)$.*

(i) *If $B = b^G$, then B is regular and covers b.*

(ii) *Suppose that B is regular. If B covers b, then $b^G = B$, and B is the only regular block that covers b.*

Proof.

(i) B is regular, as remarked at the beginning of this subsection. If $C \in \mathrm{Cl}_G(H)$, then $\omega_B^*(\hat{C}) = \omega_b^* \circ s_H^*(\hat{C}) = \omega_b^*(\hat{C})$, and hence B covers b by Theorem 5.5.

(ii) If B is regular and covers b, then it follows from (5.7) and Theorem 5.5 that $\omega_B^*(\hat{C}) = \omega_b^* \circ s_H^*(\hat{C})$ for all $C \in \mathrm{Cl}(G)$. Thus $b^G = B$. Also, if $B' \in \mathrm{Bl}(G|b)$ is regular, then $\omega_{B'}^* = \omega_b^* \circ s_H^* = \omega_B^*$, as was observed above. Therefore $B' = B$. ■

Lemma 5.14. *If $C_G(\delta(B)) \leq H$, then B is regular.*

Proof. If $\omega_B^*(\hat{C}) \neq 0$ for some $C \in \mathrm{Cl}(G)$, then $\delta(B) \leq_G \delta(C)$, and hence there is $x \in C$ such that $x \in C_G(\delta(B)) \leq H$. Thus $C \cap H \neq \varnothing$ and (5.7) holds. ■

Theorem 5.15. *If there exists a normal p-subgroup Q of G such that $C_G(Q) \leq H$, then every $B \in \mathrm{Bl}(G)$ is regular relative to H. Consequently, every $b \in \mathrm{Bl}(H)$ is covered by a unique block of G, being necessarily equal to b^G. In particular, we obtain a one-to-one correspondence between* Bl(G) *and the set of G-conjugate classes in* Bl(H) *by assigning $B \in \mathrm{Bl}(G)$ to the set of blocks of H covered by B.*

Proof. Since $Q \leq \delta(B)$, it follows that $C_G(\delta(B)) \leq C_G(Q) \leq H$, and hence B is regular by the above lemma. Consequently $\mathrm{Bl}(G|b)$ consists of a single element b^G by Theorem 5.13. The latter half is obvious. ■

Let us work with a slightly weaker condition than (5.7). A block B of G is said to be *weakly regular* (relative to H), provided there exists $C \in \mathrm{Cl}(G)$ satisfying

$$(5.8) \qquad \omega_B^*(\hat{C}) \neq 0, \qquad \delta(C) =_G \delta(B) \qquad \text{and} \quad C \subset H.$$

A regular block is necessarily weakly regular, because any defect class for it satisfies (5.8). The next theorem is of fundamental importance in the study of block covering.

Theorem 5.16 (Fong). *Let* $b \in \mathrm{Bl}(H)$ *and* $B \in \mathrm{Bl}(G|b)$.

(i) *B is weakly regular if and only if B is a block of maximal defect in* $\mathrm{Bl}(G|b)$. *If this is the case, then the following statements hold.*

(a) $\delta(B') \le_G \delta(B)$ *for any* $B' \in \mathrm{Bl}(G|b)$.

(b) *If $\delta(B)$ is chosen so that $\delta(B) \le T(b)$, then*

$$|T(b){:}H\delta(B)| \not\equiv 0 \pmod p. \tag{5.9}$$

Moreover, we have

$$d(B) = d(b) + \nu(|T(b){:}H|). \tag{5.10}$$

(ii) *Given* $B' \in \mathrm{Bl}(G|b)$, *choose* $\delta(B')$ *so that* $\delta(B') \le T(b)$. *Then*

$$\delta(B') \cap H =_H \delta(b).$$

Proof.

(i) Suppose that B is weakly regular, and take $C \in \mathrm{Cl}_G(H)$ so that

$$\omega_B^*(\hat{C}) \ne 0,\ \delta(C) =_Q \delta(B).$$

Then by Theorem 5.5, we have, for any $B' \in \mathrm{Bl}(G|b)$, that

$$\omega_{B'}^*(\hat{C}) = \omega_b^*(\hat{C}) = \omega_B^*(\hat{C}) \ne 0,$$

and hence $\delta(B') \le_G \delta(C) =_G \delta(B)$.

Suppose, conversely, that $d(B') \le d(B)$ for any $B' \in \mathrm{Bl}(G|b)$. Express $f_b = \sum_{B' \in \mathrm{Bl}(G|b)} e_{B'}$ as

$$f_b = \sum_{C \in \mathrm{Cl}_G(H)} \alpha(C)\hat{C}.$$

Since $\omega_B^*(f_b) = 1$, there exists $C_0 \in \mathrm{Cl}_G(H)$, satisfying

$$(1)\quad \alpha(C_0)^* \ne 0, \qquad (2)\quad \omega_B^*(\hat{C}_0) \ne 0.$$

Then (2) implies that $\delta(B) \le_G \delta(C_0)$. And by (1) there is $e_{B'} = \sum_C \beta_{B'}(C)\hat{C}$ with $B' \in \mathrm{Bl}(G|b)$ such that $\beta_{B'}(C_0)^* \ne 0$ and hence $\delta(C_0) \le_G \delta(B')$. Thus we have $\delta(B) =_G \delta(C_0)(=_G \delta(B'))$. Therefore C_0 satisfies (5.8), and B is weakly regular.

In order to show (b), suppose that $B \in \mathrm{Bl}(G|b)$ is weakly regular and let $\tilde{B} \in \mathrm{Bl}(T(b)|b)$ with $\tilde{B}^G = B$. Then as $\delta(\tilde{B}) =_G \delta(B)$ by Theorem 5.10, $\tilde{B}$ is also weakly regular. Thus we may assume that $G = T(b)$. In particular, $f_b = e_b$. Now let $C_0 \in \mathrm{Cl}_G(H)$ be the same as above. Then $\delta(B) =_G \delta(C_0)$, and hence we may assume that $\delta(C_0) = \delta(B) \in \mathrm{Syl}_p(C_G(x))$ for a suitable $x \in C_0$. Let

$t = |G{:}HC_G(x)|$. Then C_0 splits into the union of t H-conjugate classes. Let C'_0 be one of them with $x \in C'_0$. Since b is G-invariant, it follows that

$$0 \neq \omega_B^*(\hat{C}_0) = \omega_b^*(\hat{C}_0) = t^*\omega_b^*(\hat{C}'_0),$$

and so $t^* \neq 0$. Therefore $|G{:}H\delta(B)| = |T(b){:}H\delta(B)| \not\equiv 0 (\text{mod } p)$, which proves (5.9). From the above, we also have $\omega_b^*(\hat{C}'_0) \neq 0$, so that

$$\delta(b) \leq_H \delta(C'_0) = H \cap \delta(C_0) = H \cap \delta(B).$$

On the other hand, since $0 \neq \alpha(C_0)^* = \beta_b(C'_0)^*$, we have $\delta(C'_0) \leq_H \delta(b)$, and hence

(5.11) $$\delta(b) =_H H \cap \delta(B).$$

Therefore, we may assume that $\delta(b) = H \cap \delta(B)$. Then it follows from (5.9) and (5.11) that

$$\begin{aligned} d(B) - d(b) &= \nu(|\delta(B){:}\delta(b)|) = \nu(|H\delta(B){:}H|) \\ &= \nu(|G{:}H|) = \nu(|T(b){:}H|), \end{aligned}$$

which proves (5.10).

(ii) We may assume $G = T(b)$ by the same reason as above. As shown in (5.11), the assertion is true when B is weakly regular. Let W' be an irreducible FH-module belonging to b with $\nu(\dim_F W) = \nu(|H|) - d(b)$. We first claim that $\text{vx}(W) =_H \delta(b)$. In fact, we know that $\nu(|H|) - \nu(|\text{vx}(W)|) \leq \nu(\dim_F W)$ (cf. Chapter 4, 7.5), and hence $d(b) \leq \nu(|\text{vx}(W)|)$. However, since

$$\text{vx}(W) \leq_H \delta(b),$$

we have $\text{vx}(W) =_H \delta(b)$, as claimed. Thus we may assume $\text{vx}(W) = \delta(b)$. By Lemma 5.8, there exists an irreducible FG-module $V \in B'$ with $W | V_H$. Hence $\text{vx}(W) \leq_G \text{vx}(V) \leq_G \delta(B')$, and we have

$$\delta(b) = \text{vx}(W) \leq_G \delta(B') \cap H,$$

and $\delta(b) =_G \delta(B') \cap H$ because $\delta(B') \cap H \leq_G \delta(B) \cap H =_G \delta(b)$. However, as b is G-invariant, any G-conjugate of $\delta(b)$ is a defect group of b. Thus we have $\delta(b) =_H \delta(B') \cap H$. ■

Theorem 5.17. *Let $b \in \text{Bl}(H)$. Then there exists $\zeta \in \text{Irr}(b)$ such that*

$$\text{ht}(\zeta) = 0, \qquad |T(b){:}T(\zeta)| \not\equiv 0 \quad (\text{mod } p).$$

More precisely, let B be a weakly regular block of G lying in $\mathrm{Bl}(G|b)$ *and let* $\chi \in \mathrm{Irr}(B)$ *with* $\mathrm{ht}(\chi) = 0$. *If we write*

$$\chi_H = e\left(\sum_{x \in T(\zeta)\backslash G} \zeta^x \right) \qquad (\zeta \in \mathrm{Irr}(b)),$$

then the following holds:

$$(e, p) = 1, \qquad \mathrm{ht}(\zeta) = 0, \qquad |T(b):T(\zeta)| \not\equiv 0 \pmod p.$$

Proof. It suffices to show the latter half when $G = T(b)$. Since $\chi(1) = e|G:T(\zeta)|\zeta(1)$, it follows that

$$\nu(\chi(1)) = \nu(e) + \nu(|G:T(\zeta)|) + \nu(\zeta(1)).$$

On the other hand, by using $\mathrm{ht}(\chi) = 0$, we have

$$\begin{aligned}\nu(\chi(1)) &= \nu(|G|) - d(B) = \nu(|G|) - \nu(|G:H|) - d(b)\\ &= \nu(|H|) - d(b).\end{aligned}$$

But since $\nu(\zeta(1)) \geq \nu(|H|) - d(b)$, we conclude that

$$\nu(e) = \nu(|G:T(\zeta)|) = 0, \qquad \nu(\zeta(1)) = \nu(|H|) - d(b). \qquad \blacksquare$$

5.4. The Extended First Main Theorem (I)

Let Q be a p-subgroup of G. The first main theorem asserts that $\mathrm{Bl}(G|Q)$ corresponds bijectively to $\mathrm{Bl}(N_G(Q)|Q)$ under the Brauer correspondence. We apply here the results of the preceding subsections to $H = QC_G(Q) \lhd N = N_G(Q)$.

Since $C_G(Q) \leq H$, there corresponds to each $\hat{B} \in \mathrm{Bl}(N)$ a block b of H, being unique up to G-conjugacy, such that $b^N = \hat{B}$. If $T(b)$ denotes the inertial group of b in N, then the following holds.

Exercise 5.18. With the notation above, $\delta(\tilde{B}) = Q$ holds if and only if

(5.12) $$\delta(b) = Q, \qquad |T(b):H| \not\equiv 0 \pmod p.$$

[Hint: Use (5.9) and (5.10).]

Let $\Lambda(Q)$ denote the set of blocks b of H satisfying (5.12) and let $\Lambda(Q)/N$ denote the set of N-conjugate classes in $\Lambda(Q)$. Now the first main theorem can be extended to the following form.

Theorem 5.19. *With the notation above, there are bijections*

$$\mathrm{Bl}(G|Q) \leftrightarrow \mathrm{Bl}(N|Q) \leftrightarrow \Lambda(Q)/N \qquad (B \leftrightarrow \hat{B} \leftrightarrow \{b\})$$

such that $b^N = \hat{B}$, $\hat{B}^G = B$.

Remark. The above correspondences will be related to some classes of irreducible characters of $QC_G(Q)/Q$ of defect zero, as we shall see in Theorem 8.13 of Section 8.

If $B \in \mathrm{Bl}(G|Q)$, we mean by a *root* of B a block b of $QC_G(Q)$ such that $b^G = B$. Thus, if b is a root of B, then $Q \leq \delta(b) \leq_N \delta(b^N) \leq_G \delta(b^G) =_G Q$, and thus $Q = \delta(b) = \delta(b^N)$. Therefore $b \in \Lambda(Q)$ and $B \leftrightarrow b^N \leftrightarrow \{b\}$ with respect to the above correspondences. From this and the uniqueness of the defect group of B up to G-conjugacy, we obtain the following.

Theorem 5.20. *Given* $B \in \mathrm{Bl}(G)$, *the roots of* B *are* G-*conjugate with each other.*

If b is a root of $B \in \mathrm{Bl}(G|Q)$, we denote by $T(b)$ the inertial group of b in $N_G(Q)$, which we shall simply call the *inertial group* of b. And the integer

$$e(B) = |T(b):QC_G(Q)|$$

is called the *inertial index* of B. Note that $(e(B), p) = 1$.

We conclude this section with the following result (cf. Brauer[14] and Alperin and Broué[1]).

Theorem 5.21. *Let* $G \geq H$ *and* $b \in \mathrm{Bl}(H)$ *with* $\delta(b) = Q$. *Suppose that* b *is admissible (i.e.,* $C_G(Q) \leq H$*). Then* $D \cap H = Q$ *holds for a suitable defect group* D *of* b^G. *Moreover,*

$$Z(D) \leq C_D(Q) = Z(Q) \leq Q \leq D. \tag{5.13}$$

In particular, we have $\delta(b^G) =_G \delta(b)$ *if* $\delta(b^G)$ *is abelian.*

Proof (Okuyama). Set $B = b^G$, $L = QC_G(Q) = QC_H(Q)$, and $N = N_G(Q)$. We shall proceed by induction on $|\delta(B):Q|$. Let $b_L \in \mathrm{Bl}(L)$ be a root of b and let $b_N = (b_L)^N$. Then $(b_N)^G = (b_L)^G = ((b_L)^H)^G = b^G = B$. Since, by Theorem 5.15, b^N is regular relative to $T = T(b_L)$, the inertial group of b_L in N, there is a defect group P of b_N such that $P \leq T$, $P \cap L = Q$ by Theorem 5.16. Set $T' = T \cap N_H(Q)$. We have $|T':L| \not\equiv 0 (\mathrm{mod}\ p)$ (cf. Exercise 5.18), and it follows that $P \cap T' = Q$, since $L \lhd T'$.

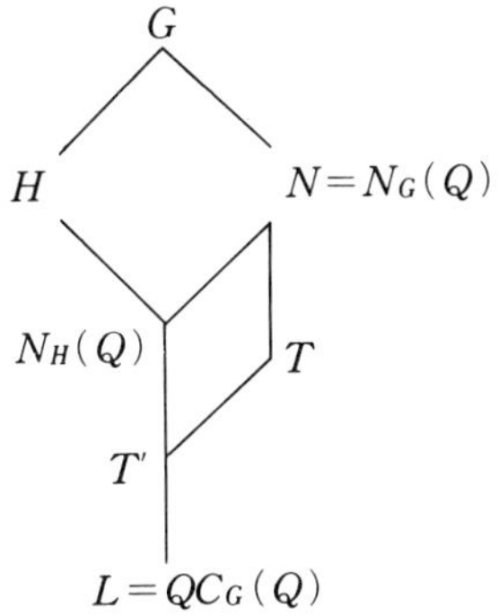

If $P = Q$, then $\delta(B) =_G Q$ by the first main theorem and the assertion follows with $D = Q$.

If $P > Q$, then $|\delta(B):P| < |\delta(B):Q|$. Thus we apply the inductive hypothesis with H and b replaced by N and b_N, respectively, and we find $D = \delta(B)$ such that $P = D \cap N$. Thus,

$$Q = P \cap T' = P \cap T \cap N_H(Q) = P \cap N \cap H = D \cap N \cap H.$$

If $Q < D \cap H$, then we have $Q < D \cap H \cap N$ by a well-known fact on p-groups. But this contradicts the above. Therefore $Q = D \cap H$.

Also, we have $C_D(Q) = D \cap C_G(Q) \leq D \cap H = Q$, and thus (5.13) follows. If D is abelian, then we have $Q = D$ because $Z(D) = D$. ■

6. The Third Main Theorem

The purpose of this section is to prove the following result on the principal block $B_0(G)$ of G, which is referred to as the *third main theorem* of Brauer.

Theorem 6.1 (The Third Main Theorem). *Let $G \geq H$, $b \in \mathrm{Bl}(H)$ and assume that b is admissible. Then we have the following*:

$$b^G = B_0(G) \Leftrightarrow b = B_0(H).$$

Before proving this theorem, we show a couple of lemmas. First we note that because the augmentation map

$$\mu_G: RG \to R \quad \left(\sum_{x\in G} \alpha_x x \mapsto \sum_x \alpha_x\right)$$

is an algebra homomorphism, $\mu_G(e_B)$ is either 0 or 1 for every $B \in \mathrm{Bl}(G)$, and there is a unique B such that $\mu_G(e_B) = 1$. But we have the following.

Lemma 6.2. $\mu_G(e_B) = 1 \Leftrightarrow B = B_0(G)$.

Proof. Write B_0 for $B_0(G)$ and let $u = (1/|G|)(\sum_{x\in G} x)$. Then, for every $a \in KG$, there holds $au = \mu_G(a)u$. In particular, $uKG = uK$ is the trivial G-module, and hence $u \in e_{B_0}KG$. Thus $e_{B_0}u = u$, and it follows that $\mu_G(e_{B_0}) = 1$. Therefore B_0 is the only block B with $\mu_G(e_B) = 1$. ■

Lemma 6.3. *Let Q be a p-subgroup of G with $QC_G(Q) \leq H \leq G$. Then we have $\mu_G(c)^* = 0$ for all $c \in R[G - H]^H$.*

Proof. It suffices to show that $\mu_G(\mathrm{Tr}^H_{C_H(x)}(x)) = |H\!:\!C_H(x)| \equiv 0 \pmod p$ for all $x \in G - H$. Suppose the contrary, then there exists $x \in G - H$ such that $C_H(x)$ contains some Sylow p-subgroup, say P, of H. Then $Q \leq_H P$, and it follows that $x \in_H C_G(Q) \leq H$, which is a contradiction. ■

Now we prove the third main theorem, following Juhász[1].

Proof of Theorem 6.1. ($\Leftarrow$). Suppose that $b = B_0(H)$ is admissible and set $B = b^G$. By Lemma 3.9, there exists $c \in R[G - H]^H$ such that $e_b e_B = e_b + (1 - e_B)c$. Since $\mu_G(e_b)^* = 1$, it follows from Lemma 6.3 that $\mu_G(e_B)^* = 1$ and hence $B = B_0(G)$.

($\Rightarrow$). Set $Q = \delta(b)$ and let $P \in \mathrm{Syl}_p(G)$ be such that $P \geq Q$. We proceed by induction on $|P\!:\!Q|$. From the assumption, we have that $QC_G(Q) = QC_H(Q) \leq H$. Let $\tilde{b}$ be a root of b, i.e., $\tilde{b} \in \mathrm{Bl}(QC_H(Q)|Q)$, with $\tilde{b}^H = b$. If we set $\tilde{B} = \tilde{b}^N$, where $N = N_G(Q)$, then $\tilde{B}^G = \tilde{b}^G = (\tilde{b}^H)^G = b^G = B_0(G)$. We first prove $\tilde{B} = B_0(N)$. Indeed, if $P = Q$, then $B_0(N)^G = B_0(G)$, as we have just proved in the above. However, since $P = \delta(\tilde{b}) \leq \delta(\tilde{B})$, we have $\delta(\tilde{B}) = P$, and it follows from the first main theorem that $\tilde{B} = B_0(N)$. If, on the other hand, $P > Q$, then we have $Q < \delta(\tilde{B})$ (note that, in general, $Q = \delta(\tilde{b}) \leq \delta(\tilde{B})$). In fact, if $Q = \delta(\tilde{B})$, then the first main theorem asserts that $Q = \delta(\tilde{B}^G) = \delta(B_0(G)) \in \mathrm{Syl}_p(G)$, which is a contradiction. Therefore $|P:\delta(\tilde{B})| < |P:Q|$. Then, by applying the inductive hypothesis to $\tilde{B} \in \mathrm{Bl}(N)$ (in place of $b \in \mathrm{Bl}(H)$), we get $\tilde{B} = B_0(N)$.

Now, for the principal block $\tilde{b}_0$ of $QC_G(Q)$, we have $\tilde{b}_0^N = B_0(N) = \tilde{b}^N$ from the implication ($\Leftarrow$). This implies that $\tilde{b}$ and $\tilde{b}_0$ are conjugate in N, and hence $\tilde{b} = \tilde{b}_0$ because $\tilde{b}_0$ is N-invariant. Therefore, $b = \tilde{b}^H = \tilde{b}_0^H = B_0(H)$ as asserted. ∎

7. The Clifford Theory of Blocks (The Stable Case)

Let H be a normal p'-subgroup of G and $\bar{G} = G/H$. Let $b \in \mathrm{Bl}(H)$ be such that $T(b) = G$. In this section, we shall prove a result of Fong, which reduces the study of the blocks in $\mathrm{Bl}(G|b)$ to the study of blocks of a certain central extension of $\bar{G}$ by a cyclic p'-group.

Following the above notation and assumption, note that $\mathrm{Irr}(b)$ consists of a single element, say ζ, and $e_b = e_\zeta$. Also, e_b is a central idempotent of RG. If we set $\mathrm{Bl}(G|b) = \{B_i;\ 1 \leq i \leq r\}$ and denote by e_i the block idempotent of B_i, then

$$e_b = e_1 + \cdots + e_r$$

is the central primitive idempotent decomposition of e_b. Equivalently

$$e_b RG = B_1 \oplus \cdots \oplus B_r$$

is the block decomposition of $e_b RG$.

First, we show

Lemma 7.1. *$b = e_b RH \simeq M_n(R)$ as R-algebras, where $n = \zeta(1)$.*

Proof. Let $\mathbf{F}: y \mapsto (\alpha_{ij}(y))$ be an R-representation that gives the character ζ. For convenience, we identify each element of $b^K (= K \otimes_R b)$ with its image in $M_n(K)$ by $\mathbf{F}^K$. Then $b \subset M_n(R) \subset M_n(K) = b^K$. Let $\{e_{ij};\ 1 \leq i, j \leq n\}$ be the set of matrix units of $M_n(K)$ and write $e_{kl} = \sum_{y \in H} \gamma_y y$. Then by Fourier's inversion formula (Chapter 3, 2.21), we have

$$\gamma_y = |H|^{-1}\left(\sum_{\theta \in \mathrm{Irr}(H)} \theta(1)\theta(e_{kl}y^{-1})\right) = |H|^{-1}\zeta(1)\zeta(e_{kl}y^{-1}).$$

Furthermore, we have $e_{kl}y^{-1} = e_{kl}e_b y^{-1} = \sum_j \alpha_{lj}(y^{-1})e_{kj}$, and it follows that $\zeta(e_{kl}y^{-1}) = \alpha_{lk}(y^{-1}) \in R$. Thus $\gamma_y \in R$ and $e_{kl} \in e_b RH$. Therefore $M_n(R) \subset b$. This completes the proof of the lemma. ■

The following theorem on the strongly G-graded algebra will be useful in determining the block decomposition of $e_b RG$.

Theorem 7.2. *Let $A = \bigoplus_{x \in G} A_x$ be a strongly G-graded algebra over R and suppose that $A_1 \simeq M_n(R)$. Then*

(i) $A \simeq A_1 \otimes_R C_A(A_1)$.

(ii) *$C_A(A_1)$ is a generalized group ring of G over R:*

$$C_A(A_1) = \bigoplus_{x \in G} Ru_x, \qquad u_x u_y = \alpha(x, y)u_{xy}.$$

Proof. We may assume $A_1 = M_n(R)$. Let $A_x = v_x A_1 = A_1 v_x$, where v_x is a unit of A (cf. Chapter 4, 6.2). Then the map $a \mapsto v_x^{-1} a v_x$ $(a \in A_1)$ gives an R-algebra automorphism of A_1, which must be an inner automorphism, by Chapter 2, 4.8. Hence there exists $m_x \in A_1^{\times}$ such that $v_x^{-1} a v_x = m_x^{-1} a m_x$ for all $a \in A_1$. Set $u_x = v_x m_x^{-1}$. Then $u_x \in C_A(A_1)$ and $A_x = A_1 u_x$.

Now we set $C = \bigoplus_x Ru_x$. Since $u_x u_y u_{xy}^{-1} \in A_1 \cap C_A(A_1) = Z(A_1) = R$, we have

$$u_x u_y = \alpha(x, y)u_{xy}, \qquad \text{with } \alpha(x, y) \in R^{\times}.$$

Thus C is a generalized group ring of G over R.

It is clear that $A = A_1 \otimes_R C$. Also, it is easy to see that an element $a = \sum_x n_x \otimes u_x$ $(n_x \in A_1)$ lies in $C_A(A_1)$ if and only if $n_x \in Z(A_1) = R$ for all $x \in G$. Therefore we get $C_A(A_1) = C$. ■

Now we return to the situation at the beginning of this section and let $A = e_b RG$. Since $bx = by$ whenever $x \equiv y \pmod{H}$, we may write

$$A = \bigoplus_{\bar{x} \in \bar{G}} bx = \bigoplus_{\bar{x} \in \bar{G}} A_{\bar{x}}, \qquad \text{where } A_{\bar{x}} = bx.$$

Thus A is a strongly $\bar{G}$-graded algebra, in which $A_{\bar{1}} = b \simeq M_n(R)$ by Lemma 7.1. Then it follows from the above theorem that

$$A = b \otimes_R C \qquad (C = C_A(b)),$$
$$C = \bigoplus_{\bar{x} \in \bar{G}} Ru_{\bar{x}}, \qquad u_{\bar{x}}u_{\bar{y}} = \bar{\alpha}(\bar{x}, \bar{y})u_{\overline{xy}},$$
$$bx = b \otimes u_{\bar{x}} \qquad \text{for all} \quad x \in G, \qquad \text{where} \quad u_{\bar{1}} = 1.$$

If $C = \bigoplus_{i=1}^{r} \bar{B}_i$ denotes the block decomposition, then it follows that

$$A = \bigoplus_{i=1}^{r} b \otimes_R \bar{B}_i. \tag{7.1}$$

However, we have $Z(b \otimes_R \bar{B}_i) = Z(b) \otimes_R Z(\bar{B}_i) = R \otimes_R Z(\bar{B}_i) = Z(\bar{B}_i)$, which is a local ring (note that Chapter 2, 4.1 remains valid if A, B are R-algebras that are R-free.) Therefore (7.1) gives the block decomposition of A.

The factor set $\bar{\alpha}\colon \bar{G} \times \bar{G} \to R^{\times}$ is a normalized one because we have assumed $u_{\bar{1}} = 1$. We shall make further "normalizations" on it, using the arguments in the subsection 5.2 of Chapter 3.

Let us identify b with $M_n(R)$. Given $h \in H$, write $e_b h = Y(h) \in b = M_n(R)$. Then $\mathbf{Y}\colon h \mapsto Y(h)$ gives an R-representation of H giving the character ζ. Given $x \in G$, we set $X(x) = xu_{\bar{x}}^{-1} \in b = M_n(R)$. Then

$$X(x)Y(h)X(x)^{-1} = xu_{\bar{x}}^{-1}(e_b h)u_{\bar{x}}x^{-1} = xe_b hx^{-1} = e_b xhx^{-1} = Y(xhx^{-1}).$$

Now we define $\mathbf{X}\colon G \to \mathrm{GL}_n(R)$ by $X(hx) = Y(h)X(x)$. Then $\mathbf{X}$ is a projective representation of G which extends $\mathbf{Y}$ (see the proof of Chapter 3, 5.7). Since $u_{\bar{x}}$ commutes with $yu_{\bar{y}}^{-1} \in b$ for any $y \in G$, it follows that

$$X(x)X(y) = xu_{\bar{x}}^{-1}yu_{\bar{y}}^{-1} = xyu_{\bar{y}}^{-1}u_{\bar{x}}^{-1} = \bar{\alpha}(\bar{x}, \bar{y})^{-1}xyu_{\overline{xy}}^{-1} = \bar{\alpha}(\bar{x}, \bar{y})^{-1}X(xy),$$

whence we have $\bar{\alpha}^{|H|^2} \sim 1$. In fact there is $\bar{\gamma}(\bar{x}) \in R$ which is expressible as a power of $(\det X(x))^{-1}$ such that

$$\bar{\alpha}(\bar{x}, \bar{y})^{|H|^2} = \bar{\gamma}(\bar{x})\bar{\gamma}(\bar{y})\bar{\gamma}(\overline{xy})^{-1} \tag{7.2}$$

(see the proof of Chapter 3, 5.7).

We assume here that R contains an $|H|^2$th root, say $\bar{\gamma}'(\bar{x})$, of $\bar{\gamma}(\bar{x})$ for all $\bar{x} \in \bar{G}$ after extending R appropriately, where we let $\bar{\gamma}'(\bar{1}) = 1$ for $\bar{\gamma}(\bar{1}) = 1$. Thus, if we set $\bar{\alpha}'(\bar{x}, \bar{y}) = \bar{\alpha}(\bar{x}, \bar{y})\bar{\gamma}'(\bar{x})^{-1}\bar{\gamma}'(\bar{y})^{-1}\bar{\gamma}'(\overline{xy})$, then $\bar{\alpha}'$ is a normalized factor set with $\bar{\alpha}'(\bar{x}, \bar{y})^{|H|^2} = 1$.

Let P be a Sylow p-subgroup of G and let ε be a primitive $|H|^2$th root of unity. If we restrict $\bar{\alpha}'$ to $\bar{P} \times \bar{P}$, we have $\bar{\alpha}'^{|P|} = 1$, hence $\bar{\alpha}' = 1$ in $H^2(\bar{P}, \langle\varepsilon\rangle)$, because $\bar{\alpha}'$ is a p'-element. Then there exists $\bar{\gamma}'': \bar{G} \to \langle \varepsilon \rangle$ such that $\bar{\alpha}'(\bar{x}, \bar{y}) = \bar{\gamma}''(\bar{x})\bar{\gamma}''(\bar{y})\bar{\gamma}''(\overline{xy})^{-1}$ whenever $\bar{x}, \bar{y} \in \bar{P}$. We now obtain the factor set $\bar{\alpha}''(\bar{x}, \bar{y}) = \bar{\alpha}'(\bar{x}, \bar{y})\bar{\gamma}''(\bar{x})^{-1}\bar{\gamma}''(\bar{y})^{-1}\bar{\gamma}''(\overline{xy})$, which satisfies the following conditions:

(a) $$\bar{\alpha}''(\bar{x}, \bar{y})^{|H|^2} = 1, \qquad \bar{\alpha}''(\bar{x}, \bar{1}) = \bar{\alpha}''(\bar{1}, \bar{x}) = 1,$$

(b) $$\bar{\alpha}''(\bar{x}, \bar{y}) = 1 \qquad \text{for all } \bar{x}, \bar{y} \in \bar{P}.$$

Therefore, we may assume that $\bar{\alpha}$ satisfies the above conditions from the beginning.

Let Z be the subgroup of $R^\times$ generated by $\{\bar{\alpha}(\bar{x}, \bar{y}); \bar{x}, \bar{y} \in \bar{G}\}$. Note that Z is a cyclic p'-group. As was already observed in Chapter 3, Section 5, the set $\tilde{G} = Z \times \bar{G} = \{(z, \bar{x}); z \in Z, \bar{x} \in \bar{G}\}$ becomes a group with the multiplication defined by

$$(z, \bar{x})(z', \bar{x}') = (\bar{\alpha}(\bar{x}, \bar{x}')zz', \overline{xx'}).$$

If we identify each $z \in Z$ with $(z, \bar{1}) \in \tilde{G}$, then $Z \subset Z(\tilde{G})$ and $\tilde{G}/Z \simeq \bar{G}$. In other words, $\tilde{G}$ is a central extension of $\bar{G}$ by the cyclic p'-group Z.

If we consider the inclusion map $\iota: Z \to R^\times \subset K (z \mapsto z)$ as a linear character of Z (over K), then the primitive idempotent e_ι corresponding to ι is expressed as

$$e_\iota = \frac{1}{|Z|} \sum_{z \in Z} \iota(z^{-1})z,$$

which, of course, is a central idempotent of $R\tilde{G}$. And the map

(7.3) $$\varphi: C = \bigoplus_{\bar{x} \in \bar{G}} Ru_{\bar{x}} \to e_\iota R\tilde{G} \qquad (u_{\bar{x}} \mapsto e_\iota(1, \bar{x}))$$

is an R-algebra isomorphism, as is easily verified. Thus we get the block decomposition $e_\iota R\tilde{G} = \bigoplus_{i=1}^r \tilde{B}_i$ with $\tilde{B}_i = \varphi(\bar{B}_i)$. Then, by combining this with the block decomposition (7.1) of A, we obtain a one-to-one correspondence $\mathrm{Bl}(G|b) \ni B_i = b \otimes_R \bar{B}_i \mapsto \tilde{B}_i \in \mathrm{Bl}(\tilde{G}|\{\iota\})$ between $\mathrm{Bl}(B|b)$ and $\mathrm{Bl}(\tilde{G}|\{\iota\})$. Before going into details of this correspondence, we show the following lemma.

Lemma 7.3. *Let $\mathfrak{o}$ denote one of K, R, and F. Let C be an $\mathfrak{o}$-algebra and $b = M_n(\mathfrak{o})$, $A = b \otimes C$. Also, let $\mathfrak{o}^n$ be the set of $1 \times n$ matrices over $\mathfrak{o}$, which we regard as a right b-module in the usual manner. Then:*

(i) *Given C-modules U, V, we have an isomorphism*

$$\psi : \mathrm{Hom}_C(U, V) \xrightarrow{\sim} \mathrm{Hom}_A(\mathfrak{o}^n \otimes U, \mathfrak{o}^n \otimes V) \qquad (f \mapsto \mathrm{id} \otimes f).$$

In particular, it holds that $\mathrm{End}_C(V) \simeq \mathrm{End}_A(\mathfrak{o}^n \otimes V)$, *where* $\otimes$ *is the abbreviation of* $\otimes_{\mathfrak{o}}$.

(ii) *Let* $\mathfrak{M}_C$ *denote a full set of representatives for the isomorphism classes of C-modules. Then the map* $V \mapsto \mathfrak{o}^n \otimes V$ *induces a bijection between* $\mathfrak{M}_C$ *and* $\mathfrak{M}_A$, *which preserves indecomposability. It also preserves irreducibility if* $\mathfrak{o}$ *is a field.*

Proof.

(i) Clearly, ψ is a monomorphism. Let $e_{ij}(1 \le i, j \le n)$ be the matrix units of $M_n(\mathfrak{o})$ and $e_i(1 \le i \le n)$ be the unit vectors of $\mathfrak{o}^n$. In particular, $e_i e_{1j} = \delta_{i1} e_j$ holds. Now, we let $g \in \mathrm{Hom}_A(\mathfrak{o}^n \otimes U, \mathfrak{o}^n \otimes V)$ be an arbitrary element and write $g(e_1 \otimes u) = \sum_i e_i \otimes v_i$. Then we have

$$g(e_1 \otimes u) = g((e_1 \otimes u)(e_{11} \otimes 1)) = g(e_1 \otimes u)(e_{11} \otimes 1) = e_1 \otimes v_1.$$

Thus, if we define $f: U \to V$ by $f(u) = v_1$, then f is a C-homomorphism. And we have $g(e_i \otimes u) = g(e_1 \otimes u)(e_{1i} \otimes 1) = (e_1 \otimes f(u))(e_{1i} \otimes 1) = e_i \otimes f(u)$. Thus $g = \mathrm{id} \otimes f$, which proves that ψ is an epimorphism.

(ii) It suffices to show that the map $V \mapsto \mathfrak{o}^n \otimes V$ is surjective. Set $\varepsilon = e_{11}$. We see that $\mathfrak{o}^n \simeq \varepsilon b$ and, for any U_A, $U\varepsilon$ is a C-submodule of U. Then we have an A-isomorphism $f: \varepsilon b \otimes U\varepsilon \xrightarrow{\sim} U$ $(\varepsilon x \otimes u\varepsilon \mapsto u\varepsilon x)$. In fact, if we choose $x_i, y_i \in b$ so that $\sum_{i=1}^m x_i \varepsilon y_i = 1$, then the map $g: U \to \varepsilon b \otimes U\varepsilon$ defined by $g(u) = \sum_i \varepsilon y_i \otimes u x_i \varepsilon$ gives the inverse of f. ■

Now returning again to our original situation, let us identify C with $e_\iota R\tilde{G}$ through the isomorphism φ. Namely, $C = e_\iota R\tilde{G} = \bigoplus_{\bar{x}} R u_{\bar{x}}$ with $u_{\bar{x}} = e_\iota(1, \bar{x})$, and $A = b \otimes_R e_\iota R\tilde{G}$. The assumption (b) on $\bar{\alpha}$ implies that $\{(1, \bar{x});\ x \in P\}$ is a Sylow p-subgroup of $\tilde{G}$, being isomorphic to P by $(1, \bar{x}) \mapsto x$.

We are now ready to prove the following theorem.

Theorem 7.4 (Fong). *Let H be a normal p′-subgroup of G and let* $\bar{G} = G/H$. *Let b be a block of H with* $T(b) = G$. *Then there exist a central extension* $\tilde{G}$ *of* $\bar{G}$ *by a cyclic p′-group Z and a linear character* ι *of Z such that* $A = e_b RG$ *is expressed as*

$$A = e_b RG = b \otimes_R e_\iota R\tilde{G}.$$

Furthermore, there is a bijection

$$\mathrm{Bl}(G|b) \leftrightarrow \mathrm{Bl}(\tilde{G}|\{\imath\}) \qquad (B \mapsto \tilde{B})$$

such that the following statements hold.

(i) $B = b \otimes_R \tilde{B}$.

(ii) *There are bijections* $\mathrm{Irr}(B) \leftrightarrow \mathrm{Irr}(\tilde{B})$ $(\chi \leftrightarrow \tilde{\chi})$ and $\mathrm{IBr}(B) \leftrightarrow \mathrm{IBr}(\tilde{B})$ $(\varphi \leftrightarrow \tilde{\varphi})$, *and with respect to these correspondences, the decomposition numbers and Cartan invariants are preserved*:

$$d_{\chi\varphi} = d_{\tilde{\chi}\tilde{\varphi}}, \qquad c_{\varphi\psi} = c_{\tilde{\varphi}\tilde{\psi}}.$$

(iii) *There is a bijection* $\mathrm{IND}(B) \leftrightarrow \mathrm{IND}(\tilde{B})$ $(V \leftrightarrow \tilde{V})$, *which preserves the vertex. Namely, if* $\mathrm{vx}(V) \leq P$, *then*

$$\mathrm{vx}(\tilde{V}) =_{\tilde{G}} \widetilde{\mathrm{vx}(V)} \qquad (\simeq \mathrm{vx}(V)).$$

(iv) *If* $\delta(B) \leq P$, *then*

$$\delta(\tilde{B}) =_{\tilde{G}} \widetilde{\delta(B)} \qquad (\simeq \delta(B)).$$

Proof.

(i) This has been already shown.

(ii) We may assume $b = M_n(R)$. The existences of the one-to-one correspondences mentioned in the first paragraph are immediate from Lemma 7.3. If $\tilde{\chi}$ (respectively, $\tilde{\varphi}$) is afforded by an $R\tilde{G}$-module $\tilde{U}$ (respectively, an $F\tilde{G}$-module $\tilde{L}_{\tilde{\varphi}}$), then the corresponding χ (respectively, φ) is afforded by $R^n \otimes_R \tilde{U}$ (respectively, $F^n \otimes_F \tilde{L}_{\tilde{\varphi}}$). Write $\tilde{U}^* \leftrightarrow \sum_{\tilde{\varphi}} d_{\tilde{\chi}\tilde{\varphi}} \tilde{L}_{\tilde{\varphi}}$. Then, since $L_\varphi = F^n \otimes_F \tilde{L}_{\tilde{\varphi}}$ is irreducible, it follows that $(R^n \otimes_R \tilde{U})^* = F^n \otimes_F \tilde{U}^* \leftrightarrow \sum_{\tilde{\varphi}} d_{\tilde{\chi}\tilde{\varphi}} L_\varphi$. Therefore we have $d_{\chi\varphi} = d_{\tilde{\chi}\tilde{\varphi}}$, and this yields the equality for the Cartan invariants.

(iii) The existence of the one-to-one correspondence described in the first paragraph is immediate from Lemma 7.3. Also, if $\tilde{V} \in \mathrm{IND}(\tilde{B})$, there is an isomorphism

$$E_R(\tilde{V}) \simeq E_{b\otimes R}(V) \qquad (\tilde{\sigma} \mapsto \sigma = \mathrm{id} \otimes \tilde{\sigma}),$$

where $E_R(\tilde{V})$ is the abbreviation of $\mathrm{End}_R(\tilde{V})$.

Since $Ve_b = V$, we have $E_{b\otimes R}(V) = E_b(V) = E_R(V)^H$. For $\tilde{\sigma} \in E_R(\tilde{V})$, $\tilde{v} \in \tilde{V}$ and $z = (z, 1) \in Z$, we see that $\tilde{v}z = \tilde{v}e_\imath z = \imath(z)\tilde{v}$, and hence $\tilde{\sigma}^z = \tilde{\sigma}$. Furthermore, for $\sigma = \mathrm{id} \otimes \tilde{\sigma}$ and $x \in G$, we find $\sigma^x = \mathrm{id} \otimes \tilde{\sigma}^{(1,\bar{x})}$ from $e_b x = m_x \otimes u_{\bar{x}}$, where $m_x \in b$.

Now, if $Q \leq P$, then $\tilde{\sigma} \in E_R(\tilde{V})$ is $Z\tilde{Q}$-invariant if and only if σ is HQ-invariant. Hence we have $E_R(\tilde{V})^{Z\tilde{Q}} \simeq E_R(V)^{HQ}$. Furthermore, if $\bar{G} =$

$\sum_{\bar{t} \in \bar{Q} \backslash \bar{G}} \bar{Q}\bar{t}$, then $G = \sum_{\bar{t}} HQt$ and $\tilde{G} = \sum_{\bar{t}} ZQ(1, \bar{t})$. Thus we have, for $\tilde{\sigma} \in E_R(\tilde{V})^{Z\tilde{Q}}$,

$$\mathrm{Tr}_{HQ}^{G}(\mathrm{id} \otimes \tilde{\sigma}) = \sum_{\bar{t}} \mathrm{id} \otimes \tilde{\sigma}^{(1,\bar{t})} = \mathrm{id} \otimes \mathrm{Tr}_{Z\tilde{Q}}^{\tilde{G}}(\tilde{\sigma}),$$

which yields that $\mathrm{Tr}_{HQ}^{G}(E_{b \otimes R}(V)) = \mathrm{id} \otimes \mathrm{Tr}_{Z\tilde{Q}}^{\tilde{G}}(E_R(\tilde{V}))$. Consequently, V is Q-projective if and only if $\tilde{V}$ is $\tilde{Q}$-projective, whence the assertion on vertices follows.

(iv) This is clear from (iii). ■

A group G is said to be *p-solvable* provided there exists a series of normal subgroups of G,

$$G = G_0 \geq G_1 \geq \cdots \geq G_m = 1$$

such that each G_i/G_{i+1} is either a p-group or a p'-group. Thus, G is solvable if and only if G is p-solvable for every prime number p.

As an application of Theorem 7.4, we prove the following result on p-solvable groups.

Theorem 7.5 (Fong-Swan). *Let G be a p-solvable group and $\varphi \in \mathrm{IBr}(G)$. Then there exists $\chi \in \mathrm{Irr}(G)$ such that $\chi = \varphi$ on $G_{p'}$.*

Proof. We proceed by double induction on $(|G|_p, |G|)$. We may assume that $|G|_p > 1$ by virtue of Chapter 3, 6.12. If $O_p(G) > 1$, then we apply the inductive hypothesis to $G/O_p(G)$ and obtain the result. Thus, assume $O_p(G) = 1$ and let $H = O_{p'}(G) > 1$. Let B be a block of G such that $\varphi \in \mathrm{IBr}(B)$ and let b be a block of H covered by B. If $T(b) < G$, then we apply the inductive hypothesis to $T(b)$ and obtain the result by Theorem 5.10.

Assume finally $G = T(b)$. With the notation of Theorem 7.4, we have $|\tilde{G}|_p = |G|_p$ and $O_p(\tilde{G}) > 1$ because $O_p(\bar{G}) > 1$. Thus the result holds for $\tilde{G}$ by the first step and therefore for G by Theorem 7.4(ii). ■

Remark. In general, it does not hold that $|\tilde{G}| < |G|$. For this reason, we have used double induction in the above proof.

8. Blocks of Factor Groups

8.1. $\mathrm{Ker}_G\, B$

We begin with the discussion of $\mathrm{Ker}_G\, B$ for a block B of RG, where B is considered to be a right RG-module.

The character χ_B given by B_{RG} is written as

$$\chi_B = \sum_{\chi \in \mathrm{Irr}(B)} \chi(1)\chi.$$

Therefore we have

$$\mathrm{Ker}_G\, B = \bigcap_{\chi \in \mathrm{Irr}(B)} \mathrm{Ker}\, \chi.$$

Theorem 8.1. *For any $\chi \in \mathrm{Irr}(B)$, we have*

$$\mathrm{Ker}_G\, B = O_{p'}(\mathrm{Ker}\, \chi) = \mathrm{Ker}\, \chi \cap O_{p'}(G).$$

In particular, if B_0 is the principal block of G, then

$$\mathrm{Ker}_G\, B_0 = O_{p'}(G).$$

Proof. Since B_{RG} is projective, it follows that $\chi_B(x) = 0$ unless x is a p'-element, whence we have $\mathrm{Ker}_G\, B \subset O_{p'}(G)$, and thus $\mathrm{Ker}_G\, B \subset O_{p'}(G) \cap \mathrm{Ker}\, \chi = O_{p'}(\mathrm{Ker}\, \chi)$.

To show the reverse inclusion, let $H = O_{p'}(\mathrm{Ker}\, \chi)$ and $u = (1/|H|)\sum_{y \in H} y$. Then u is a central idempotent with $\omega_\chi(u) = 1$ and hence $\omega_B^*(u^*) = 1$. Therefore, $ue_B = e_B$, and so $B = e_B RG \subset uRG = RGu$. But as H acts trivially on RGu, we have $H \leq \mathrm{Ker}_G\, B$.

The second half follows easily from the first, with $\chi = \mathbf{1}_G$. ■

Next we prove the following theorem.

Theorem 8.2. *Let $M = \bigcap_{\varphi \in \mathrm{IBr}(B)} \mathrm{Ker}\, \varphi^*$. Then*

$$M/\mathrm{Ker}_G\, B = O_p(G/\mathrm{Ker}_G\, B).$$

Proof. It is clear that $\mathrm{Ker}_G\, B \leq M$ and $O_p(G/\mathrm{Ker}_G\, B) \leq M/\mathrm{Ker}_G\, B$. Thus, it suffices to show that $M/\mathrm{Ker}_G\, B$ is a p-group.

If $x \in M$ is a p-regular element, then $\varphi(x) = \varphi(1)$ for all $\varphi \in \mathrm{IBr}(B)$, and hence $\chi(x) = \chi(1)$ for all $\chi \in \mathrm{Irr}(B)$. Therefore $x \in \bigcap_{\chi \in \mathrm{Irr}(B)} \mathrm{Ker}\,\chi = \mathrm{Ker}_G B$, which implies that $M/\mathrm{Ker}_G B$ is a p-group. ∎

A group G is said to be *p-nilpotent* if $G = PO_{p'}(G)$ with $P \in \mathrm{Syl}_p(G)$. The p-nilpotent groups are characterized as follows.

Theorem 8.3. *G is p-nilpotent if and only if $l(B_0) = 1$, where B_0 is the principal block of G.*

Proof. If G is p-nilpotent, then it follows from the preceding two theorems that $\mathrm{Ker}\,\varphi^* = G$ for any $\varphi \in \mathrm{IBr}(B_0)$. Therefore $\mathrm{IBr}(B_0) = \{\mathbf{1}_G\}$.

Suppose, conversely, that $\mathrm{IBr}(B_0) = \{\mathbf{1}_G\}$. Then

$$O_p(G/O_{p'}(G)) = \mathrm{Ker}\,\mathbf{1}_G^*/O_{p'}(G) = G/O_{p'}(G).$$

Therefore G is p-nilpotent. ∎

8.2. Domination of Blocks

In this subsection we let $G \rhd H$ and $\bar{G} = G/H$. As before, we denote by μ_H the natural K-algebra homomorphism

$$\mu_H\colon KG \to K\bar{G} \qquad \left(\sum_x \alpha_x x \mapsto \sum_x \alpha_x \bar{x}\right).$$

Lemma 8.4. *Let $\chi \in \mathrm{Irr}(G)$.*

(i) *Suppose that $H \le \mathrm{Ker}\,\chi$. Then χ induces the irreducible character χ' of $\bar{G}$ such that $\chi'(\bar{x}) = \chi(x)$, and we have $\mu_H(e_\chi) = e_{\chi'}$.*

(ii) *If $H \not\le \mathrm{Ker}\,\chi$, then $\mu_H(e_\chi) = 0$.*

Proof.

(i) We have

$$\begin{aligned}\mu_H(e_\chi) &= \frac{1}{|G|}\sum_{x \in H\backslash G}\sum_{h \in H}\chi(1)\chi((hx)^{-1})\bar{x} \\ &= \frac{|H|}{|G|}\sum_{\bar{x} \in \bar{G}}\chi'(1)\chi'(\bar{x}^{-1})\bar{x} = e_{\chi'}.\end{aligned}$$

(ii) If we set $\Lambda = \{\chi \in \mathrm{Irr}(G); H \le \mathrm{Ker}\,\chi\}$, then $\mathrm{Irr}(\bar{G}) = \{\chi'; \chi \in \Lambda\}$, and

$$\bar{1} = \sum_{\chi \in \Lambda} e_{\chi'} = \sum_{\chi \in \Lambda} \mu_H(e_\chi)$$

is the central primitive idempotent decomposition of $\bar{1}$ in $K\bar{G}$. By comparing this with the equation

$$\bar{1} = \sum_{\chi \in \mathrm{Irr}(G)} \mu_H(e_\chi),$$

we get $\mu_H(e_\chi) = 0$, unless $\chi \in \Lambda$. ■

We understand that $\mathrm{Irr}(\bar{G}) \subset \mathrm{Irr}(G)$ by identifying every $\chi' \in \mathrm{Irr}(\bar{G})$ with $\chi = \chi' \circ \mu_H$.

For $B \in \mathrm{Bl}(G)$, $\mu_H(e_B)$ is a central idempotent of $R\bar{G}$ unless it is zero. Thus, if $\mu_H(e_B) \neq 0$, then there exist blocks $\bar{B}_1, \ldots, \bar{B}_r$ of $\bar{G}$ such that

$$\mu_H(e_B) = \bar{e}_1 + \cdots + \bar{e}_r, \tag{8.1}$$

where $\bar{e}_i$ denotes the block idempotent of $\bar{B}_i$ $(1 \le i \le r)$.

If (8.1) holds, we say that B *dominates* $\bar{B}_i$. Note that in $Z(R\bar{G})$ we have

$$\bar{1} = \sum_{B \in \mathrm{Bl}(G)} \mu_H(e_B) = \sum_{\bar{B} \in \mathrm{Bl}(\bar{G})} e_{\bar{B}}.$$

Hence every $\bar{B} \in \mathrm{Bl}(\bar{G})$ is dominated by a unique $B \in \mathrm{Bl}(G)$.

Lemma 8.5. *If $\bar{B} \in \mathrm{Bl}(\bar{G})$ is dominated by B, then*

$$\omega_B^* = \omega_{\bar{B}}^* \circ \mu_H^*. \tag{8.2}$$

Proof. With the notation of (8.1), let $\bar{B} = \bar{B}_i$. Then $\omega_{\bar{B}}^* \circ \mu_H^*(e_B^*) = 1$, and hence (8.2) holds. ■

Lemma 8.6. *Let $B \in \mathrm{Bl}(G)$, $\bar{B} \in \mathrm{Bl}(\bar{G})$.*

(i) *$\mu_H(e_B) \neq 0$ if and only if there exists $\chi \in \mathrm{Irr}(B)$ such that $H \le \mathrm{Ker}\,\chi$.*

(ii) *$\bar{B}$ is dominated by B if and only if $\mathrm{Irr}(\bar{B}) \subset \mathrm{Irr}(B)$.*

Proof.

(i) This is clear from Lemma 8.4.

(ii) With the notation of (8.1), let $\bar{B} = \bar{B}_1$. If $\chi \in \mathrm{Irr}(\bar{B}_1)$, then $\chi(e_B) = \chi(\bar{e}_1) \neq 0$. Therefore $\chi \in \mathrm{Irr}(B)$, and it follows that $\mathrm{Irr}(\bar{B}) \subset \mathrm{Irr}(B)$.

Suppose, conversely, that $\mathrm{Irr}(\bar{B}) \subset \mathrm{Irr}(B)$. Then $\mu_H(e_B) \neq 0$ by (i). We have $\chi(e_B) \neq 0$ for any $\chi \in \mathrm{Irr}(\bar{B})$, hence $\chi(\bar{e}_i) \neq 0$ for some $\bar{e}_i$ appearing in the central primitive idempotent decomposition (8.1) of $\mu_H(e_B)$. Thus we have $\bar{B}_i = \bar{B}$. ∎

Theorem 8.7. *Let $B \in \mathrm{Bl}(G)$.*

(i) *If $\mu_H(e_B) \neq 0$, then $\delta(B) \cap H \in \mathrm{Syl}_p(H)$.*

(ii) *If $\bar{B} \in \mathrm{Bl}(\bar{G})$ is dominated by B, then*

$$\delta(\bar{B}) \leq_{\bar{G}} \overline{\delta(B)} = \delta(B)H/H.$$

In particular, we have

$$d(\bar{B}) \leq d(B) - \nu(|H|).$$

Proof.

(i) Write $e_B = \sum_C \beta_B(C)\hat{C}$. Since $\mu_H^*(e_B^*) = \mu_H(e_B)^* \neq 0$, there exists $C \in \mathrm{Cl}(G)$ such that $\beta_B(C)^* \neq 0$, $\mu_H(\hat{C})^* \neq 0$. The first condition implies that $\delta(C) \leq_G \delta(B)$, and from the second condition, we see that $\delta(C)$ contains some Sylow p-subgroup of H (cf. Lemma 2.6(iii)). Hence, $\delta(B)$ contains some Sylow p-subgroup of H, which implies that $\delta(B) \cap H \in \mathrm{Syl}_p(H)$.

(ii) Set $D = \delta(B)$. We first claim that $\mu_H^*((FG)_D^G) \subset (F\bar{G})_{\bar{D}}^{\bar{G}}$. If $a \in (FG)^D$, then $\mathrm{Tr}_D^G(a) = \mathrm{Tr}_{HD}^G(\mathrm{Tr}_D^{HD}(a))$. Therefore $\mu_H^*(\mathrm{Tr}_D^G(a)) = \mathrm{Tr}_{\bar{D}}^{\bar{G}}(\mu_H^*(\mathrm{Tr}_D^{HD}(a)) \in (F\bar{G})_{\bar{D}}^{\bar{G}}$, as claimed.

Now, since $e_B^* \in (FG)_D^G$, it follows from the above claim that $\mu_H^*(e_B^*) \in (F\bar{G})_{\bar{D}}^{\bar{G}}$. But this implies that $e_{\bar{B}}^* \in (F\bar{G})_{\bar{D}}^{\bar{G}}$, since $e_{\bar{B}}^* = e_{\bar{B}}^*\mu_H^*(e_B^*)$ from the assumption. Thus we have $\delta(\bar{B}) \leq_{\bar{G}} \bar{D}$. The remaining assertion is clear from (i). ∎

When H is a p'-group, the following assertion is true.

Theorem 8.8. *Assume that H is a p'-group. If $\mu_H(e_B) \neq 0$, then $\mu_H(e_B)$ is precisely the block idempotent of some block, say $\bar{B}$, of $\bar{G}$ and it holds that*

$$\mathrm{Irr}(\bar{B}) = \mathrm{Irr}(B), \qquad \delta(\bar{B}) =_G \overline{\delta(B)}.$$

Proof. Since $\mu_H(e_B) \neq 0$, there exists $\chi \in \mathrm{Irr}(B)$ such that $H \leq \mathrm{Ker}\ \chi$ by Lemma 8.6(i). Then it follows from Theorem 8.1 that $\mathrm{Ker}_G B = O_{p'}(\mathrm{Ker}\ \chi) \geq H$ and hence $\mathrm{Irr}(B) \subset \mathrm{Irr}(\bar{G})$, $\mathrm{IBr}(B) \subset \mathrm{IBr}(\bar{G})$. This implies that if $\chi_1, \chi_2 \in \mathrm{Irr}(B)$, then we have $\chi_1 \sim \chi_2$ as irreducible characters of $\bar{G}$, where "$\sim$" is defined in Section 6.4 of Chapter 3. Therefore, we have $\mathrm{Irr}(B) = \mathrm{Irr}(\bar{B})$ for some $\bar{B} \in \mathrm{Bl}(\bar{G})$ (and hence $\mu_H(e_B) = e_{\bar{B}}$). To show the rest, choose $\chi \in \mathrm{Irr}(B)$ so that $\mathrm{ht}(\chi) = 0$. Since $\nu(\chi(1)) = \nu(|G|) - d(B) = \nu(|\bar{G}|) - d(B)$, we have $d(B) \leq d(\bar{B})$. Then we get $\delta(\bar{B}) =_{\bar{G}} \overline{\delta(B)}$ from Theorem 8.7(ii). ■

8.3. Blocks of Factor Groups by p-Groups

In this subsection, we assume that

(8.3) *Q is a normal p-subgroup of G*

and let $\bar{G} = G/Q$.

Lemma 8.9. *Let $C \in \mathrm{Cl}(G_{p'})$, and assume that $C \subset C_G(Q)$. Given $x \in C$, we let $\bar{C}$ denote the conjugate class of $\bar{G}$ containing $\bar{x}$. Then the following statements hold.*

(i) $C_{\bar{G}}(\bar{x}) = C_G(x)/Q$.
(ii) $\delta(\bar{C}) =_{\bar{G}} \overline{\delta(C)}$.
(iii) $\mu_Q(\hat{C}) = \hat{\bar{C}}$.

Proof.

(i) If $\bar{y}^{-1}\bar{x}\bar{y} = \bar{x}$, then $y^{-1}xy = xz$ for some $z \in Q$. But since x is a p'-element with $Q \subset C_G(x)$, we have $z = 1$ by comparing the order. Therefore we have $C_{\bar{G}}(\bar{x}) \subset \overline{C_G(x)}$, whence the equality holds because the reverse inclusion always holds.

(ii) Clear by (i).

(iii) Since $|\bar{C}| = |C|$ by (i), the result is obvious. ■

Theorem 8.10. *We have $\mu_Q(e_B) \neq 0$ for every $B \in \mathrm{Bl}(G)$. Also, there is $\bar{B} \in \mathrm{Bl}(\bar{G})$, dominated by B with $\delta(\bar{B}) =_{\bar{G}} \overline{\delta(B)}$.*

Proof. Since the kernel of $\mu_Q^*: FG \to F\bar{G}$ is nilpotent, $\mu_Q^*(e_B^*) \neq 0$, and hence $\mu_Q(e_B) \neq 0$.

Write $e_B = \sum_C \beta_B(C)\hat{C}$ and let $C_0 \in \mathrm{Cl}(G)$ be such that $\beta_B(C_0)^* \neq 0 \neq \omega_B^*(\hat{C}_0)$. Then $C_0 \in \mathrm{Cl}(G_{p'})$ and $\delta(C_0) =_G \delta(B) \geq Q$. Thus, we have $C_0 \subset C_G(Q)$, that is, the assumption of Lemma 8.9 is satisfied for $C = C_0$.

Now, with the notation of (8.1), let us write $\bar{e}_i = \sum_C \beta_i(\bar{C})\hat{\bar{C}}$. By comparing the coefficients of $\mu_Q(\hat{C}_0) = \hat{\bar{C}}_0$ on both sides of the equation $\mu_Q(e_B)^* = \sum_i \bar{e}_i^*$, we find that $\beta_B(C_0)^* = \sum_i \beta_i(\bar{C}_0)^* \neq 0$, and hence $\beta_i(\bar{C}_0)^* \neq 0$ for some i. Consequently we have $\delta(\bar{C}_0) \leq_{\bar{G}} \delta(\bar{B}_i)$. Then

$$\overline{\delta(B)} =_{\bar{G}} \overline{\delta(C_0)} =_{\bar{G}} \delta(\bar{C}_0) \leq_{\bar{G}} \delta(\bar{B}_i)$$

and $\overline{\delta(B)} =_{\bar{G}} \delta(\bar{B}_i)$ follows. ∎

Theorem 8.11. *Suppose that $|G:C_G(Q)|$ is a power of p. If $B \in \mathrm{Bl}(G)$, then $\mu_Q(e_B) = e_{\bar{B}}$ for some $\bar{B} \in \mathrm{Bl}(\bar{G})$, and the map $B \mapsto \bar{B}$ gives rise to a one-to-one correspondence between $\mathrm{Bl}(G)$ and $\mathrm{Bl}(\bar{G})$. Also, the following is true for the Cartan matrices.*

(8.4) $$C_B = |Q| C_{\bar{B}}.$$

Proof. Since $G_{p'} \subset C_G(Q)$ from the assumption, it is clear that the p'-elements of G are in one-to-one correspondence with the elements of $\bar{G}_{p'}$ under the natural map $G \to \bar{G} = G/Q$. Also, we have $\mathrm{IBr}(G) = \mathrm{IBr}(\bar{G})$, since $Q \leq \mathrm{Ker}\, \varphi^*$ for all $\varphi \in \mathrm{IBr}(G)$. Thus, if $\varphi_i, \varphi_j \in \mathrm{IBr}(G)$, then

$$\begin{aligned}(\varphi_i, \varphi_j)' &= \frac{1}{|G|} \sum_{x \in G_{p'}} \varphi_i(x)\varphi_j(x^{-1}) \\ &= \frac{1}{|Q|}\frac{1}{|\bar{G}|} \sum_{\bar{x} \in \bar{G}_{p'}} \bar{\varphi}_i(\bar{x})\bar{\varphi}_j(\bar{x}^{-1}) = \frac{1}{|Q|}(\bar{\varphi}_i, \bar{\varphi}_j)'.\end{aligned}$$

Thus we get $C = |Q|\bar{C}$ for the Cartan matrices $C, \bar{C}$ of $G, \bar{G}$, respectively, (cf. Chapter 3, 6.11 (ii)). In particular, we have $|\mathrm{Bl}(G)| = |\mathrm{Bl}(\bar{G})|$, which yields that $\mu_H(e_B)$ is a block idempotent for all $B \in \mathrm{Bl}(G)$, and hence (8.4) holds. ∎

8.4. The Extended First Main Theorem (II)

In this subsection, we assume that Q is a p-subgroup of G and we let $N = N_G(Q)$, $H = QC_G(Q)$, $\bar{N} = N/Q$.

As is shown in Section 5.4, there corresponds to each $B \in \mathrm{Bl}(G|Q)$ a certain N-conjugate class of $\mathrm{Bl}(H|Q)$. Here we shall show that the latter corresponds to some classes of irreducible characters of $\bar{H}$ of defect zero.

For $b \in \mathrm{Bl}(H)$, $T(b)$ denotes the inertial group of b in N, as before. Also, for $\theta \in \mathrm{Irr}(\bar{H})$, we denote by $T(\theta)$ the inertial group of θ in N, considering θ as a character of H.

First we prove the following lemma as a corollary of Theorem 8.11.

Lemma 8.12. *Let $b \in \mathrm{Bl}(H)$ with $\delta(b) = Q$. Then we have the following.*
(i) $l(b) = 1$.
(ii) *There exists a unique $\theta \in \mathrm{Irr}(b)$ with* $\mathrm{Ker}\,\theta \geq Q$, *and we have* $\mathrm{ht}(\theta) = 0$.

Proof. Apply Theorem 8.11 with $H \rhd Q$; then there exists $\bar{b} \in \mathrm{Bl}(\bar{H})$ such that $\mu_Q(e_b) = e_{\bar{b}}$. By Theorem 8.10 $\delta(\bar{b}) =_{\bar{H}} \overline{\delta(b)} = 1$, i.e., $d(\bar{b}) = 0$. Therefore $l(\bar{b}) = 1$, and we get $l(b) = 1$ from (8.4).

Now let θ be a unique element of $\mathrm{Irr}(\bar{b})$, which is clearly of height zero as an element of $\mathrm{Irr}(b)$. If there exists $\theta' \in \mathrm{Irr}(b)$ such that $\mathrm{Ker}\,\theta' \geq Q$, then $\theta' \in \mathrm{Irr}(\bar{b})$ and hence $\theta' = \theta$. ■

We call the above θ the *canonical character* of b. If $x \in N$, then θ^x is the canonical character of b^x, and in particular, we have $T(b) = T(\theta)$.

Recall that with the notation of Theorem 5.19 there are one-to-one correspondences

$$\mathrm{Bl}(G|Q) \leftrightarrow \mathrm{Bl}(N|Q) \leftrightarrow \Lambda(Q)/N \qquad (B \leftrightarrow \hat{B} \leftrightarrow \{b\})$$

such that $b^G = \hat{B}^G = B$. Now Theorem 5.19 and Lemma 8.12 allow us to describe the first main theorem in the following form.

Theorem 8.13. *Let $\bar{\Lambda}(Q)$ denote the set of irreducible characters of $\bar{H}$ of defect zero satisfying*

$$|T(\theta){:}H| \not\equiv 0 \qquad (\mathrm{mod}\ p),$$

and let $\bar{\Lambda}(Q)/N$ be the set of N-conjugate classes of $\bar{\Lambda}(Q)$. Then there is a one-to-one correspondence

$$\mathrm{Bl}(G|Q) \leftrightarrow \bar{\Lambda}(Q)/N \qquad (B \leftrightarrow \{\theta\})$$

such that if $\theta \in \mathrm{Irr}(b)$, where $b \in \mathrm{Bl}(H)$, then $b^G = B$.

Finally, we show the following theorem in connection with Lemma 8.12.

Theorem 8.14. *Let $b \in \mathrm{Bl}(H|Q)$ and let θ be the canonical character of b. For each $\xi_i \in \mathrm{Irr}(Q) = \{\xi_1 = \mathbf{1}_Q, \xi_2, \ldots, \xi_r\}$, define the class function θ_i on H by*

$$\theta_i(h) = \begin{cases} \xi_i(h_p)\theta(h_{p'}), & \text{if } h_p \in Q \\ 0, & \text{otherwise.} \end{cases}$$

Then $\mathrm{Irr}(b) = \{\theta_1 = \theta, \theta_2, \ldots, \theta_r\}$.

Proof. Let ζ be an arbitrary character in $\mathrm{Irr}(b)$. It is clear that every $\xi_j \in \mathrm{Irr}(Q)$ is H-invariant, and so $\zeta_Q = e\xi_i$ for some ξ_i with $e \geq 1$.

Given $h \in H$ we have $\zeta(h) = 0$ if $h_p \notin Q = \delta(b)$. If, on the other hand, $h_p \in Q$, then $h \in Q \times \langle h_{p'} \rangle = L$, and ζ_L is written as $\zeta_L = \xi_i \times \alpha$ with some character α of $\langle h_{p'} \rangle$. Thus, $\zeta(xh_{p'}) = \xi_i(x)\alpha(h_{p'})$ for every $x \in Q$. In particular, it follows that $\alpha(h_{p'}) = \zeta(h_{p'})/\xi_i(1)$, and hence

$$\zeta(xh_{p'}) = \frac{\xi_i(x)}{\xi_i(1)}\zeta(h_{p'}) \qquad (x \in Q).$$

As we saw in the proof of Lemma 8.12, θ is an irreducible character of $\bar{H}$ of defect zero. Considering it as an irreducible Brauer character of H, let $d_{\zeta\theta}$ and $c_{\theta\theta}$ be the decomposition number and Cartan invariant, respectively. Now if $x \in Q$, $y \in H_{p'}(\subset C_G(Q))$, then

$$\zeta(xy) = \frac{\xi_i(x)}{\xi_i(1)}\zeta(y) = \frac{\xi_i(x)}{\xi_i(1)} d_{\zeta\theta}\theta(y)$$

$$= \frac{d_{\zeta\theta}}{\xi_i(1)}\theta_i(xy).$$

Hence

$$1 = (\zeta, \zeta)_H = \frac{1}{|H|}\sum_{x\in Q}\sum_{y\in H_{p'}} |\zeta(xy)|^2$$

$$= \frac{1}{|H|}\left(\frac{d_{\zeta\theta}}{\xi_i(1)}\right)^2 \sum_{x\in Q} |\xi_i(x)|^2 \sum_{y\in H_{p'}} |\theta(y)|^2$$

$$= \left(\frac{d_{\zeta\theta}}{\xi_i(1)}\right)^2 \frac{|Q|}{|H|} \sum_{\bar{y}\in \bar{H}_{p'}} |\theta(\bar{y})|^2$$

$$= \left(\frac{d_{\zeta\theta}}{\xi_i(1)}\right)^2 \frac{1}{|\bar{H}|} \sum_{\bar{y}\in \bar{H}} |\theta(\bar{y})|^2 = \left(\frac{d_{\zeta\theta}}{\xi_i(1)}\right)^2,$$

whence we have $d_{\zeta\theta}/\xi_i(1) = 1$ and $\zeta = \theta_i$.

Thus we have shown that every $\zeta \in \mathrm{Irr}(b)$ coincides with some θ_i, and we may write $\mathrm{Irr}(b) = \{\theta_1 = \theta, \ldots, \theta_s\}$ with $s \le r$. However, since

$$|Q| = c_{\theta\theta} = \sum_{i=1}^{s} (d_{\theta_i\theta})^2 = \sum_{i=1}^{s} \xi_i(1)^2,$$

we have $s = r$. ∎

9. Subpairs and Subsections

9.1. Subpairs

Let Q be a p-subgroup of G. If $b \in \mathrm{Bl}(QC_G(Q))$, then the pair (Q, b) is called a *subpair* of G. We say that (Q, b) *belongs* to B or is a *B-subpair* if $b^G = B$ (note that b^G is defined). In this subsection, we shall mention some basic facts on subpairs following Alperin [5].

Let (Q, b_Q) and (P, b_P) be subpairs. We write $(Q, b_Q) \lhd (P, b_P)$ if the following three conditions are satisfied:

(1) $Q \lhd P$,
(2) b_Q is P-invariant,
(3) $(b_Q)^{PC_G(Q)} = (b_P)^{PC_G(Q)}$

If this is the case, then, by noting that $QC_G(Q)$, $PC_G(P) \le PC_G(Q)$, we have $(b_Q)^G = (b_P)^G$. If this equal block is denoted by B, then both (Q, b_Q) and (P, b_P) are B-subpairs.

If (Q, b_Q) and (P, b_P) are subpairs, we say that (Q, b_Q) is *contained* in (P, b_P) and write $(Q, b_Q) \le (P, b_P)$, provided there is a sequence of subpairs such that

$$(Q, b_Q) = (Q_1, b_1) \lhd (Q_2, b_2) \lhd \cdots \lhd (Q_r, b_r) = (P, b_P).$$

If (Q, b) is a subpair and $t \in G$, (Q^t, b^t) is also a subpair, which is called a G-conjugate of (Q, b).

For a B-subpair (Q, b), we have $Q \le \delta(b) \le_G \delta(b^G) = \delta(B)$. If $\delta(B) = Q$, then we say that (Q, b) is a *Sylow B-subpair*. If this is the case, then b is a root of B.

We first show the following theorem of Sylow type on B-subpairs.

Theorem 9.1. *Let $B \in \mathrm{Bl}(G)$.*

(i) *All Sylow B-subpairs are G-conjugate to each other.*
(ii) *Every B-subpair is contained in some Sylow B-subpair.*

Proof.

(i) This is immediate from Theorems 5.19 and 5.20.

(ii) Let (Q, b) be a maximal B-subpair with respect to the partial order $\leq$. We want to show that $\delta(B) =_G Q$.

Let $H = QC_G(Q)$, let $T(b)$ be the inertial group of b in $N_G(Q)$, and L be a subgroup of G with $L/H \in \mathrm{Syl}_p(T(b)/H)$. From Theorem 5.15, we know that b^L is regular. Let $P = \delta(b^L)$. Then we have $L = PH = PC_G(Q)$ and $P \cap H = \delta(b)$ by Theorem 5.16. If $b_P \in \mathrm{Bl}(PC_L(P))$ is a root of b^L, then $(b_P)^L = b^L$, and hence $(b_P)^{PC_G(Q)} = b^{PC_G(Q)}$. On the other hand, since $Q \leq P$, it follows that $C_G(P) \leq C_G(Q) \leq L$ and hence $C_L(P) = C_G(P)$. Thus (P, b_P) is a subpair with $(Q, b) \lhd (P, b_P)$, because $Q \lhd P$ and b is P-invariant. But since (Q, b) is maximal, we have $Q = P = \delta(b)$ and hence $L = H$. This implies that $|T(b){:}H| \not\equiv 0 (\mathrm{mod}\ p)$, whence we have $\delta(b^G) = \delta(B) =_G Q$ by Exercise 5.18. Thus, by definition, (Q, b) is a Sylow B-subpair, which completes the proof. ■

Lemma 9.2. *Let (P, b_P) be a subpair.*

(i) *If $Q \lhd P$, then there exists a unique $b_Q \in \mathrm{Bl}(QC_G(Q))$ such that $(Q, b_Q) \lhd (P, b_P)$.*

(ii) *If $(R, b_R) \lhd (P, b_P)$ and $(Q, b_Q) \lhd (P, b_P)$ with $R \leq Q$, then $(R, b_R) \lhd (Q, b_Q)$.*

Proof.

(i) Since $QC_G(Q) \lhd PC_G(Q)$ from the assumption, there is $b_Q \in \mathrm{Bl}(QC_G(Q))$ covered by $(b_P)^{PC_G(Q)}$. But by Theorem 5.15, $(b_Q)^{PC_G(Q)}$ is a unique block of $PC_G(Q)$ covering b_Q, so we have $(b_Q)^{PC_G(Q)} = (b_P)^{PC_G(Q)}$. We shall show that b_Q is P-invariant in order to get $(Q, b_Q) \lhd (P, b_P)$.

Set $L = PC_G(Q)$ for brevity and let T be the inertial group of b_Q in L. By Theorem 5.10(v), $\delta((b_Q)^L) \leq_L T$. And $P \leq \delta(b_P) \leq_L \delta((b_P)^L) = \delta((b_Q)^L)$, thus $P \leq_L T$. But since $C_G(Q) \leq T$, we conclude that $P \leq T$, namely, P stabilizes b_Q. Also, we have $T = L$, and so b_Q is the unique block of $QC_G(Q)$ covered by $(b_Q)^L = (b_P)^L$. This proves the uniqueness of (Q, b_Q).

(ii) Our assumption implies that $R \lhd Q$ and b_R is Q-invariant. Therefore, we only need to show that $(b_R)^{QC_G(R)} = (b_Q)^{QC_G(R)}$.

Set $H = QC_G(R)$ and $X = PC_G(R)$, then $H \lhd X$, $R \lhd X$ and $C_X(R) \leq H$. By assumption, $(b_R)^X = (b_P)^X$ and $(b_Q)^{PC_G(Q)} = (b_P)^{PC_G(Q)}$, whence it follows that $(b_Q)^X = (b_P)^X = (b_R)^X$. Thus $((b_Q)^H)^X = ((b_R)^H)^X$, and so $(b_Q)^H$ and $(b_R)^H$ are conjugate in X by Theorem 5.15. But since b_R is X-invariant as it is

P-invariant, it follows that $(b_R)^H$ is X-invariant, and so we get $(b_Q)^H = (b_R)^H$, as desired. ■

Finally, we prove the following result, which shows a close connection between the p-subgroups of G and the subpairs of G.

Theorem 9.3. *If (P, b_P) is a subpair and $Q \leq P$, then there exists a unique $b_Q \in \mathrm{Bl}(QC_G(Q))$ such that $(Q, b_Q) \leq (P, b_P)$. Moreover, if $Q \lhd P$, then $(Q, b_Q) \lhd (P, b_P)$.*

Proof. We proceed by induction on $|P{:}Q|$. By Lemma 9.2(i), we readily find $b_Q \in \mathrm{Bl}(QC_G(Q))$ such that $(Q, b_Q) \leq (P, b_P)$. To show the uniqueness of b_Q, let $(Q, b') \leq (P, b_P)$. From the definition of containment of subpairs, there are sequences of subpairs $(Q, b_Q) \leq (R, b_R) \underset{\neq}{\lhd} (P, b_P)$ and $(Q, b') \leq (R', b_{R'}) \underset{\neq}{\lhd} (P, b_P)$. Set $S = R \cap R'$, which is normal in P, then there exists a subpair (S, b_S) with $(S, b_S) \lhd (P, b_P)$. By Lemma 9.2(ii), we have $(S, b_S) \lhd (R, b_R)$ and $(S, b_S) \lhd (R', b_{R'})$. Now there exists $b'' \in \mathrm{Bl}(QC_G(Q))$ such that $(Q, b'') \leq (S, b_S)$, as is observed above. Hence (Q, b_Q), $(Q, b'') \leq (R, b_R)$, and it follows from the inductive hypothesis that $b_Q = b''$. Similarly, we have $b' = b''$ because (Q, b'), (Q, b'') are contained in $(R', b_{R'})$. Therefore $b_Q = b'$.

The second half follows immediately from Lemma 9.2(i) and the uniqueness assertion of the theorem. ■

Remark. According to the third main theorem, a subpair (Q, b) is a $B_0(G)$-subpair if and only if $b = B_0(QC_G(Q))$. Moreover, it follows from the above theorem that

$$(Q, B_0(QC_G(Q))) \leq (P, B_0(PC_G(P))) \Leftrightarrow Q \leq P,$$

$$(Q, B_0(QC_G(Q))) \lhd (P, B_0(PC_G(P))) \Leftrightarrow Q \lhd P.$$

Thus Theorem 9.1 may be considered as a generalization of the Sylow theorem in group theory.

9.2. Subsections

If $x \in G$ is a p-element, then a subpair $(\langle x \rangle, b)$ with $b \in \mathrm{Bl}(C_G(x))$ is abbreviated to (x, b) and called a *subsection* (belonging to $b^G = B$). Thus

$(1, B)$, where $B \in \mathrm{Bl}(G)$, is a subsection. In this terminology, the second main theorem states that if $d^x_{\chi\mu} \neq 0$, where $\chi \in \mathrm{Irr}(B)$ and $\mu \in \mathrm{IBr}(b)$ with $b \in \mathrm{Bl}(C_G(x))$, then (x, b) is a subsection belonging to B.

Let (x, b) and (x', b') be subsections. We say they are conjugate in G and write $(x', b') =_G (x, b)$ if $(x', b') = (x, b)^t$ for some $t \in G$, i.e., $x' = x^t$ and $b' = b^t$. Note that if (x, b) and (x, b') are conjugate in G, then $b = b'$. Thus we obtain the following result from Theorem 4.13(ii).

Theorem 9.4. *Let $B \in \mathrm{Bl}(G)$ and let $\{(x_i, b_i);\ 1 \leq i \leq n\}$ be a complete set of representatives of the G-conjugate classes of subsections. Then*

$$k(B) = \sum_{i=1}^{n} l(b_i).$$

A subsection (x, b) is said to be a *major subsection*, provided that $\delta(b) =_G \delta(b^G)$. The next result is immediate from Theorem 5.21.

Theorem 9.5. *If $\delta(B)$ is abelian, then every subsection belonging to B is a major subsection.*

We shall describe a complete set of representatives of G-conjugate classes of major subsections. We start with the following.

Theorem 9.6. *Let x be a p-element of G, and let $B \in \mathrm{Bl}(G)$ with $\delta(B) = D$. Then there is a major subsection (x, b) belonging to B if and only if $x \in_G Z(D)$.*

Proof. If there is a major subsection (x, b) belonging to B, then we have $x \in Z(\delta(b)) =_G Z(\delta(b^G)) = Z(D)$, since $\langle x \rangle \lhd C_G(x)$.

To show the converse, we may assume that $x \in Z(D)$. Let C be a defect class for B, then there exists $y \in C$ such that $D \in \mathrm{Syl}_p(C_G(y))$. If $\chi \in \mathrm{Irr}(B)$ has height zero, then $\chi(y)^* \neq 0$ as $\omega_\chi(\hat{C})^* \neq 0$. Thus

$$\chi(y) \equiv \chi(xy) = \sum_{\mu} d^x_{\chi\mu}\mu(y) \not\equiv 0 \pmod{(\pi)},$$

and there exists $\mu \in \mathrm{IBr}(C_G(x))$ such that $d^x_{\chi\mu}\mu(y) \not\equiv 0 \pmod{(\pi)}$. If $\mu \in b \in \mathrm{Bl}(C_G(x))$, then $b^G = B$ by the second main theorem. Set $H = C_G(x)$. Then we have $D \in \mathrm{Syl}_p(C_H(y))$ from the assumption. On the other hand, it follows from $\mu(y)^* \neq 0$ that $d(b) \geq \nu(|C_H(y)|) = \nu(|D|) = d(b^G)$ (cf. Chapter 3, 6.26), whence we have $\delta(b) =_G \delta(b^G)$, as $\delta(b) \leq_G \delta(b^G)$ in general. Therefore (x, b) is a major subsection belonging to B. ∎

Actually, we have shown the following fact in the proof of the "if" part of the above theorem.

Corollary 9.7. *Let $B \in \mathrm{Bl}(G)$ with $\delta(B) = D$. If $x \in Z(D)$, then there exist a major subsection (x, b) and $\mu \in \mathrm{IBr}(b)$ such that, for any $\chi \in \mathrm{Irr}(B)$ of height* 0, $d^x_{\chi\mu} \not\equiv 0 (\mathrm{mod}(\pi))$.

For the rest of this section, let $B \in \mathrm{Bl}(G)$ with $\delta(B) = D$, and let $b \in \mathrm{Bl}(DC_G(D))$ be a root of B. If $x \in Z(D)$, then $C_G(x) \geq DC_G(D)$, and $(x, b^{C_G(x)})$ is a major subsection belonging to B.

Lemma 9.8. *Every major subsection (u, b_u) belonging to B is conjugate in G to some $(x, b^{C_G(x)})$ with $x \in Z(D)$.*

Proof. We may assume that $u \in Z(D)$ and $\delta(b_u) = D$. Let $H = C_G(u)$, then $H \geq DC_G(D) = DC_H(D)$. Let $b' \in \mathrm{Bl}(DC_G(D))$ be a root of b_u. Then, since $b'^G = (b'^H)^G = (b_u)^G = B$, b' is a root of B, and hence there exists $t \in N_G(D)$ such that $b'^t = b$. Let $x = u^t (\in Z(D))$, $(b_u)^t = b_x \in \mathrm{Bl}(C_G(x))$, then $(u, b_u)^t = (x, b_x)$. However, we have $b^{C_G(x)} = (b'^t)^{C_G(x)} = (b'^t)^{C_G(u^t)} = ((b')^{C_G(u)})^t = (b_u)^t = b_x$. Namely, (u, b_u) is conjugate to $(x, b^{C_G(x)})$. ∎

Remark. In the above proof we have used the following (easy) fact freely. Namely if, for $H \leq K \leq G$ and $b \in \mathrm{Bl}(H)$, b^K is defined, then $(b^t)^{K^t}$ is also defined for any $t \in G$ and we have $(b^t)^{K^t} = (b^K)^t$.

Lemma 9.9. *Let $x, y \in Z(D)$ and $T = T(b)$. Then the following conditions are equivalent.*

(1) $(x, b^{C_G(x)}) =_G (y, b^{C_G(y)})$.
(2) $(x, b^{C_G(x)}) =_T (y, b^{C_G(y)})$.
(3) $x =_T y$.

Proof. We only need to show the implication "(1) ⇒ (2)". Let $b^{C_G(x)} = b_x$, $b^{C_G(y)} = b_y$ and $H = C_G(y)$. By assumption, there is $s \in G$ such that $x^s = y$, $(b_x)^s = b_y$. Hence, $(b_x)^s = (b^s)^H = b_y = b^H$. On the other hand, we have $\delta(b^s) = D^s \leq_H \delta(b_y) =_H D$, thus $\delta(b^s) =_H \delta(b_y)$, and it follows that b^s and b are both roots of b_y. Therefore there exists $h \in H$ such that $b^{sh} = b$, i.e., $t = sh \in T$, and we get $x^t = y$, $(b_x)^t = (b^t)^H = b^H = b_y$. ∎

From the above two lemmas we get the following.

Theorem 9.10. *Let $B \in \mathrm{Bl}(G)$ with $\delta(B) = D$ and let $b \in \mathrm{Bl}(DC_G(D))$ be a root of B. If Λ denotes a complete set of representatives of $T(b)$-conjugate classes in $Z(D)$, then*

$$\{(x, b^{C_G(x)});\ x \in \Lambda\}$$

gives a complete set of representatives of G-conjugate classes of the major subsections belonging to B.

Remark. See Brauer[18] for the representatives of G-conjugate classes of general subsections. Subsections are useful in studying the structure of blocks. For example, blocks with dihedral defect groups are determined by Brauer[19], and blocks with quaternion and quasidihedral defect groups are determined by Olsson[1].

10. *RG* as an *R*[*G* × *G*]-Module

10.1. The R-rank of a Block

For a later use, we prove the following Theorem 10.1, following Broué's method.

Theorem 10.1 **(Brauer).** *Let $B \in \mathrm{Bl}(G)$. Then*

$$\nu(\mathrm{rank}_R B) = 2\nu(|G|) - d(B).$$

Remark. $\mathrm{rank}_R B = \sum_{\chi \in \mathrm{Irr}(B)} \chi(1)^2$.

Before proving this theorem, we show the following.

Lemma 10.2. *Let $\chi \in \mathrm{Irr}(B)$, and set*

$$e_\chi^0 = \sum_{y \in G_{p'}} \chi(y^{-1})y.$$

Then we have $e_B e_\chi^0 = e_\chi^0$.

Proof. Set $e_\chi^1 = \sum_{z \in G - G_{p'}} \chi(z^{-1})z$. Thus, $e_\chi = \chi(1)/|G|(e_\chi^0 + e_\chi^1)$, and from $e_B e_\chi = e_\chi$ it follows that

$$e_B e_\chi^0 + e_B e_\chi^1 = e_\chi^0 + e_\chi^1. \tag{10.1}$$

Apply Theorem 4.9 with $x_1 = 1$, thus $\mathfrak{S}(x_1) = G_{p'}$. Then we get e_χ^0, $e_B e_\chi^0 \in R[\tilde{\mathfrak{S}}(x_1)]$ and $e_\chi^1, e_B e_\chi^1 \in \sum_{i=2}^{r} R[\tilde{\mathfrak{S}}(x_i)]$. Thus we conclude that $e_B e_\chi^0 = e_\chi^0$ from (10.1). ∎

We record here the following fact as a consequence of the above lemma.

Corollary 10.3. *Let $\chi \in \mathrm{Irr}(B)$, $C \in \mathrm{Cl}(G)$ and $x \in C$. Then we have*

$$\chi(x) \not\equiv 0 \pmod{(\pi)} \Rightarrow \delta(C) \leq_G \delta(B).$$

Proof. Since $\chi(x_{p'}) \equiv \chi(x) \pmod{(\pi)}$ and $C_G(x) \leq C_G(x_{p'})$, we may assume that x is a p'-element. Let $D = \delta(B)$ and $C' = \{x^{-1}; x \in C\}$. We know that $(e_\chi^0)^* \in e_B^* Z(FG) \subset Z_D(FG)$ from the above lemma. But the assumption implies that the coefficient of $\hat{C}'$ in $(e_\chi^0)^*$ is not zero, and hence $\delta(C') =_G \delta(C) \leq_G D$. ∎

Now we set

$$X_R(G) = \bigoplus_{\chi \in \mathrm{Irr}(G)} R\chi, \qquad X_R(B) = \bigoplus_{\chi \in \mathrm{Irr}(B)} R\chi.$$

Lemma 10.4. *If $\theta \in X_R(B)$ and $z \in Z(RG)$, then we have*

$$v(\theta(z)) \geq v(|G|) - d(B).$$

If $z \in J(Z(RG))$, then

(10.2) $$v(\theta(z)) > v(|G|) - d(B).$$

Proof. We may clearly assume that $\theta \in \mathrm{Irr(B)}$. Then $\omega_\theta(z) = \theta(z)/\theta(1) \in R$, and it follows that $v(\theta(z)) \geq v(\theta(1)) \geq v(|G|) - d(B)$. If $z \in J(Z(RG))$, then $\omega_\theta(z) \equiv 0 \pmod{\pi}$, and therefore (10.2) holds. ■

For $\theta \in X_R(B)$, we have

$$v(\theta(1)) = v(|G|) - d(B) + \mathrm{ht}(\theta),$$

where $\mathrm{ht}(\theta)$ is a nonnegative integer called the *height* of θ.

Lemma 10.5. *Let $\theta \in X_R(B)$. Then $\mathrm{ht}(\theta) = 0$ if and only if there exists $z \in e_B Z(RG)$ such that*

(10.3) $$v(\theta(z)) = v(|G|) - d(B).$$

Proof. If $\mathrm{ht}(\theta) = 0$, then $z = e_B$ satisfies (10.3), since $\theta(e_B) = \theta(1)$.

To show the converse, let $v(|G|) = a$, $d(B) = d$, and consider the F-linear map

$$(p^{-(a-d)}\theta)^*: e_B Z(RG)/J(e_B Z(RG)) = F\bar{e}_B \to F \qquad (\bar{z} \mapsto (p^{-(a-d)}\theta(z))^*).$$

The assumption (10.3) implies that this is a nonzero map, so we have $p^{-(a-d)}\theta(e_B) = p^{-(a-d)}\theta(1) \not\equiv 0 \pmod{\pi}$ and hence $\mathrm{ht}(\theta) = 0$. ■

Let θ be the function on G such that $\theta(x) = 1$ or 0, depending on whether x is a p-element or not, and let us denote by the same letter θ the linear extension of it to RG. We know that $\theta \in X_R(G)$ from Chapter 3, 6.15(iii). Thus, if $B \in \mathrm{Bl}(G)$, then the B-component θ_B of θ belongs to $X_R(B)$.

Lemma 10.6. *For every* $B \in \mathrm{Bl}(G)$, $\mathrm{ht}(\theta_B) = 0$.

Proof. Let $\chi \in \mathrm{Irr}(B)$, with $\mathrm{ht}(\chi) = 0$. Using the notation of Lemma 10.2, we have $e_\chi^0 \in e_B Z(RG)$ and

$$\theta_B(e_\chi^0) = \theta(e_B e_\chi^0) = \theta(e_\chi^0) = \chi(1).$$

Thus, $v(\theta_B(e_\chi^0)) = v(|G|) - d(B)$ and the result follows from Lemma 10.5. ■

Proof of Theorem 10.1. We get $v(\theta_B(1)) = v(|G|) - d(B)$ from the above lemma. But since $\theta_B(1) = \theta(e_B) = 1/|G|(\sum_{\chi \in \mathrm{Irr}(B)} \chi(1)^2)$, we have

$$v(\mathrm{rank}_R\, B) = v(|G|) + v(\theta_B(1)) = 2v(|G|) - d(B). \qquad ■$$

10.2. The Vertex of a Block

We consider RG as a right $R[G \times G]$-module via the natural action as follows:

$$a(x, y) = x^{-1}ay \tag{10.4}$$

for $a \in RG$ and $(x, y) \in G \times G$. Thus, each block of RG is an indecomposable component of $RG_{R[G \times G]}$. In this section, we shall study blocks of RG from this point of view. Especially, their vertices, defect groups, and relations with blocks of subgroups of G will be considered. If B is a block of RG, we write $B_{G \times G}$, when it is considered as an $R[G \times G]$-module.

Given $a = \sum_{x \in G} \alpha_x x \in RG$, we set $a^{(-1)} = \sum_x \alpha_x x^{-1}$, then $RG \to RG$ $(a \mapsto a^{(-1)})$ is an anti-automorphism of RG. Also, the action of $x \otimes y \in R[G \times G] = RG \otimes_R RG$ on $a \in RG$ is given by $a(x \otimes y) = x^{(-1)}ay$.

We denote $(x, x) \in G \times G$ by $x^\triangle$ and write $H^\triangle = \{x^\triangle; x \in H\}$ for $H \leq G$. Now, $G \times G$ acts transitively on G under the action of (10.4), and the stabilizer of 1 is $G^\triangle$. Thus, considering R as a trivial $G^\triangle$-module, we have

$$RG \simeq R_{G^\triangle} \otimes_{RG^\triangle} R[G \times G] = (R_{G^\triangle})^{G \times G}$$

as $R[G \times G]$-modules. This yields the following lemma.

Lemma 10.7. *If B is a block of RG, then $B_{G\times G}$ is $G^{\triangle}$-projective.*

More precisely, we have the following.

Theorem 10.8 (Green). *Let $B \in \mathrm{Bl}(G)$ with defect group D. Then*

$$\mathrm{vx}(B_{G\times G}) =_{G\times G} D^{\triangle}.$$

Proof. Note that $e_B^{(-1)} \otimes e_B$ acts on B as the identity map. On the other hand, we see that $e_B^{(-1)} \otimes e_B \in (R[G \times G])_{D\times D}^{G\times G}$, since $e_B \in (RG)_D^G$. Hence B is $D \times D$-projective (cf. Chapter 4, 2.2 (2)).

On the other hand, we may assume that $\mathrm{vx}(B) = Q^{\triangle}$ by Lemma 10.7, then $Q^{\triangle} \leq_{G\times G} D \times D$, and hence $Q \leq_G D$. Therefore B is $D^{\triangle}$-projective and by Chapter 4, 7.5.

$$\begin{aligned} v(\mathrm{rank}_R B) &\geq v(|G \times G|) - v(|\mathrm{vx}(B)|) \\ &\geq 2v(|G|) - v(|D|). \end{aligned}$$

But in view of Theorem 10.1, only the equalities are possible in the above and hence $\mathrm{vx}(B) =_{G\times G} D^{\triangle}$. ■

If $P \leq G$, then $P \times P$ acts on G as in (10.4). Let us denote by $(P \times P)_t$ the stabilizer of $t \in G$ in $P \times P$:

$$(P \times P)_t = \{(x, y) \in P \times P;\ x^{-1}ty = t\}.$$

Lemma 10.9. *With the above notation, we have the following.*

(i) $$(P \times P)_t = \{(x, t^{-1}xt);\ x \in P \cap tPt^{-1}\}.$$

(ii) *If $(u, v) \in P \times P$, then*

$$(u, v)^{-1}((P \times P)_t)(u, v) = ((P \times P)_t)^{(u,v)} = (P \times P)_{u^{-1}tv}.$$

(iii) *In the following decomposition,*

(10.5) $$RG = \bigoplus_{t \in P\backslash G/P} R[PtP],$$

we have an $R[P \times P]$-isomorphism

(10.6) $$R[PtP] \simeq R \otimes_{R[(P\times P)_t]} R[P \times P].$$

Proof. Parts (i) and (ii) will be shown by direct computations, (10.5) is trivial, and (10.6) follows since $P \times P$ acts on PtP transitively. ∎

The following gives a necessary condition for a p-subgroup of G to be a defect group of a block of G.

Theorem 10.10 (Green). *Let D be a defect group of a block B of G and let P be a Sylow p-subgroup of G such that $P \geq D$. Then there exists $z \in C_G(D)$ such that*

$$D = P \cap P^z.$$

Proof. It turns out from the isomorphism (10.6) that $R[PtP]$ is an indecomposable $R[P \times P]$-module (cf. Chapter 4, 7.3 and 7.6). Hence, (10.5) gives an indecomposable decomposition of RG as an $R[P \times P]$-module, so that there is $\{t_i\} \subset P \backslash G/P$ such that

$$B_{P \times P} \simeq \bigoplus_i R[Pt_iP].$$

In particular, $\mathrm{vx}(B_{G \times G}) = D^\triangle =_{P \times P} (P \times P)_t$ for some $t = t_i$ (cf. Chapter 4, 3.8), and it follows from (ii) of the above lemma that $D^\triangle = (P \times P)_z$ for some $z \in G$. Thus we have $D = P \cap zPz^{-1}$ and $z \in C_G(D)$. ∎

If $H \leq G$ and $b \in \mathrm{Bl}(H)$, then there is $B \in \mathrm{Bl}(G)$ such that $b | B_{H \times H}$.

Theorem 10.11. *If b^G is defined and equal to $B \in \mathrm{Bl}(G)$, then $b | B_{H \times H}$.*

Proof. With the notation of Section 3, let us extend s_H to the whole RG, i.e., $s_H(\sum_{x \in G} \alpha_x x) = \sum_{y \in H} \alpha_y y$, so that s_H is an $R[H \times H]$-homomorphism from RG into RH. Now the maps $\lambda\colon b \to B(u \mapsto e_B u)$ and $\mu\colon B \to b(v \to e_b s_H(v))$ are both $R[H \times H]$-homomorphisms, and we have

$$\mu \circ \lambda(u) = e_b s_H(e_B u) = e_b s_H(e_B) u \qquad \text{for} \quad u \in b.$$

Hence by Lemma 3.9 (i), $\mu \circ \lambda : b \to b$ is an isomorphism and we have $B_{H \times H} \simeq \lambda(b) \oplus \mathrm{Ker}\, \mu$ with $\lambda(b) \simeq b$. ∎

The converse of the above theorem is true under certain circumstances (cf. Alperin[2], Green[6], Okuyama[1]). Here we prove the following.

Theorem 10.12. *Let $B \in \mathrm{Bl}(G)$ and $b \in \mathrm{Bl}(H)$. If b is admissible in G, (i.e., $C_G(\delta(b)) \leq H$), then the following statements hold.*

(i) *b appears as an indecomposable component of $(RG)_{H \times H}$ with multiplicity 1.*

(ii) $b \mid B_{H \times H} \Rightarrow b^G = B$.

Proof.

(i) Let $Q = \delta(b)$. As is noted in Lemma 10.9, we have

$$(RG)_{H \times H} = \bigoplus_{t \in H\backslash G/H} R[HtH] = RH \oplus \left(\bigoplus_{t \notin H} R[HtH] \right).$$

Hence $b \mid (RG)_{H \times H}$, since $b \mid (RH)_{H \times H}$. In order to show that the multiplicity of b is 1, it is sufficient to prove that if $t \notin H$, then $\mathrm{vx}(b_{H \times H}) = Q^{\triangle} \not\leq_{H \times H} \mathrm{vx}(L)$ for any indecomposable component L of $R[HtH]$. If on the contrary, $Q^{\triangle} \leq_{H \times H} \mathrm{vx}(L)$, then, since L is $(H \times H)_t$-projective by (10.6), there exists $(u, v) \in H \times H$ such that $Q^{\triangle} \leq ((H \times H)_t)^{(u,v)} = (H \times H)_{u^{-1}tv}$. This yields that $u^{-1}tv \in C_G(Q) \leq H$, and so $t \in H$, which is a contradiction.

(ii) From the assumption, b^G is defined (cf. Theorem 3.6), then by Theorem 10.11, we have $b \mid (b^G)_{H \times H}$, and then $b^G = B$ by (i). ■

Remark. In the module-theoretic version of the Brauer correspondence we write $b^G = B$, provided B is the only block with $b \mid B_{H \times H}$ (Alperin[5]).

For later convenience, we mention here the following fact as a by-product of the proof of the above theorem.

Corollary 10.13. *Let Q be a p-subgroup of G and $N = N_G(Q)$.* If B is a block of G, then

(10.7) $$B_{N \times N} \simeq \left(\bigoplus_{\substack{b \in \mathrm{Bl}(N) \\ b^G = B}} b \right) \oplus M,$$

and any indecomposable component L of M satisfies

(10.8) $$Q^{\triangle} \not\leq_{N\times N} \mathrm{vx}(L).$$

Proof. Since $Q \leq \delta(b)$, every $b \in \mathrm{Bl}(N)$ is admissible. Thus $b \mid B_{N\times N}$ if and only if $b^G = B$, which proves (10.7). If L is an indecomposable component of $B_{N\times N}$ such that $L \nmid (RN)$, then $L \mid R[NtN]$ for some $t \in G - N$. As was observed in the proof of the above theorem, this implies (10.8) ■

11. Lower Defect Groups

11.1. Block partitions of Cl(G)

In this subsection, we summarize Brauer's results [15] concerning block partitions of conjugate classes of G following Iizuka[2], Broué[3], and Olsson[2]. We start with the following rather general setting.

Lemma 11.1. *Let V be a free $\mathfrak{o}$-module with basis $\Omega = \{v_1, \ldots, v_n\}$. Given a direct sum decomposition,*

(11.1) $$V = W_1 \oplus W_2 \oplus \cdots \oplus W_r$$

and the projection π_i of V on W_i $(i = 1, \ldots, r)$, there exists a partition $\Omega = \bigcup_{i=1}^{r} \Omega_i$ of Ω $(\Omega_i \cap \Omega_j = \varnothing$ if $i \neq j)$ such that $\pi_i(\Omega_i)$ is a basis of W_i for each i.

Proof. Let $\{w_1^{(j)}, \ldots, w_{n_j}^{(j)}\}$ be a basis of W_j and write

$$v_i = \sum_{j,k} \alpha_{i,(j,k)} w_k^{(j)} \qquad (1 \leq i \leq n).$$

Therefore,

(11.2) $$\pi_j(v_i) = \sum_{k=1}^{n_j} \alpha_{i,(j,k)} w_k^{(j)}.$$

Let $A = (\alpha_{i,(j,k)})$ be the $n \times n$ matrix with i as the row index and (j, k) as the column index, taken in lexicographic order. Then $\det A \not\equiv 0$ $(\mathrm{mod}(\pi))$. According to the Laplace expansion formula, we can arrange the rows of A so that the first principal $n_1 \times n_1$ minors, the second principal $n_2 \times n_2$ minors, ..., and the rth principal $n_r \times n_r$ minors are all units of $\mathfrak{o}$. Corresponding to the new arrangement of the rows of A, we get a partition $\Omega = \bigcup_{j=1}^{r} \Omega_j$ with $|\Omega_j| = n_j$. It is clear that $\pi_j(\Omega_j)$ is a basis of W_j. ■

The above partition $\Omega = \bigcup_{i=1}^{r} \Omega_i$ is said to be a *partition associated with the decomposition* $V = \bigoplus_{i=1}^{r} W_i$.

Lemma 11.2. *With the same notation and assumption as in the above lemma, let U be an $\mathfrak{o}$-submodule of V with basis Λ such that*

$$\text{(a)} \quad \pi_i(U) \subset U \quad (1 \leq i \leq r), \qquad \text{(b)} \quad \Lambda \subset \Omega.$$

Then $\Lambda = \bigcup_{i=1}^{r} (\Lambda \cap \Omega_i)$ is a partition of Λ associated with the decomposition $U = \bigoplus_{i=1}^{r} \pi_i(U)$ *of U. In particular, the following holds*:

$$\operatorname{rank}_{\mathfrak{o}} \pi_i(U) = |\Lambda \cap \Omega_i| \qquad (1 \leq i \leq r).$$

Proof. The condition (a) yields that $U = \bigoplus_{i=1}^{r} \pi_i(U)$. *Let* $\Lambda_i = \Lambda \cap \Omega_i$, then $\bigcup_{i=1}^{r} \pi_i(\Lambda_i)$ is $\mathfrak{o}$-free by (b), and hence $|\bigcup_{i=1}^{r} \pi_i(\Lambda_i)| = |\Lambda| = \operatorname{rank}_{\mathfrak{o}} U$. But since $V/\langle \bigcup_{i=1}^{r} \pi_i(\Lambda_i)\rangle$ is $\mathfrak{o}$-free, it follows that $U = \langle \bigcup_i \pi_i(\Lambda_i)\rangle$, i.e. $\bigcup_{i=1}^{r} \pi_i(\Lambda_i)$ is an $\mathfrak{o}$-basis of U. Consequently $\pi_i(\Lambda_i)$ is an $\mathfrak{o}$-basis of $\pi_i(U)$. ■

Now, let us apply Lemma 11.1 to $V = Z(RG) = \bigoplus_{B \in \mathrm{Bl}(G)} e_B Z(RG)$ and $\Omega = \{\hat{C}; C \in \mathrm{Cl}(G)\}$, where the projection on $e_B Z(RG)$ is the multiplication by e_B. Before proceeding, we note the following.

Lemma 11.3. $\operatorname{rank}_R e_B Z(RG) = k(B)$.

Proof. $\operatorname{rank}_R e_B Z(RG) = \operatorname{rank}_R Z(B) = \dim_K K \otimes_R Z(B) = \dim_K Z(B^K) = k(B)$, where the last equality follows from $Z(B^K) = \bigoplus_{\chi \in \mathrm{Irr}(B)} K e_\chi$. ■

The first result is the following.

Theorem 11.4. *There is a partition of* Cl(*G*),

$$\mathrm{Cl}(G) = \bigcup_{B \in \mathrm{Bl}(G)} \Omega_B \qquad (\Omega_B \cap \Omega_{B'} = \varnothing \quad \text{if} \quad B \neq B'), \tag{11.3}$$

such that the following statements hold.

(i) $\{e_B\hat{C};\ C \in \Omega_B\}$ *is an* R-*basis of* $e_B Z(RG)$. *Hence* $|\Omega_B| = k(B)$.

(ii) *The value of* v *at* $\det(\omega_\chi(\hat{C}))_{\chi \in B, C \in \Omega_B}$ *is not larger than any other values of* v *at the* $k(B) \times k(B)$ *minors of the* $k(B) \times k$ *matrix* $(\omega_\chi(\hat{C}))_{\chi \in B, C \in \mathrm{Cl}(G)}$.

Proof. It is clear from Lemma 11.1 that there exists a partition (11.3) satisfying (i). Since $\omega_\chi(\hat{C}) = \omega_\chi(e_B\hat{C})$ for $\chi \in B$, part(ii) will be easily verified by making use of (i) ■

Remark. An analogous result to (i) holds in $Z(FG)$.

We call a partition (11.3) of Cl(*G*) satisfying (i) of the above theorem a *block partition* of Cl(*G*).

Theorem 11.5. *Let* (11.3) *be a block partition of* Cl(*G*) *and let* Q *be a* p-*subgroup of* G. *Then the following statements hold.*

(i) *Set* $m_B(Q) = \#\{C \in \Omega_B;\ \delta(C) =_G Q\}$. *Then this does not depend on the choice of a block partition of* Cl(*G*). *Indeed, we have*

$$m_B(Q) = \dim_F(e_B^* Z_Q(FG)/e_B^* Z_Q^1(FG)). \tag{11.4}$$

(ii) *Let* $\delta(B) = D$. *Then*

(a) $m_B(D) > 0$,
(b) $m_B(Q) > 0 \Rightarrow Q \leq_G D$.

(iii) *Let* $\{b_1, \ldots, b_s\}$ *be the set of blocks of* $N_G(Q)$ *associated with* B *and define* $m_{b_i}(Q)$ *in* $N_G(Q)$ *in a similar way. Then*

$$\sum_{i=1}^{s} m_{b_i}(Q) = m_B(Q).$$

(iv) *If* $\{Q_1, \ldots, Q_r\}$ *is a complete set of representatives of the* G-*conjugate classes of* p-*subgroups of* G, *then*

$$\sum_{i=1}^{r} m_B(Q_i) = k(B).$$

Proof.

(i) Since $Z_Q(FG)$ and $Z_Q^1(FG)$ are ideals of $Z(FG)$, they are closed under the multiplication by e_B^*. Hence, by Lemma 11.2, $\{e_B^*\hat{C};\ C \in \Omega_B,\ \delta(C) \leq_G Q\}$ and $\{e_B^*\hat{C}; C \in \Omega_B, \delta(C) <_G Q\}$ are bases of $e_B^* Z_Q(FG)$ and $e_B^* Z_Q^1(FG)$, respectively, and (11.4) holds.

(ii) Part (a) is clear because $e_B^* \in Z_D(FG) - Z_D^1(FG)$. If $m_B(Q) > 0$, then there is $C \in \mathrm{Cl}(G)$ such that $e_B^*\hat{C} \in Z_Q(FG) - Z_Q^1(FG)$, and so $e_B^*\hat{C} = \sum_j \alpha_j \hat{C}_j$ with $\alpha_i \neq 0$ and $\delta(C_i) =_G Q$ for some i. But since $e_B^*\hat{C} \in Z_D(FG)$, we then have $Q = \delta(C_i) \leq_G D$.

(iii) Set $N = N_G(Q)$. As is observed in the proof of Theorem 2.15, Br_Q induces an isomorphism $Z_Q(FG)/Z_Q^1(FG) \simeq Z_Q(FN)/Z_Q^1(FN)$ such that $\mathrm{Br}(e_B^*) = \sum_{i=1}^s e_{b_i}^*$. Thus (iii) follows.

(iv) This follows immediately from the definition of $m_B(Q)$, since $|\Omega_B| = k(B)$. ■

A p-subgroup Q of G is said to be a *lower defect group* of a block B if $m_B(Q) > 0$.

We next consider the R-module U spanned by the p-regular class sums: $U = \bigoplus_{C \in \mathrm{Cl}(G_{p'})} R\hat{C}$.

Theorem 11.6. *Let* (11.3) *be a block partition of* $\mathrm{Cl}(G)$ *and* $\Lambda_B = \Omega_B \cap \mathrm{Cl}(G_{p'})$. *Thus we have a partition*

$$\mathrm{Cl}(G_{p'}) = \bigcup_{B \in \mathrm{Bl}(G)} \Lambda_B.$$

Concerning this partition, we have the following.

(i) $\{e_B\hat{C};\ C \in \Lambda_B\}$ *is an* R-*basis of* $e_B U$. *Moreover*, $|\Lambda_B| = l(B)$.

(ii) *Let* $\mathrm{IBr}(B) = \{\varphi_1, \ldots, \varphi_{l(B)}\}$ *and* $\Lambda_B = \{C_1, \ldots, C_{l(B)}\}$. *If* $x_j \in C_j$, *then* $\det(\varphi_i(x_j)) \not\equiv 0 \pmod{(\pi)}$.

(iii) $\delta(C) \leq_G \delta(B)$ *for all* $C \in \Lambda_B$, *and there exists a unique* $C \in \Lambda_B$ *such that* $\delta(C) =_G \delta(B)$.

(iv) *The elementary divisors of the Cartan matrix* C_B *of* B *are given by* $\{|\delta(C)|;\ C \in \Lambda_B\}$.

Proof. Since $G_{p'}$ is the p-section containing 1, it follows from Theorem 4.9 that $e_B U \subset U$, and so $\{e_B\hat{C}; C \in \Lambda_B\}$ is an R-basis of $e_B U$ by Lemma 11.2. Now let $|\Lambda_B| = l'(B)$, $\mathrm{Bl}(G) = \{B_1, \ldots, B_t\}$, $\Lambda_{B_\mu} = \{C_{(\mu, i)};\ 1 \leq i \leq l'(B_\mu)\}$ and

$\mathrm{IBr}(B_\mu) = \{\varphi_{(\mu,i)};\ 1 \le i \le l(B_\mu)\}$. Also, we let $\eta_{(\mu,i)}$ denote the principal indecomposable character corresponding to $\varphi_{(\mu,i)}$ and let $n_{(\mu,i)} = |C_G(x_{(\mu,i)})|$, where $x_{(\mu,i)} \in C_{(\mu,i)}$. Recall that $n_{(\nu,j)}^{-1}\eta_{(\mu,i)}(x_{(\nu,j)}) \in R$(cf. Chapter 3, 6.10(ii)). Set

(11.5) $$e_{B_\mu}\hat{C}_{(\mu,i)} = \sum_{(\nu,j)} \alpha_{(\mu,i)(\nu,j)}\hat{C}_{(\nu,j)}.$$

Since $e_\chi \hat{C}_{(\mu,i)} = \omega_\chi(\hat{C}_{(\mu,i)})e_\chi$, the left-hand side of the above equation is

$$\begin{aligned} e_{B_\mu}\hat{C}_{(\mu,i)} &= \sum_{(\nu,j)} \sum_{\chi \in \mathrm{Irr}(B_\mu)} n_{(\mu,i)}^{-1}\,\chi(x_{(\mu,i)})\overline{\chi(x_{(\nu,j)})}\hat{C}_{(\nu,j)} \\ &= \sum_{(\nu,j)} \sum_{\varphi \in \mathrm{IBr}(B_\mu)} n_{(\mu,i)}^{-1}\eta_\varphi(x_{(\mu,i)})\overline{\varphi(x_{(\nu,i)})}\hat{C}_{(\nu,j)}, \end{aligned}$$

and by comparing this with (11.5), we get

(11.6) $$\alpha_{(\mu,i)(\nu,j)} = \sum_{(\mu,k)} n_{(\mu,i)}^{-1}\eta_{(\mu,k)}(x_{(\mu,i)})\overline{\varphi_{(\mu,k)}(x_{(\nu,j)})}.$$

Let $A = (\alpha_{(\mu,i)(\nu,j)})$ be the $l \times l$ matrix, where $\{(\mu,i)\}$ and $\{(\nu,j)\}$ are arranged in lexicographic order, and so $\det A \not\equiv 0 \pmod{(\pi)}$. Set $l_\mu = l(B_\mu)$, $l'_\mu = l'(B_\mu)$ for brevity.

Let $Y_\mu = (n_{(\mu,i)}^{-1}\eta_{(\mu,k)}(x_{(\mu,i)}))_{i,k}$ be the $l'_\mu \times l_\mu$ matrix, and consider the square matrix of degree l,

$$Y = \begin{pmatrix} Y_1 & & 0 \\ & \ddots & \\ 0 & & Y_t \end{pmatrix}.$$

If we set $\Phi = (\varphi_{(\mu,k)}(x_{(\nu,j)}))_{(\mu,k)(\nu,j)}$, then (11.6) gives

$$A = Y\bar{\Phi},$$

and so $\det Y \not\equiv 0 \pmod{(\pi)}$. Consequently, we have $l'_\mu = l_\mu$ and $\det Y_\mu \not\equiv 0 \pmod{(\pi)}$ for all μ $(1 \le \mu \le t)$.

Let furthermore $\Phi_\mu = (\varphi_{(\mu,i)}(x_{(\mu,j)}))_{i,j}$, then we have

(11.7) $$Y_\mu = \mathrm{diag}(n_{(\mu,1)}^{-1}, \ldots, n_{(\mu,l_\mu)}^{-1})^t\Phi_\mu{}^tC_{B_\mu}.$$

This yields

$$\det Y_\mu = \left(\prod_{i=1}^{l_\mu} n_{(\mu,i)}^{-1}\right) \det C_{B_\mu} \cdot \det \Phi_\mu,$$

and hence

$$\det Y = \left(\prod_{(\mu,i)} n_{(\mu,i)}^{-1}\right) \det C \cdot \left(\prod_\mu \det \Phi_\mu\right) \not\equiv 0 \pmod{(\pi)}.$$

Since $\nu(\prod_{(\mu,i)} n_{(\mu,i)}^{-1}) \det C = 0$ by Chapter 3, 6.32, it follows from the above that $\det \Phi_\mu \not\equiv 0 \pmod{(\pi)}$ for all μ $(1 \le \mu \le t)$, proving (ii). Part (iv) follows easily from (11.7), and part (iii) is clear from (iv) and Theorem 11.5 (ii) (b). ■

Remark. There are similar partitions for arbitrary p-sections. See the papers cited at the beginning of this subsection for details.

11.2. Lower Defect Groups and Scott Modules

If FG is considered as a G-module by conjugate action, we write $(FG)_{G^\triangle}$ for FG. It is clear that

$$(FG)_{G^\triangle} = \bigoplus_{C \in \mathrm{Cl}(G)} F[C] = \bigoplus_{B \in \mathrm{Bl}(G)} B^*_{G^\triangle}, \tag{11.8}$$

where $F[C] = \bigoplus_{x \in C} Fx$.

For a p-group Q of G, we denote by $S(Q)$ the Scott module with vertex Q, as defined in Chapter 4, Section 8. We denote here by $M_B(Q)$ the multiplicity of $S(Q)$ as an indecomposable component of $B^*_{G^\triangle}$. Now, the purpose of this subsection is to prove the following theorem.

Theorem 11.7 (Burry). $M_B(Q) = m_B(Q)$.

Let $N = N_G(Q)$ and $f = f(G, Q, N)$ be the Green correspondence. Before proving the above theorem, we show a couple of preliminary lemmas.

Lemma 11.8. *Let $C \in \mathrm{Cl}(G)$ and $\delta(C) =_G Q$. Then the following statements hold.*

(i) $F[C]_{G^\triangle} = S \oplus T$ *with* $S \simeq S(Q)$ *and* $S(Q) \nmid T$.

(ii) *If* $G = N_G(Q)$, *then* $\mathrm{soc}(S) = F\hat{C}$.

Proof. Choose $x \in C$ so that $Q \in \mathrm{Syl}_p(C_G(x))$ and put $H = C_G(x)$.

(i) Since $F[C]_{G^\triangle} \simeq (F_H)^G$, the result is clear from Chapter 4, 8.4.

(ii) We know from the assumption that S is a projective cover of the trivial G/Q-module, thus $\mathrm{soc}(S) \simeq F$. But since $\mathrm{Inv}_G(F[C]) = F\hat{C}$, we have $\mathrm{soc}(S) = F\hat{C}$. ■

Lemma 11.9.
(i) $\sum_{B \in \mathrm{Bl}(G)} M_B(Q) = \#\{C \in \mathrm{Cl}(G);\ \delta(C) =_G Q\}$.
(ii) $M_B(Q) = \sum_{b \in \mathrm{Bl}(N),\, b^G = B} M_b(Q)$.

Proof.
(i) Both sides of the equality coincide with the multiplicity of $S(Q)$ as an indecomposable component of $(FG)_{G^\triangle}$ by (11.8).
(ii) Corollary 10.13 gives

$$B^*_{N^\triangle} \simeq \left(\bigoplus_{b^G = B} b^*_{N^\triangle} \right) \oplus M^*_{N^\triangle},$$

where Q is not contained in a vertex of any indecomposable component of $M^*_{N^\triangle}$. Also, according to Chapter 4, 4.7, the multiplicity of $S(Q)$ as an indecomposable component of $B^*_{G^\triangle}$ equals that of $fS(Q)$ as an indecomposable component of $B^*_{N^\triangle}$. Thus (ii) follows, since $fS(Q)$ is the Scott module for N with vertex Q. ■

Proof of Theorem 11.7. Since $\sum_{B,Q} m_B(Q) = \sum_{B,Q} M_B(Q) = |\mathrm{Cl}(G)|$, we need only to show that $m_B(Q) \le M_B(Q)$. Also, in view of Theorem 11.5 (iii) and Lemma 11.9 (ii), we may clearly assume that $G = N \rhd Q$.

Set $m = m_B(Q)$ for brevity. And let $\mathrm{Cl}(G) = \bigcup_B \Omega_B$ be a block partition, $\{C_1, \ldots, C_m\} = \{C \in \Omega_B;\ \delta(C) =_G Q\}$ and $F[C_i] = S_i \oplus T_i$ with $S_i \simeq S(Q)$. We want to show that $S = \bigoplus_{i=1}^m S_i (\simeq mS(Q))$ is a direct summand of $B^*_{G^\triangle}$. Since $H = C_G(Q)$ is a normal subgroup of G, the map $s^*_H\colon FG \to FH (\sum_{x \in G} \alpha_x x \mapsto \sum_{y \in H} \alpha_y y)$ is a $G^\triangle$-homomorphism. Consider the projection $\pi_B\colon FG \to B^*$ $(a \mapsto e^*_B a)$ and define

$$f = (s^*_H \circ \pi_B)|_S\colon S \to (FH)_{G^\triangle}.$$

This is a monomorphism, because $\{s^*_H \circ \pi_B(\hat{C}_i) = e^*_B \hat{C}_i;\ 1 \le i \le m\}$ is linearly independent and $\mathrm{soc}(S) = \bigoplus_i F\hat{C}_i$. But $(FH)_{G^\triangle}$ may be considered as a G/Q-module, and by Chapter 4, 8.4(c), S is projective and hence injective, as a G/Q-module. Thus there is a G-homomorphism $g\colon (FH)_{G^\triangle} \to S$ such that $g \circ f = \mathrm{id}_S$. Hence the composite map $S \xrightarrow{\pi_B} B^* \xrightarrow{g \circ s^*_H} S$ is the identity map, yielding that $S \mid B^*_{G^\triangle}$. ■

12. The Glauberman Correspondence

Let A and G be groups such that A acts on G as automorphisms. Here we denote by x^α the image of $x \in G$ by $\alpha \in A$. If $\chi \in \mathrm{Irr}(G)$ and $\alpha \in A$, then we define $\chi^\alpha \in \mathrm{Irr}(G)$ by $\chi^\alpha(x) = \chi(x^{\alpha^{-1}})$. Thus A acts on $\mathrm{Irr}(G)$. We denote by $\mathrm{Irr}_A(G)$ the set of A-invariant irreducible characters of G and by $C_G(A)$ the set of A-invariant elements of G.

Glauberman[1] showed that there is a natural one-to-one correspondence, called the Glauberman correspondence

$$\pi = \pi(G, A)\colon \mathrm{Irr}_A(G) \to \mathrm{Irr}(C_G(A))$$

under the assumptions that $(|A|, |G|) = 1$ and A is solvable.

If A is a p-group, then $\pi(G, A)$ sends each $\chi \in \mathrm{Irr}_A(G)$ onto a unique $\zeta \in \mathrm{Irr}(C_G(A))$ such that $(\zeta^G, \chi)_G \not\equiv 0 \pmod p$. Later, Alperin[1] pointed out that this is also a consequence of the first main theorem.

In this section, we shall generalize the argument of Alperin[1] to show a result of Nagao[3], which treats the case where $|G|$ is not required to be relatively prime to $|A|$. The proof given here is due to Issacs.

Remark. Concerning the Glauberman correspondence, we are referred to Issacs[1] and Wolf[1], which treat the case where A is nonsolvable. For the correspondences between Brauer characters see Uno[1].

Now let $\tilde{G} \trianglerighteq G$ and suppose that $\tilde{G}/G$ is a p-group. Denote by $\mathrm{Irr}(G)^0$ the set of irreducible characters of G of p-defect zero and by $\mathrm{Irr}_{\tilde{G}}(G)^0$ the subset of $\mathrm{Irr}(G)^0$ consisting of $\tilde{G}$-invariant characters. A subgroup P of $\tilde{G}$ is said to be a *complement* of G (in $\tilde{G}$) if $\tilde{G} = GP$ and $G \cap P = 1$. Let $\mathfrak{S} = \{P_1, \dots, P_r\}$ be a complete set of representatives of the $\tilde{G}$-conjugate classes of complements of G and set $\mathfrak{T} = \bigcup_{i=1}^r \mathrm{Irr}(C_G(P_i))^0$.

Theorem 12.1. *With the above notation and assumption, there is a one-to-one correspondence*

$$\pi\colon \mathrm{Irr}_{\tilde{G}}(G)^0 \to \mathfrak{T} = \bigcup_{i=1}^{r} \mathrm{Irr}(C_G(P_i))^0$$

such that if $\chi \in \mathrm{Irr}_{\tilde{G}}(G)^0$ *and* $\zeta \in \mathfrak{T}$, *then*

$$\zeta = \chi^\pi \Leftrightarrow (\zeta^G, \chi)_G \not\equiv 0 \pmod p.$$

Moreover, we have $\mathrm{Irr}(C_G(P_i))^0 \cap \mathrm{Irr}(C_G(P_j))^0 = \varnothing$ *if* $i \neq j$.

Remark. If $\mathfrak{S}$ is empty, then $\mathrm{Irr}_{\tilde{G}}(G)^0$ is empty. However, when G is a p'-group, $\mathfrak{S}$ consists of a single element, and the above correspondence is just the Glauberman correspondence.

Before going into the proof of Theorem 12.1, we make the following remark. By Corollary 5.6, each $B \in \mathrm{Bl}(G)$ is covered by a unique $\tilde{B} \in \mathrm{Bl}(\tilde{G})$. Concerning this, we prove

Lemma 12.2. *The following two conditions are equivalent.*
(1) *B is $\tilde{G}$-invariant and $d(B) = 0$.*
(2) *$\delta(\tilde{B})$ is a complement of G.*

Proof. Since $\tilde{G}/G$ is a p-group, it follows from Theorem 5.16 that

$$T(B) = G\delta(\tilde{B}), \quad \delta(\tilde{B}) \cap G =_G \delta(B),$$

whence the assertion is immediate. ■

Proof of Theorem 12.1. Let $\chi \in \mathrm{Irr}_{\tilde{G}}(G)^0$. If χ belongs to a block B of G, necessarily of defect zero, then $P = \delta(\tilde{B})$ is a complement of G from the above lemma, where $\tilde{B}$ is a unique block of $\tilde{G}$ covering B. Thus we may assume that $P = P_i$ for some i. Let $C = C_G(P)$, then $N = N_{\tilde{G}}(P) = C \times P$. By the first main theorem, there exists a unique $b \in \mathrm{Bl}(N|P)$ such that $b^{\tilde{G}} = \tilde{B}$. And a unique $\zeta \in \mathrm{Irr}(C)^0$ corresponds to b such that $b \in \mathrm{Bl}(N|\{\zeta\})$:

$$\begin{array}{ccc}
\mathrm{Bl}(\tilde{G} \mid P) \ni \tilde{B} & \longleftarrow & \chi \in \mathrm{Irr}_{\tilde{G}}(G)^0 \\
\Big\downarrow & & \vdots\, \pi \\
\mathrm{Bl}(N \mid P) \ni b & \longrightarrow & \zeta \in \mathrm{Irr}(C)^0
\end{array}$$

Thus we obtain a map,

$$\pi\colon \mathrm{Irr}_{\tilde{G}}(G)^0 \to \mathfrak{T} = \bigcup_{i=1}^{r} \mathrm{Irr}(C_G(P_i))^0 \qquad (\chi \mapsto \zeta).$$

We first prove that if $\chi \in \mathrm{Irr}_{\tilde{G}}(G)^0$ and $\chi^\pi = \zeta$, then the following holds:
(i) $(\zeta^G, \chi)_G \not\equiv 0 \pmod{p}$,
(ii) $\mathrm{Irr}_{\tilde{G}}(G)^0 \ni \chi' \neq \chi \Rightarrow (\zeta^G, \chi')_G \equiv 0 \pmod{p}$,
which in particular asserts that π is an injective map.

In order to prove this, with the same notation as above, let

$$\eta = \zeta \times \mathbf{1}_P \in \mathrm{Irr}(C \times P).$$

Then $\eta \in \mathrm{Irr}(b)$ and $\eta^{\tilde{G}}(1) = \zeta(1)|\tilde{G}:C \times P| = \zeta(1)|G:C|$. Thus,

$$v(\eta^{\tilde{G}}(1)) = v(|G|) = v(\chi(1)),$$

because $v(\zeta(1)) = v(|C|)$.

For $\tilde{B}' \in \mathrm{Bl}(\tilde{G})$, we denote here the $\tilde{B}'$-component of $\eta^{\tilde{G}}$ by $\tilde{\eta}_{\tilde{B}'}$ for the sake of simplicity. Thus $\eta^{\tilde{G}} = \sum_{\tilde{B}'} \tilde{\eta}_{\tilde{B}'}$. In particular, if $\tilde{B}'$ covers a block B' of G such that $B' = \{\chi'\}$ with $\chi' \in \mathrm{Irr}_{\tilde{G}}(G)^0$, then, for every $\tilde{\chi}' \in \mathrm{Irr}(\tilde{B}')$, it holds that $\tilde{\chi}'_G = m\chi'$ and hence

$$(\tilde{\eta}_{\tilde{B}'})_G = e'\chi',$$

with

$$e' = ((\tilde{\eta}_{\tilde{B}'})_G, \chi') = ((\eta^{\tilde{G}})_G, \chi') = ((\eta_C)^G, \chi') = (\zeta^G, \chi').$$

Note that $\tilde{B}' \neq \tilde{B} = b^{\tilde{G}}$ if $\chi' \neq \chi$, since $B' = \{\chi'\}$ is the unique block of G covered by $\tilde{B}'$.

Now Corollary 3.2 tells us that if $\chi' = \chi$, that is, if $\tilde{B}' = \tilde{B} = b^{\tilde{G}}$, then $v(\tilde{\eta}_{\tilde{B}}(1)) = v(\eta^{\tilde{G}}(1)) = v(\chi(1))$, whence we obtain (i), since $\tilde{\eta}_{\tilde{B}}(1) = (\zeta^G, \chi)\chi(1)$. Also, it tells us that $v(\tilde{\eta}_{\tilde{B}'}(1)) > v(\eta^{\tilde{G}}(1))$ if $\chi' \neq \chi$, and we obtain (ii) in a similar way.

To see that π is a surjective map, take $\zeta \in \mathrm{Irr}(C_G(P_i))^0$ and let $P = P_i$, $C = C_G(P)$, $N = N_{\tilde{G}}(P) = C \times P$, and b be a block of N covering $\{\zeta\}$. Clearly, $P = \delta(b)$ by the above lemma, then by the first main theorem, we have $\delta(\tilde{B}) =_{\tilde{G}} P$, where $\tilde{B} = b^{\tilde{G}}$. Let B be a block of G covered by $\tilde{B}$. Again by the above lemma, we have $d(B) = 0$, and a unique irreducible character χ belonging to B is $\tilde{G}$-invariant, and thus we have $\chi^\pi = \zeta$.

Finally, if $\zeta \in \mathfrak{T}$ lies in $\mathrm{Irr}(C_G(P_i))^0$, then with the above notation, P_i is a defect group of $\tilde{B}$ determined only by χ such that $\chi^\pi = \zeta$. But since π is an injective map, P_i is uniquely determined by χ. This proves the last statement. (In other words it does not happen to be $C_G(P_i) = C_G(P_j)$ if $i \neq j$.) ■

Problems

(K, R, F) is a p-modular system. K is assumed to have a primitive $|G|$th root of unity, unless otherwise specified explicitly.

1. If $\varphi \in \mathrm{IBr}(G)$, then $\bar{\varphi} \in \mathrm{IBr}(G)$, where $\bar{\varphi}$ denotes the complex conjugate of φ.

2. Let P be a principal indecomposable RG-module and let U be a KG-module with $U|P^K$. Then there exists an R-free RG-module V such that

(a) $V^K \simeq U$, i.e., V is an R-form of U.
(b) $\mathrm{hd}(V^*)$ is irreducible and hence V^* is indecomposable.

[Remark. If U is irreducible, then there is P such that $U|P^K$. Thus there is V_{RG} satisfying (a) and (b).]

3. Let $\chi_1, \ldots, \chi_t$ be irreducible characters of p-defect zero. Then there exist p-regular elements $y_1, \ldots, y_t$ of G such that

(a) $d(y_i^G) = 0$ for all i, where y_i^G is the conjugate class containing y_i.
(b) $\det(\chi_i(y_j)) \not\equiv 0 \pmod{(\pi)}$.

4. **(Osima)** Let Λ be a subset of $\mathrm{Irr}(G)$ such that the following statement holds for all $x \in G_{p'}$ and $y \in G - G_{p'}$.

$$\sum_{\chi \in \Lambda} \chi(x)\overline{\chi(y)} = 0.$$

Then there are blocks $B_1, \ldots, B_t$ of G such that $\Lambda = \bigcup_{i=1}^t \mathrm{Irr}(B_i)$. Prove this in the following way.

(i) For $x \in G_{p'}$, set $\theta_x = \sum_{\chi \in \Lambda} \chi(x)\chi$. Then θ_x is an R-linear combination of principal indecomposable characters of G.

(ii) Let $\varepsilon = \sum_{\chi \in \Lambda} e_\chi$. Then $\varepsilon \in RG$, and hence ε is a sum of block idempotents of RG.

5. If $\chi \in \mathrm{Irr}(B_0(G))$, then any algebraic conjugate of χ belongs again to $B_0(G)$.

6. Let $\mathfrak{S}$ be a p-section of G. Let $\lambda = \sum_\chi \alpha_\chi \omega_\chi (\alpha_\chi \in K)$ be a linear function from $Z(KG)$ into K. If there is $m \geq 0$ such that $\lambda(\hat{C}) \in (\pi^m)$ for all $C \in \tilde{\mathfrak{S}}$, then $\lambda_B(\hat{C}) \in (\pi^m)$ for all $B \in \mathrm{Bl}(G)$ and $C \in \tilde{\mathfrak{S}}$.

7. With the notation and assumption of Theorem 5.10, the following hold.

(i) If I is an ideal of $\tilde{B}$, then the map $I \mapsto BIB$ gives rise to a one-to-one correspondence between the set of ideals of $\tilde{B}$ and that of B, which preserves product. The inverse map is given by $L \mapsto \tilde{B} \cap L$ for an ideal L of B.

(ii) $J(B) = BJ(\tilde{B})B$.

8. Let $x \in G$ be a p-element and let $\theta \in \mathbf{CF}(G)$. Define $\theta^{(x)} \in \mathbf{CF}(G)$ by

$$\theta^{(x)}(a) = \begin{cases} \theta(a), & \text{if} \quad a \in \mathfrak{S}(x) \\ 0, & \text{otherwise.} \end{cases}$$

If every irreducible constituent of θ lies in a given $B \in \mathrm{Bl}(G)$, then the same is true for $\theta^{(x)}$.

9. Let Q be a normal p-subgroup of G and $B \in \mathrm{Bl}(G)$. Then B dominates a unique block of G/Q, provided one of the following two conditions is satisfied:

(a) $Q \leq Z(G)$.
(b) There is a normal p-subgroup P of G such that Q is contained in the Frattini subgroup of P.

[Hint: Use Problem 19 in Chapter 1 and Problem 19 in Chapter 4.]

10. Let $G \rhd H$. Let V_{FG} and W_{FH} be irreducible modules with projective covers P_{FG} and Q_{FH}, respectively. Set $T = T(W)$ and write

$$V_H = e\left(\bigoplus_{a \in T\backslash G} W^a\right) \qquad (e \geq 1)$$

(i) If G/H is a p'-group, then the following statements hold:

(a) $(e, p) = 1$.
(b) $P_H \simeq e(\bigoplus_{a \in T\backslash G} Q^a)$.

(ii) If G/H is a p-group, then the following statements hold:

(c) $e = 1$.
(d) $P \simeq Q^G$.

11. **(Fong)** Suppose that G is p-solvable. Let P be a principal indecomposable FG-module and $V = P/PJ(FG)$. Prove the following equality

$$\dim_F P = |G|_p\,(\dim_F V)_{p'}$$

by induction on $|G|$, by making use of the preceding problem.

12. Let $B \in \mathrm{Bl}(G)$ and suppose that $D = \delta(B)$ is cyclic and normal in G. Let $|D| = q$, $I = J(FD)$ and let $U_1, \ldots, U_r$ be representatives for the isomorphism classes of principal indecomposable B^*-modules. Then the following statements hold.

(i) $\mathrm{IND}(FD) = \{FD, I, I^2, \ldots, I^{q-1}\}$.
(ii) Every indecomposable B^*-module is isomorphic to some $U_i I^j$.
(iii) If $U_i I^j \simeq U_k I^l$, then $(i, j) = (k, l)$.
(iv) $\#\,\mathrm{IND}(B^*) = rq$.

13. If $G = PO_{p'}(G)$ is a p-nilpotent group, then for any $B \in \mathrm{Bl}(G)$, we have the following.

(i) $k(B) = \#\,\mathrm{Irr}(\delta(B))$,
(ii) $l(B) = 1$.

14. (Fong) Suppose that F is a perfect field of characteristic 2. If $\mathbf{T}$ is a self-contragredient irreducible F-representation of G that is different from $\mathbf{1}_G$, then the degree of $\mathbf{T}$ is even. Prove this in the following way.

(i) By assumption, there is a nonsingular matrix $X = (x_{ij})$ such that $XT(x)X^{-1} = {}^tT(x^{-1})$ for all $x \in G$. Show that $X = {}^tX$.

(ii) Let n be the degree of $\mathbf{T}$ and $V = F^n$ be a representation module for $\mathbf{T}$. Then

$$V_0 = \left\{v = (v_1, \ldots, v_n) \in V;\quad Q(v) = vX^tv = \sum_{i=1}^{n} x_{ii}v_i^2 = 0\right\}$$

is an FG-submodule of V.

(iii) Choose $y_i \in F$ so that $x_{ii} = y_i^2$. Then

$$V_0 = \left\{(v_1, \ldots, v_n) \in V;\quad \sum_{i=1}^{n} y_iv_i = 0\right\}.$$

If $V_0 = 0$, then $n = 1$ and hence $\mathbf{T} = \mathbf{1}_G$, contradicting the assumption. Thus $V_0 = V$, and $y_i = 0$, hence $x_{ii} = 0$ for all i. It then follows that n is even. (A symmetric matrix of odd degree over a field of characteristic 2 cannot be nonsingular if its diagonal entries are all zero. To see this, recall that a skew symmetric matrix $X = (x_{ij})$ of degree n over $\mathbf{Z}$ ($x_{ji} = -x_{ij}$, $x_{ii} = 0$) has determinant 0 if n is odd, and this is also true in the polynomial ring $\mathbf{Z}/(2)[x_{ij}; 1 \le i \le j \le n]$).

15. Let $\{S_1, \ldots, S_l\}$ be the set of p'-sections of G. Then we have

$$J(FG) = \{a \in FG;\ a\hat{S}_i = 0 \text{ for all } i\}.$$

Prove this in the following way.

(i) Show that $J(FG) = \{a \in FG;\ \varphi^*(aFG) = 0 \text{ for all } \varphi \in \mathrm{IBr}(G)\}$.

(ii) Let $S_{j'} = \{x^{-1}; x \in S_j\}$ and define $\lambda\colon FG \to F$ by $\lambda(a) = \alpha_1$ for $a = \sum_{x \in G} \alpha_x x$.

Then the following holds:

$$\varphi^*(a) = 0 \text{ for all } \varphi \in \mathrm{IBr}(G) \Leftrightarrow \sum_{x \in S_j} \alpha_x = \lambda(a\hat{S}_{j'}) = 0 \text{ for all } j.$$

(iii) For $a \in FG$, we have

$$a\hat{S}_j = 0 \text{ for all } j \Leftrightarrow \lambda(x^{-1}a\hat{S}_j) = 0, \text{ for all } j \text{ and } x \in G \Leftrightarrow$$
$$\varphi^*(x^{-1}a) = 0 \text{ for all } \varphi \in \mathrm{IBr}(G) \text{ and } x \in G \Leftrightarrow a \in J(FG).$$

16. (Landrock) Let $J = J(FG)$. For $e \in \mathrm{pi}(FG)$ let $\hat{e}$ denote a primitive idempotent of FG such that $\hat{e}FG \simeq (eFG)^\wedge$. If e, $f \in \mathrm{pi}(FG)$, then the multiplicity of fFG/fJ in eJ^{s-1}/eJ^s is equal to that of $(eFG/eJ)^\wedge$ in $\hat{f}J^{s-1}/\hat{f}J^s$ for any $s \geq 1$. [Hint: Problem 5 in Chapter 3 will be relevant.]

17. (We do not assume that K contains a primitive $|G|$th root of unity.) Let V be an FG-module.

(i) Given $y \in G_{p'}$, there is an $R\langle y\rangle$-module U such that $V_{\langle y\rangle} \simeq U/\pi U$. Hence the function φ on $G_{p'}$ defined by $\varphi(y) = \chi_U(y)$ coincides with the Brauer character given by V.

(ii) φ is a $\mathbf{Z}$-linear combination of irreducible K-characters of G. [Hint: Prove by induction on $|G|$. Problem 16 in Chapter 4 will be relevant.]

18. Let x be a p-element of G. If $\chi \in B_0(G)$, then $\chi(xy) = \chi(x)$ for all $y \in O_{p'}(C_G(x))$. [Hint: Apply the second main theorem and Theorem 8.1.]

19. Suppose that G acts on finite sets Ω and Ω' as permutations. Let F be any field of prime characteristic p. Prove that

$$F\Omega \simeq F\Omega' \text{ as } FG\text{-modules} \Rightarrow \mathbf{Q}\Omega \simeq \mathbf{Q}\Omega' \text{ as } \mathbf{Q}G\text{-modules.}$$

Thus Brauer's permutation lemma (Chapter 3, 2.18) holds for an arbitrary ground field (as is noted in the footnote on p. 934 of Brauer[4]). [Hint: $F\Omega \simeq F\Omega' \Rightarrow \mathbf{Z}/(p)\Omega \simeq \mathbf{Z}/(p)\Omega' \Rightarrow \tilde{\mathbf{Z}}_p\Omega \simeq \tilde{\mathbf{Z}}_p\Omega'$ (by Chapter 4, 8.9), where $\tilde{\mathbf{Z}}_p$ denotes the ring of p-adic integers.]

20. Let $L = \mathrm{GF}(p^m)$ be a splitting field for G and let $\sigma\colon L \simeq L(a \mapsto a^p)$. Hence the cyclic group $\langle\sigma\rangle$ is the Galois group of L over $\mathbf{Z}/(p)$ and it acts on $\mathrm{Cl}(G_{p'})$ by $C^\sigma = C^{(p)} = \{x^p; x \in C\}$ for $C \in \mathrm{Cl}(G_{p'})$; this is because $C^{(p^m)} = C$ for all $C \in \mathrm{Cl}(G_{p'})$. Set $\mathrm{IRR}_L(G) = \{\mathbf{T}_1, \ldots, \mathbf{T}_l\}$. Then for any subfield $E = \mathrm{GF}(p^n)$ of L, we have the following:

(1) $\#\{\mathbf{T}_i;\ \mathbf{T}_i \text{ is realizable in } E\} = \#\{C \in \mathrm{Cl}(G_{p'});\ C^{(p^n)} = C\}$.

(2) $\#\mathrm{IRR}_E(G) = \#\{\mathrm{Gal}(L/E)\text{-orbits on } \mathrm{Cl}(G_{p'})\}$.

Solutions to Problems

Problems 1

2. (i) Expand the product $\prod_{j \neq i}(I_i + I_j) = A$.

(ii) Write $1 = v_i + w_i$ $(v_i \in I_i, w_i \in \prod_{j \neq i} I_j)$. Given $x = (a_j + I_j) \in \bigoplus_{j=1}^{r} A/I_j$, let $a = \sum_j a_j w_j$. Then $\varphi(a) = x$.

3. Since $\operatorname{Ker} \varphi \subset \operatorname{Ker} \varphi^2 \subset \cdots$, there exists k such that $\operatorname{Ker} \varphi^k = \operatorname{Ker} \varphi^{k+1} = \cdots$. Let $u \in \operatorname{Ker} \varphi$. Since φ^k is also an epimorphism, there exists $v \in V$ such that $u = \varphi^k(v)$. Then $0 = \varphi^k(u) = \varphi^{2k}(v)$, and therefore $v \in \operatorname{Ker} \varphi^{2k} = \operatorname{Ker} \varphi^k$ and $u = 0$.

4. (i) Both modules are isomorphic to $L/(L \cap V) \oplus (L \cap W)$.

(ii) (2) $\Rightarrow$ (1) is clear from (i). Suppose that there exist $V \supset V_1 \supsetneqq V_2$ and $W \supset W_1 \supsetneqq W_2$ such that $\varphi\colon V_1/V_2 \simeq W_1/W_2$ holds. If we let $L = \{(v, w) \in V_1 \oplus W_1;\ \varphi(v + V_2) = w + W_2\}$, then $L \neq (L \cap V) \oplus (L \cap W)$.

5. Note that if M' is a maximal right ideal of A', then $\varphi^{-1}(M')$ is a maximal right ideal of A.

6. Any maximal A-submodule of V contains $VJ(A)$ by Azumaya–Nakayama's lemma. On the other hand, since $V/VJ(A)$ is a completely reducible A-module, the intersection of all the maximal submodules of it is 0.

7. The isomorphisms $f_i A \simeq e_i A$ $(1 \le i \le n)$ yield an A-automorphism of A_A. Thus, there exists $u \in A^{\times}$ such that $e_i A = u(f_i A)$ for all i. Let $f'_i = u^{-1} e_i u \in f_i A$. Then $1 = \sum_{i=1}^{n} f'_i$. Deduce $f_i = f'_i$ from this.

8. From the assumption, there exists an isomorphism $\varphi\colon A_A \simeq A_A$ such that $\varphi(fA) = eA$, $\varphi((1-f)A) = (1-e)A$ and $\varphi \equiv \mathrm{id}_A \pmod{I}$. Thus, there exists a unit $u = 1 + c$ of A, where $c \in I$, such that $eA = u(fA)$, $(1-e)A = u(1-f)A$. From this we have $f = u^{-1}eu$ (see the solution to the preceding problem).

9. $(2) \Rightarrow (1)$. If W is any irreducible A-submodule of V, then $V = U \oplus W$ for some U_A. If W' is any irreducible A-submodule of U, then $V = U' \oplus W'$, and it follows that $U = (U \cap U') \oplus W'$. Thus the condition of (2) is satisfied for U, too. Hence the assertion can be proved by induction on the composition length of V. The proof of $(3) \Rightarrow (1)$ is similar.

10. Let W be an irreducible submodule of $U \oplus V$. Then by considering the images of W by the projections of $U \oplus V$ on U and V, we have $W \subset \mathrm{soc}(U) \oplus \mathrm{soc}(V)$.

11. (i) If I is an ideal of A, then $I = \bigoplus_i IA_i$, and each IA_i is either 0 or A_i.
(ii) Apply (i) to the kernel of the given epimorphism $A \to A'$.

12. (i) Remember that $A/J(A)$ is a direct sum of simple rings.
(ii) If $I \not\subset J(A)$, then $\bar{I}$ contains an idempotent $\bar{c}$ of $\bar{A} = A/J(A)$ $(c \in I)$. We can lift $\bar{c}$ to an idempotent e of A such that $e = f(c)$, where $f(t)$ is a polynomial in t over $\mathbf{Z}$ with $f(0) = 0$. Hence $e \in I$.

13. Define $\varphi\colon A_A \to A_A$ by $\varphi(a) = xa$. The assumption $xy = 1$ implies that φ is an epimorphism, hence an isomorphism by problem 3.

14. Clear by Theorems 10.11 and 10.14.

15. (i) Since $AeA = A$, there exist $x_i, y_i \in A$ such that $\sum_i x_i e y_i = 1$. Define

$$Ve \otimes_\Gamma eA \to V \quad (ve \otimes ea \mapsto vea),$$

$$V \to Ve \otimes_\Gamma eA \quad \left(v \mapsto \sum_i vx_i e \otimes ey_i\right).$$

Then they are the inverse maps of each other.
(ii) Let $f \in \mathrm{Hom}_A(U, V)$. If $f(Ue) = 0$, then $f(U) = f(UAeA) = f(Ue)A = 0$ and thus $f = 0$. On the other hand, since $U \simeq Ue \otimes_\Gamma eA$, every $g \in \mathrm{Hom}_\Gamma(Ue, Ve)$ extends to an element of $\mathrm{Hom}_A(U, V)$.

(iii) If $W'_\Gamma \subset W_\Gamma$, then the natural map $\varphi: \mathfrak{G}(W') \to \mathfrak{G}(W)$ is a monomorphism. In fact, since $\mathfrak{G}(W')e \simeq W'$ and $\mathfrak{G}(W)e \simeq W$, it follows that $(\text{Ker } \varphi)e = 0$ and hence $\text{Ker } \varphi = 0$. It is clear from the above (i) and (ii) that if V_A is projective, then so is $\mathfrak{F}(V)$.

16. For some $x_{i\mu} \in A$, $e_{i\mu}A = x_{i\mu}e_iA = x_{i\mu}ee_iA \subset AeA$. Hence $AeA = A$.

17. (i) If $P \simeq nA$, then $J(M_n(A)) = M_n(J(A))$ and the assertion is clear. Let $P \oplus Q = F \simeq nA$ and π be the projection on P. Then $\text{End}_A(P) \simeq \pi \text{ End}_A(F)\pi$ and hence $J(\text{End}_A(P)) \simeq \pi J(\text{End}_A(F))\pi$. The result is now clear from the definition of this isomorphism.

(ii) If P is projective, then the natural map $\text{End}_A(P) \to \text{End}_A(P/PJ(A))$ is an epimorphism. (See the solution to problem 14.)

18. *Necessity*. Let U be a direct summand of a free module $F = x_1A \oplus \cdots \oplus x_nA$. If $u \in U$ is expressed as $u = \sum_{i=1}^n x_i\varphi_i(u)$ with $\varphi_i(u) \in A$, then we have $\varphi_i \in \text{Hom}_A(U, A)$ for each i. Let $\pi: F \to U$ be the projection of F on U, and let $u_i = \pi(x_i)$ for $1 \le i \le n$. Then $\{\varphi_i\}$ and $\{u_i\}$ satisfy the condition.
Sufficiency. Let $F = y_1A \oplus \cdots \oplus y_nA$ be a free module, and define $f: U \to F$, $g: F \to U$ by $f(u) = \sum_{i=1}^n y_i\varphi_i(u)$, $g(\sum_{i=1}^n y_ia_i) = \sum_{i=1}^n u_ia_i$. Then $g \circ f = \text{id}_U$.

19. Let ε be any idempotent of $Z(A/I)$. Since I is nilpotent in either case, there exists an idempotent e of A such that $\varepsilon = \bar{e} \pmod{I}$. Thus it suffices to show that $e \in Z(A)$ under the assumption (1) or (2).

Assume (1). Set $f = 1 - e$. Then $eAf \subset eNAf = NeAf$ and hence $eAf = 0$, since N is nilpotent. Similarly $fAe = 0$. Then we have $e \in Z(A)$ by an easy computation.

Assume (2). Set $J = J(A)$. From the assumption, we have $eAf \subset J^2$ and $fAe \subset J^2$. Then $eAf = eJ^2f = eJ(e+f)Jf \subset eJeJf + eJfJf \subset eJ^3f$. By repeating this, we find $eAf = 0$, and $fAe = 0$ follows similarly.

20. Let $\xi: (V \otimes W)/\text{Im} f \to V'' \otimes W''$ be the map induced by $\lambda' \otimes \mu'$. It suffices to show that ξ is an isomorphism. The map $V'' \times W'' \to (V \otimes W)/\text{Im} f$ $((v'',w'') \mapsto v \otimes w + \text{Im} f$, where $\lambda'(v) = v''$, $\mu'(w) = w'')$ is well defined and gives rise to the inverse of ξ.

21. It follows from $W_2/W_1 \cap W_2 \simeq W_1 + W_2/W_1$ that $(V \otimes W_2)/V \otimes (W_1 \cap W_2) \simeq (V \otimes W_1 + V \otimes W_2)/V \otimes W_1 \simeq (V \otimes W_2)/(V \otimes W_1 \cap V \otimes W_2)$.

22. Apply problem 20 to the following pair of exact sequences:

$$0 \to I \to R \to R/I \to 0$$

$$0 \to L \to R \to R/L \to 0$$

Problems 2

1. $J = \begin{pmatrix} 0 & 0 & K \\ 0 & 0 & K \\ 0 & 0 & 0 \end{pmatrix}$, $A/J \simeq M_2(K) \oplus K$. $\quad \mathrm{IRR}(A) = \{(K, K, 0), (0, 0, K)\}$.

2. $e_{ii}A = e_{ii}M_n(K) \simeq (M_n(K)e_{ii})^\wedge = (Ae_{ii})^\wedge$. The isomorphism in the middle is an $M_n(K)$-isomorphism, hence an A-isomorphism.

3. If $a \in J$, then $b = a + \tilde{a} \in J$ because $\tilde{J} = J$. Recall that in general every matrix b satisfying $b\tilde{b} = \tilde{b}b$ is similar to a diagonal matrix.

4. $\mathbf{Q}G \simeq \mathbf{Q}[x]/(x^m - 1)$. If $f_d(x)$ is the irreducible polynomial of ζ_d over $\mathbf{Q}$, then $x^m - 1 = \prod_{d|m} f_d(x)$. Thus the Chinese remainder theorem yields the result.

5. (ii) If $J(A) = 0$, then A is semisimple and hence $B|_A A_A$. Therefore B has identity.

(iii) If $J(A) \cap B = 0$, then $A = J(A) \oplus B$ and $A = B$.

6. Since $A/J \supset (B + J) \simeq B/B \cap J$ and φ vanishes on $B \cap J$, we may assume that A is semisimple. Thus $A = Ke_1 \oplus \cdots \oplus Ke_n$ with $e_i \in \mathrm{pi}(A)$. By the above problem, B is a semisimple ring with identity, and hence $B = Kf_1 \oplus \cdots \oplus Kf_r$ with $f_i \in \mathrm{pi}(B)$. We may assume that $\varphi(f_1) = 1$, $\varphi(f_i) = 0 \quad (2 \le i \le r)$. Write $f_1 = e_1 + \cdots + e_s$, and define $\hat{\varphi}: A \to K$ by $\hat{\varphi}(e_1) = 1$, $\hat{\varphi}(e_j) = 0 \quad (2 \le j \le n)$. Then $\hat{\varphi}$ is a desired extension of φ.

7. (i) Since $Z(D) = \mathbf{R}$, we have $\dim_{\mathbf{R}} D = [\mathbf{C}:\mathbf{R}]^2 = 4$.

(ii) Let $\sigma \in D$ be such that $\sigma^{-1} i \sigma = -i$. Then $D = \mathbf{C} \oplus \mathbf{C}\sigma$ and $\sigma^2 \in Z(D) = \mathbf{R}$. If $\sigma^2 = a > 0$, then $(\sigma - \sqrt{a})(\sigma + \sqrt{a}) = 0$ and hence $\sigma \in \mathbf{R}$, which is a contradiction. Thus $a < 0$, and it suffices to let $j = \sigma/\sqrt{-a}$.

8. (i) We may assume that $[L:K] < \infty$. Then $L \otimes_k A/J(A)$ is semisimple, and hence $J(A^L) \subset J(A)^L$.

(ii) There exist a subfield E of L that is a finite extension of K and a finitely generated A^E-module U such that $V = L \otimes_E U$. Since E is separable over K, we have ${}_E E_E | E \otimes_K E$. Hence $V = L \otimes_E E \otimes_E U | L \otimes_E (E \otimes_K E) \otimes_E U = L \otimes_K U$. Thus it suffices to let $W = U$, viewing it as an A-module.

9. We may assume that $K = K(\tilde{\zeta})$, then $A = M_n(D)$ is a central simple K-algebra. If $L \otimes_K D \simeq M_l(D')$, where D' is a division algebra, then $\dim_K D = l^2 \dim_L D'$, that is, $m_K(\tilde{V}) = l m_L(\tilde{V})$, proving (i). Also since $L \otimes_K D \simeq lU$, where U is an irreducible $L \otimes_K D$-module, we have $\dim_K L = l \dim_D U$ by comparing the dimension over D of both sides. Thus $l | \dim_K L$, and (ii) follows.

11. ($\Leftarrow$). We show that $f_p \sim 0$ for every $p \| |G|$. Since $f_S \sim 0$ by assumption, we have $(f_p)_S \sim 0$ (note that $(f_{p'})_S \sim 0$). Thus $|G{:}S| f_p \sim 0$ by the preceding problem and hence $f_p \sim 0$.

12. (ii) There exists an epimorphism $\mathbf{Q}G \to \mathbf{Q}[G/\langle y \rangle] \simeq \mathbf{Q}H = \mathbf{Q} \oplus \mathbf{Q}[\zeta_5]$. Also, by (i) $M_5(K)$ is a simple component of $\mathbf{Q}G$.

13. Let $A = \bigoplus_{x \in G} u_x K$ with $u_x u_y = u_{xy}\alpha(x, y)$. If $\lambda\colon A \to K$ is a one-dimensional representation of A, then $\lambda(u_x)\lambda(u_y) = \alpha(x, y)\lambda(u_{xy})$ and we have $\alpha \sim 1$.

14. (i) Set $M = V/S_{i-1}(V)$, then $S_i(V)/S_{i-1}(V) = \operatorname{soc}(M) = \sum_\alpha M_\alpha$, where M_α ranges over all the irreducible submodules of M. Thus $(V/S_i(V))^\wedge = (M/\operatorname{soc}(M))^\wedge = (\sum_\alpha M_\alpha)^\perp = \bigcap_\alpha M_\alpha^\perp = JM^\wedge = (JV/S_{i-1}(V))^\wedge = J(J^{i-1}(V^\wedge)) = J^i V^\wedge$ (the fourth equality follows from problem 6 in Chapter 1).

(ii) Take the dual of the following exact sequence

$$0 \to S_{i+1}(V)/S_i(V) \to V/S_i(V) \to V/S_{i+1}(V) \to 0$$

and use (i).

15. $\operatorname{Hom}_A(eA/eJ, A) = \{\varphi \in \operatorname{Hom}_A(eA, A) = Ae;\ \varphi(eJ) = 0\} = l(J)e$.

(ii) ($\Rightarrow$) By (i) we have $l(J)e \neq 0$ for any $e \in \operatorname{pi}(A)$, that is, $r(l(J))$ contains no idempotent and hence is contained in J.

($\Leftarrow$) The assumption implies that $l(J)e \neq 0$ for any $e \in \operatorname{pi}(A)$, and hence $\operatorname{Hom}_A(eA/eJ, A) \neq 0$ by (i).

17. Since K is a splitting field for A, the multiplicity of $\bar{e}\bar{A}$ in A_A is equal to $\dim_K Ae$. But we have $\dim_K Ae = \dim_K eA$, since $eA \simeq ((Ae)^\sigma)^\wedge$, where σ is the Nakayama automorphism of A.

18. ($\Rightarrow$) Let $\lambda \in A^\wedge$ be regular and symmetric. Then the map $a \mapsto \lambda_a$ gives an isomorphism ${}_AA_A \simeq {}_A(A^\wedge)_A$.

($\Leftarrow$) Let 1_A be mapped onto λ under the isomorphism $A \simeq A^\wedge$. Then λ is regular and symmetric (see the proof of Theorem 8.16).

19. (i) $V^\wedge = \operatorname{Hom}_K(V \otimes_A A, K) \simeq \operatorname{Hom}_A(V, A^\wedge) \simeq \operatorname{Hom}_A(V, A)$.

(ii) Consider the map $eA \to eA (x \mapsto cx)$.

20. (i) Let $\Gamma = eAe$ be the basic ring of A. For any extension field L of K, we have $J(\bar{\Gamma}^L) = J(\bar{e}\bar{A}^L\bar{e}) = \bar{e}J(\bar{A}^L)\bar{e} = 0$, hence $\bar{\Gamma}^L$ is semisimple. Thus $\bar{\Gamma}$ is a separable K-algebra. $P = Ae$ is projective as a right Γ-module and $A \simeq \operatorname{End}_\Gamma(P)$. Let $n\Gamma = P \oplus P'$ and π be the projection of $n\Gamma$ on P, thus $A \simeq \pi M_n(\Gamma)\pi$. If there exists a subalgebra Γ' of Γ such that $\Gamma = \Gamma' \oplus J(\Gamma)$; then $A \simeq \tilde{\pi}M_n(\Gamma')\tilde{\pi} \oplus \tilde{\pi}M_n(J(\Gamma))\tilde{\pi} = \tilde{\pi}M_n(\Gamma')\tilde{\pi} \oplus J(A)$, where $\tilde{\pi}$ is an idempotent of $M_n(\Gamma')$ that is equivalent to π.

(ii) If $J^2 \neq 0$, then by induction there exists a subalgebra B' of A such that $A/J^2 = B'/J^2 \oplus J/J^2$. Also $B' = B \oplus J^2$ by induction, and it follows that $A = B \oplus J$.

(iii) We show that $B = (A' \oplus W) \cap A$ is a subring of A. If $a_i = \gamma_i + w_i \in B (i = 1, 2)$, where $\gamma_i \in A'$, $w_i \in W$, then $a_1 a_2 = \gamma_1\gamma_2 + \gamma_1 w_2 + w_1\gamma_2$. But we have $\gamma_1 w_2 = (\gamma_1 + w_1)w_2 = a_1 w_2 \in W$ and $w_1\gamma_2 = w_1 a_2 \in W$. Therefore, $a_1 a_2 \in B$.

(iv) If K is algebraically closed and $A = \Gamma$, then $A/J \simeq Ke_1 \oplus \cdots \oplus Ke_k$, where $e_i \in \text{pi}(A)$. Thus we have $A = B \oplus J$ with $B = \bigoplus_{i=1}^{k} Ke_i$.

21. We may identify $\text{End}_R(V) \otimes_R \text{End}_R(W)$ with $\text{End}_R(V \otimes_R W)$ via the natural isomorphism. $\text{End}_\Lambda(V)$ is the centralizer of $\Lambda^\circ \text{id}_V$ in $\text{End}_R(V)$, which is R-free, since R is a principal ideal domain. Hence the proof of Lemma 4.1(ii) works.

Problems 3

3. $\text{Hom}_{FG}(W, V) \simeq \text{Hom}_{FP}(W_P, F) \neq 0$, $\text{Hom}_{FG}(V, W) \simeq \text{Hom}_{FP}(F, W_P) \neq 0$.

4. (i) $U_1 \otimes W_i$ is projective and $U_1 \otimes W_i/(U_1 J(FG) \otimes W_i) \simeq W_i$.

(ii) Since $|G| = \sum_{i=1}^{l} (\dim_F W_i)(\dim_F U_i)$, the assertion follows by (i).

5. Let $A = FG$, $J = J(FG)$. Since δ is an anti-automorphism of A, i.e., $\delta(ab) = \delta(b)\delta(a)$, it follows that $\delta(e) \in \text{pi}(A)$. For $\varphi \in (eA)^\wedge$ and $u \in eA$, we have $(\varphi\delta(e))(u) = \varphi(ue) = \varphi(eue)$. Hence there exists $\psi \in (eA/eJ)^\wedge$ such that $\psi\delta(e) \neq 0$. (Note that $eAe \neq eJe$.) But since $(eA/eJ)^\wedge = \text{soc}((eA)^\wedge)$, we have $\delta(e)A \simeq (eA)^\wedge$.

6. Let G be a permutation group on Ω and C be the conjugate class containing x. Then

$$|C|\chi(x) = \sum_{y \in C} \chi(y) = \sum_{y \in C} |\{\omega \in \Omega; y \in G_\omega\}| = \sum_{\omega} |C \cap G_\omega|.$$

Since G is transitive, $|C \cap G_\omega| = m$ is independent of the choice of ω. Therefore $|C|\chi(x) = |\Omega|m = \chi(1)m$.

7. $|C_{\bar{G}}(\bar{x})| = \sum_{\chi \in \text{Irr}(\bar{G})} |\chi(\bar{x})|^2 \leq \sum_{\chi \in \text{Irr}(G)} |\chi(x)|^2 = |C_G(x)|$.

8. (i) If $\chi(x) = \sum_i \zeta_i$, i.e., the sum of roots of unity, then $\chi(x)^\sigma = \sum_i \zeta_i^\sigma = \sum_i \zeta_i^{m_\sigma} = \chi(x^{m_\sigma})$.

(iii) By Theorem 1.30 (ii), the number of the irreducible K-characters of G equals the number of the G-orbits on Irr(G). Thus, the assertion follows from Brauer's permutation lemma.

9. Note that $\prod_x |\chi(x)|^2 \in \mathbf{Z}$ by the preceding problem. Thus, if $\chi(x) \neq 0$ for all $x \in G$, then $\sum_{x \neq 1} |\chi(x)|^2 \geq |G| - 1$. Thus

$$|G| = \sum_{x \in G} |\chi(x)|^2 \geq \chi(1) + |G| - 1 \geq |G|$$

and hence $\chi(1) = 1$.

10. Suppose to the contrary that there exists an irreducible representation $\mathbf{T}: G \to GL_2(\mathbf{C})$. Then $\det T(x) = 1$ for all $x \in G$ because G has no nontrivial representation of degree 1. Since 2 divides $|G|$, G has an involution, say y. We may assume that $T(y) = -I$. But this implies that $y \in Z(G)$, contradicting the simplicity of G.

11. (i) We may assume $T(\zeta) = G$. By Lemma 5.4 (iii) every irreducible projective representation of G/H is equivalent to a linear representation and hence its degree is 1.

(ii) $\zeta^G(x) = 0$ for all $x \in G - H$, while $(\zeta^G)_H$ is a multiple of the sum of the G-conjugates of ζ. Thus $\zeta^G = m\chi$ for some $m \in \mathbf{C}$. Deduce $m = e$ from this.

(iii) χ vanishes on $G - \operatorname{Ker} \lambda$. Since $G/\operatorname{Ker} \lambda$ is cyclic, the assertion is immediate from (i) and (ii).

12. (i) If $\zeta^G = \sum_i e_i \chi_i$, then $\sum_i e_i^2 = (\zeta^G, \zeta^G)_G = ((\zeta^G)_H, \zeta)_H = |G:H|$.

(ii) $(1) \Leftrightarrow (2)$ is immediate from the proof of (i), $(2) \Leftrightarrow (3)$ is trivial, and $(3) \Leftrightarrow (4)$ is clear from Theorem 5.8.

13. By Theorem 5.11, there exists $\mathbf{X}$ satisfying (1). Let n be the degree of $\mathbf{X}$. Since $(n, |S|) = 1$, there exists a linear character μ of S such that $(\det \mathbf{X})_S = \mu^{-n}$. Therefore, $\mu\mathbf{X}$ in place of $\mathbf{X}$ satisfies the conditions (1) and (2). The uniqueness is clear by Corollary 5.9 since $(n, |S|) = 1$.

14. This is clear by Theorem 5.7 (iii) and Problem 11 in Chapter 2.

15. We may write $Z^2(G, \mathbf{C}^\times) = M_\mathbf{C}(G) \times B^2(G, \mathbf{C}^\times)$. So if $\alpha \in M_\mathbf{C}(G)$ has order n, then $\alpha(x, y)^n = 1$ for all $x, y \in G$. In particular, if α is a p'-element, then it is regarded as an element of $M_F(G)$. In this way we get a natural isomorphism $M_F(G) \simeq M_\mathbf{C}(G)/M_\mathbf{C}(G)_p$.

16. Take an extended valuation ring S of R that contains a primitive $|G|$th root of unity and let $\tilde{S}$ be the completion of S. Since the principal indecomposable $\tilde{S}$-characters are linearly independent, we get $\tilde{S} \otimes_R U \simeq \tilde{S} \otimes_R V$ from the assumption. Let E, F be the residue fields of $\tilde{S}$, R, respectively, and let (π) be the valuation ideal of R. From the above isomorphism, we get $E \otimes_F U/\pi U \simeq E \otimes_F V/\pi V$, whence $U/\pi U \simeq V/\pi V$ by Chapter 2, 3.1, and $U \simeq V$ by Problem 14 in Chapter 1.

17. By Lemma 6.13 we have $\varphi^*(x) \in F$ for all $\varphi \in \mathrm{IBr}(G)$ and $x \in G$. Hence the assertion follows immediately from Theorem 1.32 (i).

19. (i) Choose $m \geq 0$ so that $\pi^m S \in M_n(\tilde{R})$. Taking the π-adic expansions of the entries of S^{-1}, we can write $S^{-1} = T + \pi^{m+1} Y$ with $T \in M_n(K)$, $Y \in M_n(\tilde{R})$. Then $TS = I - \pi(\pi^m YS) \in GL_n(\tilde{R})$.

(ii) By assumption there are a K-representation $\mathbf{Y}$ of A and $S \in GL_n(\tilde{K})$ such that $Y(a) = SX(a)S^{-1}$ for all $a \in A$. Let $T \in GL_n(K)$ be as in (i), then $TY(a)T^{-1} = (TS)X(a)(TS)^{-1} \in M_n(K) \cap M_n(\tilde{R}) = M_n(R)$. Hence $\mathbf{X}$ is realized in R.

20. (i) This is clear from the preceding problem.

(ii) Note that $\mathbf{Z}/(p)$ is a splitting field for G. Use Problem 4 in Chapter 2.

21. If $\mathrm{Char}\, L = 0$, the assertion has been shown in Theorem 5.6. If $\mathrm{Char}\, L = p$, then the order of a representation group of G is prime to p(Chapter 2, 7.3 (i)). Hence the assertion is reduced to the case of $\mathrm{Char}\, L = 0$ by Theorem 6.12.

23. Let $\zeta \in \mathrm{Irr}(B)$ with $\mathrm{ht}(\zeta) = 0$ and let $d = d(B)$. If $x(\neq 1) \in Z = Z(P)$, then we have $\chi(x) = 0$ by Lemma 2.26, and hence $\zeta(x) \equiv 0 \bmod \zeta(1)\pi$, that is, $\zeta(x) \equiv 0 \bmod p^{a-d+1}$. Thus we have $\sum_{x \in Z} \zeta(x) \equiv \zeta(1) \bmod p^{a-d+1}$. Since $\sum_{x \in Z} \zeta(x) = |Z|(\zeta_Z, \mathbf{1}_Z)_Z$ is divisible by $|Z|$ and $\zeta(1)$ is not divisible by p^{a-d+1}, we get $\nu(|Z|) \leq a - d$ and thus $d \leq \nu(|P:Z|)$.

24. (i) The first assertion is clear from the definition. We have $a_{\chi\chi} = (\dot{\chi}, \chi)_G = p^d(\chi, \chi)' < p^d(\chi, \chi)_G = p^d$, where the inequality holds because $d > 0$ (cf. Chapter 3, 6.29(5)).

(ii) Recall that $\nu(a_{\theta\chi}) \geq \mathrm{ht}(\chi)$ by 6.33(i). If $\mathrm{ht}(\chi') = 0$, then $a_{\theta\chi'} \not\equiv 0 (\mathrm{mod}\, p)$ by 6.34(ii). Thus the result follows from (6.17).

(iii) If $\mathrm{ht}(\chi') = 0$, then $\nu(a_{\theta\chi'}) = \nu(a_{\chi'\theta}) = \mathrm{ht}(\theta) > 0$. Thus the assertion is clear by (6.17).

(iv) This follows immediately from (i) and (iii).

(v) If $\mathrm{ht}(\chi) > 0$, then $\mathrm{ht}(\chi) \leq d - 2$ by (iv). Hence $\nu(\chi(1)) - (a - d) \leq d - 2$, i.e., $\nu(\chi(1)) \leq a - 2$, which is a contradiction. Thus $\mathrm{ht}(\chi) = 0$, and hence $d = 1$.

Problems 4

1. $\mathbf{X}_n$ is an F-representation of $P \Leftrightarrow X_n(x)^q = I$, where $q = |P| \Leftrightarrow (I - X_n(x))^q = 0 \Leftrightarrow 1 \leq n \leq q$.

2. What can be said about $T \in GL_2(F)$ satisfying $T^{-1}\begin{pmatrix} 1 & 1 \\ 0 & 1 \end{pmatrix} T = \begin{pmatrix} 1 & 1 \\ 0 & 1 \end{pmatrix}$?

3. (i) If $V \in \mathrm{IND}(FG)$, then $V \mid U^G$ for some $U \in \mathrm{IND}(FP)$.

(ii) Since P is not cyclic, there exists $Q \lhd P$ such that $P/Q \simeq \langle x \rangle \times \langle y \rangle$ with $x^p = y^p = 1$.

(a) Since $\mathrm{IND}(F[P/Q]) \subset \mathrm{IND}(FP)$, we have $|\mathrm{IND}(FP)| = \infty$ from the preceding problem. Also if $W \in \mathrm{IND}(FP)$, then there exists $V \in \mathrm{IND}(FG)$ such that $W \mid V_P$.

(b) Let $\tilde{F}$ be the algebraic closure of F. If $|\mathrm{IND}(FG)| < \infty$, then it follows from Chapter 2, Problem 8 that $|\mathrm{IND}(\tilde{F}G)| < \infty$.

4. (ii) If $f(u) = \pi v$ for some $u \in U_0$ and $v \in V_0$, then by the commutativity of the diagram in (i) there exists $u' \in U_0$ such that $u = \pi u'$.

(iii) Since the composite map $V_0 \to V_0^* \to U$ is an epimorphism, it follows that $\mathrm{rank}_R(\mathrm{Im} f) \geq \dim_F U = \mathrm{rank}_R U_0$ and hence $\mathrm{Ker} f = 0$. Also $V_0/\mathrm{Im} f$ is R-free, because it is torsion-free by (ii).

(iv) Since U_0 is $\{1\}$-projective, f splits as an RG-homomorphism (cf. Exercises 2.3 and 2.4).

5. (ii) It suffices to show that $V \simeq U$ if V is an indecomposable module such that $\mathrm{Tr}_1^G(V) \neq 0$. By assumption, there exists $v \in V$ such that $v\sigma \neq 0$. Define $f: U \to V$ by $f(a) = va$. Then f is a monomorphism and hence $U \mid V$. Therefore $U \simeq V$.

(iii) $\sigma = (\sum_{x \in H} x)(\sum_{t \in H \backslash G} t) \in fFG$.

6. (i) Since $V \mid W^G$ for some $W \in \mathrm{IND}(\mathfrak{o}H)$, we have $V_H \mid (W^G)_H = \bigoplus_{t \in H \backslash G} W^t$. Note also that $(V_H)^t \simeq (V^t)_H$.

(ii) If $G \geq T \geq H$, then, for $W_{\mathfrak{o}\bar{T}}$, we have $(W_T)^G \simeq (W_{\bar{T}})^{\bar{G}}$.

(iii) Let $\mathrm{vx}(V) = Q \subset P \in \mathrm{Syl}_p(G)$. We have $V_{PH} \mid (V_{QH})^{PH}$ by assumption. But since PH/H is a p-group, $(V_{QH})^{PH}$ must be indecomposable, and hence $V_{PH} \simeq (V_{QH})^{PH}$. Thus $|PH{:}QH| = 1$.

(iv) Let $I = J(FH)$. Since $P(Y)/P(Y)I \simeq Y/YI$, we have $P(Y)^G/(P(Y)I)^G \simeq Y^G/(YI)^G$. This implies that $P(Y)^G \simeq P(Y^G)$, because $(P(Y)I)^G \subset P(Y)^G J(FG)$ and $(YI)^G \subset Y^G J(FG)$. Thus $\Omega(Y^G) \simeq \Omega(Y)^G$. The same is true for Ω^{-1} by Chapter 2, 8.25.

(v) We know that $J(FG) \supset J(FH)FG$ by Chapter 3, 1.25. The equality sign follows if every $FG/J(FH)FG$-module is shown to be completely reducible. Let X be any $FG/J(FH)\,FG$-module and Y be a submodule of X. Since X_H is completely reducible, the exact sequence $0 \to Y \to X \to X/Y \to 0$ splits as a sequence of FH-modules. Since $p \nmid |G{:}H|$, it splits as a sequence of FG-modules.

(vi) If Y is a principal indecomposable FH-module, then Y^G is a principal indecomposable FG-module, and conversely every principal indecomposable FG-module is written in this form. Now, we have ${}_{FG}(Y^G, W^G) \simeq |T(W):H|_{FH}(Y, \bigoplus_{t \in T(W)\backslash G} W^t)$. Hence, if ${}_{FG}(Y^G, W^G)$ is not zero, then Y is a projective cover of some W^t. Thus it suffices to let $U = \mathrm{hd}(P(W)^G)$.

7. Apply the Mackey decomposition theorem to $(\mathfrak{o}_H^G)_P$, where $P \in \mathrm{Syl}_p(G)$.

8. Let $P = \mathrm{vx}(U)$. It is clear that $P \leq_G \mathrm{vx}(V)$. Let $\pi_U\colon R' \otimes_R V \to U$ be the projection. Since $\mathrm{id}_U \in \mathrm{Tr}_P^G(E_{R'}(U))$, it follows that $\pi_U \in \mathrm{Tr}_P^G(E_{R'}(R' \otimes_R V)) \simeq R' \otimes_R \mathrm{Tr}_P^G(E_R(V))$. If $P <_G \mathrm{vx}(V)$, then $\mathrm{Tr}_P^G(E_R(V)) \subset J(E_{RG}(V))$, and it follows from the above that $\pi_U \in J(E_{R'G}(R' \otimes_R V))$, which is impossible.

9. (i) $E_{\mathfrak{o}G}(V_1 \otimes V_2)$ is a local ring by Problem 21 in Chapter 2.

(ii) Since $V_1 \otimes_{\mathfrak{o}} V_2 | (V_1)_{P_1}^{G_1} \otimes_{\mathfrak{o}} (V_2)_{P_2}^{G_2} \simeq (V_1 \otimes_{\mathfrak{o}} V_2)_{P_1 \times P_2}^{G_1 \times G_2}$, it follows that $Q = \mathrm{vx}(V_1 \otimes_{\mathfrak{o}} V_2) \leq_G P_1 \times P_2$. On the other hand, since $V_1 \otimes_{\mathfrak{o}} V_2 | W_Q^G$ and $(V_1 \otimes_{\mathfrak{o}} V_2)_{G_1} \simeq (\mathrm{rank}_{\mathfrak{o}} V_2) V_1$, we have $V_1 | (W_Q^G)_{G_1}$, and hence

$$P_1 \leq_{G_1} Q \cap G_1 \leq Q$$

by the Mackey decomposition theorem. Similarly we have $P_2 \leq_{G_2} Q$. Thus $P_1 \times P_2 \leq_G Q$, whence we have the equality.

10. (i) This is clear from problem 6 (ii).

(ii) In Theorem 4.7 (iii), let $W = F_P$, $H = P$.

11. ($\Leftarrow$) Since W is absolutely indecomposable, the assertion is clear from Corollaries 6.8 and 7.3.

($\Rightarrow$) We may assume $G = T(W)$. Let $E = E_{\mathfrak{o}G}(W^G)$, $E_1 = E_{\mathfrak{o}H}(W)$. Then $A = E/J(E_1)E$ is a generalized group ring of $\bar{G} = G/H$. (See the proof of Theorem 7.2.) By assumption, we have $A/J(A) \simeq F$. Thus, A has a one-dimensional representation, and hence $A \simeq F\bar{G}$ from Problem 13 in Chapter 2. But since A is a local ring, $\bar{G}$ must be a p-group (Problem 5 (iii)).

12. Let $\tilde{F}$ be the algebraic closure of F. Then $F^{(\alpha)}G \subset \tilde{F} \otimes F^{(\alpha)}G = \tilde{F}^{(\alpha)}G$ and $\tilde{F}^{(\alpha)}G \simeq \tilde{F}G$ by Chapter 2, 7.4. Thus $J = J(\tilde{F}^{(\alpha)}G)$ is nilpotent and $\tilde{F}^{(\alpha)}G/J \simeq \tilde{F}$. Set $I = J \cap F^{(\alpha)}G$. Then I is a nilpotent ideal of $F^{(\alpha)}G$ and $F \subset F^{(\alpha)}G/I \subset \tilde{F}$. This implies that $F^{(\alpha)}G/I$ is a finite extension field of F (cf. Chapter 1, 1.7).

13. (ii) It is easy to see that $C \subset A_H$. Note that A_H is an E-algebra and A_1 is a central simple over E. In particular, every E-algebra endomorphism of A_1 is inner (Chapter 2, 4.6(v)), and hence the proof of Chapter 5, 7.2 works, thus yielding our assertion.

(iii) For $a = \sum_x u_x \alpha_x \in A$, let $\mathrm{Supp}(a) = \{x \in G; \alpha_x \neq 0\}$ and $s(a) = |\mathrm{Supp}(a)|$.

Proof of $I \subset (I \cap C)A$. Let $v \in I$. We prove $v \in (I \cap C)A$ by induction on $s(v)$. Let $a \in I$ be such that $s(a) = \min\{s(y); y \in I, y \neq 0\}$ and let $a = \sum_x u_x \alpha_x$. We may assume that $\alpha_1 = 1$. Since $a\gamma - \gamma a \in I$ for all $\gamma \in A_1$, we get $a\gamma = \gamma a$ by the choice of a. Thus $a \in C$. We write $v = \sum_x u_x \beta_x$ and let $y = v - a\beta_1 \in I$. Then $s(y) < s(v)$, and we have $y \in (I \cap C)A$ by the inductive hypothesis. Thus $v = y + a\beta_1 \in (I \cap C)A$.

Proof of $LA \cap C = L$. It is clear that $LA \cap C = LA_H \cap C$. On the other hand, since $A_H = C \otimes_E A_1$, we have $LA_H = L \otimes_E A_1$, and thus $LA_H \cap C = L$.

For the last assertion, one may notice that LA is an ideal of A if L is a G-invariant ideal of C.

(iv) Since C is a local ring by the preceding problem, $J(C)$ is a unique maximal ideal of C, which is G-invariant. Hence, by (iii), $J(A)$ is a unique maximal ideal of A.

14. Let $E = \mathrm{End}_{oG}(W^G)$. It suffices to show that $E/J(E)$ is a simple ring (cf. Chapter 1, 5.4(iii)). Therefore we may assume that $G = T(W)$. Let $E_1 = \mathrm{End}_{oH}(W)$ and $A = E/J(E_1)E$. Then A is a strongly G/H-graded F-algebra and $A_1 = E_1/J(E_1)$ is a division algebra. Hence the assertion is clear by the preceding problem.

15. (i) Consider the Scott module.
(ii) This is clear from Problem 7.

16. Take a p-modular system (E, R', F') such that $(\tilde{\mathbf{Q}}_p(\chi), R, F) \leq (E, R', F')$ and E contains a primitive $|G|$th root of unity. If χ is given by an $R'G$-module V, then V^* is irreducible since $p \nmid |G|$. On the other hand, since $F = F(\chi^*)$, there exists $W \in \mathrm{IRR}(FG)$ such that $V^* = F' \otimes_F W$. And by the completeness of R, there exists $W_0 \in \mathrm{IND}(RG)$ such that $W = F \otimes_R W_0$. Thus $V^* \simeq (W_0^{R'})^*$, and it follows that $V \simeq W_0^{R'}$, because both V and $W_0^{R'}$ are projective as $R'G$-modules. Thus χ is realizable in $\tilde{\mathbf{Q}}_p(\chi)$.

17. Suppose that V is indecomposable and let $Q = \mathrm{vx}(V)$. If $Q = 1$, then V is projective and the assertion is trivial. Let $f = f(G, Q, N_G(Q))$ be the Green correspondence and $M = f(V)$, then $M^G \simeq V \oplus V'$. Recall that $\mathrm{vx}(L) <_G Q$ for all $L \in \mathrm{Comp}(V')$. Since M^G is a local module, the assertion follows by induction on the order of $\mathrm{vx}(V)$.

18. From the exact sequence $0 \to \Omega(V) \to P(V) \to V \to 0$, we get an exact sequence $0 \to \Omega(V) \otimes_F U \to P(V) \otimes_F U \to V \otimes_F U \to 0$. Hence from Chapter 1, 10.18, the assertion holds for $n = 1$. Similarly, we get the assertion for $n = -1$ from Chapter 1, 10.19. The general case can be proved by induction on n.

19. (i) Let $\{V_1, \ldots, V_m\}$ be a complete set of nonisomorphic indecomposable components of W^G that are nonprojective. Recall that $\Omega^n(W)^G = \Omega^n(W^G) \oplus$ (projectives). Thus, if $\Omega^n(W) \simeq W$, then $\{\Omega^n(V_1), \ldots, \Omega^n(V_m)\} = \{V_1, \ldots, V_m\}$, and we have $\Omega^{nr}(V_i) = V_i$ for some $r = r(i)$.

(ii) Since $\Omega^n(V)_H = \Omega^n(V_H) \oplus$ (projectives), the proof is similar to the above.

20. (i) $x^p - 1 = (x-1)^p \in J(FP)^2$,

$$x^{-1}y^{-1}xy - 1 = x^{-1}y^{-1}\{(x-1)(y-1) - (y-1)(x-1)\} \in J(FP)^2,$$

whence the assertion follows, since $\Phi(P) = \langle x^p, [P, P]; x \in P \rangle$.

(ii) It suffices to show the assertion for the case of $F = F_0$. Define $f: P \to J(FP)/J(FP)^2$ by $f(x) = (x-1) + J(FP)^2$. Then f is a homomorphism because $xy - 1 = (x-1) + (y-1) + (x-1)(y-1)$. From the equations noted in the proof of (i), we find that $\mathrm{Ker} f \supset \Phi(P)$ and obtain the induced map $\bar{f}: P/\Phi(P) \to J(FP)/J(FP)^2$. On the other hand, we define $g: J(FP) \to P/\Phi(P)$ by $\bar{a}(x-1) \mapsto x^a\Phi(P)$, where $0 \le a < p$ and $x \in P$. Then g induces the homomorphism $\bar{g}: J(FP)/J(FP)^2 \to P/\Phi(P)$, which gives the inverse map of $\bar{f}$.

(iii) $\Omega(F_P) \simeq \Omega^{-1}(F_P) \Leftrightarrow J(FP) \simeq FP/\mathrm{soc}(FP) \Leftrightarrow J(FP)$ is a cyclic module $\Leftrightarrow J(FP)/J(FP)^2 \simeq F \Leftrightarrow P/\Phi(P) \simeq \mathbf{Z}/(p)$.

21. (i) Suppose first that V is indecomposable. By assumption, there exist $f: V^{R'} \to W^{R'}$, $g: W^{R'} \to V^{R'}$ such that $g \circ f$ is the identity map of $V^{R'}$. By Chapter 1, 11.12, we may write $f = \sum_i \alpha_i \otimes f_i$, $g = \sum_j \beta_j \otimes g_j$, where α_i, $\beta_j \in R'$, $f_i \in \mathrm{Hom}_A(V, W)$ and $g_j \in \mathrm{Hom}_A(W, V)$. Thus $g \circ f = \sum_{i,j} \beta_j \alpha_i \otimes g_j \circ f_i$. But since $\mathrm{End}_A(V)$ is a local ring, some $g_j \circ f_i$ must be a unit of $\mathrm{End}_A(V)$, which yields that $V | W$. The general case can be proved by induction on the number of the indecomposable components of V. Part (ii) is immediate from (i).

Problems 5

1. Consider the contragredient representation of an F-representation giving φ.

2. Let $P^K = U \oplus M$ and $L = P \cap M$. Then, from Chapter 2, 1.8, L is a pure submodule of P, and so $V = P/L$ is a direct summand of P_R. Thus $L^* \subset P^*$, $P^*/L^* \simeq V^*$ and $V^K \simeq U$. Furthermore $\mathrm{hd}(V^*) \simeq \mathrm{hd}(P^*)$ is irreducible.

3. If $y \in G_{p'}$ and $p \| C_G(y)|$, then $\chi_i(y) \equiv 0 \pmod{(\pi)}$. Hence the assertion follows, since the modular character table is nonsingular.

4. (i) See the proof of Chapter 3, 6.15.

(ii) Let

$$a_x = \frac{1}{|G|} \sum_{x \in \Lambda} \chi(1)\chi(x),$$

then

$$\varepsilon = \sum_{x \in G} a_x x^{-1}.$$

Furthermore,

$$a_x = \begin{cases} \frac{1}{|G|} \theta_x(1) \in R, & \text{if} \quad x \in G_{p'}, \\ 0, & \text{otherwise.} \end{cases}$$

5. Note that $\chi \in \mathrm{Irr}(B_0(G))$ if and only if $\omega_\chi(\hat{C}) \equiv |C| \pmod{\pi}$ for all $C \in \mathrm{Cl}(G)$. With the notation introduced just before Theorem 4.16, if $\chi \in \mathrm{Irr}(B_0(G))$, then $\omega_{\chi^\sigma}(\hat{C}) = \omega_\chi(\hat{C}^\sigma) \equiv |C^\sigma| \equiv |C| \pmod{\pi}$.

6. Since $\lambda_B(\hat{C}_i) = \lambda(e_B \hat{C}_i)$, the assertion is immediate from Theorem 4.9.

7. Let $\tilde{e}$ be the block idempotent of $\tilde{B}$. By Corollary 5.11, we have $B\tilde{e}B = B$, $\tilde{e}B\tilde{e} = \tilde{B}$, and $\tilde{B} \cap L = \tilde{e}L\tilde{e}$ (note that if $x \notin T$, then $\tilde{e}x\tilde{e} = x\tilde{e}^x\tilde{e} = 0$.), whence part (i) follows easily. Since the analogous statement holds for FG and $J(B^*)$ is nilpotent, part (ii) follows easily.

8. From the assumption we have $(\theta^{(x)}, \chi) = 0$ if $\chi \notin \mathrm{Irr}(B)$, by Theorem 4.7.

9. Show that $B^*/I(Q)B^*$ is a block of $FG/I(Q)FG$, where $I(Q)$ denotes the augmentation ideal of FQ.

10. (a) e is the degree of some irreducible projective representation of G/H. But since G/H is a p'-group, e divides $|G:H|$ (cf. Problem 21 in Chapter 3).

(b) Let U be any irreducible FH-module. Since U^G is completely reducible by Problem 6(v) in Chapter 4, we have

$$\mathrm{Hom}_{FH}(P_H, U) \simeq \mathrm{Hom}_{FG}(P, U^G) \simeq \mathrm{Hom}_{FG}(V, U^G) \simeq \mathrm{Hom}_{FH}(V_H, U).$$

The last F-module is not zero only when $U \simeq W^a$ for some $a \in G$, in which case its dimension equals e.

(c) Since G/H is a p-group, the trivial representation is the only irreducible projective representation of G/H up to equivalence. Hence $e = 1$.

(d) By Green's indecomposability theorem, Q^G is a principal indecomposable FG-module. From this and $\mathrm{Hom}_{FG}(Q^G, V) \simeq \mathrm{Hom}_{FH}(Q, V_H) \neq 0$, we have $Q^G \simeq P$.

11. Let H be a proper normal subgroup of G such that G/H is a p-group or a p'-group. We use the same notation as in the above problem.

Case 1. G/H is a p'-group: We have $\dim_F P = e|G:T|\dim_F Q$. Apply the inductive hypothesis to Q.

Case 2. G/H is a p-group: The assertion follows easily since $\dim_F P = |G:H|\dim_F Q$ and $(\dim_F V)_{p'} = (\dim_F W)_{p'}$.

12. (i) Remember that $\#\mathrm{IND}(FD) = q$ (Chapter 4, Problem 1). On the other hand, if $D = \langle x \rangle$, then $(x-1)^{q-1} = x^{q-1} + \cdots + 1 = \hat{D} \neq 0$. Thus $\{I^j; 0 \leq j \leq q-1\}$ gives all the indecomposable FD-modules.

(ii) Let $V \in \mathrm{IND}(B^*)$. Since V is D-projective, there is j $(0 \leq j \leq q-1)$ such that $V | I^j \otimes_D B^* \simeq I^j B^* = B^* I^j$, and hence $V \simeq U_i I^j$ for some U_i.

(iii) Since $\mathrm{soc}(U_i I^j) \simeq \mathrm{soc}(U_k I^l)$, it follows that $i = k$. Let $U_i = e_i FG$ with $e_i \in \mathrm{pi}(FG)$. If $U_i I^j = U_i I^l$ for $j < l$, then $U_i I^{q-1} = 0$ and $e_i \hat{D} = 0$. Hence $e_i \in (FG)I \subset J(FG)$, which is a contradiction.

13. Let b be a block of $H = O_{p'}(G)$ that is covered by B. We may assume $G = T(b)$, thus $\delta(B) = P$ by Lemma 12.2. Also B is the only block covering b by Corollary 5.6. Let $\mathrm{Irr}(b) = \{\beta\}$. Since $T(\beta)) = G$, β lifts to an irreducible character $\hat{\beta}$ of G by Chapter 3, 5.11 (iii). Then by Chapter 3, 5.12 (ii), $\mathrm{Irr}(B) = \mathrm{Irr}(G|\beta) = \{\hat{\beta}\gamma; \gamma \in \mathrm{Irr}(P)\}$ and part (i) follows. Furthermore, since $(\hat{\beta}\gamma)^* = \gamma(1)\hat{\beta}^*$, $\hat{\beta}^*$ is a unique irreducible F-character belonging to B, which proves (ii).

15. (i) The inclusion "$\subset$" is clear. On the other hand, the right-hand side is an ideal of FG, which has no idempotent, so it is contained in $J(FG)$.

(ii) Note that if $x, y \in S_j$, then $\varphi_i^*(x) = \varphi_i^*(y)$, and that $\{\varphi_1^*, \ldots, \varphi_l^*\}$ are linearly independent over F.

16. Let $\delta\colon FG \to FG$ be the same as in Problem 5, Chapter 3, where we have shown that $\hat{f}FG \simeq \delta(f)FG$. Now we have

$$eJ^{s-1}f/eJ^s f \simeq \delta(f)J^{s-1}\delta(e)/\delta(f)J^s\delta(e) \simeq \hat{f}J^{s-1}\hat{e}/\hat{f}J^s\hat{e},$$

and the assertion follows.

17. (i) Since $V_{\langle y \rangle}$ is a projective $F\langle y \rangle$-module, it is lifted to an $R\langle y \rangle$-module.

(ii) We prove this by induction on $|G|$. The assertion holds, once it is shown for irreducible modules. Hence, by induction, we may assume $O_p(G) = 1$. In view of Chapter 4, Problem 16 applied to FG-modules, we only need to show the assertion, assuming that V is either a projective or local module. In the former case, V is lifted to an RG-module, and the assertion is obvious. In the latter case, the induction works because all of the p-locals are proper subgroups of G.

References

AKIZUKI, Y. AND SUZUKI, M.
[1] *Algebra I, II.* Iwanami Shoten, Tokyo, 1980 (in Japanese).
ALPERIN, J. L.
[1] "The main problem of block theory," *Proc. of the Conference on Finite Groups*, 341–356, Academic Press, New York, 1976.
[2] "On the Brauer correspondence," *J. Alg.* **47** (1977), 197–200.
[3] "Local representation theory," *Proc. Sympos. Pure Math.* **37** (1980), Amer. Math. Soc., Providence.
[4] "The Green correspondence and the Brauer lift," *J. Alg.* **104** (1986), 78–79.
[5] *Local Representation Theory.* Cambridge Univ. Press, Cambridge, 1986.
ALPERIN, J. L. AND BROUÉ, M.
[1] "Local method in block theory," *Ann. of Math.* **110** (1979), 143–157.
ALPERIN, J. L. AND BURRY, D. W.
[1] "Block theory with modules," *J. Alg.* **65** (1980), 225–233.
ANDERSON, F. W. AND FULLER, K. R.
[1] *Rings and Categories of Modules.* Graduate Text in Math. **13**, Springer-Verlag, Berlin, 1974.
ASANO, K.
[1] "Einfacher Beweis eines Brauerschen Satzes über Gruppencharaktere," Proc. Jap. Acad. **31** (1955), 501–503.
ASANO, K., OSIMA, M. AND TAKAHASHI, M.
[1] "Über Darstellung von Gruppen durch Kollineationen im Körper der Characteristik p," *Proc. Phys.-Math. Soc. Jap.* **19** (1937), 199–209.
AZUMAYA, G.
[1] "Corrections and supplementaries to my paper concerning Krull–Remak–Schmidt's theorem," *Nagoya Math. J.* **1** (1951), 117–124.
[2] "On maximally central algebras," *Nagoya Math. J.* **2** (1951), 119–150.

AZUMAYA, G. AND NAKAYAMA, T.

[1] *Algebra I*. Iwanami Shoten, Tokyo, 1954 (in Japanese).

BANASCHEWSKI, B.

[1] "On the character rings of finite groups," *Canad. J. Math.* **15** (1963), 605–612.

BENSON, D.

[1] *Modular representation theory: New trends and methods.* Lecture Notes in Math. **1081** (1984), Springer-Verlag, Berlin.

BRAUER, R.

[1] "Über Darstellungen von Gruppen in Galoischen Feldern," *Act. Sci. Ind.* **195** (1935), Paris.

[2] "On the modular and $\mathfrak{p}$-adic representations of algebras," *Proc. Nat. Acad. Sci. U.S.A.* **25** (1939) 252–258.

[3] "On the Cartan invariants of groups of finite order," *Ann. of Math.* **42** (1941), 53–61.

[4] "On the connection between the ordinary and modular characters of groups of finite order," *Ann. of Math.* **42** (1941), 926–935.

[5] "Investigations on group characters," *Ann. of Math.* **42** (1941), 936–958.

[6] "On the arithmetic in a group ring," *Proc. Nat. Acad. Sci. U.S.A.* **30** (1944), 109–114.

[7] "On the representation of a group of order g in the field of g-th root of unity," *Amer. J. Math.* **67** (1945), 461–471.

[8] "On blocks of characters of groups of finite order, *I*," *Proc. Nat. Acad. Sci. U.S.A.* **32** (1946), 182–186; "*II*," *Proc. Nat. Acad. Sci. U.S.A.* **32** (1946), 215–219.

[9] "A characterization of characters of groups of finite order," *Ann. of Math.* **57** (1953), 357–377.

[10] "Number-theoretical investigations on groups of finite order," *Proc. of the International Symposium on Algebraic Number Theory*, 55–62, Tokyo and Nikko, Science Council of Japan, 1955.

[11] "Zur Darstellungstheorie der Gruppen endlicher Ordnung, *I*," *Math. Z.* **63** (1956), 406–444; "*II*," *Math. Z.* **72** (1959), 25–46.

[12] "Representations of finite groups," *Lectures on Modern Mathematics*, Vol. 1, 133–175. Wiley, New York, 1963.

[13] "Some applications of the theory of blocks of characters of finite groups, *I*," *J. Alg.* **1** (1964), 152–167; "*II*," *J. Alg.* **1** (1964), 307–334; "*III*," *J. Alg.* **3** (1966), 225–255; "*IV*," *J. Alg.* **17** (1971), 489–521, "*V*," *J. Alg.* **28** (1974), 433–460.

[14] "On blocks and sections in finite groups, *I*," *Amer. J. Math.* **89** (1967), 1115–1136; "*II*," *Amer. J. Math.* **90** (1968), 895–925.

[15] "Defect groups in the theory of representations of finite groups," *Illinois J. Math.* **13** (1969), 53–73.

[16] "On the first main theorem on blocks of characters of finite groups," *Illinois J. Math.* **14** (1970), 183–187.

[17] "Character theory of finite groups with wreathed Sylow 2-subgroups," *J. Alg.* **19** (1971), 547–592.

[18] *On the structure of blocks of characters of finite groups.* Lecture Notes in Math. **372** (1974), Springer-Verlag, Berlin, 103–130.

[19] "On 2-blocks with dihedral defect groups," *Symposia Mathematica* **13** (1974), 367–393, Academic Press, London.

[20] "Notes on the representations of finite groups," *J. London Math. Soc.* **13** (1976), 162–166.

[21] "On finite groups with cyclic Sylow subgroups, *I*," *J. Alg.* **40** (1976), 556–584; "*II*," *J. Alg.* **58** (1979), 291–318.

BRAUER, R., AND FEIT, W.

[1] "On the number of irreducible characters of finite groups in a given block," *Proc. Nat. Acad. Sci. U.S.A.* **45** (1959), 361–365.

Brauer, R., and Nesbitt, C. J.

[1] On the modular representation of groups of finite order," *Univ. of Toronto Studies Math. Ser.* **4** (1937).

[2] "On the modular characters of groups," *Ann. of Math.* **42** (1941), 556–590.

Brauer, R., and Tate, J.

[1] "On the characters of finite groups," *Ann. of Math.* **62** (1955), 1–7.

Brauer, R., and Tuan, H. F.

[1] "On simple groups of finite order," *Bull. Amer. Math. Soc.* **51** (1945), 756–766.

Broué, M.

[1] "Sur l'induction des modules indécomposables et la projectivité relative," *Math. Z.* **149** (1976), 227–245

[2] "Radical, hauteurs, *p*-sections et blocs," *Ann. of Math.* **107** (1978), 89–107.

[3] "Brauer coefficients of *p*-subgroups associated with a *p*-block of a finite group," *J. Alg.* **56** (1979), 365–383.

Burry, D. W.

[1] "A strengthened theory of vertices and sources," *J. Alg.* **59** (1979), 330–344.

[2] "Scott modules and lower defect groups," *Comm. Alg.* **10** (1982), 1855–1872.

[3] "Components of induced modules," *J. Alg.* **87** (1984), 483–492.

Burry, D. W., and Carlson, J. F.

[1] "Restrictions of modules to local subgroups," *Proc. AMS.* **84** (1982), 181–184.

Cline, E.

[1] "Stable Clifford theory," *J. Alg.* **22** (1972), 350–364.

Clifford, A. H.

[1] "Representations induced in an invariant subgroup," *Ann. of Math.* **38** (1937), 533–550.

Conlon, S. B.

[1] "Twisted group algebras and their representations," *J. Austral. Math. Soc.* **4** (1964), 152–173.

Curtis, C. W. and Reiner, I.

[1] *Representation theory of finite groups and associative algebras.* Interscience, New York, 1962.

[2] *Methods of representation theory with applications to finite groups and orders, Vol. 1.* Wiley, New York, 1981.

Dade, E. C.

[1] "Compounding Clifford's theory," *Ann. of Math.* **91** (1970), 236–290.

[2] "Extending irreducible modules," *J. Alg.* **72** (1981), 374–403.

Dress, A.

[1] "Operations in representation rings," *Proc. Sympos. Pure Math.* **21** (1971), 39–45, AMS., Providence.

Fein, B., Kantor, W. M. and Schacher, M.

[1] "Relative Brauer Groups II," *J. Reine Angew. Math.* **328** (1981), 39–57.

Feit, W.

[1] "Some consequences of the classification of simple groups," *Proc. Sympos. Pure. Math.* **37** (1980), AMS., Providence.

[2] *The Representation Theory of Finite Groups*, North-Holland, Amsterdam, 1982.

Fong, P.

[1] "On the characters of *p*-solvable groups," *Trans. Amer. Math. Soc.* **98** (1961), 263–284.

[2] "Solvable groups and modular representation theory," *Trans. Amer. Math. Soc.* **103** (1962), 484–494.

[3] "On decomposition numbers of J_1 and $R(q)$," *Symposia Mathematica* **XIII** (1974), 415–422, Academic Press, London.

Gallagher, P. X.

[1] "Group characters and normal Hall subgroups," *Nagoya Math. J.* **21** (1962), 223–230.

GLAUBERMAN, G.

[1] "Correspondence of characters for relatively prime operator groups," *Canad. J. Math.* **20** (1968), 1465–1488.

GOLDSCHMIDT, D. M.

[1] "A group-theoretic proof of the $p^a q^b$ theorem for odd primes," *Math. Z.* **113** (1970), 373–375.

GOW, R.

[1] "Extensions of modular representations for relatively prime operator groups," *J. Alg.* **36** (1975), 492–494.

GREEN, J. A.

[1] "On the indecomposable representations of a finite group ," *Math. Z.* **70** (1959), 430–445.

[2] "Blocks of modular representations," *Math. Z.* **79** (1962), 100–115.

[3] "A transfer theorem for modular representations," *J. Alg.* **1** (1964), 73–84.

[4] "Some remarks on defect groups," *Math. Z.* **107** (1968), 133–150.

[5] "Vorlesungen über Modulare Darstellungstheorie endlicher Gruppen," *Vorlesungen aus dem Mathematischen Institut Giessen*, Heft **2**, 1974.

[6] "On the Brauer homomorphism," *J. London Math. Soc.* **17** (1978), 58–66.

HELLER, A.

[1] "Indecomposable modules and the loop space operation," *Proc. Amer. Math. Soc.* **12** (1961), 640–643.

[2] "On group representations over a valuation ring," *Proc. Nat. Acad. Sci.* **47** (1961), 1194–1197.

HIGMAN, D. G.

[1] "Modules with a group of operators," *Duke Math. J.* **21** (1954), 369–376.

HOWLETT, R. B. AND ISSACS, I. M.

[1] "On groups of central type," *Math. Z.* **179** (1982), 555–569.

HUNGERFORD, T.

[1] *Algebra*. Springer-Verlag, New York, 1980.

HUPPERT, B. AND BLACKBURN, N.

[1] *Finite Groups II*. Springer-Verlag, Berlin, 1982.

IIZUKA, K.

[1] "On Brauer's theorem on sections in the theory of blocks of characters," *Math. Z.* **75** (1961), 299–304.

[2] "A note on blocks of characters of a finite group," *J. Alg.* **20** (1972), 196–201.

ISSACS, I. M.

[1] "Characters of solvable and symplectic groups," *Amer. J. Math.* **95** (1973), 594–635.

[2] *Character Theory of Finite Groups*. Academic Press, New York, 1976.

[3] "Extensions of group representations over arbitrary fields," *J. Alg.* **68** (1981), 54–74.

ITO, N.

[1] "On the degrees of irreducible representations of a finite group," *Nagoya Math. J.* **3** (1951), 5–6.

IWAHORI, N. AND MATSUMOTO, H.

[1] "Several remarks on projective representations of finite groups," *J. Fac. Sci. Univ. Tokyo, Sect. I*, **10** (1964), 129–146.

IYANAGA, S.

[1] *The Theory of Numbers*. Iwanami Shoten, Tokyo, 1969 (in Japanese). (Translation: North-Holland, Amsterdam-Oxford, 1975).

JUHÁSZ, A.

[1] "A short proof to Brauer's third main theorem," *Hokkaido Math. J.* **13** (1984), 89–91.

JUHÁSZ, A. AND TSUSHIMA, Y.

[1] "A proof of Brauer's second main theorem and related results," *Hokkaido Math. J.* **14** (1985), 33–37.

KARPILOVSKY, G.

[1] *The Algebraic Structure of Crossed Products.* North-Holland, Amsterdam, 1987.

KNÖRR, R.

[1] "Blocks, vertices and normal subgroups," *Math. Z.* **148** (1976), 53–60.

KÜLSHAMMER, B.

[1] "Quotients, Cartan matrices and Morita equivalent blocks," *J. Alg.* **90** (1984), 364–371.

LANDROCK, P.

[1] *Finite Group Algebras and Their Modules.* Cambridge Univ. Press, Cambridge, 1983.

[2] "The Cartan matrix of a group algebra modulo any power of its radical," *Proc. Amer. Math. Soc.* **88** (1983), 205–206.

LORENZ, M. AND PASSMAN, D. S.

[1] "Prime ideals in crossed products of finite groups," *Israel J. Math.* **33** (1979), 89–132.

MACKEY, G. W.

[1] "On induced representations of groups," *Amer. J. Math.* **73** (1951), 576–592.

MARANDA, J. M.

[1] "On p-adic integral representations of finite groups," *Canad. J. Math.* **5** (1953), 344–355.

MATSUYAMA, H.

[1] "Solvability of groups of order $2^a p^b$," *Osaka J. Math.* **10** (1973), 375–378.

MICHLER, G. O.

[1] "Brauer's Conjectures and the Classification of Finite Simple Groups." Lecture Notes in Math. **1178** (1986), 129–142, Springer-Verlag, Berlin.

NAGAO, H.

[1] "A remark on the orthogonal relations in the representation theory of finite groups," *Canad. J. Math.* **11** (1959), 59–60.

[2] "A proof of Brauer's theorem on generalized decomposition numbers," *Nagoya Math. J.* **22** (1963), 73–77.

[3] "Some correspondences in the representation theory of finite groups," *Proc. International Math. Conf. Singapore*, 35–41, North-Holland, Amsterdam, 1982.

[4] *Algebra.* Asakura Shoten, Tokyo, 1983 (in Japanese).

NAKAYAMA, T.

[1] "Some studies on regular representations, induced representations, and modular representations," *Ann. of Math.* **39** (1938), 361–369.

[2] "On Frobeniusean algebras, I," *Ann. of Math.* **40** (1939), 611–633; "II," *Ann. of Math.* **42** (1941), 1–21.

OKUYAMA, T.

[1] "A note on the Brauer correspondence," *Proc. Japan Acad.* **54** (1978), 27–28.

OLSSON, J. B.

[1] "On 2-blocks with quaternion and quasidihedral defect groups," *J. Alg.* **36** (1975), 212–241.

[2] "Lower defect groups," *Comm. Alg.* **8** (1980), 261–288.

[3] "On subpairs and modular representation theory," *J. Alg.* **76** (1982), 261–279.

OSIMA, M.

[1] "Notes on basic rings," *Math. J. Okayama Univ.* **2** (1953), 103–110.

[2] "Notes on blocks of group characters," *Math. J. Okayama Univ.* **4** (1955), 175–188.

[3] "On some properties of group characters, *I*," *Proc. Japan Acad.* **36** (1960), 18–21; "*II*," *Math. J. Okayama Univ.* **10** (1960), 61–66.

PARKS, A. E.

[1] "A group-theoretic characterization of M-groups," *Proc. Amer. Math. Soc.* **94** (1985), 209–212.

PASSMAN, D. S.

[1] "Blocks and normal subgroups," *J. Alg.* **12** (1969), 569–575.

REINER, I.
[1] *Maximal Orders.* Academic Press, London, 1975.
REYNOLDS, W. F.
[1] "Blocks and normal subgroups of finite groups," *Nagoya Math. J.* **22** (1963), 15–32.
ROQUETTE, P.
[1] "Arithmetische Untersuchung des Charakterringes einer endlichen Gruppe," *J. Reine Angew. Math.* **190** (1952), 148–168.
ROSENBERG, A.
[1] "Blocks and centers of group algebras," *Math. Z.* **76** (1961), 209–216.
SCOTT, L. L.
[1] "Modular permutation representations," *Trans. Amer. Math. Soc.* **175** (1973), 101–121.
SERRE, J. P.
[1] *Represéntations Linéaires des Groupes Finis.* Herman, Paris, 1971.
SOLOMON, L.
[1] "The representation of finite groups in algebraic number fields" *J. Math. Soc. Jap.* **13** (1961), 144–164.
SUZUKI, M.
[1] *Group Theory I, II*, Springer-Verlag, Berlin, 1980.
SWAN, R. G.
[1] "Induced representations and projective modules," *Ann. of Math.* **71** (1960), 552–578.
[2] "The Grothendieck ring of a finite group," *Topology* **2** (1963), 85–110.
TACHIKAWA, H.
[1] "A remark on generalized characters of groups," *Sci. Rep. Tokyo Kyoiku Daigaku, Sect. A* **4** (1954), 332–334.
THOMPSON, J. G.
[1] "Vertices and sources," *J. Alg.* **6** (1967), 1–6.
TSUSHIMA, Y.
[1] "On the p'-section sum in a finite group ring," *Math. J. Okayama Univ.* **20** (1978), 83–86.
[2] "On the second reduction theorem of P. Fong," *Kumamoto J. Sci.* (*Math.*) **13** (1978), 6–14.
UNO, K.
[1] "Character correspondences in p-solvable groups," *Osaka J. Math.* **20** (1983), 713–725.
WARD, H. N.
[1] "The analysis of representations induced from a normal subgroup," *Michigan J. Math.* **15** (1968), 417–428.
WATANABE, A.
[1] "Relations between blocks of a finite group and its subgroup," *J. Alg.* **78** (1982), 282–291.
WOLF, T. R.
[1] "Character correspondence in solvable groups," *Illinois J. Math.* **22** (1978), 327–340.
YAMAZAKI, K.
[1] "On projective representations and ring extensions of finite groups," *J. Fac. Sci. Univ. Tokyo, Sect, I* **10** (1964), 147–195.

Postscript

In writing this book, we have mainly consulted Anderson and Fuller [1], Akizuki and Suzuki [1], Azumaya and Nakayama [1], Feit [2], Issacs [2], Serre [1], and Iyanaga [1]. We shall make here some supplementary remarks.

In Chapter 1, problems 15 and 16 treat a special case of the theory of Morita equivalence. Parts (1) and (2) of problem 19 are extracted from Külshammer [1] and Osima [1], respectively.

In Chapter 2, problem 16, which is due to Tachikawa, implies that every algebra is a homomorphic image of a symmetric algebra.

In Chapter 3, Theorem 5.11(iii) has been known to be true without any assumption on fields (cf. Gow [1], Dade [2], and Issacs [3]). It was conjectured by Iwahori and Matsumoto [1] that if one (hence all) of the equivalent four conditions of problem 12(ii) is satisfied, then G/H is solvable. Recently, Howlett and Issacs [1] verified this conjecture by using the classification of finite simple groups. Problem 14 is known as a theorem of Gaschütz.

In Chapter 4, the content of §6 and the proof of the Green indecomposability theorem are traced back to Conlon [1] and Ward [1]. Broué [1] generalized their arguments. Problem 13 treats a special case of a result of Lorenz and Passman [1] (see Karpilovsky [1], p. 258]). We would like to thank M. Takeuchi for drawing our attention to their result. Anyway, it is now routine to show that part (iv) remains true if more generally $A_1/J(A_1)$ is a simple algebra. We point out that our definition of strongly G-graded algebra is slightly different from the usual one (cf. Karpilovsky [1]). For the recent study of indecomposable modules that employs the Auslander–Reiten sequence, we are referred to Benson [1].

In Chapter 5, Theorem 4.1 is obtained by modifying the proof of Nagao [2] (cf. Juhász and Tsushima [1]). We thus have given a proof to the second main theorem of Brauer via the Green indecomposability theorem. A proof of Iizuka [1], which simplified the original proof of Brauer, is popular as one of the direct proofs to it. Theorem 5.16(ii) was first proved in the case in which B' is weakly regular. Later Knörr [1] removed the assumption of weak regularity. Some of typical applications of the modular representations to group theory will be found in Suzuki ([1], Chap. 6) and Feit ([2], Chap. XII).

By the way, Brauer [12] posed a number of significant problems, which greatly stimulated further development. But many problems have been left unsolved. For those and the Alperin–Mckay conjecture, see Feit ([2], Chap. V, §5) and Alperin [3]. For the relation with the classification of finite simple groups see Feit [1] and Michler [1].

List of Notations

Index

F

G

H

I